INVENTOR OF THE FUTURE

INVENTOR OF THE FUTURE

THE VISIONARY LIFE OF
BUCKMINSTER FULLER

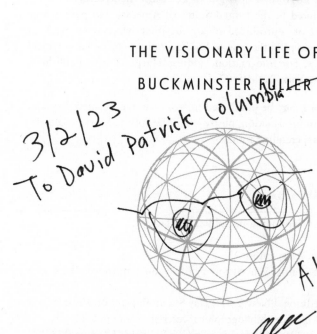

3/2/23
To David Patrick Columbia

All the best,
Alec Nevala-Lee

ALEC NEVALA-LEE

DEY ST.
An Imprint of WILLIAM MORROW

FIRST EDITION

Designed by Angela Boutin

Library of Congress Cataloging-in-Publication Data

Names: Nevala-Lee, Alec, author.
Title: Inventor of the future : the visionary life of Buckminster Fuller / Alec Nevala-Lee.
Description: First edition. | New York : Dey St., an imprint of William Morrow, [2022] | Includes bibliographical references and index.
Identifiers: LCCN 2022018416 (print) | LCCN 2022018417 (ebook) | ISBN 9780062947222 (hardcover) | ISBN 9780062947239 (paperback) | ISBN 9780062947246 (ebook) | ISBN 9780062947253 | ISBN 9780062947260
Subjects: LCSH: Fuller, R. Buckminster (Richard Buckminster), 1895-1983. | Engineers—United States—Biography. | Architects—United States—Biography. | Inventors—United States—Biography.
Classification: LCC TA140.F9 N47 2022 (print) | LCC TA140.F9 (ebook) | DDC 620.0092 [B]—dc23/eng/20220505
LC record available at https://lccn.loc.gov/2022018416
LC ebook record available at https://lccn.loc.gov/2022018417

ISBN 978-0-06-294722-2

22 23 24 25 26 LSC 10 9 8 7 6 5 4 3 2 1

To my parents

CONTENTS

Prologue: Geodesic Man 1

PART ONE: ORIGINS (1895–1927) 19

One: New England (1895–1915) 21

Two: In Love and War (1915–1922) 51

Three: Stockade (1922–1927) 78

PART TWO: THE DYMAXION AGE (1927–1947) 99

Four: The Fourth Dimension (1927–1933) 101

Five: Streamlines (1933–1942) 137

Six: The Dwelling Machine (1942–1947) 182

PART THREE: GREAT CIRCLES (1947–1967) 217

Seven: Geodesics (1947–1953) 219

Eight: Continuous Tension (1953–1959) 265

Nine: Invisible Architecture (1960–1967) 308

PART FOUR: WORLD GAME (1967–1983) 345

Ten: Whole Earth (1967–1973) 347

Eleven: Synergy (1973–1977) 386

Twelve: Equilibrium (1977–1983) 417

Epilogue: Tetrahedron
Discovers Itself and Universe 451

Acknowledgments 477

Notes 481

Bibliography 611

Index 627

I would build that dome in air,

That sunny dome! those caves of ice!

And all who heard should see them there,

And all should cry, Beware! Beware!

His flashing eyes, his floating hair!

Weave a circle round him thrice,

And close your eyes with holy dread

For he on honey-dew hath fed,

And drunk the milk of Paradise.

—SAMUEL TAYLOR COLERIDGE,
"KUBLA KHAN: OR, A VISION IN A DREAM"

INVENTOR OF THE FUTURE

PROLOGUE

Geodesic Man

The steps a man takes from the day of his birth until that of his death trace in time an inconceivable figure. The Divine Mind intuitively grasps that form immediately, as men do a triangle.

—JORGE LUIS BORGES

On October 24, 1980, a man named Taylor Barcroft drove to San Francisco. He was there to see Buckminster Fuller, the architectural designer and futurist, who was delivering a speech at a wellness conference. After the event, Barcroft headed south with Fuller and a cameraman to Cupertino, where they parked at a building on Bandley Drive. Barcroft had arrived without an appointment, and his entire plan depended on how confidently he handled himself now. Leaving Fuller in the car, he went inside and approached the receptionist. "I've got Bucky Fuller here for Steve Jobs."

The visit was a gamble, but he had reason to believe that it would pay off. Barcroft, a University of Denver graduate in his early thirties, hoped to produce a series of cable television programs featuring commentary from Fuller. A segment with one of the founders of Apple Computer would be a compelling proof of concept, but instead of calling ahead, Barcroft thought that he would have better luck by showing up unexpectedly with his famous guest. "I knew Steve was a fan of Bucky," Barcroft remembered. "Anybody like Steve would be a fan of Bucky. And I wanted Bucky to meet Steve, who was going to fulfill Bucky's dream."

It was a risky move, but it succeeded. After the receptionist passed along his message, the first person who emerged to greet Barcroft was Mike Markkula, the chairman of the company, who spoke with him for a minute as they waited for Jobs to appear. Word also reached Daniel Kottke, a mellow but bright twenty-six-year-old who had met Jobs nearly a decade earlier when they were freshmen at Reed College in Oregon. He had become close to Jobs, with whom he later shared a house, and was hired as the twelfth official employee of Apple.

Kottke was at his lab bench, which stood in a work area of cubicles and Herman Miller chairs, when someone announced that they had a visitor: "Buckminster Fuller's here." He rose immediately and hurried for the lobby, where he saw a cluster of people standing outside. His eyes were drawn at once to two men. One was Jobs, who wore his usual outfit of a casual shirt and jeans, and the other was R. Buckminster Fuller, whose face in those days was familiar across the world.

Fuller, eighty-five, was dressed in the dark suit that he favored for all of his public appearances, and, in person, he was startlingly small. His driver's license may have said that he was five foot six, but he had been about two inches shorter even in his youth, and his stature had been diminished by age. He had a huge, bald head with white hair trimmed almost to the scalp, a large hearing aid, and black, plastic glasses that magnified his hazel eyes into soft, enormously deep pools.

Joining the circle, Kottke spoke briefly with Fuller, whose work he had admired since high school. Kottke expected to talk to him further—he was often the one who showed guests around the office—but as the group headed off without him, he realized that Jobs wanted Fuller to himself.

As for Barcroft, he couldn't believe his luck. He ended up at a conference table with Fuller and Jobs, who exchanged a few words while a cameraman recorded the meeting. When it was time for a tour, however, Barcroft was left behind as well. Jobs clearly didn't want to include anyone else, and no one would ever know what he and Fuller said to each other in private at Apple, which was only months away from its initial public offering.

Afterward, Barcroft took Fuller back to his hotel. Barcroft was elated, but his plan for a cable show never materialized, and he later lost the footage of Fuller and Jobs. For his part, Fuller was unconvinced that the personal computer would enable his lifelong vision of access to information. "He didn't believe it," Barcroft recalled. "He thought that only mainframes could do that work." Fuller had devoted his career to predicting the impact of technology, but he saw nothing special in Apple: "I remember him saying that he thought the computer was a toy."

Judging from his eagerness to meet Fuller, their encounter left a greater impression on Steve Jobs, which came as little surprise to Daniel Kottke. "In my early friendship with Steve, he was interested in so many things that I was also interested in," Kottke said. "That definitely included Fuller." Since the sixties, college campuses had found an unlikely hero in Fuller, whose reputation was based on the geodesic dome, a hemispherical structure used in everything from industrial buildings to hippie communes, as well as the sculpture studio at Reed.

Fuller was renowned for his avowed optimism that technology could "make the world work" for everyone by lifting entire nations out of poverty. His message centered on the figure of a generalist known as the comprehensive designer, "an emerging synthesis of artist, inventor, mechanic, objective economist, and evolutionary strategist" capable of grasping whole systems. In an era that was skeptical of politics, his argument for a radical science of design, which emphasized universal housing and the efficient use of resources, offered a seemingly plausible program for achieving change outside conventional institutions.

To an ambitious individualist like Jobs, it was a model of what was possible, and no one embodied it like Fuller, who had been inescapable in the Bay Area of the seventies. Fuller had been given a crucial push by the *Whole Earth Catalog*, an oversized guide to books and tools for the counterculture that Jobs, who avidly read it in college with Kottke, described as "one of the bibles of my generation." Copies were kept in the Apple lobby, and its first edition, published in 1968, opened with a spread devoted to the man whom editor Stewart Brand credited as its inspiration: "The insights of Buckminster Fuller are what initiated this catalog."

At the heart of the *Catalog*, which Jobs once called "Google in paperback form," was Fuller's determination to change the environment rather than human beings. "That definitely comes directly from Fuller," Brand noted decades later. "Fuller said that a lot: that changing human nature is hard, and when you try, you mostly fail, and it's discouraging. Changing tools and technology is relatively easy." This perspective was warmly received in Silicon Valley, which Brand thought was why the microcomputer revolution happened at that specific time and place: "The stuff came out of the Stanford area, I think, because it took a Buckminster Fuller access-to-tools angle on things."

For a community that was still defining itself, Fuller was an inexhaustible source of metaphors and images, which spread through existing networks into every corner of the culture. The *Whole Earth Catalog*, for example, was born at the Portola Institute, an educational nonprofit in Menlo Park, California, that eventually included a commercial arm for books on computing. At the suggestion of Marc LeBrun—a coding prodigy who would later be one of the first four members of the Apple Macintosh team—the subsidiary was called Dymax, an homage to Dymaxion, the personal brand that Fuller used for his designs. Dymax, in turn, spun off the People's Computer Company, which held classes and potlucks that became a hangout for hackers.

One frequent attendee was an engineer named Lee Felsenstein, who had grown up captivated by one of Fuller's most famous inventions: the streamlined Dymaxion Car. In the seventies, Felsenstein joined Project One, an organization founded by Ralph Scott, who was described by *Mother Jones* magazine as "an architect and engineer who studied with Buckminster Fuller." Scott, who had built a structure inspired by Fuller's ideas for the Exploratorium children's museum, wanted to create an ecosystem of technological communes across San Francisco, beginning with an abandoned candy factory with an old mainframe that Felsenstein was eager to use.

At Project One, Felsenstein's group set up a remote terminal to allow anyone to post messages. The effort was based on a proposal by Efrem Lipkin, a veteran of the World Game workshops, which promoted Fuller's extravagant plan for a computer simulation to allocate planetary resources. Lipkin recruited two other Fuller fans, Ken Colstad and Mark Szpakowski, who together ran a World Game center out of the building. Their work on the terminal was deeply informed by Fuller, and they installed a public teleprinter and modem in nearby Berkeley as the first computerized bulletin board, which they called Community Memory.

Felsenstein dreamed of constructing a better network around a true personal computer, and, after hearing about a gathering for like-minded enthusiasts, he drove his pickup to the garage of an engineer named Gordon French, who hosted the first meeting of the Homebrew Computer Club on March 5, 1975. In his classic book *Hackers: Heroes of the Computer Revolution*, writer Steven Levy described it as "a textbook example" of Fuller's concept of synergy, in which the whole became more than the sum of its parts, and with Felsenstein as its moderator, it grew into the birthplace of an entire industry.

Fuller was there in spirit as well. Toward the back of the garage sat Steve Wozniak, who was working with Jobs on the computer that would be sold as the Apple II. When they needed an industrial designer for the housing, they hired Jerry Manock, a graduate of the legendary product design program at Stanford University. Its founder, John E. Arnold, had been a pioneer of creative engineering, with a curriculum grounded in Fuller's philosophy. His successor, Robert McKim, built a large cardboard dome in which design majors could unlock their imaginations, and he brought members of his class to see Fuller speak at the Esalen Institute in Big Sur.

One of these students was Manock, who established what became the Apple Industrial Design Group using an iterative approach that he attributed to Fuller: "He wasn't interested in solving just one tiny design problem. He would look at the next level up, and the next level up, and the next level up." Manock's team later took a leading role in developing the upright case of the Macintosh, but on the inside, its operating system resembled the graphical user interface created at Xerox PARC by Alan Kay, who also acknowledged Fuller as an influence, saying, "He was kind of an intuitive systems guy, at a time when nobody really thought in terms of systems."

Another major player was Jaron Lanier, whose father had met Fuller through friends in architecture. Lanier was raised on Fuller's books, and after his mother died in a car accident, he and his father relocated to an acre of desert in New Mexico. Although Lanier was only eleven years old, he was given permission to design their house, which he decided to base on a dome.

If his mother had lived, Lanier thought that he might have pursued a conventional career, but fate, as he later reflected, took him in other directions: "My dad was more into 'Be the Buckminster Fuller or the Frank Lloyd Wright'—be the weird outsider who becomes influential." After moving to California, where he saw Fuller speak, he entered the gaming industry. Lanier was fascinated by the concept of a computer simulation of the world, which was "very much in reference to Fuller," and he became known as the father of virtual reality, joking that it came to him naturally because of the dome: "The way everything is triangulated felt like home to me."

Lanier often met other technologists "enchanted by Buckminster Fuller," whose presence he observed in those circles long afterward: "For those who have context, of course, you see Fuller everywhere." Theodore Roszak, who coined the term *counterculture* in the late sixties, saw him as one of the "prophetical voices" of the era: "Fuller and his Bay Area disciples [were] the major spokesmen for a philosophy of postindustrial life that has done much to shape the style and expectations of the computer industry, especially as it has grown up in Silicon Valley." Its leaders, Roszak wrote, were "cut from the mold of the Bucky Fuller maverick," including Wozniak, who praised him as "the twentieth century's Leonardo da Vinci."

Given their distrust of authority, the hackers had to believe that they had independently discovered Fuller, who exerted his greatest impact from a distance. A year after his visit to Cupertino, Fuller received an Apple II as a gift, although he had little interest in learning how to use it. He was provided with a demonstration at home in Los Angeles by Janek Kaliczak, whose work as a graphics engineer led insiders to refer to him as "the Han Solo of the Apple world"—but Fuller only passed it along to his office in Philadelphia, where his staff utilized it to manage his massive itinerary.

In any event, Fuller had finally been convinced of the value of the personal computer, which he embraced as the realization of all that he had predicted. He inevitably wanted it for his projects, particularly the World Game, and his associate George Madarasz reached out to Wozniak on Fuller's behalf. Wozniak arranged for the delivery of a second computer, which Fuller and Madarasz hoped to use for "a graphical display of the world's needs and resources."

The connection to Apple resulted in one last tribute during Fuller's life. Wozniak had spent millions on the US Festival in San Bernardino, which he conceived as a Woodstock for his generation, "but maybe better." On May 30, 1983, Kaliczak and Madarasz handled the visuals for an introduction to Fuller and the World Game, which hundreds

Fuller in the television commercial that introduced the Apple slogan "Think Different."

of thousands of concertgoers watched on an enormous screen before
Stevie Nicks took the stage to sing "Dreams."

Fuller died a month later, a half year before the release of the Macintosh, but his legacy at Apple endured. On September 28, 1997, a
television commercial debuted during the network premiere of the
animated film *Toy Story*. Steve Jobs had recently returned in triumph
to Apple, and over a montage of luminaries—from Martin Luther King
Jr. to Pablo Picasso—the actor Richard Dreyfuss offered a declaration
of principles: "Here's to the crazy ones. The misfits. The rebels. The
troublemakers. The round pegs in the square holes. The ones who see
things differently."

One of the seventeen icons was Buckminster Fuller, who had been
featured at the request of Jobs himself. Fuller had crossed paths with
many of the other personalities in the commercial, including Albert
Einstein, Amelia Earhart, Martha Graham, and Frank Lloyd Wright,
and he appeared after John Lennon and Yoko Ono and before Thomas
Edison—a suitable place for a man who had been revered by both the
counterculture and the establishment.

The "Think Different" campaign was meant to sell computers, but
it also spoke to an authentic vision personified by Fuller, who thought
differently—for better or worse—than just about anyone else. "You can
quote them, disagree with them, glorify, or vilify them," Dreyfuss
concluded. "About the only thing you can't do is ignore them. Because
they change things. They push the human race forward. And while some
may see them as the crazy ones, we see genius. Because the people who
are crazy enough to think they can change the world are the ones
who do."

When Fuller visited Apple in 1980, he was famous to a degree that
can be hard to grasp today. In an indulgent biography published in the
seventies, his friend Alden Hatch wrote, "Apart from the temporary
notoriety of a few high government officials and some athletes like
Muhammad Ali and Mark Spitz, Buckminster Fuller is probably the

best-known American in the world outside of the United States." This may have been an overstatement, but in the eyes of his millions of admirers, Fuller undoubtedly stood for nothing less than the future itself.

In the last decades of his life, Fuller appeared on the cover of *Time*, lectured around the globe, and collected honorary degrees and awards that culminated in the Presidential Medal of Freedom. At Disney World, the iconic Spaceship Earth at Epcot Center was named for his most celebrated metaphor, and after his death, Fuller received an even greater accolade in a newly identified type of carbon with striking affinities to a geodesic sphere. His ideas had provided crucial insights to the scientists responsible for its discovery, who called it buckminsterfullerene, or the buckyball.

Yet Fuller's importance was only partially understood during his lifetime. From a modern perspective, he sometimes seems like a Silicon Valley visionary who was born a half century too soon, and he served as a key prototype for a myth that had yet to be named: the start-up founder. Like many of his successors, he benefited from a privileged background, which allowed him to take repeated risks without falling out of the upper class. Long before the fictionalized version of Facebook's Mark Zuckerberg, Fuller brooded over his failure to join Harvard University's exclusive final clubs, and his departure from college allowed him—like many Ivy League dropouts—to define himself as a rebel.

In 1927 Fuller underwent what he described afterward as a mystical experience, inspiring him to address the problem of housing. He implied later that his actions had been guided by first principles: "My objective has been humanity's comprehensive welfare in the universe. I could have ended up with a pair of flying slippers." The truth was that he focused from the start on architecture, which was a world to which he had access. Fuller wanted to manufacture a house that was so light in weight—using tension, not compression—that it could be produced in a factory and delivered anywhere. Just as a phone was a single node in a larger network, his house would be the front end of a dwelling service industry, which he promoted with prophetic language and conceptual art in the absence of any finished product.

By the late twenties, Fuller was a serial entrepreneur, pivoting from one disruptive invention to another under his Dymaxion brand. All of these "artifacts" fed into his plans for shelter, including the Dymaxion Car, which was a piece of the house that could drive off on its own. Fuller learned to construct arguments using trend curves, which displayed the accelerating rate of industrial progress, and infographics based on his Dymaxion Map, an unfolded polyhedron that preserved the relative sizes of the continents. He branched out into economics, sociology, and especially geometry, hoping for a moonshot breakthrough that would confer legitimacy on his other ideas.

At first, Fuller was limited by a dilemma of his own making. For all his efforts to escape the system, the rules of capitalism decreed that he had to conduct research with other people's money. He longed to operate without any obligation to investors, and while this might have been possible in a future industry that was less capital intensive, such as software development, it seemed unthinkable in shelter design.

Unbelievably, Fuller succeeded, thanks to three nearly simultaneous discoveries that he made in 1948. One was the geodesic dome. Although he emphasized its structural efficiencies, its first real benefit was that it could be prototyped for almost nothing, and its triangulated framework evoked the technology of tomorrow, as much for its aesthetic qualities as for its actual strengths. As the architectural writer Lloyd Kahn, a fan turned harsh critic, observed later of the domes, "They appear exciting and revolutionary, they promise untold advantages, the simple geometrical aspects have great appeal, and moreover, they photograph well."

Fuller's second discovery was a limitless pool of free labor. A conventional entrepreneur would have founded a business, but instead, he used the postwar boom in higher education to build a vast, invisible research operation. At colleges, he organized students into divisions inspired by the aircraft industry, like a simulation of a company that he had imagined on paper. In fact, it was closer to an emulation—a massively distributed virtual corporation that ran on the platform of the university system, in which he embedded himself like a virus.

His third discovery was his own persona. If the best way to start a movement was to target college students, it called for ideas of legitimate urgency and importance, which he learned to exemplify in himself. When a tech start-up proclaims that it wants to change the world, it follows in a long line of visionaries who used similar messages to convince disciples to devote their lives to a cause. Fuller did this better than anyone, and his vaunted relationship with the young was driven by his need for followers who were receptive and unattached enough to be recruited for a crusade.

Decades before the rise of the personal computer, a dome was a hackable machine for living that could be built in a garage by following tables of numbers known as chord factors, in a kind of physical coding that allowed users to start experimenting and debugging at once. After being embraced by industry and the military, it was claimed by individualists who brought it to the masses—in a precursor to the open-source movement—by defiantly revealing its secrets in underground publications that presented geodesic geometry as a gateway to a better life.

This view was encouraged by Fuller, whose public image foreshadowed the start-up founders who followed. To keep the focus on his message, he wore a nondescript black suit, in the midcentury equivalent of Steve Jobs's turtleneck and jeans, and he publicized his experiments with sleep and diet, which today would be called biohacking. As the shaper of his own corporate culture, he created a convoluted language that divided the world into insiders and outsiders. At his lectures, he would ask for a show of hands to see how many listeners knew the word *synergy*, and he always commented on how few there were. Today it seems like the quintessence of business jargon, which only testifies to Fuller's lasting impact.

His most enduring idea was the concept of "ephemeralization." At first, it simply meant doing more with less, but Fuller took it infinitely further. Just as compression yielded to tension, the physical would give way to the visual, which would be replaced in turn by total abstraction. When asked what would come after the dome, Fuller replied, "Nothing—

an electromagnetic field." This sequence paralleled the stages in his career, which started with concrete and ended with geometry, and its final phase sounded like a paradox: "Ephemeralization trends toward an ultimate doing of *everything* with *nothing at all.*"

More recently, the venture capitalist Paul Graham defined it as "the increasing tendency of physical machinery to be replaced by what we would now call software," noting that phones and tablets "have effectively drilled a hole that will allow ephemeralization to flow into a lot of new areas." The same trend can be seen in cloud computing, the invention of the blockchain, and the inexorable integration of technology into everyday life. Before they became part of our collective future, Fuller independently arrived at the principles of online education, remote working, and universal access to data, which he developed without any computers at all.

On a more pragmatic level, Fuller used ephemeralization to sustain his virtual company, which he ran for years as a perpetual start-up. He made most of his money from lecturing, and, as the original digital nomad, he minimized his physical needs. The chord factors for an undergraduate dome project could fit on a page in his wallet, and he was capable of lecturing extemporaneously for hours using slides and a suitcase of visual aids. His models often drew on a principle that he called tensegrity, based on lightweight sculptures where the solid parts never touched, which Fuller utilized to further reduce the bulk of his mobile operation.

As his geometry grew into a product division in itself, he often depended on words and images alone. Fuller was notorious for talking past his scheduled time, speaking in his New England accent at seven thousand words per hour, and his lectures were compared to a geodesic dome—or, more precisely, a tensegrity sphere—that made sense only after the last piece was in place. As the Swiss designer Yves Béhar has pointed out, Fuller was among the first to realize that a reputation for visionary architecture could be achieved without any tangible artifacts whatsoever: "We owe the notion of architect as statement maker to Buckminster Fuller—the idea of creating photographically accurate renderings of an architectural intervention at a gigantic scale."

Ephemeralization was also the engine of his program for change. Fuller paired technological innovations, including automation, with a universal basic income, which would distribute the benefits of efficiency to all mankind. He observed that progress tended to occur as a by-product of war, or "emergence through emergency," which he aimed to replace with a peaceful design-science revolution. Unlike a government or company, a single person could work on something that might not be needed for fifty years: "The individual can simply start to think."

For more than three decades, Fuller maintained his virtual operation in expectation of a turning point that seemed just around the corner, which explained aspects of his career that were often mistaken for personal eccentricities. To pay its expenses, he lectured nonstop, and he poured money into patents to protect his intellectual properties. The Dymaxion Chronofile, his enormous personal archive, has been called the largest of its kind in history, but it was more comparable to the files of a moderately sized company, which challenged the storage and retrieval limits of paper itself.

Fuller transformed himself into a corporation to free himself from capitalism, but his example enabled others to surrender to it completely. Although he became a conduit for the ideals of modernism—with its bold program to reimagine society—into the cult of technology, his story can make Silicon Valley seem resolutely conventional. As a practical matter, most start-ups have abandoned the built environment for what the journalist Derek Thompson has called "the ethereal world of software." Fuller's success in pulling it off in the field of housing is astonishing.

To make it possible, he found himself living permanently in the world of tomorrow, until the vision that he used to advance his goals became an end in itself. Fuller has been voted the most influential futurist of all time, but he wanted his ideas to be realized, even if it took a half century, which allowed him to insist that he was right on schedule. In reality, he survived by staking a claim in an undiscovered country, and he colonized more of it than anyone else ever would. Even the architect Philip Johnson, a lifelong enemy, granted him this grudging praise: "There hasn't been anybody since that could lead us into the land of the future."

Toward the end of his life, Fuller said that he had always avoided the temptation "to revert to saying that you are a special son of God," which was hardly a disclaimer that an ordinary person would feel obliged to make. As a teacher, he changed countless lives for the better, but his private behavior was shaped by the risks of the path that he chose. To maximize his freedom on his own terms, he had to control others, and in the absence of the usual incentives, he kept them in line with persuasion, charm, or anger, which repeatedly drove away collaborators.

A cult of personality has always existed in architecture, but Fuller took it to extremes, since he lacked both real power and obvious monuments. He was a superb choreographer of other people's lives, and as the nodes of his network grew further apart, it took on a familiar pattern. The biographer Alden Hatch thought that Fuller was "above all a mystic," and although this reflected an authentic element of his character, it was also a tool to motivate others at a distance. Mystics, like start-up founders, tend to converge on similar strategies, and Fuller's mysticism assumed a form that was appropriate for America in the age of the machine.

Above all else, Fuller needed the young. In the sixties, he saw the hippies as an extension of his research team, but his skepticism toward politics—he described the civil rights movement as "Blacks skillfully persuaded into activism"—led him to recommend design solutions instead: "I tell students to stop using their feet and start using their heads." This conveniently overlooked how technology could reinforce existing systems of oppression, and while inventions could undoubtedly change the world, the consequences were hard to plan in advance.

Even his designs were less than met the eye. Although Fuller's views on sustainability were welcomed by environmentalists, he never questioned the huge unseen infrastructure required by the project of ephemeralization. His domes were said to cover "more square feet of the earth than any other single kind of shelter," but at their peak, there were just a few thousand permanent structures based on his patents, most of which were radar enclosures, while other uses were undermined by their inherent shortcomings. To maintain his own influence and encourage publicity, he undervalued the potential of computer networks, while promoting centralized databases with gigantic displays that he could supervise himself.

Fuller's obsession with control was inseparable from the question of credit. He never operated as a licensed architect, and it was taken for granted that he needed partners who could sign off on his plans. In fact, he had minimal input on many of the structures with which he was associated, including the Montreal Expo Dome, which was seen as his masterpiece. The most generous reading was that he focused on the big picture, like Jobs, and left the details to others, but his name was frequently linked to domes in which his participation was nonexistent. His reputation as a Renaissance man had to expand in proportion to the diminution of his visible work, and he consistently understated the roles of his colleagues and students.

Like ephemeralization, Fuller's transmutation of himself into a myth had hidden consequences. His protégé Edwin Schlossberg knew this problem well—he married into the Kennedy family—and he saw it clearly in Fuller: "The way the world constructs the metaphor of being a man in the world isn't conducive to a supporting and loving relationship." One of Fuller's oldest associates, Harold Cohen, was even more blunt: "Everyone who was close to him, he destroyed." This was an exaggeration, but it hinted at a long inward struggle that was unimaginable to those who first encountered Fuller as "the planet's friendly genius."

Fuller's reluctance to share credit was part of a larger pattern. Since his death, it has become widely known that he invented elements of his biography, which the late scholar Loretta Lorance described accurately as "a public relations tool." He had to become what others believed he was, and by the end, he was the only living source for many of the facts. Fuller's writings and talks overflowed with misinformation and outright falsehoods, which he methodically built into the reality distortion field that allowed him to achieve so much in a single lifetime.

His embellishments naturally left previous biographers in a difficult position. Alden Hatch simply took him at his word, while Fuller reportedly advised Athena Lord, the author of a book about his life for young readers, to "mythologize his childhood." Lloyd Steven Sieden, whose admiring tome *Buckminster Fuller's Universe* looked at first glance like a substantial biography, undercut this possibility in the preface: "In the future, someone may well write a complete biography of Buckminster Fuller, filled with all the quirks of his personal life. This is not that book."

As a result, despite the dozens of books about Fuller, many of which are riddled with errors, he has never been the subject of a comprehensive biography that covered his full career using the best available sources. Given the vast claims that have been made for his ideas, he deserves the exhaustive, respectful, but unsentimental treatment accorded to any prominent figure of the twentieth century, which is an indispensable part of the case for his importance. Fuller's archives reveal aspects of his personality that his fans never saw, while the surviving witnesses can testify firsthand to his strengths, flaws, and contradictions.

Until now, the lack of an authoritative survey has only made it harder to take him seriously. All the pieces need to be in place before the overall shape of his life can be seen, and in the aftermath, he emerges, if anything, as even more prophetic than before. Fuller perceptively diagnosed the world's unrealized ability to care for all of its population, and he warned that capitalism favored the concentration of wealth in the hands of a tiny minority, leading to a succession of crises—from inequality to climate change—that he foresaw a generation ago.

When contextualized properly, the tools that he provided are as essential as ever. Fuller emphasized whole systems, interdisciplinary thinking, and intuition, and his fascination with structure, which was driven by his need to carry his entire operation in his head, led to insights of genuine value in engineering and the sciences. His influence on our culture is enormous, and its explanatory power only expands as we push back against the legend. In a time of unprecedented transformation, he offers a set of proven strategies for how an individual can produce change—but at a cost.

His significance was perceived long ago by the critic Calvin Tomkins, who wrote to Fuller, "It has always seemed to me that your life is not only exemplary, but altogether fascinating, and that the major biography of it could and should be one of the more important books of our period." Tomkins never finished the work that he had in mind, but in a 1966 profile in *The New Yorker*, he famously quoted Fuller on the ancient seafarers who went beyond the laws that applied on land: "When you've made things so good out there in the outlaw area that they can't help being recognized, then gradually they get drawn in and assimilated."

Fuller saw himself as a man who sailed into the unknown, like Odysseus, and even if it began as a myth, he pursued it with such unparalleled single-mindedness that it became more real than the truth, at least in its ability to affect the world and the lives of others. As Fuller himself knew well, the example that he left for the rest of us doubles as a warning. "Because I live in the frontiers," Fuller said, "what happens to me usually happens to others later on."

PART ONE

ORIGINS

1895–1927

It is one of the strange facts of experience that when we try to think into the future, our thoughts jump backward. It may well be that nature has some fundamental metaphysical law by which opening up what we call the future also opens up the past in equal degree.

—BUCKMINSTER FULLER, *SYNERGETICS*

ONE

NEW ENGLAND

1895–1915

The tactile is very unreliable; it has little meaning. Though you know they are gentle, sweet children, when they put on Hallowe'en monster masks, they "look" like monsters.

—BUCKMINSTER FULLER, *SYNERGETICS*

Richard Buckminster Fuller Jr. was born nearly blind. When he entered the world on July 12, 1895, at his family's home in Milton, Massachusetts, his eyes were crossed, but his parents were told to postpone providing him with glasses until after he turned four. No one realized that he was also severely farsighted, and the inability to see his surroundings clearly would have lasting effects on the boy with the cumbersome name whom everyone called Bucky.

His earliest memories were impressions of motion. One was of riding in a baby carriage and wiping away the mud from the wheels with his tiny fingers. Another was of being lifted into the air for a lullaby— "And down will come baby, cradle and all"—and then swung to the floor to be startled out of crying. His first fully formed thought was inspired by the landscape of snow at his house. The more distant it was, the sharper it seemed, and when he looked out at the footprints of squirrels and birds, he felt overwhelmed by the shining expanse.

Fuller and his father in the summer of 1897.

Up close, his field of vision was a blur. Faces were vague shadows, and whenever his older sister, Leslie, told him what she saw, he thought that she was making it up. In response, he invented stories of his own, and he was pleased when the adults laughed. Although he was vividly aware of color, he learned to rely on his other senses, especially his nose, and he recognized individuals by their smells.

Fuller had a limited perception of pain. When he cut himself, he knew only that the blood led to comfort from grown-ups: "So every time this red stuff showed up, I thought I would get this loving treatment." He reminded people of a little animal—he tried to lap water like a puppy—and his father's nickname for him, after a tortoise in Rudyard Kipling's *Just So Stories*, was "Slow and Solid."

His parents were concerned about his intelligence, and after his brother, Wolcott, was born in 1898, Fuller was enrolled in a Milton kindergarten based on the philosophy of Friedrich Froebel, a German educator who used toys as instructional tools. These took the form of "gifts," or sequences of play materials, including blocks in the shapes of spheres, cylinders, and cubes, which deeply influenced future architects such as Frank Lloyd Wright and Le Corbusier.

Froebel's nineteenth gift was "peas work," in which students assembled small objects out of peas and pointed rods. According to Fuller, most of the children built square structures that resembled conventional houses, which were unstable. Fuller, unable to see what they were doing, relied on touch: "When I got to the triangle . . . it had a form of rigidity, so I tried finally making a teepee, like an Indian teepee with three of them coming from a triangular base."

Working by feel, he made a tetrahedron, or triangular pyramid, followed by the eight faces of an octahedron. Fuller recalled that his teacher, Grace Parker, was so impressed by what he had done that she showed it to another instructor. In fact, such structures were an established part of the Froebel curriculum, and tetrahedra and octahedra appeared frequently in standard teaching materials—indicating that their appearance in the classroom was less unusual than he implied.

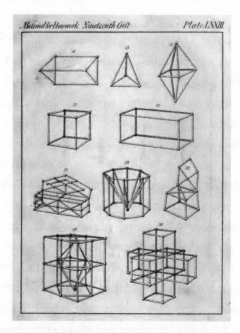

Diagram of "peas work" constructions from *The Paradise of Childhood*, a manual for Froebel educators published in 1869. The tetrahedron and octahedron appear in the top row.

At four, Fuller finally received glasses, which allowed him to see the faces of his parents for the first time. He was equally drawn to the eyes of frogs and reptiles, which seemed to say, "I love and trust you," and his nurse screamed when she found a snake that he had brought home in his shirt. After leafing through a picture book, he said that he wanted to be a cow when he grew up, because he longed "to be out in that green grass with the flowers and so forth, and go around eating."

For the rest of his life, Fuller would remove his glasses occasionally to restore his old view of reality, but his newfound clarity of sight expanded his horizons immediately. He saved pieces of paper with his name, which grew into a substantial collection, and felt more conscious of the wider world. On New Year's Eve in 1899, his father opened a window to let in a new century, and Fuller never forgot his earnest vision of "the coming years in which I would live beyond his time."

Richard Fuller Sr., who had been raised as a Unitarian but became an Episcopalian after his marriage, advised his older son to enter the ministry. They would walk holding hands to St. Michael's Episcopal Church in Milton, where the boy had been baptized and his father served as a vestryman. On special occasions, they attended services in Boston, which the son associated with the smell of incense.

Fuller's mother, the former Caroline Wolcott Andrews, was more distant and devout. As he persisted in making up stories to entertain grown-ups, which turned him into what he later called "a little liar," his mother was advised to discipline him, but he never understood why. "So to avoid what seemed to be leading up to a spanking, I would use my imagination and storytelling ability to misinform my mother," Fuller remembered. "This has built a trait into me which I've had to cope with as the years went on."

His mother showed him pictures of the Taj Mahal, which she called the world's most beautiful building because of "the love of its conceiver for his wife," and told him the ancient Greek tale of the honorable Spartan boy who kept silent as a fox devoured his entrails. She taught him to always look at the new moon over his right shoulder, but she could also be forbidding. He never forgot hearing how her sister had died when a candle burned up her nightgown, and she described seeing the Great Chicago Fire of 1871 from her childhood home in Illinois, which they visited when he was five.

Fuller asked endless questions, and his precocious appearance sometimes annoyed adults, who allegedly rejected his bids for affection. He had similar issues with his sister, who was three years older. Fuller felt that Leslie, who encouraged him to tell lies, resented no longer being the only child, and when he entered elementary school in Milton, she said, "Bucky, when I was in first grade, I was able to do the work very easily, but you will find it hard."

He responded by dominating Wolcott. With an age difference of three years, the brothers had trouble playing together, and his closest friend was the son of their neighbor, Dr. Vassar Pierce, who had delivered Fuller. Lincoln Pierce was another product of a venerable New England family, and as a natural pair—they had been wheeled alongside each other as babies—they excluded Wolcott, which Fuller called later "an act which I have regretted all the rest of my life."

On nice days, Fuller would jump down from his bedroom window to play with Lincoln in the woods. They formed an alliance against other children, including a rich boy who changed the rules of ball games as he went along. If anyone objected, Fuller recalled, "he would pick up all the playing gear, put it in his pony cart, and drive away." As they grew bolder, he and Lincoln ventured to the Blue Hills Reservation state park and the girls' dorm at Milton Academy, the nearby prep school, where they hung a flower basket as a May Day prank and were almost caught when Fuller lost his glasses there.

The backdrop to these memories was Fuller's family home, which had been the first house built on Columbine Road. Its architect was Alexander Wadsworth Longfellow Jr., his father's cousin, and although Fuller dismissed it later as "a rather large Gay Nineties Victorian," its elegant lines bore the influence of the Japanese Pavilion at the 1893 World's Columbian Exposition in Chicago. Sunlight entered the dining room through diamond panes, and Fuller was once sent upstairs without supper for cutting his initials in a window with his mother's ring.

Because of the house's narrow plot, the domestic staff lived in the attic. Before he wore glasses, Fuller would go to the top floor to see farther—he climbed trees for the same reason—and he was adopted by the servants, especially a cook named Johanna Sullivan, who treated him like her own son. He did homework each morning in the kitchen, where he watched Sullivan toss a pan of boiling water outside "to get rid of the evil fairies," and, after dinner, she would beckon him through the drawing room curtain to come to the pantry for a treat.

In the attic, Fuller threw paper planes from the window and looked out at the grounds, which included his father's garden, and the woods that sheltered "gypsy camps" in the winter. "The road curved immediately away in both directions—I couldn't see any great distance," Fuller reminisced long afterward. "I started then, when I was first allowed to, walking beyond the end of the driveway and found myself going through an almost continuous tunnel of trees."

Milton was an affluent suburb, eight miles from Boston, with a population of seven thousand. The air was filled with delicious aromas from the Walter Baker chocolate factory, and Fuller was entranced by the jars of colored water in the drugstore windows. He was warned that more distant towns were filled with violent types who carried knives and drank whiskey, but he noticed nothing dangerous when he went on his first motorcar ride.

Fuller was more intrigued by Boston. When his family took the trolley to the city, his overdeveloped sense of smell was struck by the odors of unwashed passengers, of beer from saloons, and of the island in the harbor where dead horses were incinerated. More pleasant were the murals of King Arthur at the library—Fuller was enthralled by tales of chivalry, and he was drawn both to the magician Merlin and to his father's stories of Robin Hood, who stood in his imagination as the personification of integrity.

The highlights of these trips were the visits to the waterfront offices of Howe, Fuller & Trunkett on State Street. His father imported leather for shoes and chests of Indian tea, which were lined with foil that Fuller used to create suits of armor. In 1902 his parents sailed on business to South America, leaving the children with their paternal grandmother in Cambridge, Massachusetts. While they were gone, Fuller read about the eruption of Mount Pelée in Martinique, which filled his mind with images of disaster.

A year later, after his business partnership was dissolved, Fuller's father went to India as an agent for the trading firm of Stein, Forbes & Co. Fuller headed with the rest of his family for Chicago, where they visited his uncle, Rockwell King, who was married to his mother's sister. King was president of a firm that manufactured products out of cast iron, as well as of the Western Cold Storage Company, where Fuller marveled at the enormous cooling system.

The visit kindled a lasting interest in industry for Fuller, who liked to say that he had been born in the same year as radio. In practice, his past and future were impossible to separate, and he seemed destined to play a clearly defined role in a household with generational ties to the New England establishment. His father had dropped out of Harvard, but there was no question of any other college for the son, who had been put up for the Somerset Club—another traditional gathering place for the upper class—as soon as he was born.

Certain expectations had been set into place, but his social position also exposed him to an alternate set of values. In the summer of 1904 his mother's relatives made their annual trip east. Because of a flu epidemic, instead of their usual vacation spot in Marblehead, Massachusetts, they stayed on Eagle Island in Maine, where Caroline Wolcott Andrews, his grandmother, decided to buy property in Penobscot Bay as a convenient middle ground for family gatherings. She purchased Bear Island, which had the best harbor, while the Fullers and Kings jointly bought Compass Island and Little Spruce Head Island.

The following year, they all returned to inspect their holdings more closely. They departed Boston on an evening steamer, leaving behind the city's fumes, and Fuller awoke at dawn in Penobscot Bay, which always remained his ideal of nature's beauty. From the town of Rockland, they headed for Bear Island, which was a mile long by a quarter wide, with forests, dramatic bluffs, and a natural breakwater. Arrowheads, cellar holes, and graves reflected its long history, and the family liked to share the story of how a previous owner, a Mormon, had climbed a tree to await the apocalypse, only to descend quietly when the world failed to end on schedule.

As they commenced renovations, Fuller's father planted elm trees on the weekends, while the children played in the tall grass and the red flowers of Indian paintbrush. There were seals to observe, lobsters to trap, and oysters to harvest, and the caretaker taught them rope work and sailing. Fuller lived in a separate house on the island with his wealthier and more athletic older cousins, Rockwell King's sons MacGregor and Andy, but he preferred to spend time on his own. His favorite chore was to fetch the mail from Dirigo Island in a dinghy, the *Sea Bird*, which amounted to a round trip of four miles.

This led to his first invention, which he described later as "a pair of hinged oar blades for sculling." Over time, his claims for it expanded, until he said that a jellyfish had inspired his "teepee-like, folding, web-and-sprit cone which was mounted like an inside-out umbrella on the submerged end of a pole." When he pushed it back from the stern, it opened to propel him forward, allowing him to face where he was

Fuller with his sister Rosamond in the playpen that he built for her at Bear Island (1907).

going. After acquiring a new dinghy, Fuller abandoned his "mechanical jellyfish," which survived only in his fond accounts of his intellectual development.

It was also evidently around this time that Fuller had a long talk with his uncle. His grandmother may have taught him the Golden Rule, but Rockwell King sided with the English economist Thomas Malthus, who argued that rising population led to scarcity and grief. "If you are going to survive and have a family of five and wish to prosper," King said, "you're going to have to do it at the expense of five hundred others. So do it as neatly and cleanly and politely as you know how."

In 1906 Fuller recovered from appendicitis in time to visit Bear Island with Wolcott, and they returned following the birth of their younger sister, Rosamond, in September. The boys played well together in Maine, where they explored the islands and competed at skipping stones, but they still fought occasionally, prompting their father to issue a warning: "He who calls his brother a fool shall be in danger of hell's fire." Fuller became fonder of Rosamond, and, the following summer, he built her a wooden playpen along the usual rectangular lines.

On Bear Island, Fuller wanted to prove his independence, and he was learning to listen to his intuition. One day his parents took a cousin out in their boat, leaving Fuller and his siblings with the other adults. When a storm came, he feared for their safety and convinced a grown-up to conduct a search; they found that the three of them had barely survived after capsizing. Fuller saw this as proof of his clairvoyance, but he was also gaining experience on the water, mostly in a sloop, the *Ursa Minor*, that he had innocently proposed calling the *Cuspidor*.

He was growing up in an era when science and technology had displaced the frontier in the American imagination, and he was thrilled back home by his first chance to drive a car. On one occasion, when Fuller was sent to his room at the impressionable age of twelve, he stole a book from downstairs about a man who built a house with rounded corners to make sweeping easier, as well as automatic fountains for cleaning. "These two ideas seemed the quintessence of common sense as applied to house building," Fuller recalled, "and I informed the family of the benefits they were missing as soon as I was released from my room."

In 1907 the catastrophes that had haunted his imagination—from Mount Pelée to the storms of Bear Island—finally came true. Fuller's father suffered a massive, incapacitating stroke caused by an undiagnosed brain tumor. Overnight, Fuller saw the man he had associated with travel and freedom reduced to a wreck—he had to lead his father by the hand, read to him, and fan him for hours.

With their income reduced, his mother laid off six of their servants, except for Johanna Sullivan, which forced Fuller to undertake most of the chores, including beating rugs and filling barrels with ash. He disliked the upkeep required in an old house without electricity, and when a fire destroyed much of the nearby town of Chelsea in the spring of 1908, it suited his darkening mood.

His family spent that summer in the newly constructed Big House on Bear Island. It had been built from lumber ferried over by schooner on the southeast promontory, where it was anchored with iron stanchions. Like the house in Milton, it was designed by Alexander Longfellow, who filled it with Japanese touches, and since it was never wired for power, the glow of candles and kerosene lamps suffused it by night. Wolcott and Fuller occupied the attic, where they held races in the early morning that often ended with their mother spanking them with a hairbrush.

Fuller had never forgotten his uncle's lecture on the unforgiving laws of survival, and MacGregor and Andy King taught their younger cousin a similar lesson in their own way. The Big House had a Victrola for playing light opera and Enrico Caruso, but while Fuller had built a special cabinet to store records upright, Andy flung the disks into the sea "when they were played too frequently for his taste." Fuller enjoyed drawing, painting, and carving houses from driftwood, but his artistic inclinations, which became more important to him as his father's strength declined, were dismissed by his mother and mocked by his cousins.

To keep him in line, the older boys administered beatings. Although Fuller pretended not to care, he sometimes feigned illness to avoid their "desissifying" treatment, which was encouraged by his mother, who had been annoyed by his refusal to defend Wolcott from bullies at home. "She wanted me to get angry about it and fight," Fuller recalled, "and, finally, to gratify her, I would find compassion for my younger brother if somebody were brutalizing him, but I did not go out and battle the other kid."

Fuller never forgave her failure to protect him, and it led him afterward to avoid his family, who, he wrote later in the third person, "would undoubtedly be astounded at this revelation and find it hard to conceive how important a part it has played in his life." At the time, however, he arrived at a practical solution. Working from an article in *St. Nicholas*, a children's monthly magazine, the boy built a cabin out of leftover wood and shingles, allowing him to live apart from the others. In retrospect, the simple shelter was a rudimentary version of a strategy—based on mobility and access to construction plans—that would one day become central to his career.

Escape was less possible at school. After the third grade, Fuller had spent two years in public schooling, where the "horrid smells" of the toilets and the "straitjacket" of his desk shaped his views on both plumbing and education. From there he enrolled in Milton Academy as a day student. At the time of his father's stroke, he had been there for a year, and everyone assumed that it was a stepping-stone to Harvard, which had been attended by four generations of Fullers.

His friend Lincoln Pierce was a pupil at the day school as well, but Fuller's life there was miserable. At first, he was eager to impress the older boys, which prompted what he later saw as a betrayal of his values. When the headmaster resigned, Fuller joined in "a town-wide riot of joy," despite knowing nothing about the man: "I whooped and danced and ran with the others."

Milton's motto, based on a poem by the seventeenth-century British author George Herbert, was "Dare to Be True," which Fuller rephrased as follows: "Dare to tell the truth as you see it, and you'll find yourself in trouble." Because of his natural oddness, he became a target for the other students, who contrived to have noises come from his direction in class. Without looking up, the teacher would usually say, "One mark for Fuller." For each offense, Fuller had to remain in detention for fifteen minutes.

As the school "antifavorite," Fuller was repeatedly told that he wasn't very bright, which filled him with a desire to prove the others wrong—although it also allowed him to overlook his own considerable privilege. Brooding over the Scottish poet Robert Burns's famous lines on seeing ourselves as others see us, he reached an unpleasant conclusion: "I assumed that I was just a freak."

Fuller developed a reputation for insubordination. After devoting weeks to fractions, his teacher informed the students, "I am now going to teach you a better way. It's called decimals." Fuller wasn't sure why she hadn't mentioned this first, and he was confused by how repeating decimals could go "out the window and over the hill." On another occasion, the teacher drew a point and a line on the chalkboard, explained that both were imaginary, and then asserted that a cube was real. Deciding that this sounded like nonsense, Fuller asked, "How old is it?" He followed up with questions about how heavy or hot it was, but when she grew irritated, he shut his mouth. "I thought she was very pretty and appealing," Fuller recalled, "and if that's the way she wanted to play the game, I'd play it her way."

At fourteen, he saw his first airplane, which left the greatest impression of all his encounters with technology, and in the spring of 1910 he used his father's camera to photograph Halley's comet. Later that year, he won a footrace, which gave him "great confidence," but the most peaceful period of his boyhood was ending. Grandmother Andrews had died in March, dividing the ownership of Bear Island among her grandchildren, and Fuller's father was losing his mind as well as his sight. For the previous two years, he had been essentially incommunicative, and a visiting relative vividly remembered him "on his deathbed screaming."

Caroline was overwhelmed by the demands of caring for her husband, and they left him under the watchful eye of Dr. Pierce next door when they went away for the summer. On July 12, 1910, while the rest of the family was on Bear Island, Fuller's father died of "apoplexy." He was forty-nine. It was Fuller's fifteenth birthday, and he was devastated that he had been unable to say good-bye to his father, whose place he was expected to take as the man of the house.

The tragedy coincided with a seemingly minor event. Bear Island's caretaker, who had left to become a preacher, was replaced by Jim Hardie, a Scottish immigrant who had rowed over to ask for the job. Hardie had a frightening appearance—his eyes were bloodshot from years at sea—but he was a quietly remarkable man who left his mark on everyone, especially Fuller.

At first, their relationship consisted of a friendly competition for Caroline's approval. Fuller thought that Hardie, who worked outdoors in a vest and tie, had "a most extraordinary and uncanny intuition and logic, as well as a gargantuan will and muscular strength to accomplish." Hardie had taught himself to read from the newspapers, and, like Johanna Sullivan, he eventually became a surrogate parent to Fuller, although he was only eleven years older.

Away from the island, life went on. His sister Leslie was a recent debutante—Fuller studied her boyfriends' cars with interest—and he was confirmed at an Episcopal church in Boston in 1911. The following summer, he surprised everyone by heading alone to Bear Island, where he and Hardie used horses and oxen to lay down stones for a road and worked on a clay tennis court.

Fuller took pleasure in directing his frustrations into labor, and he later described the island as the fountainhead of his education, pointing to its lessons in conserving resources and in "such tension systems as seines, trawls, weirs, scallop drags, lobster pot heads, and traps." It was the beginning of his interest in shaping the environment with tools, as well as the formative experience of his youth, and as he grew older, he imitated the local sailors who called children "darling" or "dear."

His growing independence was also expressed in less constructive ways. Although his father had left an estate of around $150,000, or the equivalent today of several million dollars, Caroline kept their money under tight control. Since Fuller lacked an allowance, he secretly charged purchases to his mother: "I'd forge her name, hide bills, be deceptive in every way I could." She lamented his behavior—Rosamond remembered "shivering" in the next room as their mother scolded him over his spending—and asked the husbands of her friends to lecture him, which only made the young man more aware of his own father's absence.

At Milton Academy, he remained unpopular with the faculty, with the exceptions of biology instructor William Lusk Webster Field and Homer LeSourd, a physics teacher who sensed that Fuller had a head for numbers. LeSourd tutored him in math during his frequent detentions, and Fuller enjoyed a talk by the Milton alumnus Starling Burgess, a yacht designer and aviator, who showed slides of his biplane, the *Flying Fish*.

As his confidence grew, Fuller sang in the glee club—he would always dream of being "a song and dance man"—and looked for ways to prove himself athletically. He was handsome but sensitive about his height, which would never be much greater than five feet, four inches. As a result, he cherished the memory of modest physical feats, including the time that he impressed his classmates by rowing a kayak up the Charles River to Boston.

Improbably, he made it onto the football team as an upperclassman, although he was ashamed by the cheap quilted helmet that his mother had bought for him. In his senior year, Fuller played quarterback without glasses, even though he was legally blind without them. Relying on his speed, he successfully aimed passes at the outlines of his teammates, but he broke a knee in the last game of the season.

He said later that he received good grades by repeating whatever his teachers wanted, but in actuality, his written record was mediocre. In his junior exams, he earned a B in physics, a B- in plane geometry, and a failing E in Latin. His next round was worse, with a C+ in geography and elementary history, a C- in English, a D+ in German, a D in solid geometry, a D- in French, and another E in Latin.

Despite this unimpressive showing, Fuller was admitted to Harvard in 1913, and his relatives agreed to pool money for his tuition. That summer, he signed the family guest book at Bear Island—in which he had experimented with various signatures, unsure about whether to include his "too pretentious, too heavy" middle name—as "Richard B. Fuller," dropping the *Jr.* for the first time.

When Fuller headed for Cambridge in the fall, he knew that he was entering a world in which the past would be inescapable, but he also saw a chance for liberation from the oppressive forces—his teachers, his peers, and his own family—that had caused him so much pain. Toward the end of his life, Fuller compared himself to a genius against whom he constantly measured himself, but he drew a key distinction. "I think Einstein was fortunate in that he probably never lost his childhood," Fuller reflected. "I temporarily had mine taken away from me."

During what turned out to be his difficult first term at Harvard, Fuller often thought of his family history, and he was consoled by the knowledge that his father's college years had been just as unconventional. Richard Buckminster Fuller Sr. had dropped out as a junior to travel to the silver mines in Chihuahua, Mexico, where he became sick with typhoid. He worked afterward for the merchant firm of Howe & Goodwin in Boston and New York, writing in one alumni report that he had held "no offices of profit and none of honor worth mentioning."

In fact, Dick Fuller fared well in business, and he successfully courted Caroline Andrews. Their letters hinted at real affection—he called her Sandy, she called him Dicky—but her family's fortune was also a consideration. On the maternal Wolcott side, her ancestors included a signer of the Declaration of Independence and several governors of Connecticut. Another branch headed for Ohio, where her mother married a wealthy Yale University graduate named Martin Andrews. From there they moved to Chicago, where Caroline was born in 1867.

Dick and Caroline's son heard many of these stories growing up, but the enterprising Wolcott and Andrews families rarely seized his imagination. He was much more invested in his father's family—he later paid genealogists to trace back his lineage eighteen generations to England—and he naturally focused on the "freedom loving and fearless" colonists who immigrated to America, their lives filled with portents and warnings of challenges to come.

In his mind, their history began with Thomas Fuller, who sailed in 1638 from the Isle of Wight to the Massachusetts Bay Colony. A sermon in Cambridge led him to convert to Puritanism, although another family tradition credited "the black eyes of a certain Miss Richardson," who was not one of the three women he later married, with his decision to settle as a blacksmith in what became Middleton, Massachusetts. He served as a selectman and in the local militia, and, near the end of his life, he and his son were among the accusers in the Salem witch trials.

After two uneventful generations, a more active element returned in the Reverend Timothy Fuller, a Harvard graduate whose wife, Sarah Williams, was an educated woman descended from the Buckminsters. He was dismissed by his parish over his skepticism toward the American Revolution, but he represented the town at the Massachusetts Constitutional Convention, where his opposition to the Fugitive Slave Clause prompted him to oppose ratification. After forcibly moving his family to a New Hampshire farm, he died in poverty in his sixties.

This marked the end of any inherited wealth among the Fullers, which deeply shaped the life of his eldest son, Timothy Jr. At Harvard, he was an early member of the prestigious Hasty Pudding Club, and he was allegedly demoted to second in his class for participating in "a rebellion of the students," who objected to the rules that undergraduates receive a blessing at meals and sit in "an upright and decent posture." While working as a teacher to pay for law school, he became known for flirting with his pupils, which distressed his wife, Margarett Crane, whom he married in 1809.

Timothy's talent for speaking carried him to the United States House of Representatives, but he was "too indolent or unenterprising" to succeed in politics, and his failure to secure a diplomatic posting prompted him to relocate to a farm in Groton, Massachusetts. In abruptly uprooting his family, he was repeating his father's actions, and he alternated between working in the fields and writing an unfinished history of America before dying of cholera at fifty-seven.

His most lasting impact came through two of his children. One was Arthur Buckminster Fuller, who became a Unitarian minister after graduating from Harvard. His second wife, Emma Lucilla Reeves, bore him two sons, including Richard Buckminster Fuller, who was born in 1861. Arthur was an ardent abolitionist, and despite his poor health, he was commissioned as a chaplain of the Sixteenth Massachusetts Volunteer Infantry during the Civil War.

On December 11, 1862, Arthur volunteered to cross the Rappahannock River at the battle of Fredericksburg, Virginia. Within five minutes of taking up arms, he was killed by bullets in his chest and hip, and in the ensuing slaughter, his body was bayoneted and robbed. Arthur's grandson Buckminster, who inherited his sword, often had reason to brood over his last reported words: "I must do something for my country. What shall I do?"

But the most remarkable Fuller of all was Arthur's older sister, whom the writer Susan Sontag would later describe as "the first important American woman of letters." Margaret Fuller was born in 1810, and as Timothy Jr.'s eldest child, she became the repository for all of his hopes. He taught her Latin at age six, while demanding constant accuracy in thought and speech, and she was praised as a "wonderful child" who could discourse at length on mathematics.

Margaret paid the price for her precocity at school, where she felt that none of her teachers ever bothered to ask the important questions about life. A friend thought that she suffered from "nearsightedness, awkward manners, extravagant tendencies of thought, and a pedantic style of talk," and she was often ridiculed. From an early age, the girl felt torn between love for her father and resistance to his authority, which had left her with "no natural childhood."

All of this culminated on Thanksgiving 1831, as her father was about to move the family to Groton. After church, Margaret wandered alone until she came to a pool in the woods. "And, even then, passed into my thought a beam from its true sun, from its native sphere, which has never since departed from me," Margaret would write. "I remembered how, a little child, I had stopped myself one day on the stairs, and asked, how came I here? How is it that I seem to be this Margaret Fuller?" All her sufferings, she concluded, came from the illusion that the self was real.

She resolved to honor her father's wishes, and after his death in 1835, she wrote in the family Bible, "We shall go to him, but he can-

The allegorical frontispiece to Margaret Fuller's *Woman in the Nineteenth Century* (1845).

not return to us." To earn money, she worked in Boston for the educator Amos Bronson Alcott, whose daughter Louisa May would grow up to write the novel *Little Women*. In 1839 Margaret inaugurated her famous "conversations" for women in the Boston bookshop of Elizabeth Peabody, an advocate for the Socialist experiment at Brook Farm. One of her listeners recalled, "I felt that the whole wealth of the universe was open to me."

Margaret Fuller became known as the most widely read person in New England, and she had a tumultuous friendship with the poet and essayist Ralph Waldo Emerson, to whom she said, "I now know all the people worth knowing in America, and I find no intellect comparable to my own." Emerson was "astonished and repelled," but he granted that she was the country's finest conversationalist. Nathaniel Hawthorne, by contrast, had reservations: "It was such an awful joke," said the author of *The Scarlet Letter*, "that she should have resolved—in all sincerity, no doubt—to make herself the greatest, wisest, best woman of the age."

In recognition of her talents, Emerson gave her the editorship of the *Dial*, the journal of the transcendentalism movement, a spiritual philosophy of individualism that he defined in part as "the tendency to respect the intuitions." Margaret took the unpaid position, and during her tenure—which included the first appearance in print of Henry David Thoreau—she shaped the literary voice of the entire nation, particularly in her essay "The Great Lawsuit," which she expanded into the book *Woman in the Nineteenth Century*.

When this landmark feminist work appeared, it immediately secured Margaret's reputation. Its frontispiece was an allegorical image of two conjoined triangles, one black and one white, which would later figure prominently in her grandnephew's symbolic vocabulary. She wrote of women, "But if you ask me what offices they may fill; I reply—any. I do not care what case you put; let them be sea-captains, if you will." According to Margaret, a reduction in labor would enable this new era, which she saw in the rising generation: "At present I look to the young."

In 1844 Margaret accepted an offer from Horace Greeley to become the literary editor of his *New-York Daily Tribune*, for which she went abroad as the first American woman to serve as a foreign correspondent. Although she was apparently attracted to both men and women, she had never been in a romantic relationship until she fell in love in Rome with a marquis named Giovanni Angelo Ossoli, who was ten years her junior and spoke no English.

Within the year, she became pregnant, and she evidently married Ossoli in secret. In Rieti, she bore a son, Nino, and returned to Rome to witness France's assault against the republican forces of General Giuseppe Garibaldi. Margaret cared for dying soldiers, but after the French took the city, she and her husband decided to sail for America. While anchored off Gibraltar, the ship's captain died of smallpox, and an inexperienced first mate took command.

After crossing the Atlantic, the ship missed its New York port in the night and was grounded off Fire Island on July 19, 1850. Margaret, who had celebrated her fortieth birthday during the voyage, was awakened by the impact. Going on deck, she saw by the light of dawn that they were three hundred yards from shore. As waves washed over

the listing ship, the cabins broke apart, and Margaret was swept overboard to drown with her husband and infant son.

Throughout his career, Buckminster Fuller spoke of his great-aunt
as an influence, but although her books were in the library of the Milton
house, he was unaware of her work until he was in his thirties. Fuller
became interested in her while searching for validation for the life that
he had chosen, and he took comfort in knowing that she had seen herself in the context of history, writing in the late twenties that he hoped
to accomplish in his own field "what Margaret Fuller did for American
literature."

In 1845 Margaret told a friend, "The destiny of each human being
is no doubt great and peculiar, however obscure its rudiments to
our present sight, but there are also in every age *a few* in whose lot
the meaning of that age is concentrated. I feel that I am one of those
persons in my age and sex. I feel *chosen among women*." Fuller had
similar intuitions about himself, and he identified with his great-aunt's
reputation long before achieving anything comparable.

Daguerreotype of Margaret Fuller by photographer John Plumbe (1846).

Like her grandnephew, who would spend his life seeking to advise the wealthy and powerful, Margaret felt that she was made "to fill the ear of some Frederick or Czar Peter with information and suggestions on which he might reflect and act." The theologian James Freeman Clarke recalled that her acquaintances often looked back on their interactions as a turning point "in which we took a complete survey of great subjects . . . and were led to some definite resolution or purpose which has had a bearing on all our subsequent career[s]."

Fuller approvingly quoted Margaret's lines on American genius, which she believed would be realized only after "the physical resources of the country being explored . . . talent shall be left at leisure to turn its energies upon the higher department of man's existence." In his 1902 book *The Varieties of Religious Experience*, William James recorded her most famous aphorism: "I accept the universe." Fuller misremembered it, in a small but telling variation, as "I must start with the universe."

As an adult, Fuller was also a close reader of Emerson, who wrote in similar terms about the origins of civilization: "Chiefly the seashore has been the point of departure, to knowledge, as to commerce. The most advanced nations are always those who navigate the most." Emerson even anticipated Fuller's thoughts on design efficiency: "The line of beauty is the result of perfect economy. The cell of the bee is built at that angle which gives the most strength with the least wax; the bone or the quill of the bird gives the most alar strength, with the least weight."

Emerson and Margaret Fuller were both shaped in turn by Emanuel Swedenborg, the eighteenth-century Swedish mystic who began as an engineer—he sketched an early concept for a flying machine—and was transformed by a prophetic vision in middle age. "Swedenborg approximated to that harmony between the scientific and poetic lives of mind, which we hope from the perfected man," Margaret said. "The links that bind together the realms of nature, the mysteries that accompany her births and growths, were unusually plain to him. He seems a man to whom insight was given at a period when the mental frame was sufficiently matured to retain and express its gifts."

If Fuller ever read these lines, he would have seen their relevance to his own story. As Emerson wrote memorably of Swedenborg, "The genius which was to penetrate the science of the age with a far more subtle science; to pass the bounds of space and time; venture into the dim spirit-realm; and attempt to establish a new religion in the world—began its lessons in quarries and forges, in the smelting-pot and crucible, in ship-yards and dissecting-rooms." And a few pages earlier: "This man, who appeared to his contemporaries a visionary, and elixir of moon-beams, no doubt led the most real life of any man then in the world."

When Fuller joined his Harvard class of seven hundred freshmen in the fall of 1913, the eighteen-year-old's attention was initially caught by another prodigy. William James Sidis had entered the college several years earlier at the age of eleven—the youngest student ever to enroll—and famously lectured at the Mathematical Club on four-dimensional bodies. Fuller noted that most students regarded Sidis as "a freak," but the talk, which had taken place before his arrival, inspired his interest in the fourth dimension.

In Cambridge, Fuller himself was blinded by a "mythical Harvard of my own conjuring." His mother placed him in Randolph Hall, a building for undergraduates located in the fashionable district called the Gold Coast, but he was disappointed when his longtime friend Lincoln Pierce roomed instead with a Milton graduate named Thomas Whitall. Like all freshmen, Fuller wanted to make an impression, and he dreamed of joining the football team, where Milton quarterbacks traditionally did well. As a boy, his hero had been Waldo Fuller, a first cousin once removed, who had been a star player for the Crimson.

At tryouts, however, Fuller "busted" his knee again, ending his football career. The injury failed to keep him from leaping from one dormitory roof to another to see a friend when the building's door was locked, in another gratuitous display of physical daring, or from joining the cross-country team in October. Its coach, the distance runner Al Shrubb, advised Fuller not to run with his fists clenched: "Take it easy and go faster." His running improved, and he said afterward that the admonition "helped me through many difficult moments when I was worrying and trying too hard."

During his first term, this quality was less than evident. Fuller recalled that he avoided studying science and math because he dismissed them as games: "I thought I'd take something really difficult, like government or English." In fact, his schedule consisted of introductory classes in English, French, chemistry, art, and government, and he did poorly in them all. He had the most trouble in a required rhetoric course that assigned short themes, four of which he skipped. The following month, he broke his arm, and he turned in no assignments before winter break, leading one professor to conclude that his work was "very neglectful."

Fuller was more preoccupied with the social scene. He claimed later, implausibly, that he was unaware of class divisions before arriving at Harvard, and he was undoubtedly unprepared for the exclusive final clubs that dominated student life. By his own account, he was discouraged when his sister Leslie, who was dating a man variously described as a member of either the illustrious Porcellian Club or the A.D. Club, told him that he wouldn't be accepted into any clubs without a rich father. Fuller's more popular friends informed him that they could no longer be seen together, and the last straw was when the first girl he ever loved "jilted" him for a club man.

In practice, the scene at the college was evolving—Fuller's class was the first in which a majority of students came from public schools—and Leslie was seeing a tennis player named Edward Larned, who graduated from Princeton University, not Harvard. The most prestigious clubs didn't even take freshmen, and although Fuller's cousin and classmate John P. Marquand, who became a best-selling novelist, resented his own lack of social success, many students, including the future theater critic Brooks Atkinson, rarely gave it a passing thought. Even the biographer Alden Hatch, who generally accepted his subject's claims, thought that Fuller was just "too much of an oddball" for this elitist world, which he proceeded to reject preemptively.

A sense of inferiority was common enough among Harvard freshmen, but Fuller's response only reflected his strangeness. As a gesture of rebellion, he established a society of his own, the HCKP Club, after the four closest stops on the subway—Harvard, Central, Kendall, and Park—and claimed to have "attracted some excellent members." He had read the college diary of his great-grandfather Timothy Fuller Jr., who had needed a full day to reach Boston "for his rum," and the fact that the train took only seven minutes supposedly struck him as "a harbinger of an entirely new space-time relationship of the individual and the environment."

As a practical matter, the Boston nightlife was more alluring. The city was an enticing destination, and in an era when automobiles were still a novelty—only three of his classmates owned cars—the subway opened up enormous possibilities. Fuller's preferred routine was to get off at Park and walk south to the theater district, where undergraduates congregated at The Nip on Tremont Street. After one drink, he felt "magically converted into a real, live millionaire."

In December the Shubert Theatre across the way was featuring *The Passing Show of 1913*, a satirical revue that had concluded a run at the Winter Garden in New York. According to Fuller, his sister Leslie had left her Russian wolfhound at a kennel during her honeymoon, and he paid a messenger to bring him the dog. When Fuller walked it to the stage door, the chorus girls loved it, especially Marilyn Miller, who later became a famous performer in the Ziegfeld Follies. Fuller allegedly asked her to join him at the Hotel Touraine with her "hired mother," or chaperone, and although the date ended at dinner, it impressed his friends.

The truth was rather different, and the incident became one of the first of many in his life that Fuller fictionalized after the fact. Leslie wouldn't marry Edward Larned until the following November, and the revue at that time didn't include Miller, who would make her debut in 1914. Either Fuller had taken out another chorus girl entirely, or the event occurred one year later, which required an even more drastic revision of his account of his college days.

What was indisputable, however, was the change in his habits. His mother provided him with an annual allowance of $1,000, and at first, he had lived frugally. By winter break, though, Fuller had grown increasingly irresponsible, and the situation came to a head after Christmas. While his family vacationed in Bermuda, he returned to college for his exams, and when he realized that he lacked enough money to make it through the year—at least not in style—he withdrew all of it at once.

In the most celebrated version of the story, Fuller accompanied the cast of *The Passing Show* by train to New York, where he treated them all to a lavish dinner at Churchill's restaurant, burning through most of his bank account. After checking into a hotel, he ran out of funds, forcing him to call a rich relative in Boston for help. Depending on the telling, one or more detectives from the William J. Burns agency was dispatched either to rescue him or to bring him to the manager, and Fuller slunk back to college to face the consequences.

His true actions were more complicated and less colorful. On January 17, 1914, Fuller paid for dinner at the Hotel Westminster in Boston using a bad check—a far more serious offense than spending all of his personal allowance. In a report on Fuller, Henry A. Yeomans, the assistant dean of the college, wrote, "He secured a considerable sum of money upon worthless checks, drawn and cashed, when he knew he had no funds in the bank to meet them and when he had no reason to believe that funds would be available to meet them when presented."

By the time Fuller left for New York, *The Passing Show* was already in Toronto, so he would have traveled there alone. According to the dean's report, he lived "riotously" and spent money "on one or more chorus girls," perhaps from *The Whirl of the World*, which was playing at the Winter Garden Theatre. "I have been told that one such woman at least was quartered with him at a New York hotel," Yeomans wrote. "The evidence, however, is mere hearsay, and I am unable to substantiate it. Fuller denies that he went further with any woman than eating and drinking."

After this encounter, which was evidently his most extensive romantic experience to date, Fuller was stranded at the Imperial Hotel, and he cabled the superintendent at Randolph Hall on February 4 to ask for thirty dollars. On the same day, Yeomans sent him a stern letter: "Your failure to take several of your midyear examinations places you in a very serious position. I am sure that the best thing you can do is to talk the whole matter over with me fully and without delay."

On February 13 the freshman was asked to withdraw from Harvard. "They can't fire you for spending your money—it's your money—but they could fire me for not going to my classes," Fuller recalled long afterward, understating his actual misdeeds. He departed abruptly, abandoning most of his possessions, as creditors threatened to sue his mother, who had sold the Milton house to reduce the family's expenses. Caroline spoke of sending Fuller to a reformatory, but a paternal cousin, Lindsley Loring, offered a possible alternative.

Loring, the treasurer of the Merrimac Chemical Company, assisted in contacting Edward W. Atkinson, a friend of Fuller's late father who owned a brokerage firm for cotton machinery. Atkinson, Haserick & Co. had contracted to equip a mill for the Canadian Connecticut Cotton Mills Company in Sherbrooke, Quebec, and Atkinson proposed that

Fuller (second from left) in Sherbrooke, Quebec (1914).

Fuller could work there under "practical men who had made good without the advantages of an education." To teach the young man the value of college, he told the foreman "to show him no favors, to pay him what he was worth, and to report to me."

It was a chance to redeem himself for nine dollars a week, and Fuller had no choice but to accept. At the end of February, he took the train to Sherbrooke, which was a railway center at the confluence of two rivers. He checked into the relatively comfortable Château Frontenac, a short walk from the mill, which he reached by trudging along the tracks past houses with snow up to their roofs. The temperature was below zero, and he arrived to find an empty shell of a building that needed to be filled with equipment.

Fuller knew that he was there because of his connections, and he craved the respect of the working men, but the task ahead was challenging. The pulley shafts had to be connected to a central powerhouse driven by a water wheel. In the finished mill, bales would be opened at track level, with the cotton blown to the top to travel downward through pickers, combers, twisters, and spinning and weaving machines. The machinery was imported from England and France, and the English components often arrived mangled in transit.

For his first assignment, Fuller was given "a burlap bag of busted parts" to repair with the help of local foundries. To everyone's surprise, he took to it avidly. The chief engineer encouraged him to keep a sketchbook—Fuller claimed that he sometimes built better components than the ones that were broken—and he relished his exposure to the highly capable British and German machinists.

He had yet to regain the trust of his mother, who had moved into a small farmhouse to pay his debts, but she expressed cautious approval. What she failed to anticipate was that his "self-tutored course of engineering exploration" might introduce him to a system of values very different from the ones at Harvard. If the plan had been to underline the importance of an Ivy League education, it backfired spectacularly, and he identified with the machinists instead.

After a short visit home before Easter, Fuller labored in Sherbrooke until July, finishing the installation and teaching workers to use the machinery. Later that summer, he sailed to a costume ball thrown by the businessman Herman Oelrich in Bar Harbor, Maine. Fuller spent the night in the cabin of his sloop at the dock, only to be awakened early the next morning by a piercing ship's whistle. Going above deck, Fuller saw the *Kronprinzessin Cecilie*, a German liner carrying gold in search of a neutral port. It was August 1914.

Europe had plunged into a war that would upend any notion of progress or change, but for now, Fuller only wanted to return to Harvard. In a letter to Dean Byron Hurlbut, Edward Atkinson wrote on his behalf, "Fuller now appreciates how other men have to live, and what a great opportunity it is to secure an education, as well as what a fool he made of himself last year in Cambridge." Lindsley Loring, speaking by phone with Dean Yeomans, acknowledged that Fuller had left behind unpaid bills in Canada, but he emphasized that everyone understood that this was his final chance.

Fuller was readmitted in the fall. When he returned to Cambridge, still bearing the residual glamor of his misadventures, the narrative of his redemption seemed clear. With most avenues for social recognition gone, he would have to justify his benefactors' trust by devoting himself to his studies. His schedule consisted of courses in English literature, German, government, music, and rhetoric, and Loring insisted on hiring a law student as his academic supervisor.

Once again Fuller struggled. In his literature class, he was warned never to guess the meaning of a word but to check the dictionary—a sign of the carelessness that led a teacher to call his work "pretty bad on the whole." In his rhetoric course, which he had to retake, Fuller cut lectures and finished only seven themes out of fifteen, earning Ds for his efforts. When his professor asked for a conference, the nineteen-year-old never showed up. An instructor recommended his dismissal, and his overdue bills made Caroline afraid that her son was deceiving her again.

Looking for a way out, Fuller enlisted in the Massachusetts Volunteer Militia, although he quickly had second thoughts. He stayed in college, but while he submitted a few more themes, he continued to miss lectures. After he entered the infirmary for an unknown illness, his government teacher wrote, "I do not believe that the fellow is right physically and that this affects his mentality." Fuller's music instructor was equally critical: "He . . . seems to make little effort to get at the subject which may be hard for him to understand. The chief trouble appears to be a lack of mental energy."

On January 22, 1915, Yeomans informed Fuller that the Administrative Board would once again ask for his withdrawal: "It hardly seems after my talk with you that further words are of use. You understood clearly when you returned to college at the beginning of the current year that you were to attend strictly to your business and do your very best in every way. . . . Indeed, I understand that you yourself feel that you have not kept your agreement."

Yeomans concluded on a compassionate note: "You have really tried to overcome the weaknesses which brought you to grief last year. Because I believe this, it is especially hard for me to write this letter." He invited Fuller to visit him before he left, and he advised, "Take the first decent job that comes your way, do it to the best of your ability, and while you do it, be considering the question of just what work you would like to make your business in life."

Fuller later attributed his second departure to "lack of sustained interest," although he also said that Harvard might have forgiven him again if the war hadn't intervened, since he had earned honors in biology, mathematics, and physics. In reality, he failed to complete any courses, and he never earned the "high scholastic grades" that he often claimed. During his more philosophical moods, he said that colleges were pressured by their donors to produce specialists, not generalists, which allowed him to implicitly characterize his actions as a kind of rebellion.

At the time, however, Fuller felt only "anguish and shame." Instead of becoming the man of the family, he had broken his mother's heart, and his atonement would take the form of another apprenticeship in the real world. In response to the dean's offer to meet, Fuller scrawled a hasty reply:

Dear Dean Yeoman [*sic*]

I am leaving in quite a hurry as I have a job open for me in New York which Mr. Loring wishes me to come down for immediately on the three o'clock. I had not expected to leave till tomorrow when I intended to see you to say good-bye. I am very sorry that I missed you, as I wanted to thank you for the interest you have taken in me and all that you have done. However, let me thank you just the same by this note.

Most sincerely yours,
Richard B. Fuller

T W O

IN LOVE AND WAR

1915–1922

Life and Universe that goes with it begins with two spheres: you and me . . . and you are always prior to me. I have just become by my awareness of you.

—BUCKMINSTER FULLER, *SYNERGETICS*

On February 15, 1915, Fuller reported to the 527 West Street branch of Armour & Company, which stood near the bank of the Hudson River in the Hamilton Heights neighborhood of New York. As the nation's leading meatpacker, Armour was a massive industrial operation, founded on harsh labor practices and the refrigerated railway car, that extended its reach through a network of local distributors to wholesale and export clients, including hotels and restaurants.

Armour was an ideal place for a young man to learn the logic of supply and demand, but Fuller began as a lowly "beef lugger," earning ten dollars a week for loading trucks, checking freezers, and making deliveries to stores and ships. Instead of the trained machinists of Sherbrooke, his colleagues were uneducated men whose vocabulary, he recalled, consisted of about one hundred words, half of them obscene: "They were sure that my speech pattern meant that I was some kind of sex pervert, and they whistled at me as I came along."

Yet he owed the job to his privilege. Fuller had landed it through a family friend, William A. Hazard, a salt magnate and polo expert who wrote to obtain a reference from Arthur Meeker, an Armour executive in Chicago with a shared interest in horses. Meeker predicted that Fuller would end up in sales: "It is where almost everyone commences who eventually amounts to anything in the business. In other words, it is what we call in the trenches or on the firing line."

The work was hard on his sensitive nose, but Fuller knew that he was being watched, and he put in long hours. He quit his college smoking habit, cursed to fit in with the others, and welcomed the chance to explore the area around the Chelsea Piers. Fuller claimed that he checked the cargo on the luxury passenger liner *Lusitania* before it sailed for England on what would be its final voyage. He also recalled the disgust he felt when Germans in the market "exulted" after hearing the news that a U-boat had torpedoed the ship off the Irish coast. According to Fuller, he had spoken German well until then—although he had scored a D+ on his last exam—but now he deliberately pushed it out of his mind.

Fuller impressed his supervisors as "an unusually ambitious young man," and his relatives struck an encouraging but careful tone, with Wolcott writing to his brother that he was lucky to have such "big men" on his side. When his mother and siblings left for Bear Island, he stayed in his residential hotel on East Thirty-First Street. Later that summer, he visited Leslie in Lawrence, Long Island, where he donned a tuxedo for a party on Ocean Avenue. Most of the attendees were teenagers, including Alden Hatch, a local boy who met Fuller for the first time.

Both of them were more interested in the ladies, whom Hatch fondly remembered in their "ankle-length chiffon or voile dresses." Fuller, having just turned twenty, especially enjoyed dancing with a nineteen-year-old with a sensitive face and dark eyes. Anne Hewlett, the "best girl" of his friend Kenneth Phillips, had recently returned from a trip to Colorado and the West Coast, and she was accompanied by her younger sister Anglesea, who was known as Anx. They were a world apart from Fuller's chorus girls, and he was strongly drawn to them both.

Out of deference to his friend, he focused on Anx, who was sixteen and pretty, with hair that she wore in golden ringlets. Over the summer, he and Phillips took the sisters on double dates to the movies and the beach, but Fuller sensed that he had sunk his chances with Anx "by being the dumbest, gloomiest old fool that ever existed." He proceeded to transfer his affections to Anne, who was closer in age, as well as more suited for the partner that he needed.

Anne, the oldest of ten children, was born in Brooklyn on January 9, 1896, to James Monroe Hewlett, a socially prominent architect and muralist, and the former Anna Willets, who was described by Alden Hatch as "a semi-invalid," which reflected only that she was almost constantly pregnant. After preparatory school at the Packer Collegiate Institute in Brooklyn, Anne attended the pioneering New York School of Applied Design for Women, and she proved to be a talented artist, particularly at needlework, sketching, and watercolors.

Left to herself, Anne thought that she might have become an illustrator of children's books. Her father encouraged his eldest daughter's

Photo studio portrait of Anne, Fuller, and Anx (1916).

artistic side, but by the time she met Fuller, she had settled into the diverting but conventional existence of a girl of her class. She appeared frequently in the society pages, and in the fall of that year, she had her debut at her paternal grandmother's house, Rock Hall, a colonial estate in Lawrence built by enslaved workers. Anne got along poorly with the other debutantes, and she engaged in mild forms of rebellion—she could occasionally be seen wearing a necktie or even smoking a cigarette.

Fuller was attracted to Anne, but he contented himself for the moment with advising Phillips on how to court her. He had moved by then to the accounting department at the Adams Brothers Company, an Armour branch in Fort Greene, Brooklyn, where he worked as an assistant cashier, weighing meat, cataloging loads, and calculating columns of figures. It taught him the trick of "casting out nines" to look for errors, as well as the concept of the trial balance, a preliminary review of the books that he would later use as a metaphor for the initial survey of any situation.

His room at the YMCA was about two miles from the Hewlett house in Brooklyn's affluent Columbia Heights neighborhood. Once Anne returned to the city in October 1915, Fuller visited her there often. When he tried to clarify her romantic status, he was relieved by what he heard: "She made it perfectly clear to me that she was *not* engaged to Kenneth, and she thought it was perfectly appropriate for me to call on her." Anne was intrigued by Fuller's scandalous backstory, but otherwise his past didn't seem to matter.

Fuller, typically, moved too quickly. Soon after their first kiss, he said that he wanted to marry her, and Anne reacted badly. She confessed that she had feelings for him but revealed that she had entered into a brief engagement at sixteen, which she had ended by breaking the boy's heart. In a note that she handed to Fuller, she concluded sadly, "You can come and see me if you want to—just as you think best—but you mustn't ever try to act as you did tonight."

She closed on a devastating last line: "I can't ever belong to you." After Fuller read it, his work supervisor noticed that he seemed ill. Taking the rest of the day off, he wrote an emotional letter to Anne that he finished at four in the morning: "You *are* the wife of my heart and soul, and nothing can stop me from fighting and fighting up till my last day just for you." They reconciled, and he courted her somewhat less persistently, but he was unable to resist proposing again in December.

Their relationship remained a secret from her family, and they shared private tokens of affection. When he asked for her picture, Anne allegedly gave him a photo that she had taken with Mark Twain, whom Fuller cut out and threw away. He spent much of his paycheck on roses and violets, and although he later dismissed flowers as "a bribe or a lever" for funerals or "to make the girl feel good," the passion behind his gestures was real.

For Anne's twentieth birthday on January 9, 1916, Fuller wrote "a little story" in the third person that opened with his father's death:

> Now his eldest son, whom one would have expected to come forward and take the reins and help his mother, seemed to lose all that sense of responsibility which had theretofore been characteristic in him. . . . Utterly selfish, he sought after all these things which were bad and led a riotous and unhealthy life, until things came to such a pass that he had to leave college and his father's friends all loyally came to his aid and gave him a piece of work to do. He seemed to do it well and they all pleased sent him back to college again and then the fool went right back to his old habits, throwing away practically his last chances.

Fuller wrote that he had been saved by "the most wonderful girl in the world," who had given him a reason to live. She was constantly on his mind as he walked alone for miles, and every night, he prayed feverishly, "Gracious God, if ought heed can be taken of the prayer of a repentant sinner, make me strong, that some day thou mayest give me Anne Hewlett for my wife."

At times, he overwhelmed her with his guilt, which led Anne to speak of their relationship more cautiously. When Fuller informed her that his progress at Armour would benefit them both, she warned him not to call it "good news for us, because of course nothing connected with your own affairs has anything to do with me at all." She also scolded him for speaking poorly about himself: "I wouldn't even like you a speck if all the 'complimentary' [sic] remarks you make about yourself were true—and you know how I love you."

As the nature of their relationship became clear to others, they had to navigate a series of obstacles. One was Edward Willets, Anne's devoted maternal grandfather, who ran a fur business across from her family's home in Brooklyn. At his urging, Anne reportedly "sort of dismissed" Fuller for a time, but she took him back after Willets passed away in January 1916.

Anne's parents were more favorably inclined. Her mother often asked questions about Fuller's frequent visits, and Anne decided to sound her out one day. "Oh, I think he's *very* attractive, and I think he has the nicest manners," Anna Hewlett responded. She liked how Fuller helped with the dishes and got along with the servants, and she said that he had "a very strong face," although she noted one flaw in his appearance: "I don't think I like those big glasses."

To save money, Fuller was living with his mother, who had moved into the Hotel Wentworth in Midtown Manhattan, and he took advantage of any chance to see Anne's family. Her father had a background in stage design, and the children—Anne, Jim, Anglesea, Willets, Carman, Laurence, Arthur, Hope, Hester, and Roger—were naturally theatrical. They performed in pageants and household skits, and they opened themselves up to Fuller, whom they nicknamed "the Butcher Boy."

But Anne's other relatives were wary of his reputation, and, according to family legend, they tried to rescue her from Fuller one last time by sending her out west. In fact, the separation lasted for just six days in May, with Anne going to Washington while Fuller was in Atlantic City, New Jersey. Before they parted, Anne symbolically gave him one of a pair of gloves, keeping the other for herself, and they arranged to meet "unexpectedly" when she came home.

On her return, Fuller escorted her back to Lawrence, where they climbed the garden wall to be alone, and she finally accepted his marriage proposal. After his mother heard the news, however, she warned Anne's father that her son was "irresponsible, thoughtless, and selfish." In response, Hewlett said that he trusted Fuller, who angrily moved out of his mother's hotel suite to return to the YMCA. Caroline eventually relented enough to give them a ruby ring that had been a gift from her late husband, and they made their engagement official in June 1916.

Fuller had landed a position as an assistant salesman in Harlem, selling "small stock, beef, provisions, and specialties," and he felt optimistic for the first time since college. He also had one eye on the Plattsburg military camp in upstate New York, which had been organized by the founders of the Preparedness Movement, including former president Theodore Roosevelt, as a volunteer program in anticipation of America's entry into the ongoing fighting in Europe. Fuller told William Hazard that his patriotic feelings were running high: "I'm so sick of listening to all these men that I meet here in the markets, most of them foreigners, who are enjoying American citizenship and yet talking like dogs about our country."

Plattsburg had been established on the assumption that America would enter the war on the side of the Allies—Great Britain, France, Russia, Italy, Romania, and Japan—and its trainees were drawn from the upper classes. On a practical level, Fuller hoped this would put him in line for a commission, and he signed up for July. Before he left for training, he and Anne spent one last night together, as he remembered to her in a letter: "We lay there, saying good night, and I could just feel my whole heart leaping out of my body and into yours for you to keep forever."

At Plattsburg, Fuller joined more than seven thousand volunteers, including four hundred Harvard men. They roughed it outdoors, but weekends and evenings were spent at the Hotel Champlain, where Fuller—who was trying to stop drinking—avoided the wilder parties. At home, Anne was called "the war bride," but they were still getting to know each other, and when she reassured Fuller in a letter that he looked like a Greek god, he said that he felt more like "a sour old fool."

He continued to live in the shadow of his misdeeds. Caroline repeatedly reminded him to make connections, and he wrote back at length "so she would not get any more foolish ideas about wasting my time up here." He also told Anne that he had never given money to chorus girls: "I have taken them to supper and danced with them but nothing more." Anne replied, "Did you really think I was really worried about your chorus girls and things? My own darling silly Bucky!"

Fuller showed a surprising talent for marksmanship, and on the rifle range, wearing his glasses, he earned a ranking as a sharpshooter. In a nod to the stories of chivalry that they both cherished, he imagined that he was a knight in a tournament shooting for the honor of "Lady Anne," a nickname that she carried for the rest of her life. He was named a corporal in his squad, and the summer concluded with a march through neighboring farms, punctuated with mock battles that had little in common with the realities of trench warfare.

After moving back into his hotel with Caroline and his sister Rosamond, Fuller arranged to rent a house on Pearsall Place in Lawrence, where he and Anne would live after the wedding. Through James Hewlett, he entered the artistic circles exemplified by the Beaux Arts Ball of the Architectural League of New York, which he attended with Anne in February 1917. At work, he was promoted to an assistant sales role at 120 Broadway, and he made his rounds "with a suitcase of grape juice" to sell a portfolio of canned fruit, bouillon cubes, and soda fountain supplies.

His longtime friendship with Lincoln Pierce pulled him in another direction. Like Fuller, Pierce was unhappy at Harvard, and, with the war looming, young men of the right social status had resources that allowed them to enlist without serving on the front lines. Ships were badly needed, and Pierce proposed that they volunteer the power cruiser from Bear Island. The *Wego* was only forty feet long, and after the US Coast Guard rejected it for being too short, Fuller decided to try the Naval Reserve.

The impending war had provided a convenient outlet for his ambitions, and he was under the impression that Armour would continue to pay his salary in his absence. On the day after President Woodrow Wilson requested a declaration of war, Fuller and Pierce headed for Rockland, Maine. With the help of "an old Owl Club man" from Harvard, they had an encouraging meeting with the head of the naval command, and Fuller worked up the nerve to ask his mother for the *Wego*, which they were fixing up with Jim Hardie when America finally declared war on April 6, 1917.

After their return to Rockland, they became local celebrities in town, but a complication soon arose. Fuller's uncorrected eyesight was so poor that he was unable to pass a mandatory vision test, and he appealed to his contacts in the service. He was advised to see a doctor in Bar Harbor, who was "a good sport," and was once again saved by his connections. As Fuller recalled, "The doctor looked at my eyes and said he had never seen better ones, and he put me right through."

Fuller was made a chief boatswain, an unusually high rank for such a young man, and the *Wego* was assigned to patrol Bar Harbor and Frenchman's Bay. As they overhauled the craft, he served as an aide to the section commander, while Wolcott Fuller and Pierce were detailed to his crew. Hanging a picture of Anne in the cabin, Fuller exercised his prerogatives as captain by pointedly taking his meals alone. He claimed later that as the only ship in the region, they were called "the State of Maine Navy," but several larger vessels had joined them by May, and their duties consisted primarily of stopping boats, taking photos, and recording the names of those on board.

The threat of U-boats attacking the East Coast was real enough, and in Rockland, they heard rumors that the Germans were planning to establish a secret naval base nearby. On the evening of June 21 they sailed to Machias Bay to look into reports by fishermen who claimed to have seen "German submarines on foggy days, when they were out tending their lobster pots." The *Wego* was supposed to lead seven ships, but one of its lines became fouled in the propeller, forcing Fuller to jump overboard to untangle it, and they were left bringing up the rear.

Anchoring in the channel, Fuller stood watch on the bow with a rifle. At two in the morning, Wolcott and a crewmate saw a series of blinking lights that they interpreted as Morse code messages between Machiasport and Cross Island, although they couldn't decipher any of the signals. Shortly afterward, a sardine boat passed by without seeing the *Wego* and vanished into the night. On their return, they reported the incident, and the fleet was sent to investigate.

The details of this encounter changed from one telling to the next—Fuller said elsewhere that it happened in the afternoon—but some kind of incident evidently did occur. After the war, Fuller alluded to an acquaintance who had conducted "secret service work" at Bar Harbor and "knew all about the *Wego*'s part in the Machias episode," which Wolcott suggested later had been a case of mistaken identity: "I have never been so much on edge or excited as I was that night in Machias Bay, when we thought that weir was a submarine."

Fishing weirs, or low dams that often carried warning lights, were common in Machias Bay, but Fuller offered a different explanation. He said that Anne's uncle, who served in naval intelligence, told him that German subs had refueled at Machiasport, and he went on to make an even more startling claim: "It was not until after the war that the *Wego*'s crew learned that it had uncovered the point of landing of Boy-Ed and von Papen and the refueling point of German submarines."

Fuller was referring to Franz von Papen and Karl Boy-Ed, who had organized a sabotage ring responsible for a bridge bombing across the Canadian border in 1915. The government expelled both of them that year, but Fuller alleged that the two merely relocated to Maine, which was home to a large population of German immigrants. According to Fuller, a Chicago businessman bought a sardine cannery as a secret refueling port, trusting that it would be disguised by the oil that it used in routine operations.

The story was accepted by several of Fuller's biographers, but no independent evidence for it ever emerged. Papen—who would later usher Hitler into power while serving as chancellor of Germany—had collaborated with an officer who was jailed in Machias, but the town had no other connection with any foreign intrigue. In 1916 the explosion of a fishing shack had led to allegations of submarine activity, which had lasted long enough to greet Fuller in Maine.

In reality, Fuller's account may have been inspired by the exploits of Miss Dixie Mason, a wartime spy who had abandoned a promising stage career to enlist in the Secret Service. After following a German submarine captain to a clandestine meeting in Newport, Rhode Island,

Mason had overheard none other than Captain Boy-Ed himself out-
lining his scheme to conduct raids on ships: "Von Papen will join us at
East Machias—after we have touched there for supplies."

It was the plot of an episode of *The Eagle's Eye*, a 1918 silent serial
billed as "a story of thrilling, heart-stirring romance." Dixie Mason
and the German conspiracy were both utterly fictional, but Fuller, an
avid moviegoer, may well have combined it with the rumors in town
to spin the story of his own adventure in Machias Bay, which was
absorbed without question into the narrative of his life.

Back in Lawrence, Long Island, Anne was overwhelmed by the stress
of wedding planning. After Caroline revealed her son's modest salary,
James Hewlett said that it was "ridiculous" to marry on so little, and
the Armour company informed Fuller that it had no intention of paying
him during his service after all. Anne impulsively applied to be a lady's
maid in New York, but after landing the post, she was so flustered that
she ran out without a word. A doctor warned that she was "on the
verge of nervous prostration or breakdown," but they finally mailed
out invitations for a ceremony to be held on Fuller's birthday.

As their wedding approached, the pressure continued to take its toll
on Anne. On June 22, 1917, she wrote melodramatically to Fuller, "I
wish I was dead. What has happened? Are you just tired of me? It's
so hard not to be able to understand, and I can't believe you don't love
me. I can't write or think or anything. I've gotten so sort of dazed by
it all. I haven't even cried, and I don't think I will cry. I'll wait until you
tell me you don't care for me, and then I'll kill myself."

Finding himself placed in the unfamiliar role of a stabilizing force,
Fuller said that his feelings for her were unchanged, and he confirmed
that he would be coming down soon with Wolcott and Pierce. When
they arrived on Long Island, Alden Hatch arranged a celebratory
banquet for the wedding party. As they dined by candlelight, Hatch
joked about Anne's silence: "It must be quite something to think
about—going off with a strange man in the morning."

Anne glanced across the table at Fuller. "You know, he isn't as strange as he looks."

On July 12 the wedding was held at Rock Hall, which one local paper described as a mansion that had hosted "the lieutenants of George Washington's day and maids of that time, with a background of dusky slaves." Five hundred guests took their seats outdoors to the strains of a harp and strings. As a grandfather clock chimed five, Anne appeared on the verandah, wearing organdy and carrying a bouquet of tea roses, and walked around to the porch to stand with her bridesmaids.

Fuller joined her nervously in his dress whites. Pierce was his best man, with Thomas Whitall, Willets Hewlett, and Wolcott among the ushers, and the officiant was the Reverend William A. Sparks of St. John's Church of Far Rockaway, where Anne's family attended services. Anne cut the cake with Fuller's sword, which he thrust back into its scabbard, still covered in crumbs, and as the musicians played a folk tune on the lawn, the newlyweds left in a flurry of confetti.

A steamship carried them to Bar Harbor, where Anx and Pierce joined them in their cottage on Snow Street, along with Jim Hewlett and his prospective fiancée, Marjorie Beard. The six of them made a lively household, and they passed the time by inventing stories about Spongee, an imaginary explorer who had supposedly discovered America before Christopher Columbus.

Fuller and his crew resumed their patrols. While investigating another report of submarine activity, they passed the filming of a Fox Film Corporation silent epic, *The Queen of the Sea*, which featured a replica ship that struck Fuller as a combination of a galley, a longboat, and the *Santa Maria*. Watching the extras straining to row it forward, he felt grateful for his own modern craft.

In August, however, its engine failed, leaving it out of commission for weeks. During the break, Fuller passed the ensign exam at the Boston Navy Yard, and after a trip to Bear Island with Anne—whose husband later noted that she found the rustic setting "less inspiring and refreshing" than he did—it became clear that the *Wego* was unfixable. He was ordered to Boston for a new assignment, and he returned the boat to the family's summer home, where he had to pay a fisherman to tow him part of the way.

On his arrival, Fuller was briefly given command of the USS *Whistler*, which he brought to the shipyard for renovations, and then the *Inca*, a larger boat with a one-pounder gun. Fuller felt that it was "the best in the fleet," and he was excited to be assigned to the naval operating base in the Hampton Roads region of Virginia. On the way down, he anchored in Philadelphia on Halloween, and he later recalled being kissed for the entire evening "by beautiful girls."

At the beginning of November, Fuller reported in Norfolk, Virginia, to Lieutenant Commander Hugh McLean Walker. To his delight, he was posted to the aviation detachment, where his commanding officer would be Lieutenant Edward McDonnell, a recipient of the Medal of Honor. Fuller was tasked with convoying airplanes on training runs and assisting in tests, which fed into his fascination with flight, and he remembered afterward how "struggling designers, builders, and fliers of the early planes became men of marked importance" during the war.

Fuller aboard the USS *Inca*.

Fuller was pleased to join their ranks in his lodgings at the Norfolk Country Club. Anne was there by Thanksgiving, and although they had to sleep in separate quarters, she snuck over to see him at night, which she thought was "awfully naughty." Fuller made a point of befriending officers, especially McDonnell's successor, Lieutenant Commander Patrick Bellinger, who roomed next door. Bellinger was a daring pilot—he would be the senior air officer present at the Japanese attack on Pearl Harbor in 1941—and he grew fond of Fuller on their shared rides to base.

Until the ice on the bay broke up, activity was limited. The most interesting work that winter was by Lee de Forest, the inventor of the first practical amplifier for broadcasting, who conducted range tests at nearby Langley Field of a radiophone for pilots. Fuller said later that de Forest spent two months aboard the *Inca*, where he "established the first successful voice communication ever heard between a ship and an airplane." This wasn't entirely true—planes had received transmissions from ships as early as 1911—and it left no trace on Fuller's orders or correspondence.

After coming down with a mild illness, Anne returned to New York in January 1918. As he watched her leave by boat, Fuller felt like crying, but he channeled his emotions into securing Walker's recommendation for a reserve officer program at the US Naval Academy in Annapolis, Maryland. Having forgotten most of his geometry and trigonometry, he took night classes at the local high school to prepare for the exam, and this may have encouraged an existing train of thought. The year before, while looking at the white wake of the *Wego*, he had estimated the "fantastically absurd number" of bubbles in one cubic foot, which he later connected to a major insight:

I said to myself, "I've been taught at school that to be able to design a model—because a bubble is a sphere—you have to use π, and the number, π, 3.14159265, on and on goes the number. We find it cannot be resolved because it is a transcendental irrational." So I said, "When nature makes one of those bubbles, how many places did she have to carry out π before she discovered you can't resolve it? And at what point does nature decide to make a fake bubble?" I said, "I don't think nature is turning out any fake bubbles; I think nature's not using π."

Fuller gave no indication of any of these thoughts in his letters at the time. In his more leisurely moments, he played with Spaghetti, a fox terrier that had wandered aboard the *Inca* one day, while his working hours were spent on ordinary duties, including conveying passengers around the harbor. He was allegedly faulted for refilling his water tanks at a convenient ferry slip, and Lieutenant Commander Bellinger had to intervene when an officer threatened to charge Fuller with piracy.

As the ice cleared, test flights resumed, and Fuller took command of six boats that were assigned to rescue downed planes. On February 13, 1918, he witnessed an accident that occurred after a seaplane was unable to recover from a nosedive in time. It crashed where the water was only three feet deep, and the pilot was "all cut up" but otherwise unharmed.

Exactly one month later, the base suffered its first fatality. Three planes were caught in an "air pocket" and nearly collided, with one aircraft losing a wing and plunging into the sea. One man was rescued unconscious, but the pilot, Leslie M. MacNaughton, was trapped under the surface. Three men, including a lieutenant named Clifford L. Webster, dove repeatedly into the frigid water to free him, as Fuller's crew stood by to pull out the rescuers. After finally extricating MacNaughton, they tried for an hour to revive him, but failed.

Fuller was haunted by the incident, which purportedly inspired him to develop a device for his boat: "I invented and had installed a powerful winch to pull the airplanes out of the water in a hurry." What he described as "a seaplane rescue mast and boom" supposedly proved its value in a dramatic fashion. When another plane went down, the *Inca* was there in ninety seconds, and after the winch lifted the aircraft, the pilot was stretched out on deck and saved. It was exactly like the Mac-Naughton crash, but with a happy ending.

If this rescue ever happened, it went unmentioned in any official records. Neither was there evidence of the claim that winches were

mounted in all the boats, saving "hundreds of pilots' lives," or Fuller's
statement that the base had been losing "an aviator a day." In reality, there
were only two fatalities during his time in Hampton Roads. He never
discussed any rescues in his correspondence or earliest recollections, and
although he said later that his invention "won me an appointment" to
Annapolis, it was conspicuously absent from his application materials.

Fuller found other ways to cultivate Bellinger, his commanding officer,
for whom he carved a wooden map to plot flights. He claimed later that
he flew an airplane himself at the base, and he had long conversations
with Bellinger about aviation, in which he shared his concept of "jet
stilts"—a propulsion method that used thrusters to power planes upward
like wild ducks. His separation from Anne remained difficult, and he
grew jealous when she mentioned other men in her letters. She returned
to Virginia in March and, shortly afterward, she became pregnant.

After placing a respectable tenth out of one hundred on the Annapolis
exam, Fuller headed for the Naval Academy in June. Fuller was worried
about his eyes—he was so nervous that he threw up upon his arrival—but
felt reassured by the sight of others with glasses, and he immersed himself
in what he called a rigorous course in "navigation, ballistics, strategy,
handling of ships, everything necessary to become an officer-of-the-line
familiar with the duties of all officers of all ranks and levels of responsibility
and able, if all other officers are killed, to take over the fleet."

This was more than slightly exaggerated. Fuller was participating
in a special three-month program with narrowly defined objectives,
as Admiral Edward W. Eberle, the superintendent, made clear at the
time: "Our plan has been to equip the reserve officer for a limited and
particular task to be performed during the period of the war, and no
attempt has been made to qualify him for all the duties at sea required
of a graduate of the Naval Academy."

The experience still made a lasting impression on Fuller, who would
refer to Annapolis long afterward when he championed the ideal of the
generalist. An officer, he said, had to handle any situation that arose
when he was out of touch with home, which might include establishing a naval base halfway across the world. Such work demanded "very
comprehensive men," and Fuller regarded himself as a member of the

last generation to receive this education, which was rendered unnecessary after radio made it possible to transmit orders from a distance.

Meanwhile, Anne was having a harder time. She had moved back in with her parents on Long Island, only to have her mother treat her like a chambermaid, and she was upset at the prospect of having a "horrible baby" so soon after their marriage. Anne felt that it was awful "to have to sacrifice everything you love for something you hate and despise and detest," prompting Fuller to respond cautiously: "Please try to get used to the baby enough not to hate it anyway."

Fuller was distracted by an investigation of some unpaid mess accounts left by the crew of the *Inca*, which felt like a return to his problems with money, and the pregnancy remained a sensitive subject. Alluding to women friends who had suffered miscarriages, Anne hinted that she longed for a similar outcome. "It doesn't matter how sick it makes me," Anne pleaded. "Please find out if there isn't a way to bring it on." For the moment, Fuller quietly filed these words away, and he arranged for Anne to join him in July.

After passing his exams, Fuller graduated from Annapolis on September 18, 1918. He was posted as a navy ensign to the USS *Finland* in New York, but when he arrived, he found that the ship had been delayed until November. In the interim, he would serve as an assistant communications officer to Vice Admiral Albert Gleaves, the head of the US Navy's Cruiser and Transport Force, which conveyed troops to France. Gleaves's flagship was the USS *South Dakota*, but Fuller worked out of the information office in Hoboken, New Jersey, compiling the official statistics of the transport service.

In late October he was ordered to the USS *Great Northern*, which was carrying five hundred men to French naval bases, along with Rear Admiral Alexander S. Halstead. As Fuller boarded, he couldn't bear to look back at Anne, but he was excited by the scale and danger of his first overseas trip. The young ensign worked on decoding watch, and he was impressed by the refrigerators that allowed ships to enjoy "the finest cream," while the infantry had to drink canned milk—an observation that led him to reflect on the impact of technology on the everyday standard of living.

In France, they transferred the men to shore in Brest, which was alive with rumors of an imminent end of the war. They left the next day, and on November 11 a radiogram announcing a cease-fire agreement was released by Captain Stafford Doyle, who reminded his men "to always remember the necessity for a great navy." Fuller took these words to heart—he came to see the naval defense of supply lines as a key element in the balance of power—and he plunged back into his work tracking ships as an information and statistical officer.

Anne was living with her parents in anticipation of the delivery. When she went into labor, she was joined by her husband, as well as Lincoln Pierce, who comforted her by cracking jokes on the ride to the hospital. Alexandra Willets Fuller, called variously Sandra, Xandra, and Zandra, was born in Brooklyn on December 12, 1918. Pierce was her godfather, and although Anne had a slow recovery, her negative feelings toward the baby—whom she named after the queen she had once glimpsed in England—vanished at once.

The Spanish flu pandemic was raging, and despite the family's precautions, Alexandra became sick in Lawrence. In January she came down with a fever, followed by spinal meningitis and what Fuller called "infantile paralysis"—another term for polio—although it did not appear in her medical records. After receiving antiserum injections in her spine, Alexandra recovered, but she suffered from hemiplegia, or weakness on one side of her body. By spring, the only trace of illness was on the left side of her face, which was slightly paralyzed.

Fuller was working hard at the office in Hoboken, where he had come up with an idea for a magazine, the *Convoy*, devoted to the Cruiser and Transport Force. It was modeled on existing service publications that were funded by advertising, and he received permission to edit it alongside his regular duties. The work consisted largely of writing profiles of senior officers, allowing him to meet many of the important men in a unit that he had joined only recently, and it became his first successful business enterprise, with a peak circulation of fifty thousand copies.

In the spring, Fuller went on his last voyage. At the Paris Peace Conference, President Wilson unexpectedly called for the USS *George*

Washington, which left New York on April 11, 1919. After Anne saw him off, Fuller prepared to assist Vice Admiral Gleaves's chief of staff while continuing as an assistant communications and information officer. His position gave him access to secret messages, shaping his ideas on clarity in "large teamwork maintenance of colossal dynamic patterns" and kindling an interest in keeping a chronological record of his life. In Brest, he was granted five days of leave, which he spent in Paris and Italy before heading back at the end of the month.

On the way over, Fuller had written to Anne that he had spoken by radio with New Brunswick, New Jersey, and he later made a notable claim: "The human voice was transmitted trans-oceanically for the first time in history, as man was heard through the receiving instrument in Arlington, Virginia, speaking over the transmitter in the radio shack atop the USS *George Washington* at anchor in Brest, France." Elsewhere, he said definitively, "I was there when the first transoceanic telephone call was made by Wilson from the steamship *George Washington* to Arlington."

The latter statement was demonstrably incorrect—the ship returned with Secretary of War Newton D. Baker, but without Wilson—and the former had to be carefully qualified. Transatlantic radio messages had been sent as early as 1915 from Arlington to Paris, and experiments on Fuller's trip produced mixed results. According to one eyewitness, "the New Brunswick radiophone was heard, but not understood" on the way home from Brest, and the most successful call took place when the ship was just two hundred miles away from Washington.

In years to come, Fuller would describe World War I as "an enormous metallurgical war" that forced nations to modernize, especially in the fields of alloys and radio, which inadvertently yielded permanent improvements to society. It led to a brief flurry of radical fervor across America, followed by an equally intense period of reaction, but Fuller's concerns for the moment were more personal. Gleaves was about to take command of the US Asiatic Fleet, and there was a strong prospect that he would request Fuller as his aide.

Fuller said later that he was reluctant to be separated from his family by a role that might take him to the Philippines, but his finances were the real consideration: he was confident that he could earn more back at the Armour company, where his wartime experience would be an advantage. He proposed that he rejoin his former manager at the branch in Rio de Janeiro, Brazil, but he was offered his old position instead. Although Gleaves was reluctant to let him go, Fuller's resignation was accepted on August 28, 1919.

He had left his beloved navy, in which he had suffered only some minor hearing loss, to support his wife and daughter. Behind the wheel of their first car, a Chevrolet that he had bought with Anne's money, Fuller went to join them in their Pearsall Place rental in Lawrence. He was justifiably proud of his record, which had redeemed him even in the eyes of his mother, who wrote that she was pleased to hear "that little Alexandra is doing so well and is so cunning."

The rest of the year passed pleasantly. Alexandra seemed healthy, and they often saw the Hewletts at Martin's Lane, the family's elegant shingle house, where Fuller wrote and performed light verse for skits. James Hewlett introduced them to friends from artistic circles, including the Italian-born sculptor Antonio Salemme and his wife, Betty, both fixtures in Greenwich Village. One night the Salemmes brought them to a tavern on Christopher Street, and Fuller had his first encounter with its owner, Marie Marchand, a woman in her thirties universally known as Romany Marie.

Fuller would be closely associated one day with the Romanian restaurateur and former anarchist, who oversaw a thriving salon for bohemians and intellectuals. For now, he was a tourist in this world, and he spent most of the evening in conversation with Eugene O'Neill, who could frequently be found staring into the tavern's fireplace. Fuller and the playwright never met again, but James Hewlett collaborated with O'Neill the following year on the sets for *Beyond the Horizon*.

In the fall, Fuller went to Chicago to see Arthur Meeker of Armour, hoping for "bigger work," and landed a sales role in the West Indies export department at 120 Broadway. Money was still tight—his salary remained stuck at forty dollars a week—and Fuller felt that the New York office manager failed to appreciate how hard he was working. He confessed to William Hazard, "I have not told Anne how badly we are off, for I don't want her to worry."

His professional concerns were interrupted by a succession of tragedies. On January 11, 1920, Anne's brother Willets was killed when his car collided with a telephone pole in Lawrence, and two months later, her mother passed away of acute kidney disease. Anne and Fuller spent the first part of the summer with the grieving James Hewlett, who wrote gratefully that they had been "a comfort and joy."

They went afterward to join Fuller's mother at Bear Island, where Anne stayed with the baby while Fuller returned to Armour. Alexandra was looking strong, and Anne would wrap her up warmly to bring her to the beach. She wrote to her husband, "I seem just to go around in sort of a dream remembering everything when we were together." He replied, "Dearest little girl, take very great care of yourself and our little baby. . . . She is God's manifest of his blessing upon *our* love."

In December Alexandra caught pneumonia. She was only two years old, and her fever, which lasted for three weeks, left her with empyema, in which the pleural cavity of the lung fills with pus. A case of abdominal paralysis persisted for two months, leaving her badly constipated— Anne administered an enema every morning, which was difficult for them both. Alexandra was confined to her bed, and Fuller thought that her lack of mobility led her to experience the world in other ways, just as his poor vision had heightened his sense of smell.

For Alexandra, Fuller said, the effects of her condition were even more profound: "We discovered she was extremely sensitive to other persons, their feelings, and so forth. My wife and I would be about to communicate—I would be about to say something to my wife—and the words would come out of my daughter's mouth as she lay over there in her bed. My wife and I would both turn, for she wouldn't be using her kind of words, but ours." He concluded that Alexandra was reading his mind, and it encouraged his belief in telepathy.

Alexandra recovered enough to walk normally, but the cost of her care forced Fuller to ask Armour for a loan, which he received, along with a promotion to assistant export manager. On a typical day in what he later called "one of the most material and highly organized of necessary industries," he might correspond about pineapple sales in Martinique, tallow prices in Suriname, and beef advertisements in British Guiana, and the sheer number of minor transactions wore him down.

During this period, he reconnected with Edward McDonnell, who had briefly been his commanding officer at Norfolk. McDonnell directed a reserve battalion in the Third Naval District, for which he asked Fuller to establish a unit in Nassau County. It was a welcome source of money and contacts, and he asked Gleaves for help in securing an Eagle Boat—a steel patrol craft built by the Ford Motor Company—for a training cruise in June 1921.

Alexandra had come down with tonsillitis in the spring, and she suffered fever for a month. In July she seemed well enough to accompany her mother to Bear Island, but she continued to have digestive problems. After one particularly difficult day, Anne wrote to Fuller, "I thought of how you and she and I would be doing things together when she's older and has outgrown all these things."

When Alexandra experienced another fever and a rash, they brought her to Boston Children's Hospital, where she was diagnosed with kidney inflammation. After she was successfully treated, the three of them returned to a new apartment on Pierrepont Street in Brooklyn, but when their doctor advised that Alexandra should spend the winter in a warm climate, they decided that Anne would take her to Bermuda.

Anne and her daughter departed on January 4, 1922, accompanied by Anx and their friend Gladys Bedford. Fuller saw them off at the pier, waving his red handkerchief as they sailed away. He wrote later to his wife, "My heart certainly jumped into my throat and is still sticking there at seeing you and Zandra standing there and slipping away from me."

Almost at once, Fuller was distracted by an opportunity from his network of contacts. Edward McDonnell had taken a position as general manager of the Kelly-Springfield Motor Truck Company, and he offered Fuller a chance to handle its national sales for a weekly salary of more than seventy dollars, along with commissions in his territory. It would draw on his experience at Armour, which was a large prospective buyer of trucks, and he was excited enough to accept the job without discussing it with his wife.

In the meantime, Anne had arrived in Bermuda after a calm voyage, which Alexandra spent "tearing up and down the deck all day long." On their arrival, their rented cottage in Hamilton turned out to be green with damp, with an "unspeakable" bathroom, and Anne spent the night on a sodden straw mattress. When she saw Alexandra playing on a patch of ground covered in weeds, instead of the beach that her husband would be picturing, she could only laugh.

With the help of a friendly American couple, they arranged for a better rental. Anne hired an eighteen-year-old named Susan Napier, "a mixture of colored and Portuguese," to see after housework and watch Alexandra, and the travelers began to explore the area around the capital city of Hamilton. When they visited the Crystal Cave, with its beautiful rock formations, Alexandra kept murmuring, "Isn't that lovely."

Later that month, Fuller finally told Anne about his new job, which appealed to him more than anything since the navy. "As you know," Fuller wrote, "I am really more interested in motors and mechanical things than in any other work." Anne was supportive, and after visiting the firm's factory in Springfield, Ohio, he settled into the sales department in New York. It seemed like a promising position. Trucking sold itself as the "big brother" to the railways, since it extended the transport network established during the war from the rail system to the roads, and Fuller would focus on national clients that were standardizing their trucking equipment.

In Bermuda, Anne settled into a comfortable routine. On a typical morning, Alexandra played on a neighbor's verandah, gathering stones in a pail or tying silk snippets around her waist. Anne bicycled into town for fish and vegetables for lunch, followed by beachcombing with her daughter, who held tea parties in the sand. In the afternoon, Napier took Alexandra to see goats or a cow being milked, and after she went to bed, Anne read and dealt bridge hands to herself.

Alexandra loved Napier, who prompted the girl to announce one day, "When I'm bigger, Daddy's going to buy me a bicycle and a pony and a sailboat and an umbrella." On hearing the story, Fuller wrote back, "I am so happy that I am getting into a place where I can really take care of you and protect you. . . . We may even be able to arrange for the bicycle, pony, sailboat, and umbrella. That was wonderful." When he promised to give up smoking and alcohol, Anne responded, "I don't mind your drinking or playing around with other girls."

On May 1, 1922, Anne returned from Bermuda, along with Alexandra and Napier, and rented a house with her married sister Laurence on East Rockaway Road in Hewlett, New York. Fuller remained visible in social circles, and he applied to the Police Reserve Aviation Division, a group of volunteers—led by Fiorello H. La Guardia, the future mayor of New York City—that used seaplanes for fire and harbor patrol. He attended drills in uniform, but he was fonder of producing his badge to get out of speeding tickets.

At work, Fuller's only sale consisted of a single truck to Armour, so he tried to stand out in other ways. After witnessing a display of "dishonesty" at two distributors—they used parts from new trucks to repair old vehicles—he reported it, and he suspected a conflict of interest with Kelly-Springfield's bank, J. P. Morgan Co., which was steering business to the Mack Trucks company. He went to a prospective client's office in Boston to investigate, and although no deal ensued, he thought that he had impressed the bankers as a "bright young man."

But all his efforts came to nothing. In June Kelly-Springfield's president, Charles Young, abruptly closed the national sales department: "National users are not now, and probably for many months in the future will not be, seriously in the market for additional truck equipment. Of course, you realize, as we all do, that the Kelly Company must be operated as economically as possible, and so it is with sincere regret that we must deprive ourselves of Mr. Fuller's efficient services."

It was his most crushing setback since his forced departure from Harvard. Much later, Fuller implied that the entire company had folded under pressure from J. P. Morgan, and he put the blame squarely on

McDonnell: "Eddie gave me a job very high in a corporation he had just been made president of, not telling me he might be discontinuing it, which he did." In fact, McDonnell had been his most consistent supporter, and Fuller's references to the "termination" of the whole firm were untrue.

His unemployment left him on strained terms with Laurence's husband, Robert Burr, who dismissed Fuller as "a ne'er-do-well." Alexandra was hospitalized again, and as the life that he had been building seemed to crumble, Fuller started to drink. He had been counting on a great success, and he began to show a side of himself that Anne had never seen.

There was one ray of light. After Fuller unsuccessfully sought a position at the International Merchant Marine, its general operating manager, Kirkwood H. Donavin, offered him the command of another Eagle Boat for the summer. In July the USS *Eagle 15* left on a training cruise to Connecticut, and between the water races and baseball games, Fuller's mood improved. Anne wrote him letters, and Alexandra

Alexandra Fuller.

scribbled on one page with ink, telling her mother, "Daddy will be crazy about this!"

A second cruise was led by Vincent Astor, who had inherited a fortune after his father, the business magnate John Jacob Astor IV, died on the *Titanic*. He and Fuller had a common friend in Herman Oelrich, and they spent the evenings discussing "our mutual interest in everything to do with boats and mechanical developments." The squadron hosted several distinguished guests, including Rear Admiral William Sims, who had commanded the naval forces in Europe. Fuller was reminded that he was still part of a privileged world, and he felt comforted by these signs of his social position.

In September Fuller spent the night on Astor's private yacht, the *Nourmahal*. Astor was heading on a trip overseas, and he offered to lend his monoplane, a Loening S-1 Flying Yacht, for Fuller to exhibit "the efficiency of the New York Naval Reserves air force and its ability to give to the War Department information which can be obtained only by the use of the seaplane."

Fuller saw the trip as a respite from his problems. Leaving Alexandra with her grandfather, he flew with Anne to weddings in New York, Boston, and Long Island. It felt like a belated honeymoon, as well as the life that they might have had under other circumstances. Their travels attracted extensive press, with one article stating that Astor had provided the plane "after they found themselves with so many wedding invitations that they couldn't attend all of them any other way."

In later years, Fuller claimed incorrectly to have been "the first man ever to land an airplane in the state of Maine," but the pilot throughout the trip was Clifford L. Webster, whom the *Inca* had assisted during the failed rescue of Leslie MacNaughton in Hampton Roads. Hoping to impress his family, Fuller flew the plane to Bear Island, but Jim Hardie had doubts about its safety: when they went on a flight over the water, he brought along an oar.

After returning the airplane, Fuller and Anne attended Anx's wedding in October, where Alexandra struck the guests as "big and cunning." Fuller joined the Police Reserve at that year's World Series between the New York Giants and the New York Yankees at the Polo

Grounds, ordering spectators to extinguish their cigarettes, and in early November he reunited with Astor, whom he entertained with a visit to the bohemian world of Romany Marie's.

On November 11 Fuller went to Harvard Stadium to see the Crimson football team take on the Princeton Tigers. He carried a walking stick, which was waved at games, and as his wife and daughter accompanied him to the train, Alexandra asked her father, "Daddy, will you bring me a cane?"

Fuller promised that he would, but he promptly forgot all about it. After Harvard lost, he spent the night with college friends and returned to New York the following afternoon. When he phoned the house from Penn Station, Anne sounded haggard: "Alexandra is very sick. She has pneumonia. She is in a coma."

Rushing back to the Hewletts' home at Martin's Lane, Fuller found his daughter still unconscious. The doctor and nurses could do little for her, and her parents sat at her bedside throughout the agonizing days that followed. On November 14, 1922, a month short of her fourth birthday, Alexandra died in her father's arms.

Fuller later attributed her illness to the "contiguous displacements" of the war and "the messy world of ignorance and superficiality in houses." Alexandra had actually been living in comfort with Anne's family, but he was correct in blaming overcrowding and poor hygiene for the flu pandemic that caused her earliest health problems, and his ensuing views on decentralization and sanitation would be profoundly affected by this devastating loss.

For now, these considerations were far from his thoughts. Fuller had associated his daughter with the love of God, and her death was the greatest tragedy that he had ever faced. He was not alone in his grief, but he felt isolated by the knowledge that he had been away during her final decline, and he was filled with shame for years afterward by one of the last things that she said. At some point in the night, Alexandra had opened her eyes, looked up at her father, and smiled slightly. "Daddy, did you bring me my cane?"

THREE

STOCKADE

1922–1927

It is a basic principle that you only discover what you had had
by virtue of losing it.

In his standard account of his life, Fuller always placed the turning point
in 1927. He retroactively credited that year with nearly every important insight that he ever had, contrary to all available evidence, and even
backdated his sketches to fit the chronology. In reality, the hinge moment
had occurred five years earlier. Alexandra's death, which haunted him
forever, was his single most traumatic loss, and in the aftermath, he
encountered many of his lifelong obsessions for the first time.

If he downplayed the significance of this period, it was partly because
it was too painful to remember for long, but he was also unable to discuss
it without emphasizing the contributions of his most significant mentor.
"Mr. Hewlett was the first man ever to tell me that my own ideas were
valid," Fuller said, which was true—but it would have been even more
accurate to acknowledge that many of these ideas had arisen through
James Hewlett himself.

Fuller and his father-in-law were united at first in mourning. Alexandra's funeral was held on November 16, 1922, at St. John's Church

in Far Rockaway, five years after its rector had officiated at Anne's wedding. After they buried their daughter with a tiny headstone in a Quaker cemetery in Westbury, New York, Anne took refuge with her father and siblings, who suffered another tragedy when her sister Laurence's husband, Robert Burr, died from pneumonia.

Anne kept her faith, writing in response to a letter of condolence, "I believe truly in the immortality of the soul—not in a dread merciless and punishing power but in one most wise, most tender." As for Fuller, his mind often returned to the opening stanza of a poem by the writer Christopher Morley, "To a Child," that he had read shortly before Alexandra died:

> The greatest poem ever known
> Is one all poets have outgrown:
> The poetry, innate, untold
> Of being only four years old.

The poem took up permanent residence in his memory, especially a phrase that evoked how children, with their otherworldly innocence, accepted the mystery of life with all its contradictions: "Your strange divinity still kept."

He was also drinking, which he played down in his recollections: "I was never what you would call an alcoholic. The drinking side in those first years was really for companionship. . . . Drinking did not make me testy as it does some people, but rather very warm and optimistic; overoptimistic. This was a false kind of manner, which I did not like, for it made me undertake to do more than I could carry through and would often get me into trouble."

Fortunately, this energy was diverted at once into a project envisioned by Anne's father. Until then, Hewlett had served as Fuller's guide to the world of art and fashionable society, which was a prelude to his initiation into the crafts of building, engineering, and design. Fuller was about to discover that James Hewlett had remarkable ideas about housing, and he might never have found his life's calling if he had married into any other family.

James Monroe Hewlett was born on August 1, 1868, in Lawrence, Long Island, and studied at Columbia University and the École des Beaux-Arts in Paris. Following a stint at the legendary architectural partnership of McKim, Mead & White, he started his own firm. Hewlett was best known as a muralist, but he designed many buildings, including the City Club in Manhattan, and presided over the Brooklyn chapters of the American Institute of Architects and the Architectural League of New York.

Hewlett occasionally worked on the stage—he designed sets and costumes for a production of the Metropolitan Opera—but his grandest achievement would be seen by millions of commuters. In 1912 the French painter Paul César Helleu conceived of the celestial ceiling of Grand Central Station, which was entrusted to Hewlett and artist Charles Basing to execute. After it was unveiled, however, an attentive observer noted that the constellations had been painted backward.

According to Hewlett, the mistake had been caused by a miscommunication. An architect's plan was normally drawn with a viewpoint from the ceiling, leaving an inverted design on the page, so Hewlett provided two drawings—one reversed, one correct—with instructions. When he left on another assignment, the contractor assumed that he had forgotten to reverse the plans, which were changed yet again. It was too expensive to fix, but despite the confusion, the idea of a vault with a view of the stars would linger in Fuller's imagination.

Hewlett was an avowed generalist—he was one of the founders of the Digressionist Society, which held up Leonardo da Vinci as a model for architects—and Fuller came to see his father-in-law as an example of how an unconventional background, with experience in painting and stage design, could foster creative solutions in architecture. At the moment, however, they were connected most profoundly by loss. By 1922 Hewlett had suffered the deaths of his wife and a son, which freed him to embark on a venture that might otherwise have seemed excessive for a man in his fifties.

His objective was to reduce the inefficiency of conventional housing, which was remarkably like the goal that Fuller would one day set for himself. During World War I, residential construction had slowed to a standstill, and the migration of workers from rural areas into cities resulted in dangerous overcrowding. The housing shortage had only worsened after the war, due to rising prices for materials, and commercial builders concentrated on houses for the high end of the market, which neglected buyers with lower incomes.

In Europe, the challenge of rebuilding from the ground up produced the philosophy of the Bauhaus school in Germany, which called for the construction of worker housing using modern industrial methods. The situation in the United States, which had survived the war with its cities intact, required a more incremental approach. Rather than proposing a radical redefinition of architecture, Hewlett wanted to build traditional houses with greater efficiency. Masonry, while durable, was wastefully heavy, so he decided to find a substitute.

His solution was based on an organic fiber, such as wood shavings, mixed with plaster or another binding agent. When dried in a mold, it yielded a lightweight block the size of a standard brick—sixteen inches long by eight wide and four high—that weighed just two pounds. Its defining feature was a pair of holes in each of the blocks, which were laid in courses on a temporary scaffold, using metal clips like elongated croquet wickets. When aligned correctly, the holes made a vertical shaft in which concrete was poured, and the result was virtually indistinguishable from brick.

After developing the system on his own, Hewlett filed for a patent on May 26, 1922. He also brought it up with Fuller, who hesitated at first to get involved, although he mentioned it to Vincent Astor on one of their training cruises. After Alexandra died, Fuller considered participating more seriously, but he framed it as a favor to Hewlett: "He didn't seem to know what to do with his invention, so I decided I would do something about it."

It was really the lifeline that he needed, and, like Hewlett, whose personal losses had given him the courage to assume a professional risk, Fuller was ready to reinvent himself. In January 1923 they founded

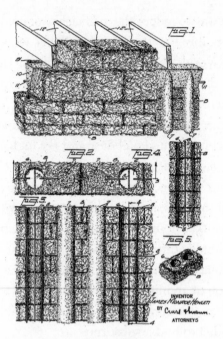

Patent illustration of the Stockade Building System by James Hewlett.

Stockade Building System—a name that was meant to evoke a protective barrier, not a military prison—to raise money to acquire patents "for concrete stanchions."

To perfect the formula for the blocks, Fuller set up shop in the barn at Rock Hall. It would be a messy task, but it was the kind of clearly defined mechanical problem that he liked, and it offered his first opportunity in years to work directly with machines. Fuller dove into it, but he was still drinking, and the time that he spent on the project took him away from Anne, who was more invested in emotionally supporting her own large family.

After experimenting with straw fibers, peat moss, cornstalks, and even "seaweed from the Sargasso Sea," Fuller settled on wood excelsior bound with lime plaster and sugar. At first, he packed it tightly, which drew moisture from the concrete, causing it to set incorrectly. He switched to coating the fibers with a minuscule amount of binder, working it with a pitchfork, "the way you'd pull spaghetti out of a bunch," and tamped it down just enough to fill out the mold.

Fuller successfully used these blocks to construct a garage on the family property in Lawrence, and to refine them for mass production, he moved his operation to Hewlett's studio in a converted roller rink in Brooklyn. For his first foray into production engineering, he had to build a track and kiln himself, and he developed an improved system in which excelsior was fed into an ensilage chopper, blown into a tub of cement, routed into a second cutter, and packed into a mold that was rotated into place on a turntable.

It was a pivotal chapter in his education. Fuller brought on the mechanical engineer Robert McAllister Lloyd, the father of a friend, and for advice on fibers, he visited a felt manufacturer and dined with Scott H. Perky, the son of the inventor of shredded wheat cereal. He was insecure about his qualifications, and whenever he had an idea, he would run it past a colleague first: "If he was enthusiastic enough about it, then I told the man with money that the mechanic had invented the thing, because I was pretty sure he trusted the mechanic's judgment."

They opened a plant in a converted lumber mill in Summit, New Jersey, but it had barely begun to produce in volume when it caught fire and collapsed in the early morning of August 31, 1923. Although insurance covered the loss of machinery, any profit was at least two years away. In the language of future generations, Stockade was a start-up, and it faced the same challenges that other founders confronted in securing investment capital. If it was positively received from the beginning, it was because it had genuine attractions: it drew on Hewlett's reputation, could be easily explained to investors, and was perfectly timed to address the postwar demand for construction.

Not surprisingly, their first round of support came from friends and relatives. Edward Larned, Leslie's husband, had been a star tennis player two decades earlier alongside his oldest brother, William, and his family owned extensive real estate holdings in Summit. Larned provided a loan for another New Jersey plant, but when the company was unable to repay him on schedule, he closed down the factory.

Despite this setback, Stockade continued to expand. Fuller was optimistic about their ability to undersell competitors, but only if the price compared favorably to that of brick, and to cut transportation costs, they needed a plant in each potential market. Accordingly, the Stockade Blue Ridge Corporation was founded in Baltimore, while a factory in Brookline, Massachusetts, was financed by Anx's husband, George Abbot, most likely with Anne's encouragement.

After a test at the Massachusetts Institute of Technology, the system was approved for use in Boston, and Fuller oversaw the installation of equipment in every plant. He became especially interested in the vats, which were assembled out of wooden staves wedged together by steel bands, leading him to think at length about the forces of tension and compression. Fuller also continued to tinker with the pneumatic forming process, and he filed a patent in his own name—the first of many—on December 31, 1924.

The Stockade system debuted publicly in April 1925, when it withstood another load test at the Grand Central Palace exhibition hall on Lexington Avenue. As its president and chief salesman, Fuller was in charge of promoting it, and he was granted a free hand. The previous summer, Hewlett had married his second wife, Estelle Rodgers Wilbur, and adopted her five-year-old son. With a spouse and small child back in his life, Hewlett was glad to delegate practical matters to his younger partner.

After raising money to repurchase the New Jersey factory from Larned, Fuller began to sell the system to builders, which tested all the skills that he had acquired at Armour. He relied at first on contacts from Hewlett, who provided letters of recommendation, and he implemented a clever cosmetic change. The blocks were resistant to fire, but the wood fibers made them appear flammable, so Fuller simply used a carbon black powder to turn the color gray.

It was a pioneering application of an insulating concrete form, and many of the benefits that Fuller emphasized in selling it became central to his future work. The system was based on unconventional materials, and its standardized modules could be combined in countless ways. Because they were light, the blocks for an entire house could be transported on a single truck; they could be assembled quickly by relatively unskilled workers; and they could be held easily in one hand—a convenience that he would often seek in later projects.

The real difficulty lay in persuading contractors, and in the mean-time, Fuller had to build the houses himself. At every stop, he found himself repeating the same procedure: "I would have to, then, find a rather prominent client, an architect, and they would be able to get me somebody they knew in the town council. . . . And then they would say, 'We will give you a special permit, provided you have tests made at our particular university.'" After the authorities were satisfied, the resulting order would often amount to just a few truckloads.

By January 1926 the New England branch was finally showing a profit, and the company prepared more patents, including a joint filing by Fuller and Hewlett for a refinement of the underlying wall structure. Negotiations were under way for subsidiaries in three states and six foreign countries, including Canada, but the most promising region was Chicago, where they opened a new office. Fuller was renting an apartment on East Ninety-Fifth Street in Manhattan, while Anne focused on life with her family in Lawrence, but business commitments often took him away from New York.

He was feeling pressure at work, and tensions at home sometimes erupted in jealousy. One night at Martin's Lane, he became angry when Anne flirted with Alrick H. Man Jr., a charming Yale graduate and tennis star. After glowering in silence, Fuller left the house and walked twenty miles to the city, crossing the Queensboro Bridge at dawn. Alden Hatch speculated that Fuller stayed "perhaps in speakeasies or empty warehouses, perhaps in some girl's room" for the next two days. When he returned, Hatch said, Anne took him back without any questions.

The truth was that their marriage had been tested by Alexandra's death, and it seemed best for them to spend some time apart. In April 1926 Fuller visited the Chicago office, which was headed by Frederick V. Burgess, a family friend from New York. Burgess was a Yale war hero, nicknamed Bunny, who moved in fashionable circles—his background provided material to F. Scott Fitzgerald for the character of Nick Carraway in The Great Gatsby. He was a poor manager, however, and Fuller decided to oversee Stockade Midwest himself.

At that point, Fuller had been working on Stockade for over three years. He had made significant advances on the mechanical front, and he was starting to show results in sales, but he had yet to prove that he could build a successful subsidiary in the absence of his partners. This was his next great challenge, and he would do it largely by himself. Anne decided to stay behind in Lawrence, and when Fuller departed for Chicago in June, he was on his own.

Upon arriving in Illinois, Fuller had reason to think that it would be a defining chapter in his career. Until then, he had been frustrated by the slow pace of progress at Stockade. He had built many houses, but the hassles of dealing with regulations, banks, and insurers meant that no one project ever led to others: "There was absolutely no momentum from the thing you'd done before."

Fuller saw Chicago, by contrast, as "pretty much the center of the business and building material world," and he was drawn to it by many of the same factors that had attracted the young Frank Lloyd Wright exactly four decades earlier. Geography had formed the city into a thriving commercial hub with a river through its heart, and after the Great Fire of 1871, which Fuller's mother had witnessed, it had been rebuilt into a metropolis like nothing else at the time in America.

Because its downtown was constrained by water on three sides, the city had been forced to rise. Chicago had perfected the skyscraper, while a demand for new techniques produced the balloon frame for wooden structures and the skeleton frame for steel. It was a major market for building materials and machinery, as well as the home of many architectural firms, and Fuller understandably viewed it as a stage on which he could put his business plans into practice.

He focused at first on capitalizing Stockade Midwest. After Bunny Burgess left the office, Fuller found an unlikely ally in Andy King, the cousin who had once bullied him on Bear Island. King, a stockbroker, agreed to provide backing and seek subscriptions from his affluent contacts in Lake Forest. They were joined by none other than Wolcott Fuller, a recent graduate of the Harvard School of Engineering, while Anne recruited her brother Carman in New York.

Fuller had been entrusted with the family concern, and he needed to show results. As he sought approval to operate in Chicago, he closed a deal with the developer Arthur T. McIntosh, who wanted the Stockade system for a housing tract south of the Morton Arboretum in nearby Lisle, Illinois. To amortize their costs, Fuller was convinced that they had to scale up dramatically, and he put up two model homes that were inspected by potential buyers in August.

He was emboldened by his distance from New York, but it also caused logistical difficulties. Checks had to be mailed for his signature, and he was slow to respond, which revived familiar doubts over his reliability with money. After learning about the terms of the stock options that he granted to investors, the board of directors was annoyed that the parent company had been left with a smaller percentage than usual. Carman Hewlett complained that his brother-in-law had been "soft," which prompted Fuller to push back: "The Chicago crowd are a very strong and powerful group."

To consolidate his position, he decided to open a new factory. Until then, his houses had been built using blocks from the East Coast, and a local plant would cut costs in the long run. Fritz E. Schundler, "a funny little Dutchman" who owned the nation's leading producer of magnesite cement, agreed to provide space in Joliet, which Fuller saw as his best chance to establish the massive operation that he envisioned. He withdrew from his involvement with the other offices, which were losing money, and closed the Long Island branch entirely.

When James Hewlett had to liquidate some of his holdings to manage his personal finances, Fuller seized the opportunity to further strengthen his control. Hewlett advised him to sell to a Chicago buyer who would support his efforts, and Fuller approached Farley Hopkins, a Stockade Midwest investor with a background strikingly similar to his own. Hopkins, a Yale graduate who had known Lincoln Pierce at summer camp, had volunteered a patrol boat in the war and served in the transport service. He had recently moved from wholesale coal into advertising, and his expanded ownership share brought an outsider into a key position for the first time.

Anne was still living with her relatives in Lawrence. Fuller saw her whenever he went back to New York, although he apologized after one visit for "that darned mad rush and to have you cross at me, and no chance to say good-bye, decently." He confessed that he was lonely in his room at the Chicago Club, and Anne hoped that her husband wouldn't feel so anxious once the business side was settled. Fuller was paying his expenses himself, which strained their budget, and when Anne began working as a shopgirl to earn extra money, he disliked the implication that he was unable to provide.

To deal with the stress, Fuller told Anne that he had taken up running to reduce his "bulging tummy," and he promised to quit drinking and smoking: "I still ride high on the wagon and with increasing pleasure and good health." In practice, entertaining clients and potential investors carried certain expectations, and not just regarding alcohol, which was widely available even at the height of Prohibition. Fuller asserted later that he patronized "over a thousand" brothels in Chicago, and although this was undoubtedly an exaggeration, it reflected an aspect of his life that went unspoken between him and Anne.

The city was full of temptations. On his drives, Fuller often passed through Cicero, Illinois, a regional center of organized crime, including prostitution. He may have been amused that the Chicago outfit's most notorious "vice resort" was also known as the Stockade, and he cast his visits to these establishments in an incongruously romantic light, recalling, "They seemed to be the only place where people really talked straight to me—those girls. Many of them had babies. I wanted to see them as human beings, to know how they got there. They were terribly interesting as people."

While patronizing one speakeasy off Joliet Road, Fuller reportedly encountered its owner, the mobster Al Capone, whom he remembered as "not a very attractive man." Fuller wrote nonchalantly to Anne of "the latest machine gun gang murders" outside Holy Name Cathedral, claiming later to have been disembarking from the trolley as the bullets began to fly. His real interaction with the underworld was limited, but he used it frequently to convey his range of experience, which included "contacts with all sorts, down to Capone and up to heaven."

In Lisle, which had been his greatest success so far, buyers seemed satisfied with their homes—one praised it as "the warmest house I have ever lived in"—but Fuller had committed to the project before materials and labor were available, and the pressure to reduce costs reflected poorly on the system. The houses were advertised for under $6,000, forcing the company to cut corners, and an engineer who conducted an inspection noted that the construction seemed "cheap."

Although Fuller was elected president of Stockade Midwest, his partners felt that he had closed deals prematurely and overpromised on growth. He believed that he needed to keep up a good front for investors, but DeCoursey Fales, a board member and friend from the Cruiser and Transport Force, warned him to stick to the terms of his contract. After Fuller canceled an order for blocks to be made by the New England branch, George Abbot announced that he was ending all of their personal relations outside of work, while the attempts to improve manufacturing at the Joliet factory had taken longer than expected, bringing about "a bitter fight" with Andy King.

Another confrontation was looming back east. In November Hopkins acquired a controlling interest in Stockade, which he intended to restructure to concentrate on national promotion. The plan was in line with Fuller's goal of establishing an industry on a large scale, and he told Anne that he was determined to prevail: "I am going to have the hardest directors meeting we have yet had to put over Farley Hopkins's reorganization plan, which is very fair and *absolutely* essential." After it passed over Abbot's objections, Fales resigned.

Fuller felt that he had won an important battle, and he welcomed the chance to focus on the Joliet plant as it finally began production. He consulted a former engineer from the Quaker Oats Company, whose expertise in puffed rice could be repurposed for the molding process, and he noted that the factory was based on "the elevation of materials in concentrated form to the starting point with a descent involving the expansion of the product to the final block." Like the cotton mill in Sherbrooke, it took full advantage of gravity, which became one of his core principles of design.

By then, he had reconciled with Anne, who was pregnant by the end of the year. Fuller reassured her that he would eventually earn a real salary, which would allow them to avoid living in "a horrid, cheapish way," but in the same breath, he revealed his concerns: "I don't think it would be safe for me to try to get away on a trip very soon, as I have to watch everything like a hawk, or else I believe everyone would steal all my interest and control of Stockade away from me."

His misgivings were justified. The reorganization had left hard feelings, and even minor transgressions came back to haunt him. While waiting on hold on the telephone, a board member allegedly heard Fuller and Carman singing a dirty song, and his drinking, which he tried to conceal by keeping his reflexes under control, was more obvious than he thought. There were lingering concerns over his desire to expand, and Hewlett warned him about his use of what outsiders would see as "tricky methods."

Fuller had misjudged the political situation, and his earlier fights had cost him valuable goodwill. Early the following year, he was asked to resign as president of the parent company, while remaining the head of Stockade Midwest. On February 10, 1927, he wrote tersely: "Being informed that it is desired by its Board of Directors, and deemed necessary to the welfare and success of the Company that I take the following action, I herewith tender my resignation as President of the Stockade Building System Inc., to take effect immediately."

He felt betrayed by Hopkins, the new president, who turned out to be an opportunistic and unimaginative businessman. Fuller had hoped to sustain investor enthusiasm until the company could compete on price, but his more pragmatic partners concluded that he had made promises he couldn't keep. He was undermined by his reliance on outside support and funding for an idea that needed time, and it would be decades before he found a way to avoid these constraints.

At Stockade, the humiliation was both professional and personal, and it had been witnessed by his entire family. Anne happened to be in town when the situation at the office fell apart, and their reunion could hardly have been worse. Their marriage had already been tested by his long absences, which now seemed to have been for nothing, and Fuller began to drink heavily again. Anne noted later that her husband's problems with alcohol had resulted in "the only times we've been dreadfully upset," and she alluded in a letter to "that nasty night in Chicago when you stayed out so long and then sort of banged me around when you came in."

Shortly after his resignation, Fuller appeared at the Own Your Own Home Show at Madison Square Garden, which was part of a campaign by the United States Housing Corporation to encourage small buyers to invest in residential developments. According to Fuller, as he was putting together his exhibit, the president of the Common Brick Manufacturers' Association "wantonly approached and destroyed" part of the display in a gesture of contempt.

Apart from this incident, the show seemed like a success. Fuller left with the names of around fourteen thousand attendees, and he worked to guide many prospects through the buying process. Out of all these leads, however, not a single one resulted in a house. It was another setback for his hopes of building on a grand scale, and he continued to devote all of his energy to isolated deals.

Back in Chicago, Fuller assured his contacts that his resignation was in line with his plans. He told a potential backer in Detroit that he didn't want Stockade to suffer "due to any derogatory remarks about myself," and his colleagues acted as though they had generously granted him another chance. Andy King assumed that his cousin had given up any idea of running the business on a national level, and he patronizingly advised Fuller to "keep your temper, your head, and your confidence."

It was all vaguely condescending, but it provided breathing room for Fuller, who told Anne, "Things are black sometimes, but are going to get better and better." He diligently improved manufacturing, arranged for product tests, closed small deals, and promoted the system without any sign of strife. On March 8, 1927, he overcame his nervousness during a radio appearance by picturing Anne. As he told her in a letter:

"I pretended to myself that I was reading out loud to you, which I love so to do, and it went beautifully."

In Joliet, Fuller was thinking intensely about cost per pound, which would dominate his thoughts for decades, but private matters especially weighed on his mind. On March 20 Edward Larned died of the flu in Florida, five months after his brother William, the famous tennis player, committed suicide at his club in New York. Although Larned had divorced Leslie, he had proposed partnering with Fuller on the New Jersey branch, and his death closed the door on any prospect of a triumphant return back east.

A shake-up of the Manhattan office led to the firing of its secretary, who had supported Fuller over criticisms of his "misstatements or overstatements," and he felt sidelined when Hopkins postponed adding him to the company expense account. When Fuller finally raised the issue, the new president said that he had put him on the account a month ago. The implication that he had misunderstood only added insult to injury, and Anne agreed that Hopkins was a "skunk."

Their separation was hard on both of them, but Anne insisted that she didn't want to distract Fuller in Chicago: "I could just stay on here . . . and be sort of mildly happy, with no heights of joy or depths of despair." After reading about the adventurous nineteenth-century British explorer Sir Richard Francis Burton, she drew a revealing comparison to her own marriage: "Isabel [Burton] had a much worse time trying to keep track of her Richard than I have. . . . Finally, she couldn't bear to be left anymore and went with him on all his expeditions, which was very daring for a lady of those times."

As Fuller drove between sites in his truck, he found himself crying aloud for Anne, and, in one letter, he came close to a full confession: "Darling, I am going to go to church a lot the way I used to when I was trying so hard to win you. . . . I have been in no way as devout or good as I used to be." In March he began attending the Chicago Sunday Evening Club at Orchestra Hall on Michigan Avenue, which offered a place of worship in the business district. It was an apparently modest change in his routine, but it would play an important role later that year.

When Fuller visited Anne in May, they decided that if she rented out their New York apartment, she could join him at the end of the summer. Money was a constant concern: Fuller had trouble keeping up his dues for the elite Somerset Club back in Boston, which represented all the establishment values that he felt obliged to maintain, and he was advised to resign. As a fresh start, he considered moving to a California office, but he worked instead with the engineer Martin Chamberlain to overhaul the Joliet plant.

While the world marveled at Charles Lindbergh's solo flight across the Atlantic, which Fuller later said could never have occurred "if he brought along a board of directors," he brooded over his precarious position at the company. When Leslie married her second husband, John Nelson Borland II, Fuller felt unable to take time off to attend the wedding. He told Anne that it made him want to cry, but he had no choice: "It would be absolutely ruinous for me to leave here now, as business is critical every minute, and [I] just *have* to make it go."

In August Fuller overcame his fears of leaving the city to travel to Rock Hall, where his wife—having broken the lease on their apartment after failing to find another tenant—was living with Anx. Anne wrote in her diary, "I dressed and went to the end of the lane and met him, and we are going to stay together always, as we miss each other too much." Fuller packed a trunk with his navy pictures and papers, and they took the train from Pennsylvania Station to Chicago, moving into the elegant Virginia Hotel on the corner of Rush and Ohio Streets.

The hotel, where Fuller had been staying for several weeks, seemed like a reasonable place for a small family. It offered butler and maid service, and their room on the ninth floor looked out on Lake Michigan. Fuller admired its view of the recently completed Tribune Tower—he had met one of its architects, Raymond Hood, through James Hewlett—and Anne sketched the neo-Gothic building from their window. Before her arrival, Fuller had written, "I don't believe business will ever go entirely smoothly. It never does, but it will go smoothly enough for me, and I guess I like a certain amount of excitement in it anyway."

Before long, everything fell apart. Fuller had been right to worry about his situation, and he abruptly lost his managerial position in

Stockade Midwest on August 18. According to Fuller, the subsidiary was in solid shape for the first time, which encouraged his partners to make "the big grab," and he complained later that they branded him "a bad man" to push him out. "The thing that hurt most," Fuller said later, "was false statements to undermine me. It was very unnecessary, as shown a proper reason, I would always concede for the good of my backers."

For the moment, he kept his dismissal a secret from his wife. Fuller hoped to salvage the situation—he was still involved in manufacturing—and he wanted to spare her any anxiety in advance of her delivery date. A week later, Anne felt labor pains, and they headed for the Chicago Lying-in Hospital. Fuller set her up with flowers and books, but the baby refused to come. The following evening, they returned to the maternity ward, where their daughter was born early on August 28, 1927.

She weighed seven pounds, with light-brown hair and blue eyes, and they decided to call her Anna Allegra. Her middle name, which was how they always referred to her, was inspired by the English poet Lord Byron, who had sent his illegitimate daughter, Allegra, to a convent in Italy. With the help of the nuns, she had written her father a famous letter asking him to visit, but Byron never returned before the child died of a fever at the age of five. It had obvious resonance for Anne, although Fuller preferred to note only that the name meant "happy girl."

Fuller saw Allegra's birth as his second chance. As Anne recovered, he kept a growth chart for the newborn, who was left in the care of Nettie Bonenkamp, a chambermaid at the hotel. Bonenkamp asked him for advice on an investment property that she had bought from a company in Indiana, which refused to repurchase it when she needed to cover her medical expenses. The real estate market took advantage of ordinary buyers, and Fuller began to consider their predicament more seriously.

When Allegra was two weeks old, Fuller finally told Anne what had happened at work. At that point, he had no good idea of what to do next. He was still a shareholder in Stockade, but he declared that his future involvement would be "impersonal." In collaboration with Martin Chamberlain, he discreetly prepared a pair of patents—one on a wall system, the other on mixing fibers with cement—and he contemplated entering the housing business on his own.

After Chamberlain was laid off, Fuller felt vulnerable as well. He admitted later to having loaned $700 to a sales manager who promptly skipped town, and—in a crowning humiliation—he was mugged in October. In Fuller's version of events, he was walking on Wabash Avenue when "a colored taxi driver" asked for the time. As he was about to reply, another man broke his cheekbone with brass knuckles, and the pair escaped with his watch and all the money that he was carrying.

In November Fuller was removed from his manufacturing role, severing his last practical connection to Stockade. He devoted a day to unauthorized work with Chamberlain, who still hoped to file a stealth patent, but Fuller was well aware that he had formally signed the rights to all his ideas over to the company. After overseeing 150 homes, he had been left with nothing.

On Sunday, November 20, 1927, Anne made a seemingly ordinary entry in her diary: "Cloudy, drizzly day. Stayed in all morning. I went for a walk in the late afternoon. Nettie stayed with Allegra during dinner, and then Bucky went to a service at the auditorium and walked for a while." In his own journal, Fuller merely noted that the Harvard football squad had lost to Yale. Eventually, however, he came to regard that evening as the turning point of his entire life, and while the specifics evolved over time, the night in question was very real.

At seven o'clock Fuller attended a Thanksgiving service at the Chicago Sunday Evening Club. After a choir of one hundred singers performed, it concluded with an organ postlude. The rainy weather was less than promising for a stroll, but he liked to walk and think, and it was only ten minutes across the park to Lake Michigan.

According to the version that Fuller recalled endlessly afterward, he found himself looking out at the darkened water for a long time, much as Margaret Fuller had left church to contemplate her own existence by a pool on Thanksgiving 1831. Although the details varied, Fuller claimed consistently that he intended to drown himself, which would have been a rather difficult way to die: "I had adequate courage to swim out into the lake until I became exhausted and sank."

Even if suicide only crossed his mind, Fuller questioned the value of his existence. He was thirty-two, with little to show for it, and he wondered whether his wife and daughter would be better off without him. Anne's family could support them more comfortably than he could, and additional money would come from his life insurance. It would be the ultimate abdication of responsibility—the definitive proof that his relatives had been right about him all along.

In his account of that night, Fuller supposedly posed himself a series of questions. He began by asking whether he believed in God, which he defined as a "greater intellect than the intellect of man." After concluding that he did, he moved on: "Do I know best or does God know best whether I may be of any value to the integrity of universe?" He decided that no man could know his true worth, and that faith told him only that God was aware he existed.

This led to what Fuller characterized later as a blinding revelation, which arrived in a rush by the lake: "You do not have the right to eliminate yourself; you do not belong to you. You belong to the universe. The significance of you will forever remain obscure to you, but you may assume that you are fulfilling your significance if you apply yourself to converting all your experience to highest advantage of others. You and all men are here for the sake of other men." And he resolved from that moment onward to "work out his fate."

Instead of suicide, he had arrived at an "egocide," but despite the parallels to Margaret Fuller's resolution, the outcome was very different. She had accepted her father's values, at least for a time; in contrast, her grandnephew allegedly decided to live for the benefit of all humanity, which meant no longer placing himself at the mercy of others. Rather than seeking approval from his family, he would frame his life around his daughter, although he would use his own judgment to determine what she would want him to do. The very next day, in his first concrete step to distance himself from Stockade, Fuller started looking for a new job.

It was less than the decisive move that the profound nature of his breakthrough might have implied. Fuller was determined to trust his intuitions, but he also hoped to benefit from his ties to his former company and social circle, and his desire to transform himself would always be tempered by human weaknesses. The night of change by the lake would never truly end, and it resulted shortly afterward in a second experience, often conflated with the first, that was even more striking.

On February 1, 1928—to slightly anticipate the narrative—Fuller went to a series of business meetings. He decided to stop at the Ford Industrial Exposition on Michigan Avenue and Sixteenth Street, which he had attended on its opening day earlier that week. It was advertised as "a miniature picture of the work of the Ford Industries," and on his previous visit, Fuller had been granted an uncanny preview of his own future.

The exposition included cars that were cut away to show their insides; a monoplane with room for twelve passengers; exhibits of precision gauges, manufacturing, and trade education for children; and a display of engineering methods, including the use of the assembly line in "the Ford method of continuous production." In his diary, Fuller referred to the show as a "marvelous exhibition," and it amounted to another kind of religious awakening.

As he headed to the 131st Regiment Armory for the second time, he saw a familiar figure south of the river on Michigan Avenue. It was Farley Hopkins, whose face filled Fuller with the impulse "to kill him." Instead, he resolved to never mention the circumstances of his firing again—although not before he pointedly announced this on the street to Hopkins himself. From there, he continued on his walk, and he was passing by the Art Institute, with the Wrigley Building visible in the distance, when he was blindsided by what he described as another moment of insight.

Fuller characterized it decades afterward as nothing short of a mystical epiphany: "I found myself with my feet not touching the pavement; I found myself in a sort of sparkling kind of sphere." It was the dramatic appearance of a geometric shape that would dominate his life, and as he felt it surround him, a voice spoke with all the force of divine authority: "From now on, you need never await temporal attestation to your thought. You think the truth."

Whatever the nature of this alleged visionary experience, as soon as it ended, Fuller kept walking. He paid his second visit to the Ford exposition, attended another business meeting, and then headed back on the bus, pausing to buy candy at the Fannie May shop at the Tribune Tower. His mind had been opened to the possibility of further messages, however, and on the way home, he supposedly heard another voice: "From now on, write down everything you think."

After dinner with Anne, Fuller felt "much in need of mental diversion." He went by himself to a showing of *The City Gone Wild*, a silent gangster melodrama starring Louise Brooks, and when it was over, he returned to the hotel. Anne was working on a new dress, and, at one point, feeling sentimental, she took out some of her old party clothes. In his diary, Fuller devoted a single line to what had happened: "After much philosophical thought while walking about, worked out theory of spheres."

Life went on, but regardless of the true details of the incident, Fuller came to see it as a fundamental transformation. The circumstances were ordinary, but no more than in the case of the mystic Emanuel Swedenborg, whose most famous spiritual encounter took place at an inn, where a mysterious figure warned him to watch his diet before unfolding the secrets of heaven and hell. Fuller genuinely believed that he had been changed, and he felt temporarily liberated from the doubts and desires that had caused him so much suffering.

Decades later, a friend claimed that Fuller had revealed an even more extreme version of the vision, in which the voice from the light said, "Bucky, you are to be a first mini-Christ on earth." Elsewhere, he put it more simply, and perhaps the most honestly of all. When asked how it had felt to return to earth after such an experience, Fuller confirmed that the worst part of him had remained in orbit, and that he had been left, he said, with "all that my mother would like."

PART TWO

THE DYMAXION AGE

1927–1947

You cannot get out of Universe. Universe is not a system.
Universe is not a shape. Universe is a scenario. You are
always in Universe. You can only get out of systems.

—BUCKMINSTER FULLER, *SYNERGETICS*

FOUR

THE FOURTH DIMENSION

1927–1933

We start playing the game, the most complicated game of chess that has ever been played. We start to play the game Universe, which requires absolute integrity.

—BUCKMINSTER FULLER, *SYNERGETICS*

Of all the elements in Fuller's origin story, one of the most persistent was the notion that he gave up speaking after his transformative experience in 1927. "I scarcely spoke at all for two years," Fuller said. "I couldn't be completely free of words, but my wife had to talk to people for me. I didn't want to say anything, make any sounds, until I was pretty sure what those sounds meant and why I wanted to use them." He repeated this claim throughout his life, leading Susan Sontag to refer much later to "the voluntary experiments in silence that some contemporary spiritual athletes, like Buckminster Fuller, have undertaken."

Even at the time, this account was questioned. "He did not seal up his mouth," Anne said dryly to the biographer Alden Hatch, who allowed Fuller to stick to his basic story: "Anne made all my oral contacts with the outside world." In fact, this assertion was contradicted by his letters, his diaries, and the official chronology of his projects. If anything, after his epiphany, Fuller never stopped talking. His ability to spin a grandiose narrative had cost him his job, but now he needed it to survive, and he even wrote to a correspondence school to request a booklet titled *How to Work Wonders with Words.*

After Fuller returned in November from his "blind date with principle," he dove at once into an ambitious undertaking spurred by his frustration at Stockade over his lack of momentum and his inability to operate at scale. The solution, he decided, was a house designed for mass production, with components that could be built in a factory, delivered as a package, and assembled rapidly on-site. It demanded nothing less than a complete industry, but it would center on a scientific housing company of his own, which he called Fuller Houses.

Before he could launch his new venture, however, he had to find work. On November 21, the day after his revelation by the lake, Fuller met with a vice president at the Celotex Corporation, which made fiberboard out of sugarcane. No positions were available, but his contact expressed interest in his business plans. Through his Stockade associates, he also reached out to Franklyn R. Muller, a flooring firm. The manager there offered Fuller a sales job at fifty dollars a week plus commissions, fully expecting him to turn it down, but at thirty-two, with a wife and a child to support, he needed the income.

The Virginia Hotel was too expensive, and Fuller relocated his family to the more affordable Lake View Hotel on Belmont Avenue on December 5. Anne wrote in her diary, "I felt awfully depressed. . . . It is a cheap, tiny place, but clean, so I think we can stand it for the sake of getting straightened out financially, but it does seem frightful to have to come here with Allegra."

Fuller subsequently described it as a tenement in a slum, and Anne reminisced that one neighbor, who helped to load their garbage into the incinerator, was a gangster who carried two pistols in shoulder holsters. It was really a new building in a fashionable part of the city, but it lacked the Virginia's amenities, and Anne spent hours on chores at what they called the headquarters of Fuller Industries. There was sometimes enough money for just one meal a day, and Fuller later wrote a poem about his wife returning from the store with "a nice red apple that she stole."

As Fuller reinvented himself, he withdrew from his old friends, and he was content to let Anne look after their everyday needs, which may have been the truth behind his alleged vow of silence. On nice days, he took Allegra to Lincoln Park, where he also went running alone, and he often paused to admire the statue by Augustus Saint-Gaudens of Abraham Lincoln, whom he saw as a symbol of integrity. His exercise regimen included pulling himself up onto tree branches, which prompted him to marvel at the strength of their "slender single supports."

Fuller carried a bundle of pencils for jotting down notes on his walks "like a man with ague," and he started to outline his thoughts more systematically after attending a dinner for Eliel Saarinen at the Architects Club in December. The event included an exhibition of the Finnish architect's unused plans for the Tribune Tower, and Fuller was told that that he would have been invited to speak had they known he was coming—a reminder that he had access to a potential audience through his contacts in the architectural world.

In January 1928 Fuller bought a typewriter to set forth the specifications of his proposed house in more detail. Like James Hewlett, he was responding to what he saw as obsolete methods of construction, and he began with the concept of a centralized operation that could deliver houses anywhere. His desire for a housing industry led him to the design and materials that he needed, not the other way around, and it called above all else for lightness and standardization.

Although his plan was driven by his frustration with a business model "in which each house is a pilot model for a design which never has any runs," it also reflected legitimate social concerns. A postwar rise in construction costs had forced builders to cut corners, build smaller houses, or specialize in homes for upscale buyers. The working class could afford only older houses that had fallen in value, usually because of deterioration or obsolete construction. Government regulations to protect public health inadvertently discouraged investment in residential buildings, which resulted in overcrowded tenements for the poor.

As a result, Fuller was more than justified in advocating for an affordable house that could be progressively improved by industrial methods. He began by thinking about the inadequacies of "the house of his childhood," but his first sketches depicted a conventional square plan and pitched roof. It had concrete supports, stabilized by piano wire, and a transparent ceiling for conveying sunlight through the interior with prisms or mirrors. For the walls, he specified duralumin, a lightweight but expensive aluminum alloy used in airplanes. He said that the engineers he consulted thought that he was joking, and he recognized that it clearly made sense only for mass production.

Fuller bought a drawing board and drafting supplies and set to work at the kitchen table from midnight until early morning, brooding over design philosophy, sales psychology, and organization. At Stockade, he had been constrained by his partners and the technology at hand, but now he could afford to dream, although he also took the practical step of meeting with a patent attorney.

His belief in patents would never fade, but a fundamental change in his strategy was just around the corner. On January 30, 1928, shortly before he attended the Ford Industrial Exposition for the first time, Fuller met with Russell Walcott, an architect who had joined the board of Stockade Midwest after building a house with its blocks. Walcott was interested in Fuller Houses, and he also lent Fuller a copy of the book *Towards a New Architecture* by the Swiss-French architectural theorist Le Corbusier, which had been translated into English the year before.

Fuller devoured it until late at night, writing in his diary that he was "startled at coincidence of results." Le Corbusier, born Charles-Édouard Jeanneret in 1887, was both the role model that he had unknowingly needed—an architect who had established his reputation with drawings and aphorisms rather than finished buildings—and a source of specific precepts:

> The lesson of the airplane lies in the logic which governed the statement of the problem and its realization.
> The problem of the house has not yet been stated.
> Nevertheless there do exist standards for the dwelling house.

Machinery contains in itself the factor of economy, which makes for selection.

The house is a machine for living in.

Towards a New Architecture advised the serious architect to seek "the majesty of solutions which spring from a problem that has been clearly stated." Fuller adopted this as a guiding principle, and he often paraphrased it without attribution: "A problem adequately stated is a problem solved."

Elsewhere, Le Corbusier spoke of mass production in housing; of drawing inspiration from airplanes, cars, and ships; of reducing the weight of construction to lower transportation costs; and of designing a house from the inside out "like a soap bubble." He described World War I as a client that could never be satisfied, punishing mistakes and rewarding efficiency, and casually threw out ideas that Fuller would pursue at length—including a "parallel table" of inventions from history and a streamlined car with an ovoid body.

They also differed in meaningful ways. After reading the book for the first time, Fuller wrote in his diary that Le Corbusier "misses main philosophy of home as against house." Fuller drew a distinction between the house, a material problem that could be solved by efficient design, and the home, which was brought into being by "the spiritual and abstract." Le Corbusier treated the engineer as an example for the architect, not as a direct collaborator, and Fuller took the idea of the house as a machine for living further than his predecessor ever would.

All the same, Fuller absorbed many strategic lessons immediately. He redirected his efforts overnight toward writing a book that would attract similar attention, and he began to read widely in sociology and economics. In this new phase, Anne served as his creative partner; she produced conceptual art, clipped articles, did library research, accompanied him to a meeting with a potential backer, and brainstormed "the philosophy of the business end."

At night, Fuller worked on drawings with Clair Hinkley, a draftsman in Walcott's office, and he resolved a lingering dispute with Stockade.

After Farley Hopkins warned him of trouble if he tried to profit from the inventions that he had developed there, they signed a mutual release of all claims. Although Fuller complained of the "garbled facts," Stockade was no longer relevant to his plans, and it would cease operations several years after his departure.

The settlement removed his last obstacle to starting a business. He had been making sales calls and overseeing a flooring commission for Muller, but in the spring, he quit to focus exclusively on Fuller Houses. Fuller presented it "as a matter of integrity," since he was devoting so much energy to his own ideas, but it was also a gesture of fanatical faith in himself. It was the first time that he voluntarily left one job before having lined up another, and he expected to secure funding quickly.

His proposed house still had a rectilinear plan, supposedly on his attorney's advice, and he placed all of its utilities in the center, allowing the roof and walls to be engineered as a separate lightweight shell. A mast rose from a concrete foundation, like a tree trunk, to support the outer framework and protect the utility core, which included a ventilator and a forced air system for cleaning. Beams and tension rods radiated outward for the floor, and to minimize the transport of heavy compression materials, aluminum tubes and inflatable panels served as rigid members.

Prefabricated housing was far from a new idea—house kits were available from the Sears, Roebuck & Co. catalog—and comparable designs had been explored by other architects. Fuller had seen drawings for the Moscow Palace of Labor by two Russian brothers, Leonid and Viktor Vesnin, with tensile structures that resembled radio antennas, while another pair of brothers, Bodo and Heinz Rasch of Germany, had envisioned an apartment tower stabilized with wires on a mast. By reasoning backward from mass production and delivery, Fuller arrived at an analogous solution, and his nautical training inspired him to draw on the principle of tension.

Throughout the history of architecture, most buildings had counteracted the force of gravity with sheer bulk, using construction materials with high compressive but low tensile strength, such as stone or brick. Steel made it possible to take advantage of tension components that were stretched rather than compressed, as seen in suspension bridges, bicycle wheels, and dirigibles. By employing tensile elements that were stronger than conventional compression structures, weight could be reduced drastically, but usually at a higher price per pound.

Building houses out of these materials would depend on economies of scale, which demanded an industrial operation that vastly exceeded the cost of a single dwelling. To seek the necessary funding, Fuller decided to emulate Le Corbusier as a prophet of scientific and architectural trends. Since he wanted to present the resulting business model as an extension of natural law, the name Fuller Houses struck him as excessively personal, so he changed it to Cosmopolitan Homes.

In an unpublished essay of the same title, he framed his sales pitch around a sweeping argument: "The living abodes of humans will become more and more decentralized, and industry more and more centralized and mechanicalized." At a time when shifts in transportation, communications, and power transmission encouraged urban theorists such as Lewis Mumford to explore the notion of decentralization, Fuller wisely positioned his original objective—houses distributed from a factory—as part of the inevitable expansion of the geographical freedom of all mankind.

A truly decentralized dwelling had to be light and autonomous. Using a term that he claimed inaccurately to have coined himself, Fuller spoke of "debunking" excess weight using materials and methods that were economical only in mass production. In a reversal of his actual train of thought, he indicated that he had deduced the need for a housing industry from the basic requirements of shelter itself. Like an airplane or the human body, Fuller wrote, a house had discrete tasks that could be segregated and solved in "one best way."

In April he wrote a much longer unpublished piece, "Lightful Houses," that took the tone established by Le Corbusier in even stranger directions. Fuller opened by observing that people were less demanding of the objects that they used every day: "It is the subjects farthest removed from personal contact which have received the most honest and unbiased attention and design." Airplanes and cars had benefited from this detachment, which was long overdue in housing.

Fuller offered a thought experiment of how a car would look if it were purchased like a house. A prospective buyer would hire a designer, take bids from garages, and apply to the town council for permission to build the resulting eyesore, which would cost $50,000. If houses were produced instead with industrial techniques, the visionaries who put these principles into practice would be rewarded, as expressed in what he called Fuller's Law of Economics: "As time is saved by individuals for other individuals, so shall that individual control capital in the direct equivalent of men hours saved."

Several pages were devoted to a critique of conventional geometry, which he saw as an obstacle to science. Fuller contended that there was no such thing as a perfectly flat plane—when two people were standing on the earth's surface, their heads were farther apart than their feet. He also discussed the "theory of spheres" that he had developed in February: "All matter in unforced state is spheroidal not cubistic, and these spheres are expanding for the life of their existence at a fixed rate."

Along with his main argument, he touched on telepathy, education through television, the message of Jesus Christ, his unhappy childhood, and the importance of looking to the young: "Figure out what they are talking so glibly of that seems so unaccustomed to you. This is a new law of prognostication." He outlined every department of his proposed corporation, from engineering to advertising, claiming to have run it entirely on paper, and concluded that there was no time to lose, since it was barely possible to predict the future before "the demand be upon us."

Until then, Fuller had been living off gifts from friends and family, but on May 4, 1928, he secured an investment from Russell Walcott and a stockbroker named John W. Douglas, who would provide $500 in advance and up to $5,000 to cover expenses for what Fuller had renamed the 4D Company. He saw mechanics as "the application of time or the fourth dimension to the other three dimensions," and the name of his company doubled as a tribute to Henry Ford, "the greatest time saver of the world," whom he honored with the coded pun of "4D."

Under an arrangement that would often recur in the future, Anne served as treasurer and majority owner of any patent rights, and she

Fuller, Allegra, and Anne at the shore of Lake Michigan in Chicago on April 8, 1928.

assumed control of Fuller's bank accounts—a responsibility, he noted, that she took on "with unexpected enthusiasm." After receiving a small inheritance, however, Anne thought twice about entrusting it to her husband, and when Fuller learned that she had given it to her brothers to invest, he felt betrayed by her lack of faith.

He was more concerned by a looming deadline. On May 16 the American Institute of Architects would hold its annual convention in St. Louis, making it his best chance to reach an audience of potential supporters in one place. With Anne serving as a proofreader, Fuller feverishly assembled his drafts and letters into a revised prospectus, removing most of the religious references from the manifesto that he ultimately called *4D Time Lock*.

Its title evoked a safe that could be opened with the correct combination of symbols, and he packed its forty dense pages with ideas that he would mine for the rest of his life. Fuller began by stating that if houses were designed like aircraft or cars, the outcome would inevitably be standardization, which represented the best way of doing things. Perfection could be approached but never achieved, while progress demanded continuous "cycles of research, analysis, design, and performance."

The 4D House would be built from the mast outward, producing a "circular progression" for the most efficient enclosure of space, featuring appliances previously available "only in hotels, hospitals, liners, and other points where mass purchase has made them possible." Like all his subsequent houses, it had two bathrooms, which reflected his personal fastidiousness, and a unit for washing and drying laundry, although not ironing. He hoped that clothes could be designed to eliminate "ironing drudgery," which later led John Marquand, his cousin, to express amusement toward his ideas: "The trouble is the shirts never come out pressed."

A generator and sewage tank would allow the house to be located off the utility grid, with telecommunications equipment for what we now call remote working and education. Fuller wrote that this decentralized approach would revolutionize the housing industry itself, which would use calculating machines to route orders between departments: "The operator of all this may be anywhere in the world." Fuller was largely silent on cost, but he combined ideology and engineering in a memorable aphorism: "Philosophy to be effective must be mechanically applied."

Many sections were close to unreadable, however, and he was left with the problem of selling it. Deciding to avoid publishers entirely, he struck a deal with a stationery shop to use its mimeograph machine, which at the time was a groundbreaking tool for independent publishing. With copies in hand, he prepared to attend the AIA convention with Pierre Blouke, a member of the Chicago chapter, who had written an article on housing that inspired Fuller to pay him a visit. According to Blouke, "This chunky little fat boy, looked like a Buddha, walked in, and asked if I'd seen what he was working on."

Blouke was impressed enough to offer Fuller a ride to St. Louis. Although Fuller implied later that he spoke there as a sponsored guest, he paid his own way, and he never delivered the speech that he had prepared—he instead sought out listeners on the bus and in the lobby. He boasted that his work "was greeted with the most astonishing remarks as to scholarly ability," but after it failed to make any impact, he reversed himself, complaining that he had been misunderstood by men who were "complacently befuddled with liquor."

Fuller said afterward that his presence prompted the convention to issue a proclamation rejecting "peas-in-a-pod-like reproducible designs." The statement was actually finalized before his arrival, and its emphasis was very different: "Local characteristics are fast disappearing in this era of common thought and mechanical advancement. Communities are coming to look more and more like peas of one pod." It was about the loss of regional style, not industrial housing, and while Fuller took credit for inspiring it, he had made no real impression at all.

He rebounded by pursuing a remote network of supporters. After adding new pages to the existing copies of his manifesto, he mailed it out to a long list of notables, including Henry Ford, Christopher Morley, the British mathematician and philosopher Bertrand Russell, the author Thornton Wilder, and the architect Claude Bragdon. He also wrote to James Hewlett to offer a controlling interest in his patents to the AIA. Hewlett advised him to send a proposal to the appropriate committee, but Fuller quickly abandoned the effort, which he later held up as proof of the establishment's indifference to his work.

The feedback from other relatives was disheartening. Fuller had invested in the stocks of industries that he expected to benefit from his housing revolution, such as radio and plastic, and he urged his mother to exchange her railway bonds for aviation shares. He also told her to buy more islands, which he predicted would rise in value after houses could be delivered by air. Fuller asked her to treat his recommendations, which he insisted were a "gold mine," as though they had come from a stranger, but Caroline replied only that his ideas were "over my head."

Wolcott was even more skeptical. The younger of the two brothers was working for General Electric, which Fuller hoped would invest in the 4D House, but after finishing the book, Wolcott said bluntly, "I honestly cannot make head or tail out of it." Fuller was discouraged, but Anne wrote a thoughtful response to Wolcott. After reading it, Fuller professed to be amazed that his wife, who had assisted him for months, grasped his ideas so well. He later told Alden Hatch, "That was the only time I had seen her show that she understood what I'd been thinking."

Anne was doing what she could to stabilize their home life, which was strained by Fuller's experiments with sleep. After noticing how long it took him to catch a second wind on his runs, he decided to rest as soon as his energy levels began to fall, and he began taking a short nap every six hours or so, followed by a period of relaxation—usually by dancing to records—before returning to work. He claimed that it allowed him to function on much less sleep than average, which became part of his legend, although Anne soon persuaded him to abandon it.

This irregular schedule was obviously incompatible with associates who kept regular hours, and there were promising signs of interest from outsiders. The architect Paul Nelson, a native Chicagoan then living in Paris, volunteered to represent Fuller in France, and he found support closer to home. Although he had trouble recruiting established professionals, he enjoyed better luck with unattached young men, and he attracted a handful of followers, including Leland Atwood, who would become an important collaborator.

His younger admirers were in no position to provide funding, but they served as useful sources of unpaid labor, and they agreed to disclose their ideas to Fuller "for his exclusive ownership." Much of their time was spent on illustrations, supplementing the pictures that Fuller, like an aspiring Leonardo, had been turning out all year. One sketch depicted a building delivered by an airship that dropped a bomb for the foundation, which Fuller compared to planting a tree.

Fuller's other drawings ranged from esoteric visual allegories—often bordering on outsider art—to designs for furniture, and many were incorporated into the second printing of his manifesto. He mailed the revised edition, which included a "chronofile" of his positive and negative correspondence, to enough additional recipients to bring the total to two hundred—a number that Fuller believed, mistakenly, was required for copyright. The new picture section was even more provocative than the text, with renderings of towers like vertical zeppelins, which Fuller admired as a model of structural efficiency.

Fuller's One Ocean World Town Plan, from the second printing of *4D Time Lock* (1928).

Although he was still focusing on a house for a single family, as it would be easier to prototype, he knew that futuristic images could impress potential investors. Other pictures contained the seeds of ideas that would emerge down the line. Instead of using a circular layout, the towers were hexagonal, with a mast that provided a mooring point for dirigibles, while another illustration presented an "auto-airplane" with collapsible wings.

His most memorable drawing rendered the earth in such a way as to show its maximum land area, which Fuller later called the One Ocean World Town Plan. The caption warned that humanity would need two billion more homes within eighty years, and the issue wasn't a lack of space but inefficient housing. He envisioned towers sprouting across the globe, connected by polar air routes, with the Eastern pagoda and Western skyscraper united over "the forgotten graves of mysticism and doubt."

Like many start-up founders to come, Fuller was promoting an untested technology by making huge claims for its importance, but he also wanted the backing of practical men, such as Vincent Astor. Not surprisingly, the tension between these impulses surfaced occasionally. In August he sent the banker George N. Buffington a proposal that he promised would be "the greatest letter which you will ever have received." Instead of delivering a straightforward pitch, however, he vowed to make Buffington the leading financier in the world, and his language reached new heights of fervor: "4D is my own creation, or rather has been revealed through me."

As Fuller continued to revise *4D Time Lock*, he took a similar approach to his model of the house, which would present his case even more dramatically. The first version, consisting of two structural rings suspended from a central mast, was based on the hexagon rather than the circle. This was easier to execute, but it also marked an ideological shift. As a compromise between the ideal circle and the square, the hexagon resulted in a symmetrical distribution of stresses, and it allowed the floor to be divided into triangles, which he embraced as the most stable structure.

This model was first displayed in September 1928 at Le Petit Gourmet restaurant on North Michigan Avenue, in an exhibition arranged by the poet and novelist Jean Toomer, whose address Fuller had obtained from Russell Walcott. "We might arrange supper, or lunch, or a walk," Fuller wrote in an introductory letter. "I walk the length of Lincoln Park at least once a day." Toomer responded enthusiastically: "I have an increasing interest in your whole idea and work."

It was an intriguing connection. Toomer was mixed race, and his interest in Eastern philosophy was compatible with Fuller's mystical side, which had been drawn out by the prophetic mode that he had adopted. At the restaurant, the attendees would have included followers of George Gurdjieff, the Russian guru who urged his disciples to awaken to a heightened state of consciousness. Toomer was Gurdjieff's leading advocate in Chicago—he had studied with the master in France—and he ushered Fuller into this world for the first time.

As he began to make strides in bohemian circles, Fuller experienced setbacks along more practical lines. After a search uncovered nine related patents, several claims for his house were rejected, and he gave up the effort entirely. For this, he blamed his attorney, claiming, "I did not know that subsequent resubmission of the patent was possible." This explanation was inconsistent with his previous experience of patent law, and he said elsewhere that he abandoned the applications "to prevent their being exploited by any selfish interest."

Deciding to proceed as though the house already existed, Fuller stated publicly that it would weigh six tons and cost $3,000, which he based on the assumption that it could be produced on an industrial scale at the same price per pound as a car. Over the winter, his associates built an improved model out of cardboard, with three small nude figures on the inside to illustrate its advanced climate control.

After an exhibit at the studio of the painter Rudolph Weisenborn, Fuller heard from Ray Schaeffer, the head of advertising for Marshall Field's, about displaying the model at the department store chain's flagship location on State Street. According to Fuller, the retailer had received a consignment of modernist furniture from Europe, which it hoped would seem less radical in comparison with his house. In fact, it was part of a long-standing program by department stores to publicize modern architecture, and Fuller was chosen partly because he was a Chicago resident rather than a member of the foreign school dominated by the Bauhaus in Germany.

Before the exhibition, he met with Waldo Warren, the former advertising manager of Marshall Field's. Fuller mistakenly believed that Warren coined the term *radio*—it was actually popularized by Lee de Forest—but he was unquestionably a skilled wordsmith, and he thought that "4D" sounded like a hotel room or a failing grade. Over two listening sessions, Warren and a colleague used the vocabulary in Fuller's talks to assemble a list of invented terms that they invited him to pair off successively, picking his favorite until one was left.

After the final bracket, the name that remained was *Dymaxion*. This word, which would follow Fuller throughout his life, was later described as a combination of *dynamism*, *maximum*, and, implausibly, *ion.* It was more credibly derived from his notion of "the dynamically balanced home," and he loved it: "It seemed at first mouthing a bit uncomfortable but has improved constantly with use." After Fuller was granted the rights, it became his personal brand, which he defined retroactively as the most efficient use of available technology.

His lectures at Marshall Field's were equally formative. On April 6, 1929, he took up residence in the Interior Decorating galleries on the ninth floor to speak six times a day for the next two weeks. It was Fuller's first extended attempt at a lifelong performance that he learned to adapt to different audiences. At the store, he emphasized that his ideas would cut housework to just fifteen minutes a day, and he claimed that the architects Howard Fisher and George Fred Keck had been inspired by his presentation to start housing projects of their own.

Another attendee, the Polish-American philosopher Alfred Korzybski, struck the model with his cane and asked, "What's this?" As the founder of the arcane cognitive discipline known as general semantics, Korzybski identified a principle that he called "time binding," which allowed humans to build on the discoveries of previous generations, and said that it demanded a language that was structured to precisely reflect the events that it described. Korzybski extended Fuller an invitation to a lecture that he was giving at the University of Chicago, and the pair met on several other occasions.

Fuller's next major showing was at the American Institute of Architects convention in Washington, DC. He brought a suitcase with his house model, which he managed to repair after it was accidentally damaged, and met Harvey Corbett, the head of the architectural committee for the upcoming 1933 Chicago World's Fair. Its theme, "A Century of Progress," offered a prominent platform for the concepts that Fuller was promoting, including prototypes of experimental homes, and he was confident that Corbett would agree to display the Dymaxion House.

In May Fuller accepted an invitation from the Harvard Society for Contemporary Art, which had been founded by the undergraduates Lincoln Kirstein, Edward Warburg, and John Walker. It was a meaningful return to Cambridge, where he reportedly presented his model to "many famous Harvard professors," as well as to a senior named Philip

Johnson, who was equally intrigued by its use of aluminum—the basis of his family fortune—and by Fuller's showmanship. Although he disliked the house itself, Johnson granted that it left a strong impression: "I learned vast amounts of the potentialities of architecture that I never forgot."

More events followed, including a dinner held for the Chicago Architects Exhibition League by the art collector Rue Winterbotham Carpenter. Fuller had been experimenting with "absolutely comfortable clothes," including sneakers and khakis, and he arrived in a rebellious mood: "I showed up and rudely announced, 'I don't eat that kind of food,' and was in every way obnoxious." Afterward, he felt badly, and he began to emulate Le Corbusier in wearing a plain black suit, or "bank clerk's clothing," in order to highlight his ideas, not his eccentricities.

He rarely achieved this goal, but his efforts were paying off. Henry Saylor, the editor of *Architecture* magazine, was interested in a book on his work, potentially to be written by Lewis Mumford for Scribner, for which Fuller recorded a lecture at the Architectural League of New York. After an introduction by Corbett, who confirmed that he wanted the house for the World's Fair, Fuller explained its features, including photoelectric sensors to eliminate germs from doorknobs. It was a far-fetched notion at the time, and Wolcott worried that it made his brother "a liar."

The book never appeared, but Fuller received a modest amount of funding from a few sympathetic architects, which encouraged him to return to New York permanently. Before the move, he and Anne brought Allegra on her first trip to Bear Island. On their vacation, he amused his daughter with Goldilocks stories based on scientific ideas from his reading—which he described later as rehearsals for the "thinking out loud" that became his trademark—and bought a boat that he called the *Lady Anne*.

When Fuller headed for Manhattan at the end of the summer, Anne moved to Woodmere, Long Island, with Allegra, who soon became accustomed to her father's extended absences. Fuller visited occasionally on weekends and sent whatever money he could, which consisted mostly of a small allowance from his mother, while Anne said that she was managing to live without a maid by working out "different labor-saving things *a la* 4D."

A few years later, however, he would tell Anne in a letter that the separation reflected deeper fissures: "All our troubles started back in Chicago—both our faults; mine for a stupid notion of a martyristic monk's life which would atone for my past sins and make possible true thought and service, which had a horrible reaction eventually on both of us and nearly wrecked everything." Fuller added that her fault was "over money."

These issues worsened in New York. When Anne visited her brother Jim, Fuller saw it as a betrayal that turned their families against him: "It was flight. It was our first separation. The side-taking then began, based on a lack of understanding of my integrity, my determination to look out for you in the biggest, most complete, and ultimately certain manner being mistaken for an utter carelessness for your welfare. Those who had grudges or jealousies fanned the flame."

This perceived hostility from his relatives encouraged him to seek support elsewhere, and the obvious place was Greenwich Village. In the fall of 1929 Fuller showed up with a suitcase at the apartment of James Hewlett's friend Antonio Salemme, who agreed to let him crash on his wicker couch. Fuller lived with the sculptor for six months, and from there it was a small step to Romany Marie's—the perfect venue for cultivating the connections that he needed.

The tavern had recently moved to 15 Minetta Street, and Fuller offered to redecorate it. Decades before the artist Andy Warhol lined his loft, the Factory, with silver foil, Fuller worked with a young sculptor named Isamu Noguchi to cover the walls in aluminum paint, and he built tables and a steamer bench out of metal and oilcloth. He used colored filters to alter the light based on the mood—blue for business, pink for love—and stated that a room "should lend itself to change so that its occupants may play upon it as they would upon a piano."

Romany Marie had mixed feelings about the result. Her patrons didn't care for it, and she found herself explaining it to everyone rather than discussing her own ideas. The tables wobbled whenever dishes were set down, and, finally, the bench collapsed. Marie placed an order for wooden furniture the next day, but Fuller waved away the disaster by saying that it had only been an experiment.

Fuller was still a charismatic figure, and the artist Hugh Ferriss, who met him through Jean Toomer, advised Marie, "If you will give a hand to Bucky Fuller, you will go down in history." Seeking a more productive outlet for his talents, Marie appointed him the tavern's "official talker." She was among the first to recognize his appeal to the avant-garde circle exemplified by her clientele, whom she saw as "a core of people who are operating for new ideas," and it amplified a visionary streak in Fuller that would have been out of place at Marshall Field's.

The role came with free servings of chorba, a rich Romanian stew, along with prunes and black bread, although Fuller—who was receiving breakfast and lunch from Salemme—claimed that he took advantage of it only every other day. He was more interested in the fact that "every intellectual in the world" passed through Romany Marie's. It was there that he encountered poet E. E. Cummings, actor Paul Robeson, sculptor Alexander Calder, writer Theodore Dreiser, dancer Martha Graham—who recalled Fuller as "always exciting"—and countless other luminaries.

From this ideal perch, Fuller made an impression on many influential contacts, two of whom became especially important. One was the explorer Vilhjalmur Stefansson, whose ethnological research in Canada and Alaska led him to experiment with an Inuit diet consisting mostly of meat. After attending an exhibit of the Dymaxion House at Salemme's studio, Stefansson became friendly with Fuller, who shared his interest in the geography of the poles.

Fuller's other lasting friendship was with Isamu Noguchi, born in Los Angeles to the Japanese poet Yone Noguchi and the American writer Léonie Gilmour in 1904. His father abandoned them before he was born, and he lived with his mother in Japan before attending school in Indiana, where he boarded with a Swedenborgian pastor. At Columbia, he studied medicine, but he showed a greater gift for sculpture, for which he won a Guggenheim Fellowship.

The grant enabled him to study art in Paris as an assistant to the influential modernist sculptor Constantin Brancusi. When Noguchi decided to return to New York City, Brancusi encouraged him to look up Romany Marie. The restaurant owner introduced Noguchi to Fuller, and the two of them hit it off at once, both creatively and personally. A show of their work was held at the tavern on October 27, 1929, two days before the stock market crash that heralded the beginning of the Great Depression. According to Stewart Brand, Fuller remembered it afterward "fondly as a period of trial and of abundant learning and creativity," but it complicated his attempts at fund-raising, and he focused instead on the newly energized bohemian world.

Noguchi earned money by doing portrait sculptures, and they grew closer during sittings for a bronze bust of Fuller, which he advised the sculptor to plate in chrome to create "a fundamental invisibility of the surface." He proposed that Noguchi use aluminum paint on the walls of his studio on Madison Avenue and Twenty-Ninth Street, where

Fuller with the third model of the Dymaxion House. A small figure of a nude woman on the inside illustrates the structure's climate control.

Fuller slept on the floor for two months, and they walked together for miles to admire the unfinished George Washington Bridge.

Although nine years apart in age, they were joined by a sense of difference that Fuller described as Noguchi's envy of those who "belong to their respective lands," which others found appealing. Kay Halle, an heiress and journalist, invited them to parties held by composer George Gershwin, who posed for a bust by Noguchi. Halle would sometimes tell them to drop by her apartment when she expected a gentleman caller, whom she would usually persuade to treat them all to a fancy meal.

Noguchi respected "Mr. Fuller," as he called him, but he was cautious about his influence: "You can't be submerged by one person. Bucky rescued me from Brancusi in the same way that Brancusi rescues me from Bucky." Fuller's own feelings were heightened by his fascination with people of mixed race, and he was enthralled by Noguchi: "It was amazing that he came out like a Neapolitan boy. He looked like some of the figures you see in Michelangelo's work." Years later, he said that they had been "in love," although he was careful to issue a clarification: "Neither of us was homosexual."

Despite their shared impoverishment, Fuller was becoming a minor celebrity. Harvey Corbett invited him to Princeton; he lectured on his house in a newsreel for Fox Movietone; and Phil Nowlan, creator of the science-fiction hero Buck Rogers, reportedly said that he took inspiration from Fuller's ideas. In March 1930 he and Noguchi loaded the latest aluminum model of the Dymaxion House, along with busts of Fuller and Martha Graham, into a station wagon and headed up to Cambridge, Massachusetts, for an exhibition of their work at the Fogg Museum and the Harvard Society for Contemporary Art.

From New England, they drove through Indiana—where they organized motorists to move stranded cars during a snowstorm—to a show by Noguchi in Chicago. The following month, Noguchi sailed for France, leaving Fuller "quite sad today at thought of your leaving." Over the summer, Noguchi sculpted what his biographer Hayden Herrera described later as "a spiritual portrait of Buckminster Fuller," in the form of a small nude with large genitals in the pose of Leonardo's

Vitruvian Man. Fuller gave it a place of honor in his collection, although his relatives continued to express confusion over "who that Jap was Bucky was always hanging around with."

Left to himself, Fuller made strides in design circles, but when a student organization invited him to Yale, Everett V. Meeks, the dean of the School of the Fine Arts, allegedly declined to attend. Fuller's technological approach to architecture had grown more controversial, and at the next American Institute of Architects convention, a walkout was led by the architect William Delano, who dismissed the Dymaxion House as "halfway between a gas tank and a greenhouse on the wing."

His complicated relationship with the architectural world was underlined at an event in the fall at the Women's University Club in New York. The attendees included Frank Lloyd Wright, Harvey Corbett, and William Van Alen, the latter of whom had designed the Chrysler Building in Midtown Manhattan. Fuller was given only ten minutes to speak, and, during his presentation, he felt a sudden inspiration. Turning to Van Alen, Fuller asked, "Can you tell me what the Chrysler Building weighs?"

Van Alen confessed that he couldn't provide an estimate within a million tons. None of the men present, Fuller recalled, could answer this question about their own buildings, which he would ask for decades whenever he wanted to catch a conventional architect off guard. Wright was amused, although he noted that Fuller had perspired until he became "a dripping rag," and he finally rose to deliver a verdict: "My young friend, Buckminster, is as bad a speaker as he is good as a designer."

Fuller urgently wanted Wright's approval. Earlier that year, he had written proudly to Anne, "Am to visit Frank Lloyd Wright at Talileassen [*sic*]—his famous Wisconsin 'love nest' this weekend." It was a rather insensitive reference to the 1914 scandal at Taliesin, where Wright had lived with his married lover before she was murdered with six others. Although the visit never occurred, the prospect testified to Fuller's growing reputation. Two years earlier, he had written of Wright, "I know that he will no sooner hear of 4D than he will claim that he invented it, in his infancy."

For now, Fuller was confident that his house would appear at the 1933 World's Fair, but he lacked a sponsor, and his personal issues remained unresolved. His annual visit to Bear Island had ended with a cruise with his wife and daughter on the *Lady Anne*, but they continued to live apart. Left to his own devices, he grew more involved in the esoteric world, which seemed more receptive than the establishment. An exhibit of his work was held at a museum devoted to the Russian artist and mystic Nicholas Roerich, and the author Muriel Draper finally introduced him to George Gurdjieff himself.

Although Fuller later dismissed Gurdjieff as a fraud, they met several times at Romany Marie's. As the tavern's resident visionary, Fuller was wary of the guru, but he also praised his "beautiful expressions," remembering how Gurdjieff offered individual toasts to the thirty-three kinds of idiots, causing everyone to become intoxicated, before asking his followers to "go to work writing his life history." Frank Lloyd Wright's third wife, Olgivanna, was among his admirers, which led Fuller to remark afterward that the architect "let Gurdjieff take over."

Fuller was more impressed by an article in the *New York Times Magazine* by Albert Einstein, whose work he saw as confirming his own conclusions. In the essay, which appeared on November 9, 1930, Einstein affirmed that religions were driven by fear or longing, while "the cosmic religious sense" gave heretics—including scientists—the courage to endure failure. Fuller embraced it as a description of himself, and he came to believe that the piece marked the dawn of a new era in history.

A few days later, a key player appeared in his life. Evelyn Schwartz was born in 1913 into a wealthy family of Hungarian Jews in Brooklyn, where her father designed fur coats. After he died of a heart attack, her mother stopped speaking, which left Schwartz to make her own way. She was attractive and intelligent, with an interest in art and theater, but she was still only seventeen when a high school classmate brought her to Romany Marie's.

On November 12 Schwartz listened intently to one of Fuller's talks. Although he was twice her age, Fuller was struck by her resemblance to the absent Noguchi, which he called a type "to which my very nature and ideal must automatically respond." After she became a regular, he started to drive her home to Bensonhurst, which was considerably out of his way, and she soon sensed that Fuller—who later described a car as a "porch with wheels" for courting—wanted to become more than friends.

Schwartz fended him off, but she respected him as a teacher, and he even brought her to meet the Hewletts. "Anne's siblings included a set of female twins with turned-up noses and adenoidal speech," Schwartz wrote in her memoirs. "Anne shared this family trait and, wearing no makeup, looked 'old' to a teenager like me." Roger Hewlett promptly fell in love with her, and she had to gently turn him down when he impulsively proposed marriage.

Fuller eventually made his intentions more explicit. "Being an expert con man," Schwartz said, "Bucky explained smoothly that Anne didn't like sex and encouraged his having other girls. Later in life, I discovered that 'my wife doesn't like sex' was a typical opening line of married men interested in extracurricular sex." Schwartz was a virgin, and they didn't sleep together yet, but the terms of their relationship had been established. If Anne refused to believe in him, Fuller was willing—like Wright—to risk disgrace with another partner.

Despite his bohemian lifestyle, Fuller remained active in high society, and he escorted Schwartz to the Beaux Arts Ball on January 23, 1931. Its theme was "The World of the Future," which inspired Fuller to dress her in a wearable version of his ideas. "The top, made of sheet aluminum lined with cork, resembled armor with two pointy breastplates," Schwartz recalled. "Bare skin showed between the breastplates and the short skirt, which consisted of a bright silk underslip covered by alternating chains of silver and gold, like a hula skirt, which swung out when I danced."

Before the ball, they called on Caroline and Rosamond. Schwartz was surprised by the "freezing reception," and she understood for the first time how they looked to others. She assumed that Fuller refused to see this, which made her wonder at his lack of empathy, but her misgivings were swept away by the excitement of her first ball, where architects came costumed as their own work. William Van Alen

wore a hat shaped like the Chrysler Building, while Fuller transformed
Schwartz into a stealth advertisement for the Dymaxion House.

Afterward, Schwartz began to assist with Fuller's models, while keeping what he described as "valuable notes of lectures and discussions." She
became the first in a series of women outside his marriage to provide
the creative support that he needed at each stage, and they were openly
seen as a couple. "Our chaste good-bye kisses grew into long, passionate encounters," Schwartz remembered. "He would drive me home to
Bensonhurst, and while everybody in my big house was asleep, we
would neck. Bucky urged me to go 'the limit,' but this was not to
be contemplated." Schwartz may have been outwardly confident, but
she was still young and vulnerable, and Fuller was more than willing
to use this to his advantage.

By 1931 two distinct themes had emerged in the coverage of Fuller's
ideas. One was exemplified by the magazine *The Nation*, which listed
him on its honor roll for "developing the potentialities of mass production, new materials and new engineering principles for housing that
is practical, cheap and of good design." The other was typified by
a caption that had appeared more than a year earlier in the New
York *Daily News*, under a photo of Fuller alongside his model of
the Dymaxion House, "which he predicts will be all the rage in the
twenty-first century."

Fuller had never wanted to be seen as a futurist, and he saw his
house as a plan of action, not a prediction. He sought funding from
the Metropolitan Life Insurance Company, which he believed would
embrace his house's potential to increase longevity, and secured an invitation through Vilhjalmur Stefansson to a dinner held by the Soviet
trade office in New York. The engineers there allegedly said that they
needed to keep the house a secret from their people—who had to focus
on their five-year plans—or otherwise "they would all want it." It still
represented a vast market, and Fuller left dreaming of an order from
the Russians.

His house had made an impression on several architects, including the urban planner Henry Wright, but his best hope for a prototype remained the upcoming Chicago World's Fair. Fuller worked on sketches in the office of Harvey Corbett, whose architectural partner introduced him to the banker Frank A. Vanderlip. At a dinner at Vanderlip's apartment, Fuller met Clarence Woolley of the American Radiator and Standard Sanitary Corporation, which operated a foundation in Buffalo—named for its founder, John B. Pierce—that was designing a new kind of bathroom.

Vanderlip was separately developing a real estate investment in California, and he invited Fuller to discuss the project at lunch the following day. Fuller brought along the Austrian architect Richard Neutra, a recent arrival to the city, who had called him after seeing his name in *The Nation*. Neutra thought that the Dymaxion House represented "a distinctly advanced segregation of functions," which led them to consider a collaboration, and since he had the practical background that Fuller lacked, he seemed like a promising candidate for Vanderlip.

Two very different accounts would appear of what happened at lunch. In one version, Fuller stated that Vanderlip hired Neutra to work for him in California, while in another, he remembered that the meeting "went dead" after his Austrian guest casually stirred ice cream into his finger bowl. Vanderlip reportedly closed the meeting by giving Fuller fifty dollars and advising him to focus on the structure of the house, not the utilities, which would only cause him trouble.

Fuller ignored the warning. He had kept in touch with Woolley about the Pierce Foundation, and on April 11, 1931, he received a formal offer from Robert L. Davison, a former *Architectural Record* magazine writer he had met at the AIA convention, to join the bathroom program. It was a chance to prototype a key part of the house, but on his arrival in Buffalo, Fuller was dismayed to learn that Davison wanted him to work on an existing design for a conjoined bathroom and kitchen.

Because Davison was often in New York, Fuller was left on his own in the workshop. "I have never been to a place where I was quite so devoid of friends," Fuller wrote to Anne, whom he sometimes visited on the weekends—although, in another letter that he mailed on the same day, he also invited Evelyn Schwartz to join him at his hotel. She never made it to Buffalo, but Fuller met her for a rendezvous in New Jersey, where he was equally taken by the sleek lines of the new Pullman railway coach. It represented how Davison would approach the bathroom, he told her, "if he weren't such a nut."

Feeling that there had been no improvement in sanitary engineering for more than a century, he quietly expanded Davison's bathroom, which concealed the plumbing in the walls, into his own vision. In his sketches, the sink blended smoothly into its surroundings, with water flowing directly from the basin and away from the user to minimize splashing. The interior surfaces were seamless aluminum and enamel, and the toilet emerged from the wall like the bow of a ship.

Fuller suspected that the indecisive Davison, whom he dismissed as "an annoying bore," was presenting it as his own work, so he began to keep aspects of his design a secret. He also met privately with a representative of the Ford Motor Company who pledged to provide a chassis for his proposed vehicle, the 4D Transport, which he optimistically expected would soon be on the road.

In June, according to Fuller, as they prepared to show the bathroom to executives in an Eleventh Avenue warehouse, Davison, "blustering and blushing," suddenly asked him not to attend. Fuller responded that he would agree only if Davison confessed that he had taken credit for their ideas. Davison allegedly acknowledged this, and Fuller, by his account, promptly resigned.

The true crisis came one month later. Fuller had been in contact about his bathroom with Neutra, who offered to provide a draftsman to "ripen and harvest" it. Word reached Knud Lönberg-Holm, the technical editor for *Architectural Record*, who asked to see the drawings, and Davison was angered by what he saw as a betrayal. Fuller shot back that he had the right to promote his own concepts, and he quit the Pierce Foundation at the end of July.

He returned to the city under a cloud, and he later told Schwartz that he had suffered "a nervous breakdown." From there, he left for

Bear Island, and although he promised to write to her, he never did. Enraged by his "unfeeling cruelty," Schwartz endured "two long months when silence, distance, and cold reality told me that Bucky would not divorce his wife."

They eventually reconciled, however, and what Schwartz called their "intense and frustrating" sexual attraction left them in need of a place to meet in private. Fuller found it in the Starrett-Lehigh Building, a recently constructed warehouse and freight terminal in the West Chelsea neighborhood with a vacant storeroom on the top floor. Its rooftop view of New York was breathtaking, and Fuller surprised the rental agent by asking if it was available.

It was his first permanent address in the city in years, and it immediately affected his relationship with Schwartz, who had turned eighteen. "I finally gave in to his promises of divorce, and we began to sleep together," she recalled. "Bucky was an experienced and tender lover, and I was gloriously happy." Almost at once, however, she became convinced that she was pregnant, and after hearing the news, Fuller went to tell Anne that their marriage was over.

When he returned to propose to Schwartz, she told him that the pregnancy had been a false alarm, but he decided to proceed with the divorce: "Having gone this far," he wrote to his young lover, "I did not wish to retract offer or mend gap at home which later seemed wise for its own sake." Schwartz responded by breaking off contact, and Fuller effectively began to stalk her. He drove past her house at night and parked where she could see him, insisting that he only wanted to show her that she hadn't been abandoned.

Schwartz was troubled by his behavior, but she agreed to meet him on Christmas Eve 1931. As a gift, Fuller inscribed a copy of Margaret Kennedy's *The Constant Nymph*, a novel about a teenager's love for a married man: "To the dearest nymph my life shall experience—may the world never injure her laughing heart or break the spontaneous inclusion of every beautiful thing." Nevertheless, Schwartz regarded him coldly. She accused him of only wanting an affair, and he noted in a diary entry that she was "in almost hysterics over our difficulties."

In a distraught letter, Fuller told Schwartz that he had confirmed his separation with Anne, and he repeated his resolve to marry her, even though marriage was "a direct contradiction of my ideals." The following night, however, he recorded in his diary that she had been "waiting to torture me" at Romany Marie's, and it was clear that any prospect of a divorce was an illusion. Their relationship was over, and although they renewed their friendship a few years later, Fuller was left for the moment with nothing but what Anne called their "partial estrangement."

While Anne devoted herself to her family on Long Island, Fuller channeled his energies elsewhere. Two months earlier, he had organized a circle of young architectural editors and critics to discuss his ideas. They were united by their interest in "tension theories, liquid compression-members, standardization of utility components, manufacturing and merchandizing methods," and Fuller agreed to serve as their "peripatetic guide." At first, the group was known as 4D Dymaxion, but it ultimately settled on the more neutral Structural Study Associates, or SSA.

The core members, who met at a trattoria on MacDougal Street, included Knud Lönberg-Holm and the architects Frederick Kiesler, Ted Larson, and Simon Breines. They were inspired by the Bauhaus in Germany—they intended to start a similar school of their own—as well as the domestic technocracy movement, which argued that economic policy should be set by scientists and engineers. Fuller encountered Howard Scott, its founder, at Romany Marie's, but he felt that the technocrats neglected intuition, and he later compared their uniforms to the "colored shirt movements of the Nazis."

All the same, he understood the value of such a network, and although his ideas were theoretically available to everyone in the SSA, there was no doubt that he was in charge. Their goals included a lecture series by Fuller, whose house they planned to fund for the World's Fair, and the other members covered a percentage of the rent at his "penthouse" at the Starrett-Lehigh Building, where they held gatherings. It became the scene of wild parties attended by the likes of Mexican painter Diego Rivera, whom Fuller led upstairs after the elevators had shut down for the day. "He was quite large and corpulent," Fuller remembered, "so we had to stop at every floor to rest."

Rivera had contributed art to *T-Square*, an architectural magazine based in Philadelphia that had published an article on the Dymaxion House. After its editor, Max Levinson, asked Fuller to submit a piece of his own, the result was the first part of "Universal Architecture," a feverish essay that warned of a conspiracy of "capitalists, industrial leaders, and technologists" who had joined forces to take over the housing industry.

The article coincided with a landmark show on contemporary architecture in February 1932 at New York's Museum of Modern Art, where Philip Johnson and the architectural historian Henry-Russell Hitchcock set out to curate the defining statement of the International Style. With its modular forms, industrial materials, and lack of ornamentation, it had obvious affinities to Fuller, but Johnson rejected the Dymaxion House on the grounds that it had "nothing at all to do with architecture."

Hitchcock and Johnson were invited to edit an issue of *T-Square*, which was forced to find a new name after a copyright dispute with the publisher Scribner. Fuller proposed that it rename itself *Shelter*, which he hoped to turn into another platform for his work. He had been carefully ranking his possible endeavors in order of the capital that they required. With enough money, he could enter business for himself; with limited funding, he could develop models or designs on paper; with less, he could write a book; and if he had nothing at all, he could fall back on conversation.

For three years, he had focused on the last alternative, but now he made a decisive return to print. The magazine had been funded largely by the architect George Howe, who approached Johnson and Fuller about taking it off his hands. Johnson was rich enough to put up his share without any trouble, while Fuller scraped together $800 and cashed out an insurance policy for the rest. "Bucky Fuller and I always disliked each other," Johnson recalled decades afterward, and a clash of wills seemed inevitable, but they postponed it by never meeting in person.

Fuller returned to New York to find himself evicted from the Starrett-Lehigh Building. He claimed later that his nemesis Robert L. Davison had been so impressed by the penthouse that he had stolen it, offering a higher price after Fuller forgot to pay the rent for a month while he was gone. The Pierce Foundation did put an experimental house on the roof, but Fuller had also failed to cover rental fees and other expenses for the year to date, which was more than enough reason to throw him out.

He recovered with an arrangement at the Hotel Winthrop, where he agreed to display the Dymaxion House in exchange for a room. It wasn't much—Fuller and Isamu Noguchi, who was back in town, slept on air mattresses and survived on doughnuts and coffee—but he was spending most of his time in Philadelphia assembling the May issue of *Shelter*. Raising the price to two dollars, Fuller announced that it would no longer take advertising, and he described the publication grandly on its masthead as "A Correlating Medium for the Forces of Architecture."

Not surprisingly, Fuller's philosophy was visible throughout the magazine, which was beautifully designed by Lönberg-Holm. A caption for a photo of the latticed radio towers erected by the League of Nations, representing a technological solution using the minimum amount of material, featured the debut of Fuller's single most powerful idea: "Progressing towards ephemerality—as emergency evolves into emergence." Elsewhere in the magazine, he wrote, "Industrially reproducible compositions and instruments will get lighter and lighter by selection and refinement. SSA terms this phenomenon the *factor of evolutionary ephemeralization*."

Johnson felt these ideas were unintelligible, and he resigned in protest: "Bucky Fuller was no architect, and he kept pretending he was. He was annoying. We all hated him." After George Howe asked to be removed as associate editor, Fuller was left on his own. He blamed the situation on his enemies, whose attacks on his character had led potential supporters to withdraw, but he conceded to Howe, "At the same time, I offer no pretense that I have not personally participated in almost every form of venture or even ethical prohibition imaginable."

While Fuller's personality caused investors to hesitate, his competitors had taken advantage of the rising demand for affordable housing. His most irritating rival, General Houses, planned to build prefabricated steel homes conceived by Howard Fisher, with a list of suppliers that included American Radiator. Fuller was positive that Fisher, who had once arranged for him to deliver a talk, was copying his work, and he grew convinced that big business—including the forces behind the World's Fair and the Pierce Foundation—was conspiring to steal his ideas.

On August 8, 1932, an SSA exhibition opened at the Winthrop. The following day, Fuller met there for four hours with Starling Burgess, the yacht and aircraft designer who had given a lecture to his class years ago at Milton Academy. At fifty-three, Burgess was bankrupt, and after studying a model of a tower with a streamlined windshield, he agreed to do calculations for the mast. Fuller saw him as a valuable ally, although he lamented to Anne that whenever he began to get his hopes up, the result was "some kind of abstract abortion."

Despite his bleak mood, Fuller was receiving more coverage than ever. A profile in *Time* opened with the unlikely image of Fuller dancing the Lindy Hop at Harlem's Savoy Ballroom "for black men and women," and it concluded with a reference to a mobile dormitory for Soviet farms that would be built "in some shipyard now being kept secret." Fuller planned to use wood and rope to make up for a shortage of metal in Russia, but it never went past the daydream stage.

The article impressed his friends and put him in a reflective state of mind. After it appeared, Fuller took Noguchi on a tour of his personal landmarks, including his "cathedral," the Brooklyn Bridge. Hoping to reconcile with Anne, he told her that it reminded him of their shared past: "War days. Armour, Kelly, and Stockade days. Days before we were engaged, days before we were married. All our joys, sorrows, perplexities, inspirations seem to wait for me at the center of that bridge."

As his public profile increased, Fuller divided his time between multiple ventures, including a homeless shelter in a building owned by the Guaranty Trust Company, which volunteers refurbished into housing for two hundred men before it was closed as a fire hazard. He hired Burgess for a model house project in Wiscasset, Maine, which he was unable to fund, and after losing his arrangement at the Winthrop, he shifted his operations to the Carlyle.

Most of his energy during this period was devoted to *Shelter*, which was about to release its most radical installment. Fuller had announced that the upcoming issue would be edited by the poet and critic Archibald MacLeish, who had hailed the Dymaxion House as possibly "the prototype of a new domestic architecture." After MacLeish was dropped, perhaps due to pressure from Johnson, Fuller edited it himself. Sensing that the magazine's days were numbered, he decided to put everything into one enormous issue in November.

The cover featured a Noguchi sculpture of the dancer Ruth Page, transformed into a streamlined shape that Fuller titled *Miss Expanding Universe*. In the text, he attacked Johnson, criticized the Pierce Foundation, and warned of a trend in economic charts, commencing on his last birthday, that pointed to martial law: "A winter residence outside of New York City, or other big cities should be planned. A month by month rental is also advisable. Be sure that they have a complete winter's coal supply *in the bins* of the place you select to stay in."

His tone was often apocalyptic, but it alternated with flashes of clarity and insight. He opened with a definition of ecology—"The study of human relations, particularly as pertains to the home"—and laid out his plan for the "conning tower," a corporate conference room that used teletype, radio, and television to share information. Many pages were visual marvels, and Le Corbusier's influence was visible in the collages of tension structures and hexagons in nature, as well as a gallery of disasters captioned with what became a famous precept: "Don't fight forces, use them."

Fuller's most notable article was "Streamlining," which built on Le Corbusier's observation that a car could be based on an ovoid shape with "the greater mass in front" to minimize air resistance, as seen in dirigibles and birds. His concept for the 4D Transport was illustrated with models, by Noguchi, that evoked flying fish, and he added that similar rules could be used to streamline society itself: "A designer may, through scientific contemplation and articulation, reorder resistances and thus demonstrate true mind over matter."

The issue was exhausting to read, however, and he was right to suspect that the magazine was nearing its end. He said later that although *Shelter* had been profitable, he shut it down to keep it from being "exploited for high gain," but then stated elsewhere that he closed it after Franklin D. Roosevelt's election as president allowed liberals to put their ideas into practice. In reality, while subscriptions had risen, the magazine's lack of advertising drained its funds, and it was sued for costs by its printer.

Fuller was grappling separately with the realization that the Dymaxion House was unlikely to appear at the World's Fair. In the November *Shelter*, he wrote that it had been canceled because it "entailed too much scientific research and industrial organization," but he never gave up entirely. As late as December, he asked Leon Levinson, his managing editor, to tell the exhibition director that it would be "a simple matter to arrange for its display," but no funding ever materialized.

The clouded circumstances led to another famous myth. According to Fuller, he received an offer from the nephew of the fair's president, Rufus C. Dawes, to build a prototype of the Dymaxion House, but he declined out of concern that it would be "another piece of design novelty, superficial, costly, and impractical in every respect." In yet another version, when the younger Dawes asked how much funding a house would require, Fuller calculated the capital needed for an entire industrial operation. "The basic cost today," Fuller claimed, "is a hundred million dollars."

One of Fuller's later advocates, Robert Marks, treated this as an example of his ambition: "Certain that he was dealing with a lunatic, Dawes turned and left the room. He was aggrieved. All he asked for was a house; Fuller offered him an industry." Fuller had actually stated elsewhere that the business could scale up gradually, and when even a modest amount of money proved to be unobtainable, he framed his retreat as an uncompromising insistence on mass production.

In spite of these setbacks, he continued to draw positive attention. The editor Clare Boothe Brokaw, who met him while he was at the Pierce Foundation, included him in the annual *Vanity Fair* magazine hall of fame, and he was invited to a private party for Frank Lloyd Wright. Wright asked him to promote his Taliesin Fellowship for architectural apprentices as one of "ten architects I consider leaders," and at another meeting, he raised the possibility of Fuller and Noguchi teaching at his side.

"I would certainly like to come out to the Fellowship," Fuller responded, "if not for permanent association, at least for a period long enough to meet and work with your family." It never happened, partly because Fuller—who now had a following of his own—knew that Wright would relegate him to a subordinate role. Soon afterward, Wright publicly chastised Fuller, whom he called "my friend," for "exclusively booming architecture for the benefit of private industry."

Fuller was left with nothing but his reputation as what Noguchi called "a messiah of ideas." Such veneration often became a burden: "I found myself being followed by an increasing number of human beings, particularly women, who were beginning to make me into some kind of messiah." He claimed that he resolved on New Year's Day to discourage the cultists: "I premeditatedly began to drink, and having known the patterns of the brothel, I made myself extremely offensive, doing things that I ordinarily would never do, drunk or sober. As a result, they soon began to avoid me."

In fact, his actual typewritten resolution of January 1, 1933, was filled with high ideals "to streamline man's competitive volition, unbeknownst to him, in the direction of least resistance." If Fuller drove away fans who saw him "stumbling out of bars or coming out of a brothel and things like that," it was less than deliberate. He believed that he was in control, but he had never given up drinking—he filled entire bathtubs with spiked eggnog at parties—and he later remembered this period as "my stupid time."

A new opportunity came from a surprising direction. The highlight of the last issue of *Shelter* had been Noguchi's models of the 4D Transport, which were displayed in an architectural bookshop in New York. After they drew an enthusiastic response, Fuller was invited by Alfred Reeves, the general manager of the National Automobile Chamber of

Commerce, to set up a booth at its annual show at the Grand Central Palace. He also spent time in Philadelphia to shop around a book proposal and earn what he could through lecturing.

After one event, he was approached by Anna "Nannie" Biddle, a Philadelphia socialite he had met years earlier in Maine. Biddle had wanted to study "the customs and living conditions of Alaskan women," and Fuller introduced her to Vilhjalmur Stefansson, who agreed to consult on the expedition. Biddle spent several months in Alaska, mostly in a snowbound cabin, and, on her return, she visited Fuller at the *Shelter* office. The two of them talked about "all sorts of things," he recalled, "including whether or not she should leave her husband."

In December the thirty-year-old Biddle, a fashionable member of cultural circles, had hosted a reception for Fuller. Among the attendees was famed conductor Leopold Stokowski, who brought him home to talk until three in the morning and quickly became another supporter. Now Biddle envisioned an even more exciting relationship. Using her own money, she offered to fund a working prototype of the 4D Transport, which she was willing to finance on Fuller's terms. It was the best chance that he had ever been given to realize one of his designs, and he prepared to disrupt an entire industry by building the car of his dreams.

FIVE

STREAMLINES

1933–1942

We think of ourselves as individuals with gravity pulling us Earthward individually in perpendiculars parallel to one another. But we know that in actuality, radii converge. We do not realize that you and I are convergently interattracted because gravity is so big. The interattraction is there, but it seems so minor we dismiss it as something we call "aesthetics" or a "love affair." Gravity seems so vertical.

—BUCKMINSTER FULLER, *SYNERGETICS*

"The first motorcars were constructed, and their bodies built, on old lines," Le Corbusier wrote in *Towards a New Architecture*. "This was contrary to the necessities of the displacement and rapid penetration of a solid body." In his earliest sketches of the 4D Transport, which dated from 1928, Fuller borrowed the ovoid contours featured in the architect's book but added three wheels on a triangulated frame. It was steered from the rear, like a boat, and came with inflatable wings.

Fuller claimed to have discussed its underlying principles years earlier with Patrick Bellinger, his commanding officer in Norfolk. The fishlike vehicle would rise into the air on "jet stilts," with a pair of gas turbines as thrusters, in a process that Fuller compared to a duck

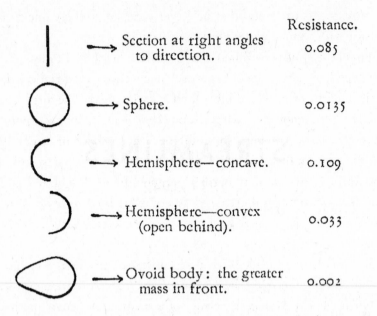

		Resistance.
	Section at right angles to direction.	0.085
	Sphere.	0.0135
	Hemisphere—concave.	0.109
	Hemisphere—convex (open behind).	0.033
	Ovoid body: the greater mass in front.	0.002

Table of wind resistance of various shapes, including "Ovoid body: the greater mass in front," from the English translation of Le Corbusier's *Towards a New Architecture* (1927).

beating its wings repeatedly to gain altitude, while its properly timed bursts would translate into forward motion, like a pole vaulter handed another pole at the peak of each jump.

Since a flying version was currently beyond his means, he decided to focus on the vehicle's "ground taxiing" phase, which would take the form of the car that his young daughter called the Zoomobile. He later spoke of it as a part of the home that had broken off, "like hydra cells going off on a life of their own," liberating people from cities. Building it was consistent with his strategy of prototyping subsidiary elements as the opportunity arose, and it seemed easier to produce than a house.

When Fuller accepted Nannie Biddle's funding, he was careful to establish control up front. Remembering his frustration at the Pierce Foundation, he stipulated an "ice cream cone" clause: "If I want to use all of it to buy ice cream cones, that will be that, and there will be no questions asked." On these incredible terms, Biddle advanced him $1,000—with the promise of more to come—in February 1933. It was the most access to capital that he'd ever received, and it came from an investor who seemed willing to follow him anywhere.

Fuller promptly recruited the penniless Starling Burgess, whose knowledge of patents and expertise in nautical and aviation design would lend the project credibility, and Max Levinson, the former editor of *Shelter*, as employees of what became the 4D Dymaxion Corporation. The firm's logo, a flying fish, combined the concept of streamlining with the name of Burgess's first biplane. For automobile parts, they spent $450 on a Ford Tudor that they plundered for its engine, running gear, chassis frame, and gearbox. To complete the prototype, Burgess estimated that they needed another $1,500, which Biddle agreed to provide.

In March they headed for Bridgeport, Connecticut, conveniently close to Burgess's home. The decline of the local auto industry during the Great Depression had left plenty of infrastructure, enabling them to commence operations at the Boudreau Machine and Tool Company on the morning of Franklin D. Roosevelt's inauguration. The new president promptly declared a bank holiday two days later. Because 4D Dymaxion was one of the few local businesses with cash, a thousand workers reportedly applied for a handful of positions, and Fuller allegedly prioritized men with families, who sometimes wept with gratitude.

Privately, Fuller thought that the rent was too high, and the tools were unusable, which forced them to subcontract much of the construction of the chassis. Worrying that Burgess was overspending on custom parts, he dealt with the pressure by drinking with his workers and eating cake at his boarding house in Darien: "I had been starving for so long." The stress also strained his already tenuous marriage. Anne wrote from Lawrence to complain of their tense visits: "We waste so much of the little time we have together by your scolding and fussing about what you consider to have gone wrong the time before."

On April 19 a test revealed drawbacks with the prototype's single rear wheel, which was fixed in place, causing it to lean into corners while the front tires remained perpendicular. At high speeds, the result was a "death wobble." They eliminated the defect with a strategically placed hinge, however, and Fuller was encouraged enough to allow

Allegra to take a ride in it. He received another $5,000 from Biddle, who was in the middle of a painful divorce, and he proposed that Burgess become a partner, although he had no intention of sharing his voting power.

Since the shop often flooded, they moved to the Automatic Machine Company on East Washington Avenue. Fuller seethed when Burgess took more interest in building a yacht for a prominent lawyer, Elihu Root Jr., but despite the distraction, they successfully remodeled the chassis with parts from Ford, Studebaker, and Chevrolet. The staff grew to seventeen, including what the writer Robert Marks described as "Polish sheet metal experts, Italian machine tool men, Scandinavian woodcraftsmen, and former Rolls-Royce coach makers," as well as an artist named Emma Lu Davis, who later became a celebrated sculptor.

As usual, Fuller was already making outrageous claims. He stated falsely to one contact that the prototype had been driven a hundred thousand miles, that a hundred cars were being built, "most of which have been sold at $2,500," and that the price would fall to $200 as

Dymaxion Car under construction in Bridgeport, Connecticut (1933).
Fuller is kneeling to examine the engine compartment at the far left.

production rose. In fact, the one car that they were actually assembling was proceeding slowly. Because of a lack of space, in June they moved yet again, this time to a Bridgeport factory that had once belonged to the automaker Locomobile Company of America, where Burgess— again to Fuller's frustration—spent most of the next two weeks working on the yacht.

Fuller's mood improved when Anne and Allegra joined him in Connecticut, and he hoped to arrange for another reunion by enlisting Isamu Noguchi. "It seems to me," Fuller wrote to the sculptor, "that this is the chance that you and I have always looked forward to in the matter of your executing the best of design." Noguchi, who was working out of a studio in London, never came, so Fuller reached a deal with Burgess instead on future house, boat, aircraft, and writing projects, stating in a letter that he hoped that they could spend their lives developing ideas "vested in us by the sagacity of the Universal mind."

In practice, Fuller became annoyed when Burgess put both of their names on the sign in front of the building, and a more personal conflict was brewing. At their first meeting in Bridgeport, Burgess was charmed by Biddle, who commended Fuller for having hired an expert instead of a "nobody." On another visit from Biddle, Fuller grew suspicious when Burgess led their benefactor away by the arm. Shortly afterward, Burgess declared his love for the wealthy socialite, and the feeling turned out to be mutual. According to Fuller, Biddle claimed that Burgess "aroused her maternal instinct more than any-one else had ever done."

Fuller foresaw a struggle for power, but he concentrated on the task at hand. On July 12, 1933, his thirty-eighth birthday, they put the finishing touches on Dymaxion Car #1. It was nineteen feet in length, weighed about 2,700 pounds, and had cost a steep $8,000. Under its canvas roof, a body of lacquered aluminum was built over a wooden framework, with lightening holes drilled in the chassis, and the Ford transmission was reversed to put the engine at the rear.

Like Fuller's sketches, the car had three wheels, two in front and one in back. It was steered by its rear wheel, using steel cables in sheaves,

Fuller with Dymaxion Car #1 (1933).

and could turn in a circle with a radius only slightly greater than its wheelbase. The roomy interior was cooled by dry ice, with a single headlamp and a crush zone of balsa wood installed in front. Its wrap-around windscreen lacked wipers, on the faulty theory that its shape would shed water, and instead of a rear window, it had a periscope.

Fuller implied later that it was the first streamlined automobile, which was untrue. He liked to contrast it with a standard sedan, which resembled a horse and buggy, but Chrysler had conducted wind tunnel tests for years, and cars with similar profiles included the Arrow Plane and the Alfa Romeo Castagna Aerodinamica. Other vehicles had been designed with three wheels, and an existing German patent became a source of particular concern at the company, since any competitor could simply purchase the American rights.

According to Fuller, the car's ovoid lines allowed it to achieve 40 miles to the gallon and 120 miles per hour, although that was debatable. Burgess had wanted an "air-cooled aviation engine," but the Ford V-8 that they used was less powerful, and to arrive at the quoted figure,

Fuller may have simply multiplied its specifications with the top gear ratio. It evidently peaked around 90 miles per hour, which was the fastest that it ever went in public.

Fuller claimed that the Dymaxion Car was stable enough to drive across fields and railroad crossings, but it had real problems with handling. Its low gearing made it hard to steer, and its long suspension and the concentration of weight at the rear led to a pronounced twisting effect when turning. As a result, cornering was difficult, and it suffered from severe tire wear. A later observer found that it was plagued even on smooth surfaces by "uneasy, oscillating swivels."

Even worse, at high speeds, its tail came off the road, leading to a dangerous lack of control. In his original plans, Fuller had proposed steering by rudder when the rear wheel rose, but on a practical level, his vision of a flying machine made it worse as a car. Its three wheels were justified by a forced analogy to landing gear, which had very different requirements than a car on the ground, and Fuller admitted that it was even less stable in crosswinds: "I had to practically fly her along the highway on a northwest gusty day; I wouldn't allow anyone else to do it."

None of these flaws was evident, though, when it was unveiled on July 21 at the Locomobile plant, where it was demonstrated before three thousand observers, including Diego Rivera. Fuller was filmed behind the wheel of the car, which he accelerated to seventy miles per hour at Seaside Park. After the mayor of Bridgeport spoke, Fuller drove him to city hall, stopping twice at intersections to touch the shoulder of a traffic officer and drive around him in a tight circle without removing his hand.

Vilhjalmur Stefansson, serving as a consultant, had introduced Fuller to Alfred J. T. Taylor, a mechanical engineer from London who was interested in a license for sales overseas. Taylor agreed to pay $5,000 for a demonstration car, but Burgess put off starting work on the second vehicle for weeks, which only deepened Fuller's growing distrust of his partners. On a visit to Maine, he learned that Biddle had told his mother that Fuller "was tired out and needed someone to look after him." Biddle was seen opening his mail at his desk in his absence, and a mutual friend warned that she was spreading rumors that Fuller was "going mad."

Feeling pushed out, Fuller brought the prototype to New York, where he was cited for speeding and supposedly set a record at a "midget racing car track" in the Bronx. On one occasion, he drove editors from *Fortune* and *The New Yorker* down Fifth Avenue and impressed a policeman by turning around him in a tight circle. At every intersection, officers asked for a repeat performance, to the point that it took an hour to drive one mile. The Dymaxion Car drew attention wherever it went, and after causing a traffic tie-up in the financial district, Fuller was reportedly asked to stay above Canal Street.

At Roosevelt Field airport, the car was tested at ninety miles per hour by the company's sales manager, Frank T. Coffyn, a former dare-devil pilot. Coffyn was close to Alford F. Williams Jr., the head of the aviation products division of the Gulf Refining Company. Williams wanted to buy the prototype as a promotional vehicle for air shows, and although his offer of $2,500 was far below its cost, Fuller took it. After an argument over control, Biddle had cut off additional funding, and they badly needed the cash.

Hoping for an investment from the Curtiss-Wright aviation company, Fuller drove to Buffalo, only to be struck by a woman turning left in her Plymouth. He settled to avoid negative press—he acknowledged that the Dymaxion's appearance could have a "startling effect"—but the deal was scuttled by financial concerns. Although he continued to dream of a domestic manufacturing arrangement, meetings with Walter Chrysler and the designer William B. Stout led nowhere.

In October Fuller filed a patent that treated the vehicle's three wheels as optional. A pair of rear wheels would have fixed a "bug," as he called it, in which the back tire often became stuck in slippery ruts, but Fuller was reluctant to revise the design, especially when the future of the company itself seemed unclear. Burgess had been acting erratically: suffering from an ulcer and pain from an old operation, he was taking daily shots of morphine. After consulting a psychiatrist, Fuller tried to put him into a sanitarium, but Burgess fled with Biddle to Reno, Nevada.

The European market remained their most promising lead, and toward the end of October, Stefansson heard from Colonel William Francis Forbes-Sempill, a British aviator who had expressed interest in an investment. Sempill, who was friends with Alfred Taylor, called by radio from the *Graf Zeppelin*, on which he was traveling as an official observer. The German airship would be in Chicago for less than a day, he explained, and he wanted to see the car while he was there on October 26.

Stefansson arranged for the Dymaxion to be driven to Chicago by a Gulf employee named Francis T. Turner. On his arrival, Turner took Sempill around in the car and ran passengers from the zeppelin, including its captain, to the World's Fair. He spent all night chauffeuring Gulf employees who were in town for a convention, and when he returned to the Stevens Hotel, he found a note under his door from Sempill requesting a lift to the airfield in the morning.

Sempill offered a ride to another passenger, aeronaut Charles Dollfus, an attaché with the French Air Ministry. When the car turned up at the hotel on the morning of October 27, Dollfus was surprised by its unusual appearance, but he obligingly climbed into the backseat on the driver's side, while Sempill took the seat next to Turner. They departed shortly after eight. Five minutes later, they passed the Field Museum, heading south on Lake Shore Drive. To the east was the entrance to the World's Fair, which was the worst imaginable place for what happened next.

The five lanes of southbound traffic were moving at around forty miles per hour. As they approached the light island, where there was a slight bend, Turner turned abruptly to the left. Dollfus testified afterward that he swerved to avoid an oncoming vehicle, but that was implausible: they were in the second outermost lane, far from any cars coming the other way. The asphalt had patches of water and grease, and Turner may have braked sharply when one wheel was in a spot where the road was slick, causing him to skid and turn broadside.

Whatever the reason, he turned in front of a car being driven in the same direction by a man named Meyer Roth. The Dymaxion rolled, smashing its canvas top, and Dollfus instinctively tucked his legs and covered his head. On the second roll, Roth struck the car, sending it tumbling farther, but he managed to steer clear. Sempill and Dollfus, whose belts were unfastened, were flung out. Dollfus hit his head on the roof, suffering cuts to his chin and right eye, and landed on his feet behind the car. Sempill fell in front, breaking his skull in two places.

Turner was trapped by his seatbelt. Losing its grille and windows, the car stopped after three or four rolls in the second northbound lane. Two police officers ran to help, along with a passing motorist, and found that Turner's face was crushed, although he was still alive. Dollfus regained consciousness, but the others remained unresponsive as they were rushed to Mercy Hospital, where Turner died of fractures to his skull. He was just thirty-three, and he left behind a wife and two sons.

Fuller heard the news in a call from the editor of the *Bridgeport Post*, followed by a telegram from Samuel Halsted, a consulting engineer for the company in Chicago. Flying out the next day to contain the crisis, Fuller immediately went with Halsted to the hospital, where a nurse brought them to see Sempill. Fuller thought that the colonel looked all right—"not disfigured, color good," he said in a telegram to Vilhjalmur Stefansson—and the doctor said that he was expected to recover.

From there they went to a garage associated with Gulf, where the car had been taken after the crash. Fuller wanted to examine it, hoping to exonerate his design, but the Dymaxion was under the jurisdiction of the South Park Commission, and the officers there refused to let them inside. Fuller's cable to Stefansson later that day contained an implicit admission about the absence of safety features: "Steel bows would apparently have saved all hands."

The following day, Fuller and Halsted visited the scene of the accident, making sketches and interviewing an officer who had witnessed the crash. Fuller met afterward with Al Williams to advise rebuilding the car with metal bows over the top, which supposedly had been part of his original plan. Williams agreed, although he wanted to avoid any changes that would imply a lack of safety. He also reconfirmed his support, saying, "I believed in the car, and I still do."

After deciding not to attend Turner's funeral, Fuller accompanied Williams later that day to inspect the car at the garage. He checked the steering cables, which looked fine, and returned to the hospital to visit Sempill. While he was there, the colonel received a call from the British consul, whom Fuller would later magnify in his recollections to the king of England himself.

Although Fuller implied afterward that the crash irretrievably damaged the car's reputation, the word *Dymaxion* was absent from most of the news stories, and the positive coverage continued for a year and a half. Gulf repaired the vehicle, which appeared at events, and later sold it to an engineer who had tested it at the National Bureau of Standards. After changing hands again, it was used to advertise soft drinks, and it was destroyed in 1943 when it caught fire after being refueled.

The company had avoided bad press, but its real problems were financial. Alfred Taylor felt that the vehicle needed an American distributor before anyone in Europe would manufacture it, and he declined to advance any funds beyond his advance payment for his car, which had yet to be delivered. Fuller turned to his friend Philip C. Pearson, a stockbroker who was "not rich but not broke." Pearson, who was part of the same social circle as Nannie Biddle, was willing to come on board with an art investor named Aurel Lupu. When Burgess and Biddle couldn't be reached, Fuller guaranteed the new shares out of his own stake.

An inquest into the crash, which had been postponed after Turner's widow collapsed on the first day, resumed on November 16. After hearing from Roth and Dollfus, it returned a verdict of accidental death, with the coroner's certificate stating that Turner had died while "driving odd type car for publicity." Halsted attended the proceedings and sent a detailed account to Fuller, who learned for the first time that another car had been involved.

Witnesses had stated unequivocally that the second driver struck the Dymaxion Car only after it had begun to roll, but Fuller seized on the new information, telling Stefansson that Turner and Roth had been "apparently having a race," which was a total fiction. He found it suspicious that the second car had been rushed away from the scene, and, in subsequent retellings, he transformed Roth—who evidently worked as a shoe salesman—into a powerful South Park commissioner who arranged to have the entire incident covered up.

Fuller also claimed that Roth had caused the crash by striking the Dymaxion's bumper, although Dollfus recalled "no such collision," and he alleged that the two drivers had been racing at seventy miles per hour instead of forty. In reality, Roth was an innocent victim, but Fuller, eager to blame any convenient target, even considered leaking the allegations to *Time* magazine.

His new partners, however, were more interested in pushing out Burgess, who had married Biddle in Reno. The latest Mrs. Burgess agreed to a buyout, although she complained afterward that it took advantage of "the pressure of finances and our personal life." Another casualty of the restructuring was Max Levinson, whom Fuller later accused of conspiring with the others to seize control of the company. In response, Levinson upbraided his former employer for "the thousands of dollars that you dissipated on insane contraptions" during his disputes with Burgess: "Your life in Bridgeport was diametrically opposed to everything you wrote and preached in *Shelter*."

By the end of the year, Sempill had recovered fully. Surprisingly, he told Vilhjalmur Stefansson that he still had a favorable opinion of the Dymaxion Car, saying it was no riskier than sports model test vehicles he had driven before, although presumably not along city streets. Sempill sailed on the HMS *Majestic* back to England, where he became infamous years later for passing wartime secrets to the Japanese. For now, though, he reportedly joked to his friends, "In future, I'll stick to flying. It's safer than the road."

On January 6, 1934, Dymaxion Car #2 was finished, with Burgess's name conspicuously absent from the plans. A modified chassis made it more stable than its predecessor, but instead of delivering it to Taylor immediately, Fuller drove the vehicle to the New York Automobile Show at the Grand Central Palace, where he hoped to erase all memory of the accident. Fuller later claimed to have been excluded from the event by Walter Chrysler, who was exhibiting his streamlined Airflow sedan. According to Fuller, he rebounded by offering a ride to New York City Police Department Commissioner John F. O'Ryan, who then gave him permission to park his car in front of the hall.

It attracted more attention than any of the exhibits inside, Fuller recalled, and he said that even Chrysler himself offered him a word of praise: "You produced the exact car I wanted to produce." He also received interest in a third automobile from Evangeline Stokowska, the wife of conductor Leopold Stokowski, an ardent Fuller admirer. Stokowska was wealthy in her own right—her family had founded the medical products company Johnson & Johnson—and she finalized her order for a car on January 22.

On the same day, Caroline Fuller died of pneumonia at sixty-six. Fuller's mother had been living in New York with Rosamond and her husband, an Englishman named Harold Edwards-Davies, and left a sizable estate of $114,797, along with her holdings in Maine. It was divided equally among her four children, but Fuller's impatience with the proceedings caused tensions with Wolcott, who lived in Cambridge with his wife, the former Persis Wellington, and their daughter.

Fuller had never overcome his feeling that he had disappointed his mother, and he was also drifting apart from Anne, who had moved with Allegra to Philip Pearson's property in Noroton, Connecticut. When Pearson offered her the chance to design her own guest home, Anne produced an attractive clapboard cottage. This allowed her to say later that she had built a house long before her husband did—although Fuller occasionally took credit for it himself.

During this fraught time, Fuller spent a memorable weekend in February with Noguchi and two remarkable women. Noguchi's girlfriend, the actress Dorothy Hale, had proposed taking Car #2 to the premiere of the experimental opera *Four Saints in Three Acts*. With Noguchi at the wheel, they were joined by Clare Boothe Brokaw, the managing editor of *Vanity Fair*. Fuller met them in Hartford, Connecticut, for the performance, which was filled with luminaries from the art world. He reconnected with Noguchi's friend Alexander Calder—

Fuller and the sculptor kept in touch intermittently afterward—and visited playwright Thornton Wilder on the way home.

Fuller was even more captivated by the glamorous Brokaw, who sardonically described herself at age thirty as "a smart young society matron with a literary flair." After divorcing her alcoholic first husband, she had risen from a caption writer at *Vogue* magazine to her present editorial role, and she easily matched wits in the car with Fuller, amusing him when she said that men found only the love that they deserved. She told Fuller that he should have been a minister, and by the end of the trip, he evidently concluded that Brokaw—with her beauty, connections, and vast ability—would be a better partner for him than Anne.

On a cold night in Bridgeport, he surprised her by declaring his love, and in a subsequent phone call, he was filled with "a sudden longing to take you in my arms and a temporary desperation at the impossibility to do so." She fended him off by teasingly asking for a Dymaxion House for two, but she later told him in a letter that she was incapable of loving anyone: "I am a tangle of vanity and fear and ambition and envy. . . . When the pyramids were built, I was not there living my life, and when your Dymaxion world comes, my life will be gone."

Brokaw closed on a melancholy note: "Anyway, Bucky I am going away, so good-bye." Before she could mail it off, she received a passionate letter from Fuller, although he misspelled her name "Claire." Since meeting her, Fuller wrote, he had felt renewed excitement in his work, and he said that their love would become "the most thrillingly beautiful, physically, mentally, and spiritously that either of us could ever have knowledge of." He followed up with a Valentine's Day poem that rhapsodized over "the lovely love I'll make to thee."

Reading her original letter, Brokaw felt that it was "very silly and hysterical and feminine." Since they had last seen each other, she had quit journalism to become a playwright, and on a train to North Carolina, she wrote to Fuller, "I do not like the mutability movability locomobility of a coming Dymaxion world. To move implies a goal and an objective. Deeply I have none." Yet she agreed to meet again: "I must get quite clearly the picture of what you are and what you hope to be so that I may compare it with the unfolding picture of what you become. And you in turn must find this out from me."

In anticipation of their next encounter, Fuller bought a new suit, and after she asked him to phone on her return, he spent a solitary dinner wondering how to express his love. When he called, however, Brokaw was out, and after it happened again, he admonished her for treating him like one of her "hungry telephoning beaux." Fuller had actually met his match, and he even granted the possibility that she might become "many times greater" than he would. Nevertheless, they stayed friends, meeting up occasionally with Dorothy Hale, but Brokaw was busy chasing her own destiny.

On the business front, Taylor withdrew his order for Car #2, which Fuller had never delivered. After refunding him from his mother's estate, Fuller took the car to New York, settling into the Racquet and Tennis Club on Park Avenue and earning press coverage by driving celebrities around town. Through his engineering contacts, he met Frank Morley, a board member of the British publishing house Faber & Faber. Frank, in turn, introduced him to his brother Christopher, the versatile writer who had once moved Fuller with the poem "To a Child." At their first meeting, Fuller recited it from memory, touching Morley deeply.

Christopher, who was five years older than Fuller, brought him to the Three Hours for Lunch Club, which he had founded as a gathering place for assorted wits, and he praised the Dymaxion Car in the *Saturday Review of Literature*. Fuller continued to court the famous, including H. G. Wells, whom Frank Morley arranged for him to meet. Wells had heard of him through Stefansson, and, according to Fuller, the prolific English novelist became "pleasantly high" with him on mint juleps before taking a ride in the car, which he promised to consider for his futuristic movie *Things to Come*.

A newsreel was filmed of what *The New Yorker* called Fuller's "astonishing automobile," and he had even greater plans for Evangeline Stokowska's car, which would feature a metal roof and a tail fin to improve stability. He was looking ahead to a completely revised design that would incorporate the lessons of previous models, but although plans were drafted for a Dymaxion Tudor Sportster—with four seats and a shorter length of fourteen feet—it was never built.

In the meantime, Stokowska was having second thoughts, and after reviewing her expenses, she asked Fuller to sell Car #3 for her. After it was finished in June 1934, he headed for the World's Fair in Chicago. Fuller claimed that it took him only five hours to cover the final stretch of the drive from Detroit, for an average speed of close to sixty miles per hour.

At the fair, then in its second year, Dymaxion Car #3 was displayed outside George Fred Keck's "Crystal House"—a model prefabricated home of glass and steel—thanks to an invitation from their mutual associate Leland Atwood. Fuller participated twice a day in a pageant of transportation, delighting in accelerating before the crowd and executing tight turns. "Everybody in the stands went up," he recalled. "They were sure that I was going over, you see."

Feeling encouraged, Fuller wrote to Henry Ford, who reportedly called to say that he was "immensely interested." When Ford offered a discount on parts, Fuller used it to purchase an engine for thirty dollars, in what turned out to be his only contact with the man he praised as "the greatest artist of the twentieth century." An overture to the army foundered, along with a demonstration for the Soviet trade office in Washington. Assuming that the Russians wanted the vehicle for military use, Fuller drove them through open fields, only to have the car start making a grinding noise. After sending his passengers home in a taxi, he checked one of the tires, which came off in his hands.

This debacle resulted in the cancelation of a Soviet deal for two cars, and Fuller began to show signs of stress. One day, he was speeding along with Frank Henry, an editor at the book publisher J. B. Lippincott & Co., and a pair of New Jersey troopers when the car began ticking loudly. On pulling over, he found that one of the wheels was hanging by threads—he had forgotten to bolt it securely. During a race in Long Island Sound, he sailed the yacht that Burgess had built, the *Little Dipper*, but he mishandled it in high wind, confused the crew, and finished at the rear of the pack.

At the end of the year, the lawyer William M. Parkhurst, who was married to Anne's sister Hope, assisted in a reorganization of the entire company. As bills went unpaid, their factory windows were smashed, and Fuller said afterward that the sheriff of Fairfield offered to hold an auction to save him further trouble. The sheriff really ordered two property seizures to cover outstanding debts, while a liquidation sale raised another $1,000.

Fuller's last chance for success came in March 1935, when Amelia Earhart—who had taken a ride in the car with Sir Charles Kingsford Smith the previous June—asked to use it in Washington, where she was being honored for the first solo flight from Honolulu to Oakland. Fuller opened their trip with a drive through upstate New York, where he gave a demonstration to the publisher Ogden Reid, and hoped that Earhart would be impressed enough to buy Car #3 herself. On one straightaway, he recalled, he let go of the steering wheel for a solid mile.

In Washington, he chauffeured Earhart and First Lady Eleanor Roosevelt down Executive Avenue to the White House. According to an article on the car's visit, Mrs. Roosevelt "was not yet convinced that she thoroughly liked its appearance," but she steered it capably herself, and she even asked Fuller to take her grandchildren for a spin. After hearing a siren on the road, he reportedly pulled over to find that President Roosevelt and his wife had caught up to take a look.

It was the best publicity imaginable, which Fuller subsequently tried to leverage with a letter to the First Lady, but his luck was running out. He had suffered a minor accident in February, and a more serious incident followed on May 27. In Milford, Connecticut, as he was driving Car #2 to his Harvard reunion with Anne, Allegra, and a friend with three children, a steering pin broke and a cable came loose. Fuller veered up an embankment and overturned, fracturing Anne's finger, while Allegra was left with a scar on her forehead.

Allegra came to believe that the incident shook him: "I think he thought that if the car did this to his wife and child, then maybe it wasn't the thing to do." After returning to the damaged vehicle, Fuller managed to push it upright, fixed the pulley, and drove to a garage in Bridgeport. In later accounts, he said that he gave it to his employees as a gift, but he really signed it over to the repair shop, J. Mezes and Sons, when he was unable to pay for its services.

Having failed to interest Earhart or the designer Norman Bel Geddes in Car #3, he finally sold it to his mechanic, Harold Browall, "for advertising hire." By the time the company folded, Fuller had burned through another $19,000 from his mother's estate. He had pledged his stock to his patent attorneys, who seized it, while the Waterhouse Company of Massachusetts, which had handled much of the bodywork, received a judgment against his share of the family properties, including Bear Island, until Rosamond repurchased it years afterward.

Fuller never discussed the true extent of his failure, and he even claimed that he had paid back his investors, whom he sometimes erased from the story entirely. His official biographies removed Biddle, implying that the funds had come from Philip Pearson alone, and he once stated that he had financed the car himself "with the little money I had from lecturing and from *Shelter* magazine." He also said that his partners refused to understand that he just wanted to produce one prototype.

In fact, the project had been undermined by interpersonal conflicts, funding shortfalls, and persistent design issues that he was unwilling to acknowledge. Fuller was far from the last technologist to underestimate the risks of the automotive business, where the cost of error was measured in human lives, but he always spoke as though the Dymaxion Car had been destroyed by nothing but public ignorance and bad luck. Decades later, he still insisted, "She was the most stable car in history."

In June 1935 Isamu Noguchi drove to California behind the wheel of his first car, a Hudson that Fuller had been forced to sell. He enjoyed the open road—he was arrested for speeding—and survived a minor crash after falling asleep and colliding with a steel fence in Philadelphia. "America becomes even more wonderful," Noguchi wrote to Fuller. "It's hard to believe that this also is America—but then this is the old America, black and mysterious."

Later that summer, Noguchi arrived in Mexico City entirely unaware of a private drama that was unfolding in his absence. Before his departure, he had introduced Fuller to Francise Clow, a twenty-one-year-old heiress from Lake Forest, Illinois, who was known to everyone as Baby. Clow was active in artistic circles—she belonged to a wealthy family that manufactured plumbing valves—and she and Noguchi had recently become lovers.

As his fortieth birthday approached, Fuller continued to be drawn to younger society women with artistic inclinations, and he began an affair of his own with Clow, a devotee of yoga and Eastern philosophy who seemed to combine Evelyn Schwartz and Clare Boothe Brokaw into one enticing package. "Our wedding gives cosmic insight and inclusion," Fuller wrote, and he told Clow elsewhere, "You should never have met Bucky, for his sake. Why? Because he just loves you so much."

Clow had once looked on the Dymaxion House "with animal fear and hatred," but she was charmed by Fuller's concept of the random element that led to change. The young woman found herself drawn to both him and Noguchi, writing in a letter to Fuller, "This route is perhaps not a circuit but a triangle." After returning to Lake Forest, Clow volunteered to organize his papers—she had nightmares of clippings "cramming my mouth with print"—and Fuller mused in a note to himself that monogamy tended to last for a period of eighteen years, which he and Anne had passed.

Over the summer, an outside factor intervened. Gilbert Stearns, nicknamed Gee, was a longtime friend of the Hewlett family. He had contemplated working at Stockade and sailed with Fuller on the *Little Dipper*, and his brother would later marry one of Anne's sisters. Stearns, a banker just a few years younger than Fuller, was fond of Anne—Fuller had grown jealous when they took a trip upstate without him—and he alluded to even deeper feelings after drinking too much at a party in July.

Allegra was shocked, and when Stearns argued that Anne would be happier with him, Fuller said that they should let her decide. He wrote to his wife accordingly, confessing that his sins had been "myriad," and promised that if she gave him a second chance, it would lead to "happiness and contentment and growth for you dearest heart such as are far beyond your most hopeful dreams."

Anne chose to stay with her husband, but she had hurt Fuller, who retaliated with an astonishingly cruel letter that compared her style of kissing to "a dog avoiding medicine." He accused her of rejecting his ideas, reminded her that she had failed to save her late mother and brother with her wishful thinking, and ruthlessly reopened old wounds: "You did not want Xandra—in advance hated her even."

Although he said later that his marriage at this time was "at its most perilously feeble state," Fuller had no intention of ending his own infidelities. Shortly after sending his letter to Anne, he saw Clow in New York, rekindling his passion. In September he wrote to her of his "two vital affairs, eight major enthusiasms, hectic jealousy, somewhere near a thousand prostitutions," but concluded, "I adore you, long for you, thrill to you, all completely."

While Clow was out of town, Fuller directed his energy elsewhere. He was avoiding Romany Marie's, in part because contact with leftists—the tavern's clientele overlapped considerably with the nearby John Reed Club—might be socially inconvenient, and he grew closer to mainstream figures such as Christopher Morley, whom he drove to speaking engagements, and the Irish writer George William Russell. According to Fuller, Russell once told him that his compatriot James Joyce had admired his essays in *Shelter*.

Fuller busied himself with new projects. One was an "air map" that he had sketched the year before, basing it on the meridian that ran longitudinally down the center of the Western Hemisphere, rather than the equator. Instead of figuring it out mathematically, he drafted it by eye, placing the North Pole at its heart, and felt that it showed the continents more accurately than standard projections.

His other undertaking, which he and Clow discussed as though it were a child that they would raise together as parents, returned to the notion of a book for a general audience that in turn could change the course of industry. By providing an inventory of relevant statistics and tools, it would enable readers to analyze current events, invest in promising lines of exploration, and guide the course of the future. He called it *Eight Chains to the Moon*, reflecting his calculation that if all the people in the world stood on one another's shoulders, they would reach the moon eight times. By the time he was done writing it, the number had gone up to nine.

Fuller was living that fall with William H. Osborn and his wife, Peggy, a patron of the arts whom he had met through Noguchi. He often approached powerful men through their wives, and he found a strong advocate in Osborn, a member of a rich railroad family who served as a research director for the copper company Phelps Dodge. Upon reading Fuller's manuscript, Osborn proposed to its president, Louis S. Cates, that they hire him as an advisor at their refinery in Laurel Hill, Queens.

The position had been created specifically for Fuller, who saw it as a chance to advance his own ideas. In a memo, he argued that instead of chasing markets, the research department should analyze technological trends to find new applications for copper. The reasoning was difficult for Osborn to follow, and so Fuller responded by reading it to him aloud—pausing after "each verbal dose" to check for comprehension. Osborn asked him to write it out that way, and Fuller resubmitted it with the paragraphs broken up into short lines. "This is lucid," Osborn said of the revision, "but it is poetry, and I cannot possibly hand it to the president of the corporation."

When Fuller objected, Osborn replied, "I am having two poets for dinner tonight, and I will take this to them and see what they say." He announced their verdict the following morning: "It's too bad—it's poetry." The critic Hugh Kenner later questioned the story, asking, "Who were those poets? Where did the research director find them?" But Osborn's connections to literary circles were real enough, and, in the end, he approved Fuller's unconventional proposal.

Fuller began by reviewing copper consumption across industries, including electronics. He noted that companies learned to use materials more efficiently over time, describing this as a trend toward ephemeralization. At Bell Telephone, an engineer told him that the number of messages able to travel through a cross section of wire was increasing so rapidly that they no longer had to buy metal for upgrades, which prompted Fuller to look at recycling.

As he dove into his work, his personal life settled down. Eight-year-old Allegra was bright but "quite dyslexic," as she recalled years later, so her father enrolled her in Manhattan's prestigious Dalton School, with the help of a reference from Evangeline Stokowska, who served on its board. Renting an apartment around the corner, on East Eighty-Seventh Street, Fuller began to live with his family in a more regular fashion than he had in years.

Clow was back in New York, along with Noguchi, who finally learned about their affair. After Clow found herself seated next to both of them one night, she told Fuller, "One of you was the new world, and one of you was the old, and I was between you, and I was both. And to you, I am the new world, but to him, I am the old, and, as he loves me, it makes very strong in him his nostalgia for the old—and you love me, and it makes you more ardent for the new."

Noguchi returned to Mexico, leaving what Clow called a "veil" between the two men. The sculptor dismissed this afterward as a rumor, insisting, "We never had any conflict with girls. We went our own way. And there was the age difference; nine years was sort of a protection." This understated Fuller's weakness for younger women, as well as Noguchi's sense of betrayal.

They reconciled in an unusual fashion. With the help of Diego Rivera, Noguchi was granted permission in Mexico City to paint a sculptural frieze, which would incorporate Albert Einstein's equation for the equivalence of mass and energy. He cabled a request to Fuller: "Please wire me rush Einstein's formula and explanation thereof." Fuller replied in December with a massive telegram:

Einstein's formula determination individual specifics relativity reads quote energy equals mass times the speed of light squared unquote speed of light identical speed all radiation cosmic gamma x ultra violet infra-red rays etcetera one hundred eighty six thousand miles per second which squared is top or perfect speed giving science a finite value for basic factor in motion universe. Speed of radiant energy being directional outward all directions expanding wave surface diametric polar speed away

from self is twice speed in one direction and speed of volume increase is square of speed in one direction approximately thirty
five billion volumetric miles per second.

The message, which went on for another 160 words, cost him ten
dollars. Noguchi included $E = mc^2$ in the frieze, and the incident became famous, although Fuller's explanation had more to do with his
own theories than with any real knowledge of Einstein: he associated
c^2 with the surface of a sphere as radiation expanded outward, when
it was actually a conversion factor that related an object's energy at
rest to its mass. Clow was relieved by Noguchi's request: "You cannot
imagine what joy it gives me to have him ask for it," she wrote to Fuller.
"I think it is a sign."

Fuller was spending more of his own time pitching products to
Phelps Dodge. One was a proposal to use copper in automotive brake
linings, which he later claimed "established the metallurgical principle
of the disk brakes now used on the wheels of heavy bombers," but it
was never patented or produced. Another was a centrifuge for processing tin, of which he recalled, "They were really very scared of the
centrifuge, and that it might really kill a lot of people."

Neither project was particularly exciting, and Fuller briefly explored
starting another corporation to sell the Dymaxion Car. Fortuitously, a
more promising alternative soon appeared. For his research on scrap circulation, Fuller studied the rates at which copper moved through various
sectors as old technology was recycled into improved equipment. This
led him to the concept of "lag," or the gap between the development of
an invention and its widespread application. In telephones, the delay was
two years, while in housing, it was a half century.

Averaging the lag times across different industries, Fuller arrived at
an overall recycling period of twenty-two and a half years for copper,
which indicated that there would be a spike in the scrap supply around
1939. He prepared attractive charts of copper output and usage, along
with other industrial trends, and concluded that the most promising
application for the future was "the highly mechanized dwelling for
ultimate mass production."

Interior of the Phelps Dodge bathroom (c. 1937).

All his research had resulted in the most obvious answer imaginable, and he just happened to have the exact background that it needed. William Osborn bought it, and as a pilot program, they agreed that Fuller would build a new version of the bathroom that he had developed at the Pierce Foundation, which Phelps Dodge saw as an outlet for copper in an otherwise slow market.

In May 1936 Fuller contracted to put together a prototype in designer William Stout's workshop in Detroit. For the rest of the year, he worked diligently on his bathroom, which consisted of four pieces sized to fit through an ordinary door. When bolted together, they formed two compartments—one with a tub and shower, the other with a sink and toilet—that homeowners could take with them when they moved. A more refined prototype in Queens, made of copper and aluminum, weighed four hundred pounds. At five feet square, some found it cramped, but it was easy to clean and could be installed by two men in three hours.

Fuller was photographed washing himself in his bathroom for an issue of *Architectural Record*, while the second prototype was installed in Morley's cabin on Long Island. Their real goal was a government contract, which required a price of no more than $400. As Fuller tried to lower the cost, he placed it in houses by prominent architects, including Richard Neutra. Another pair went to the yacht designer Jasper Morgan, who offered him a word of advice: "You are always trying to make things simple, when the first rule of success is to try to make things complicated."

He was having the opposite problem with his book, which after years of writing was still hard for readers to follow. Fuller was occupied elsewhere—he may have had a brief affair around this time with the socialite Ellen Tilton—and he hired a professional editor and travel author, Viola Irene Cooper, to make it more accessible. With the revised manuscript in hand, Fuller pitched it to Frank Henry at J. B. Lippincott & Co., accompanied by Morley and the former Evelyn Schwartz, who had recently ended her brief marriage to the innovative puppeteer Bil Baird.

Morley's endorsement led to a book deal in August 1937. Delivery was expected in a few months, and Cooper was granted a percentage of the royalties in exchange for drafting several chapters herself. They received additional feedback from Walter Gropius, the founder of the Bauhaus, who had relocated to the United States. His arrival was a major event—"The Silver Prince himself was here," Tom Wolfe wrote decades later—and Fuller obtained his notes toward the end of the year.

An even more unlikely encounter occurred after the book was submitted to Lippincott. Not surprisingly, the editors were unsure of what to make of it, and their most significant reservations involved the discussion of Einstein. According to Fuller, an editor informed him that only ten people in the world were said to understand the theory of relativity—and his name didn't appear on any published list of experts. "So I rashly wrote back and said that Dr. Einstein has come to America; he's in Princeton," Fuller recalled. "Why don't you send him my typescript?"

What happened next became a cornerstone of his personal myth, although the true details were never revealed in his lifetime. At Cooper's invitation, on Sunday, February 27, 1938, Fuller attended a party hosted by Dr. Isidore W. Held, a physician for whom she had done some free-lance editorial work. Held—whom Fuller confused in his recollections with the medical editor Morris Fishbein—was actively involved in the rescue of Jewish doctors from Nazi Germany. Through his efforts, he had become friends with Einstein, who was expected at the party that night.

When Fuller arrived at Held's Fifth Avenue apartment, he saw Einstein—who seemed to possess a "mystical aura"—surrounded by admirers in the living room. In the standard version of the story, Einstein ushered Fuller privately into the library, where the manuscript of *Nine Chains to the Moon* was lying on the desk, and lavished praise on the chapter that related his equation to the "horsepower" of everyday life: "Young man, you amaze me. I cannot conceive of anything I have done ever having practical application."

Fuller later described these words in a note to himself as "Mr. Einstein's statement to me," which indicated that the conversation did take place, although his other accounts of it differed dramatically. In another telling, Fuller said that the two of them spoke for forty-five minutes but that Einstein had not looked at the draft. After Fuller summarized the relevant section for him, Einstein replied, "Whether or not I shall be convinced by the logic of your book's derivations from my work I do not know, but I want you to understand that I understand how arduous and important is the work which you are doing."

Although this version was considerably more plausible, the encounter was never mentioned in Fuller's correspondence, which also lacked the exchange of letters with his editor at Lippincott. In any case, the section survived, but Fuller's tangled prose, customized symbols, and fondness for capital letters caused other problems, and publication—scheduled originally for his birthday as the "big book of the summer"—was delayed until the fall.

In the meantime, Fuller concentrated on his bathroom, driving it on a trailer to the Bureau of Standards in Washington for testing. When the prototype was exhibited in Noguchi's courtyard in New York, curious viewers lined up in the rain, and Ruth Green Harris of the *Times* was enthusiastic: "It may serve as a snare to lure us ultimately into the house itself, by feeding our American passion for extravagant cleanliness."

In May Fuller applied for a patent, which was assigned to Phelps Dodge, but tensions within the company were rising. Since a metal surface would be too cold, Fuller intended to make the finished version out of plastic. This removed any internal justification for the bathroom, and the latest model cost $700, which was too high for a government contract. Cates halted work after twelve units had been built, and by June, Fuller was out of a job. He later said that it had been doomed by concerns over the plumbers union, which in fact had reacted positively, but he had really been unable to solve the cost issues that had already undermined the Dymaxion House.

Although Fuller retained the rights to sell the bathroom, his first move was to leave on a trip with Christopher Morley on June 15, 1938. Riding on the Pennsylvania Railroad's new Broadway Limited passenger train, they stopped in Philadelphia to pick up the typeset galleys for *Nine Chains*. In Chicago, they entertained the heiress Narcissa Swift, met the artist László Moholy-Nagy, and broke the bed at a party for twenty in their hotel room. Fuller felt invigorated, and on his return, he seemed less interested in looking for work than in spending time with Morley, who affectionately called him "Buckling."

It became the male equivalent of the partnership that Fuller had previously sought in women. Later that summer, Fuller and Morley traveled throughout New England, where they discussed extrasensory perception and the mystical properties of the moon. In New Hampshire, they lounged by the Ammonoosuc River "in stupor, sweating, prone," and at his cabin on Long Island, Morley comforted Fuller after he awoke at three in the morning from a nightmare about a plane.

Another excursion took them to Fuller's old plant in Bridgeport, where he hoped to build the Dymaxion Bathroom. He had kept three units for himself, and Alden Hatch recalled riding in a truck through Lawrence "with Bucky at the wheel, assorted Hewletts in the bathroom, one of them sitting on the john, and streamers of toilet paper flying out behind." Fuller planned to market it with Jasper Morgan, but when he was unable to source the necessary plastic, he shut it all down.

He returned from a visit to Bear Island in time for the September publication of *Nine Chains to the Moon*. In exchange for an investment in the bathroom, Fuller had assigned all his royalties to William Osborn, who would return the rights only after having been repaid in full. Bil Baird had built a beautiful relief version of Fuller's air map for the cover, but it finally bore a copper-colored jacket that did nothing to indicate the contents, which would have been hard to describe in any case.

Despite much hard work by Cooper, *Nine Chains* was often grotesque and digressive, and it suffered from the harshly satirical tone of Fuller's early writing—he had yet to find his true voice, and the result read like a rough sketch of things to come. It was an attempt, he said, to convey a thought process that could streamline society itself, encouraging mankind to move in the direction of least resistance. "The reward," Fuller explained, "is the ability to perceive and to analyze motivating forces and trends."

His argument was based on a handful of themes. One was industrialization, which inspired his famous definition of a man: "A self-balancing, 28-jointed adapter-base biped; an electro-chemical reduction-plant, integral with segregated stowages of special energy extracts in storage batteries, for subsequent actuation of thousands of hydraulic and pneumatic pumps, with motors attached . . . guided with exquisite precision from a turret in which are located telescopic and microscopic self-registering and recording range finders."

The full passage, which filled two more paragraphs, was intended to alienate readers from the body, encouraging them to see it as a mechanism that could be extended by other machines. If the body was governed by an intangible "phantom captain" hooked up to an organic communications network, it was a small step to connect it to the rest of the world through technology. America itself was a composite mechanical organism, he stated, and the house was "the universal extension of the phantom captain's ship into new areas of environment control."

Another theme was ephemeralization. Fuller associated compression materials with immobility, and he advocated a return to the mobile culture symbolized by triangular tension structures such as tents and sails. In the next phase of design history, ephemeralization would pass into the visual and abstract. All trends culminated in the electrical stage, going from track to "trackless," and industries that reasoned abstractly would be rewarded: "This is a practical key which may be applied to market speculation."

According to Fuller, society would become increasingly decentralized and efficient, with intuition guiding inventors along the most promising avenues of experimentation. In a crucial chapter, he calculated the horsepower that the average person could generate using muscular strength alone. After converting it to kilowatts, he claimed that the utilization of existing hydroelectric power at maximum efficiency would provide the equivalent of twenty-nine inanimate servants—which Fuller later called "energy slaves"—for every American.

As a result of these factors, civilization was entering a period of accelerating change, leading to an economy in which industries "rented" raw materials that were progressively recycled into improved applications. All businesses would become service industries, including housing, which could innovate radically once residents rented their houses like phones. As long as people weren't forced to buy their homes, he wrote, dwellings could look "utterly different."

Fuller argued that competition should be confined to scarcity items such as mechanical inventions, while "plenitude" goods like food should be socialized, with basic needs provided "through the effort of one man out of every five working but one day a month." He blamed the resistance to this system on capitalism, saying that it discouraged innovation that it was unable to monetize. Capitalists built skyscrapers that remained vacant instead of sheltering the homeless, and they had assumed wrongly that radio would be used only by amateurs for "publicity."

His sarcastic critique of the financial sector was the weakest part of the book, and he devoted many pages to attacking old enemies, especially the American Institute of Architects and American Radiator. He was more convincing when he took a longer view. *Nine Chains* featured his trend charts, which he claimed to have researched with a team of engineering students at Columbia University, and a chronology of inventions from Leonardo da Vinci's "perfect whore house" to his own "scientific dwelling machine."

Fuller bracketed the book with a pair of considerations of tomorrow. It opened with a list of predictions that he expected to be fulfilled by his birthday on July 12, 1948. Many were plausible, such as a mechanical stock exchange, education through broadcast media, the socialization of leisure, and widespread recycling. Others—power transmission using gold, a vast migration into Canada, the renaming of New York as "Radio City"—were more far-fetched, and he closed by stating that there would be no change in the "way of a man with a maid."

He ended the volume with a science-fiction story about an interstellar visitor, modeled on Clow, from a planet that deliberately caused a crisis in its own society to encourage original thinking. By using the principle of "emergence through emergency," it had obtained scientific shelter, radio education, and a "receiving and sending apparatus" to disseminate information. Fuller also offered rules for innovators: never show unfinished work; introduce artifacts only when the world is ready; and share ideas freely, since each one led to more in "geometrical progression."

Inevitably, the reviews were mixed. "The author has sound knowledge of one thing and mere opinion on a thousand things," William Marias Malisoff wrote in the *Times*. "In the one thing that he knows well, namely architecture, he is a revolutionary. In the other things, he is contradictory, self-contradictory, and just as likely to be reactionary as progressive." Fuller had greater hopes for a write-up in the *Saturday Review of Literature* by Frank Lloyd Wright, whom he had encouraged to take the assignment. "I know it will be a hell of a chore," Fuller wrote in a letter, "but I need your hand."

In his review, Wright praised Fuller as "the most sensible man in New York," although he criticized many of the book's predictions, as well as the concept of industrial housing. It was a testament to the serious attention that had been granted to Fuller, who told Wright, "I still get a little chokey over your review. It is swell, provoking, amusing, and the performance of a real friend." On a visit to Taliesin, he marveled at the community of apprentices that Wright had created, and the image of the architect surrounded by disciples lingered in his mind as "something like heaven."

Nine Chains was recommended by the Book of the Month Club, mentioned in Eleanor Roosevelt's syndicated newspaper column, and covered in *The New Yorker*, which jokingly referred to its author as "R. Buckrogers Fuller." Within months, however, sales fell, and, despite assertions that it sold "by the thousands," it moved only around 1,300 copies. All the same, *Nine Chains* increased Fuller's public profile significantly, and he sought immediately to take advantage of it.

To commemorate this professional turning point, Fuller made the symbolic gesture of privately binding his Chronofile—the huge scrapbook that he had saved throughout his life—into a uniform set of more than sixty volumes. In a more practical move, he wrote to the publishing magnate Henry Luce that he had spoken to friends at *Time* and *Fortune* about the possibility of a position on his staff.

He was skillfully leveraging his contacts, but another connection carried even greater weight. Three years earlier, Luce had married Clare Brokaw, who would be known ever afterward as Clare Boothe Luce. As a playwright, she had achieved a huge success with the satirical Broadway show *The Women*, which included a line that Fuller might have seen as a comment on his own life: "A man has only one escape from his old self: to see a different self—in the mirror of some woman's eyes."

By October 14, 1938, Fuller had landed a role as a technical advisor at *Fortune*. The following week, he accompanied Isamu Noguchi, who was back in the city again, to a party thrown by their friend Dorothy Hale. In another gesture of reconciliation, Fuller and Noguchi were collaborating on the design for the *Chassis Fountain*, a sculpture—based on automobile components and plated in magnesite—for the Ford Pavilion at the upcoming New York World's Fair.

Hale was less content. Although she had appeared in another play by Brokaw, other roles were scarce, and her personal life was a disaster.

According to Noguchi, Hale mentioned that night that she was going away: "I remember very well. She said, 'Well, that's the end of the vodka. There isn't any more.' Just like that, you know." Sometime around daybreak, long after all her guests had left, Hale threw herself out the window of her small sixteenth-floor apartment on Central Park South, plunging to the pavement below.

Her suicide marked a tragic close to Fuller's bohemian phase. He attended her funeral with Clare Boothe Luce, and, in a note to a friend, he expressed how he and Noguchi had been affected: "Dorothy has given us through brave love new clarity and significance of our kinship and its responsibilities." Clow was equally devastated. She had become engaged to the Italian-born pianist and composer Mario Braggiotti, which Fuller learned about through the papers. After hearing of Hale's death, she wrote to him, "I have been terribly saddened by it because it was such an unhappy life."

A week and a half later, Fuller and Luce attended a preview of an exhibit at the Julien Levy Gallery by a relatively unknown painter from Mexico named Frida Kahlo, who wore a dress that inspired *The New Yorker* to call her "a beribboned bomb." Luce asked Kahlo for a memorial portrait of Hale, but when it was delivered, she nearly fainted. In a series of images, it showed Hale standing on her balcony, plunging through the air, and lying in a starburst of blood. Luce was shocked enough to consider destroying it, but, in the end, she only asked Noguchi to paint out the part with her own name.

In the fall of 1938 Fuller reported to editor Russell Davenport of *Fortune* in the Time & Life Building in Rockefeller Center. Henry Luce's business magazine was a gorgeous, oversized monthly with two million readers, and it offered an ideal platform for a vision of the future. Fuller began by working on an issue about New York, which he described as "the idea market of the world," and after overseeing a housing conference early the following year at Yale University, he became convinced that the time was ripe for a dwelling industry.

With America on the verge of a new stage in its economic and technological development, Fuller proposed a series of articles about science and business, which he called "The New US Frontier." In underlining the accelerating rate of progress, he expanded on his arguments in *Nine Chains*, as well as a manuscript, *The Curve of Fate*, by James Lonsdale-Bryans, a British writer and white supremacist. Fuller had no evident qualms about the author, and he even told Max Schuster, cofounder of the publisher Simon & Schuster, that "this inadvertent declaration of Nazi philosophy by an Englishman would make an easily promotable book."

Lonsdale-Bryans was one of the earliest writers to compare the curve of change to a hockey stick, and Fuller explored the implications in "Ballistics of Civilization," an unpublished call for a predictive science to guide humanity's "trajectory through the ages of space." The tools included his charts and his air map, which revealed a northwest spiral of migration along the isotherm, a band of moderate temperature coinciding with the central line of world population. According to Fuller, it indicated that the headquarters of the British Empire was destined to move to Canada.

He shared his theories with Henry Luce, who may not have known about Fuller's past history with Clare—although one of her biographers later alleged that Fuller said that she had married her new husband "on the rebound from him." Fuller took advantage of his access to the nation's most influential publisher, for whom he wrote a speech on ephemeralization: "The sewing or calculating machines of this year, for instance, are far more able and easy to work than their predecessors, yet are but half the weight and volume of their prototypes of but three years ago."

Technology was constantly on his mind. For two articles on the debut of regular television service, Fuller interviewed the cultural arts critic Gilbert Seldes, whose daughter, Marian—later an acclaimed stage actress—happened to be friends with Allegra at Dalton. Seldes had recently been named director of the Columbia Broadcasting System's experiments with the nascent medium of television. He was impressed enough to arrange televised talks by Fuller, who promoted this new form of broadcasting as a tool for "the learning of geography, ethnology, commerce, and even languages."

Fuller had moved with Anne and Allegra to an apartment at 105 East Eighty-Eighth Street. His friends from this period included Al Hirschfeld, the well-known caricaturist of Broadway stars and other celebrities; sculptor Gertrude Vanderbilt Whitney, who had once made headlines for her custody fight over her niece Gloria Vanderbilt; abstract expressionist painter Arshile Gorky; and Alexander King, a *Life* magazine editorial assistant and playwright. In September King performed a reading of a new play, acting out all thirty roles in hopes of attracting producers, which Fuller remembered as a lesson in asking listeners to use their imaginations.

He received another piece of valuable advice from Allegra, who had become interested in dance at the Dalton School. She told her father, whose naval background sometimes made him seem stiff in public, to gesture more freely in his lectures: "I believe that you could communicate with your audiences better if you would only *let yourself go*." Fuller agreed, and the hint influenced his speaking style, which gradually evolved into a performance that he often conducted, like King, to seek financial support from possible patrons.

As usual, he was preoccupied by side projects. Models of his unrealized designs for the Dymaxion Car were exhibited at the New York World's Fair, and he considered writing a book on plastics. He took up running again, and he kept in touch with Vilhjalmur Stefansson, who later married Fuller's former flame Evelyn Schwartz. It may have been through Stefansson that he met L. Ron Hubbard, a pulp author and enthusiastic sailor who had talked his way into the exclusive Explorers Club. Stefansson became friends there with Hubbard, whom Fuller encountered in passing in New York.

Fuller's series of articles in *Fortune* was announced by a note in the September 1939 issue. America, it said, was a natural innovator, but the devastating economic downturn had left investors reluctant to take risks. As a corrective, the magazine would map the frontier unfolding in research departments and laboratories, with inventors and entrepreneurs as its prime movers. The text was by others, but Fuller— who saved clippings about science on shirt cardboards—took the lead in selecting subjects. "I read *Patent Gazette*," Fuller recalled, "and I could literally spin the pages, and when my eyes saw something I wasn't familiar with, they would stop me."

He had a staff to check his figures, which left him with a habit of being careless with facts, and acquired a stock of miscellaneous information that served him well. An article on drug development reproduced a molecular model made of balls and rods, which anticipated his use of similar tools to explore geometry. Another profiled the radio electrical engineer Edwin Howard Armstrong, hailing him as a pioneer who had triumphed by having "money enough to wage his own war."

Other articles informed Fuller's work in more obvious ways. One was on the energy sector's efforts to devise transmission methods to send electricity over long distances, which would preoccupy him for decades. A visit to the Glenn L. Martin aircraft company planted the idea of using aviation manufacturers for housing, while a piece about plywood described it as an industry "that wants to revolutionize the construction of *everything*," including prefabricated houses.

Life acclaimed Fuller as an "expert on America's future," and he became known as a useful source of story ideas. Russell Davenport praised him in private as a genius with "the scope and daring of Leonardo," and when asked to justify Fuller's place on the payroll, he supposedly responded: "I send for Bucky, and Bucky comes in and talks for about two hours. By the time he leaves, I have more ideas than I can publish in the magazine in ten years."

Fuller's work at *Fortune* culminated in a special tenth anniversary issue in February 1940. Its articles were meant to position America as "the most important nation in the world," with contributions by a writing staff that included Peter Drucker, the future management guru. As Drucker was backing out of an office one day, he collided with Fuller, who was standing on a scaffold to draw economic charts on the wall and ceiling. After Fuller fell, he rose and said with a straight face, "You have set back the industrial development of South America by at least ten years."

At the heart of the issue was the article "US Industrialization," which presented a series of trend lines on America's economic expansion. Davenport supplied the text, Philip Ragan handled design, but the concept belonged entirely to Fuller, who returned to his experimental sleep habits to finish the piece on schedule. Its most memorable illustration was a world map, based on his cartographic projection, with white dots for population and red dots for "energy slaves." At a glance, it was evident that the United States controlled more than half the globe's power—a sign of the expanding role that it was destined to take in what Henry Luce would soon call the American Century.

On April 9, 1940, Fuller followed up with "No More Secondhand God." The prose essay, broken up into short lines, argued that, in wars, democracies were disadvantaged against dictatorships. "God is a verb," Fuller wrote, and the German blitzkrieg of Europe was proof that change was happening "in a hell of a hurry." Drafts were read by Frank Lloyd Wright and E. E. Cummings, but neither it nor his "Untitled Epic Poem on the History of Industrialization" was published at the time.

It was clear that the United States might eventually be drawn into the war raging overseas. As Fuller imagined various roles for himself, he concentrated on researching a pair of articles about the Sperry Corporation, a manufacturer of gyroscopes and navigational equipment. He came up with an elaborate analogy for the principle of precession, or movement at right angles to a force, which became his favorite metaphor for unforeseen side effects, and he made a point of befriending Sperry president Reginald E. Gillmor.

Fuller often used his articles to make valuable connections, and the first visible result came from Chrysler. After contributing to a piece on the company's hydraulic drive, he was offered $1,000 to write a book about its research department. Taking a leave of absence from the magazine, he traveled to Detroit in April to gather material for *New Worlds in Engineering*.

The project gave him a chance to meet scientists and executives, as well as an excuse to discuss his favorite subjects, including inventory control, streamlining, and the "extensions of the sensorial spectrum range." While the book had a large print run, Fuller thought that it had been "butchered" by publicists into corporate propaganda. After reading a section on employee productivity, his contact at Chrysler allegedly told him to remove it, saying, "It'll just make those fellows ask for more money."

Back in New York, Fuller channeled his frustrations into the poem "Machine Tools," writing that the conversion of one hundred tons of materials into five tons of aircraft required exceptional talent, although few were "qualified for such mastery." At night, he lay awake in the darkness, "with some part rigidly pressed against another, teeth against teeth, or fingernails into flesh," and fought the urge to kneel and pray to God to make him worthy of employment.

He was dealing instead with minor irritations. At Rockefeller Center, his house model was accidentally destroyed, while Bear Island was rented out to guests to pay the bills. As he contemplated leaving *Fortune* to teach at Dalton, he occupied himself with his design for the Mechanical Wing: a trailer with the basic requirements for living—including a bathroom, kitchen, and generator—that could be hitched to a car and driven to campsites. After it was featured in Luce's *Architectural Forum*, Fuller hoped to manufacture it, but his plans were superseded by a more practical venture in the fall.

In November Fuller left on a leisurely ramble through the Midwest with Christopher Morley, who wanted to track down some letters by Edgar Allan Poe, and an English professor named Ralph Sargent. While driving near Galesburg, Illinois, Fuller noticed that the fields were full of round steel bins measuring up to eighteen feet in diameter. They were manufactured by Butler Manufacturing of Missouri to store grain for the New Deal, and Fuller seized on the idea that they could be converted into prefabricated shelters.

After Morley offered to underwrite the project with the proceeds from his recent best-selling novel, *Kitty Foyle*, Fuller went to Kansas City to meet E. E. Norquist, the president of Butler, who brought him to a poker game. According to one story, Fuller won $500 and then deliberately gave it all back, prompting Norquist to agree to a deal. He also secured a separate investment from Robert Colgate, the vice president of research at the firm that bore his family's name.

With the architect Walter Sanders, Fuller worked out a design based on the grain bin, which he positioned as a solution to the worldwide housing shortage that had contributed to the rise of international tensions. In his pitch, he argued that conventional industry was unable to meet demand, while Butler could immediately begin production of what he named the Dymaxion Deployment Unit. The proposal succeeded, and Fuller resigned from *Fortune* early in 1941 to focus on the DDU.

As he prepared to file a pair of patents, he flew out to examine the first version at Butler, where it had been assembled in just three weeks. At twenty feet across, it was a handmade prototype for the production model, which would be built from the top down using a temporary mast. Portholes were set into rounded sides that were corrugated for stiffness, and the bin's conical top was replaced by a roof of curved panels and a ventilator.

In contrast to the futuristic Dymaxion House, the result was a kind of minimum viable product—the simplest structure that met the necessary requirements. The flooring was unfastened, and Fuller proposed using straw or newspaper for insulation. It had no bathroom, although an optional second cylinder was added with a kitchen and bath, and a deal was struck with Sears, Roebuck to provide basic furnishings, including a kerosene icebox and stove. Assuming a run of 350 units, the price, with amenities, came to $1,390.

As a potential client, Morley provided an introduction to Marie Macneil, whose husband, Robert, chaired the Inventions Board of the British Purchasing Commission, the organization that arranged for the procurement of armaments from North America. Fuller felt that Macneil, who oversaw a shelter program, was enthusiastic, but he was finally told that the British no longer wanted "any portable buildings at all." Undaunted, he founded the Dymaxion Company, with Colgate as president and himself as vice president, and the prototype was installed at Hains Point in Washington, DC, which was a convenient spot for official visitors.

In May Fuller oversaw construction of a second version in Kansas City. According to one famous anecdote, Norquist insisted on entering it on a scorching day, and when the engineers followed—reluctantly— they found that it was perceptibly cooler on the inside. Lighting cigarettes, they saw that the smoke was pulled down from the roof

Display model of the Dymaxion Deployment Unit in Kansas City, Missouri (May 1941).
Photograph by Marion Post for the US Office of War Information.

opening, sucked under the walls, and replaced by air from the top. They concluded that the white exterior of the bin reflected the sun, creating a circulating effect that lowered the internal temperature by around six degrees.

Although the product was clearly promising, Fuller had trouble moving forward. Defense housing was a real prospect, but access to steel was restricted, and they lacked the government contract necessary for a high priority rating. The navy ordered two houses, but negotiations with federal contacts went nowhere, and Fuller and Anne were distracted by grief when James Hewlett passed away in October 1941. "More than any other man in my life," Fuller said later, "[Hewlett] gave me the courage to break over into as revolutionary an architectural idea as the Dymaxion."

A more positive development came through the Museum of Modern Art. It had already exhibited his bathroom and house model, and director Alfred Barr felt that the DDU would make a case for the museum's importance in wartime. Two conjoined units were displayed in the garden, where they were photographed for *Vogue*, and the heat had to be turned off to discourage young lovers. Fuller said that the house would cost just $700—he would worry about the details later—and could survive a bomb: "It might jump off the ground, but it would come back."

As he searched for a large order, he teamed with David Cort, an editor at *Life*, to form an informal discussion group to brainstorm strategies for the war. Their proposal applied the logic of industrialization and ephemeralization to combat, with the goal of delivering the most striking power using weapons "as light and as terrible" as possible. It called for replacing naval and land traffic with polar air routes, with assaults conducted by gliders from Soviet bases, and power based on "liquid oxygen or uranium or anything under the sun."

Cort wanted to publish it in *Life*, but their plans were disrupted by the Japanese surprise attack on Pearl Harbor on December 7. On a visit the following day to Rockefeller Center, Fuller ran into John Fistere, a former colleague at the magazine. When Fistere proposed that they head for a bar, Fuller replied, "John, there's a war on. And if I'm going to be of any use to my country, I'm not going to take another drink."

Fuller saw the war as an opening for his ideas, and he became frustrated when other inventors received greater attention. In January 1942 he wrote to New Jersey governor Charles Edison—whose father, Thomas, was perhaps the most famous inventor in American history—to pitch "several technical devices which might, I believe, be developed into important weapons." One was the Crocodile, an armored vehicle for a single man with a gun on his back, while the other was a supercharged Dymaxion Car, which indicated that he was still confident enough in his designs to entrust them with the lives of soldiers in battle.

His greatest breakthrough originated at the intersection of defense and his lifelong fascination with geometry. During his discussions with Cort, he often thought about air routes, which, ideally, took the shortest distance between two points on the globe. In geometrical terms, they were geodesics, or arcs of the great circles that split the sphere into equal halves.

Fuller was evidently tracing geodesic lines on a globe by hand when he made a pivotal observation: when four great circles were arranged in a certain way, they divided the earth's surface into a spherical polyhedron with fourteen faces. If he drew a circle around the equator, for example, followed by three more that crossed to make an equilateral triangle at the pole, the result could be unfolded into a map with eight triangular and six square pieces.

This polyhedron had caught his attention before. It had emerged from his interest in the hexagon, which had appeared in the plan of the Dymaxion House. As the shape produced when a circle was enclosed by six others—as in the bundle of metal tubes in his model of the house's mast—the hexagon was often found in nature, where it could be seen in honeycombs and clusters of bubbles.

When the principle was extended to three dimensions, the closest packing consisted of a central sphere surrounded by twelve more. Fuller had been working around this time with constructions glued together from "beautiful little clear crystal balls," and he may have seen the concept in the classic 1917 book *On Growth and Form* by the Scottish naturalist and mathematician D'Arcy Wentworth Thompson, who noted that one ball could nest inside twelve others, "six in its own plane, three below, and three above."

If the spheres were connected with straight lines, they made a polyhedron with fourteen faces—eight triangular and six square—and a hexagonal cross section. This polyhedron, which had been described by the ancient Greeks, was called the cuboctahedron, although Fuller was unaware of this. It had appeared in his sketches years earlier, and he was fascinated by its properties. Not only were all its edges identical in length, but each vertex was also the same distance from the center, resulting in the only polyhedron that was radially equilateral.

Fuller realized that this polyhedron could produce a superior version of his air map, with fourteen pieces that could be configured in many different ways, which was a line of thought that had been explored by previous cartographers. A map by Leonardo da Vinci had divided the globe into triangular octants, like an unfolded octahedron, while Bernard J. S. Cahill patented a projection in 1913 that arranged the triangles like a butterfly. Fuller himself had once considered a version based on the regular icosahedron, which had twenty triangular faces.

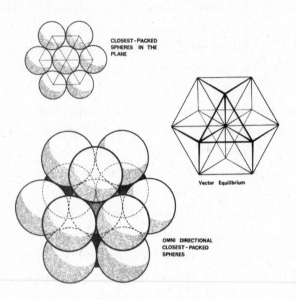

The construction of the cuboctahedron (or vector equilibrium) from the closest packing of spheres.

For now, Fuller sketched a map based on an unfolded cuboctahedron, which he later called the vector equilibrium, although he originally referred to it—in what he conceded was a "presumptuous" gesture—as the Dymaxion. The real question was why he decided against the icosahedron. He said subsequently that a vector equilibrium produced slightly less distortion, but he overlooked the obvious advantages of simplicity in a polyhedron with faces that were all the same shape, and his choice was driven by his interest in the concept of the cuboctahedron itself.

Because Fuller lacked confidence in his skills in spherical trigonometry, in order to subdivide the faces into the grid necessary for a map, he had to work by eye. The squares of the vector equilibrium were easy enough, but the triangles were challenging. He decided eventually to divide each edge into smaller segments and draw perpendicular lines at every intermediate point, yielding a triangular grid that made a geodesic pattern on the globe.

Fuller relied on purely mechanical techniques, taking care to keep the continents intact, and his earliest map divided the edges of each face into only four segments. This generated a grid that was simple enough for him to copy geographical features by hand. The "three-way great circle grid" within the triangles looked uniform at first glance, and he came to believe that it was a discovery unknown to mathematics. In fact, the grid was invisibly imperfect, with unseen gaps whenever the lines failed to cross exactly, which would later cause him considerable trouble.

It wasn't a conventional projection, since no math was involved, but it was a deeply pleasing design. By the time he spoke at Bennington College in Vermont in April 1942, he had a fairly finished version, which he presented to the son of the architectural professor Edwin Park. Fuller thought that it had commercial potential at a time when interest in wartime geography was rising, and he kept the pieces in his pocket to show off at the office and at parties.

The map also fed into Fuller's and Cort's proposed war strategy to replace ocean routes with air traffic, and he continued to promote their

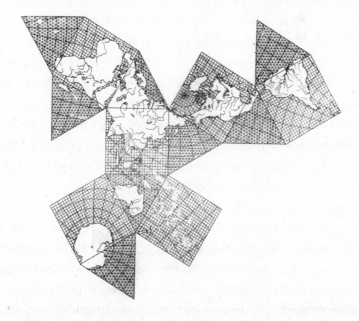

The vector equilibrium "projection" of the Dymaxion Map. Fuller's earlier version divided each face edge into only four segments. A three-way grid is visible in the triangular faces.

plan whenever he could. Through Park, he was introduced to James L. McCamy, the head of the Board of Economic Warfare (BEW), which was in charge of shaping policy for the crucial international trade in materials and equipment. After an evening of conversation in New York, McCamy offered Fuller a position.

Fuller was willing to move on. At the beginning of the year, he had built another prototype of the Dymaxion Deployment Unit for the US Army Signal Corps at Fort Monmouth, New Jersey. Although he had to discreetly fix leaks and other problems "that they never knew were wrong," he was offered a contract for five of his "igloos," followed by fifty more, which looked at first like the substantial purchase that he needed.

In the end, however, the Signal Corps never went beyond its initial order, which was used to house aircraft crews in the Persian Gulf. The shelters were installed using a custom tool kit that was usually stolen afterward—Fuller thought that this would encourage the troops to build the units quickly—and most were painted dark, without a gap at the base, so that the cooling effect went unused. Another DDU ended up at Bennington, where it was admired by Martha Graham, but although Fuller later said that hundreds were made, the real number was probably fewer than seventy.

As his first venture into mass production, it was a qualified success, and he was eager for new challenges. In November Fuller received the details of his appointment to the Board of Economic Warfare. As head mechanical engineer, he would work on designs for steam and automotive engines, conduct tests on fuel alcohol, research plant layouts and mechanization, look into potential savings in energy and resources, and develop his map. It gave him wide latitude to explore other areas, and it came with a salary of $6,500.

He was ready for it. In an unpublished article, Fuller wrote that the war—the ultimate emergence through emergency—might result in a true housing industry at last. Survival, he said, required "immaculate discipline of person and premises," and he prepared to exemplify this himself. Despite repeated efforts, he had never managed to stop drinking and smoking, but now he was determined to quit in order to be taken seriously. He also framed the resolution as an anniversary present for Anne, with whom he had celebrated their "respective twenty-five-year martyrdom."

Fuller was seen, rightly or otherwise, as an expert in industrialization, and his legitimization by the government would give him the chance to test his ideas in wartime. After closing a deal with *Life* to publish the Dymaxion Map, he moved to Washington, DC, where he began work on December 7, 1942. The new position represented the fresh start that he had wanted, but it would soon be complicated by the arrival in his life of yet another woman.

SIX

THE DWELLING MACHINE

1942–1947

Physical points are energy-event aggregations. When they converge beyond the critical fall-in proximity threshold, they orbit coordinatedly, as a Universe-precessed aggregate, as loose pebbles on our Earth orbit the Sun in unison, and as chips ride around on men's shoulders.

—**BUCKMINSTER FULLER,** *SYNERGETICS*

As he prepared for his government role, Fuller became characteristically distracted by new opportunities almost at once. Gerard Piel, the *Life* science editor who was overseeing the article on the Dymaxion Map, had introduced Fuller six months earlier to the shipbuilder and pioneering industrialist Henry J. Kaiser. After hiring Fuller to ghostwrite a speech on postwar industry, Kaiser expressed his willingness to fund a prototype of Fuller's latest automobile, which was formalized as a project on the same day that he began working at the Board of Economic Warfare.

Fuller was more excited by his arrangement with Kaiser, whom he hoped to persuade to move into housing. After reaching an agreement with his supervisor, the automotive engineer Alex Taub, he commuted on weekends from Washington to New York, where he drafted plans in the architectural office of Walter Sanders and John Breck. The grueling schedule led Fuller to develop anemia, but he knew that the chance might not come again, and he recruited Bil Baird to help with the model of what became the D-45.

If the Dymaxion Car had symbolized decentralization, this was a car for city driving. At nine feet in length, it was noticeably shorter than its predecessor, although still wide enough to accommodate a row of four passengers. Fuller designed it to be steered either in front or through a single rear wheel, which could be extended on a boom like the tail of a horseshoe crab. In early 1943 Kaiser approved the model at a meeting at the Waldorf-Astoria Hotel, and Fuller dreamed of a "three-cornered deal" that would include a housing program.

Fuller was equally busy at the BEW, which offered a promising stage for his ambitions. The board, chaired by Vice President Henry A. Wallace, focused on economic relations with foreign countries, especially the procurement of essential wartime resources. Fuller reported to Taub in the Industrial Engineering Division, working out of the Department of Commerce building at Fourteenth Street and Constitution Avenue.

His family had remained in New York, where Allegra would finish school, so Fuller took a room on Garfield Street with Ken Beirn of the Office of War Information. He often saw Clare Boothe Luce, now a freshman Republican congresswoman representing Connecticut, and reconnected with Thornton Wilder, who was then serving as an intelligence officer in the US Army Air Forces. The two of them lunched with Frank Lloyd Wright, and Fuller tried to interest Wilder, whom he saw as an accomplished "amateur mathematician," in his experiments with geometry.

Fuller's closest companion during this period was a secretary in his own office. Twenty-one-year-old Cynthia Lacey, from Occoquan, Virginia, graduated with a secretarial degree from Strayer College. After working as a senior typist at the Washington Navy Yard, she was assigned to Albert C. Shire at the Federal Public Housing Authority, whom she joined in a transfer to Fuller's division. She was an attractive, intelligent brunette, and after Fuller took her to dinner, their discussions of workplace politics blossomed into an affair.

Lacey was very different from the socialites he had courted in the past, and she soon became closer to him than either Evelyn Schwartz or Francise Clow had ever been. On a typical Sunday, Fuller would cook breakfast at her home and read Emerson aloud to her, while their evenings together opened with dinner, followed by a movie or a stroll around Georgetown. Above all else, Lacey became an invaluable sounding board for his official activities, which resembled no one else's in Washington.

Along with delivering indoctrination lectures on his theories of geography, Fuller was asked to rethink the government's approach to substitutions. Because the ongoing war had cut off the usual sources of materials, he suggested sensibly that instead of replacing specific goods, they should concentrate on uses—on fuel rather than oil, for instance—and seek efficient alternatives.

To explore this "functional" strategy, which called for novel industrial methods, cutbacks in commodities, and reducing the weight of supplies so that they could be transported by air, Fuller held a conference in January. The location was the Cosmos Club in Lafayette Square, where he attended weekly lunches organized by the chemist Lyman Chalkley. With its talks on technical subjects, the club was a prestigious meeting ground for scientists and engineers, whom Fuller quickly identified as a crucial audience.

At the conference, Fuller told the receptive crowd, "Where a man is sinking, you will throw him the first thing that will float, whether it is a particularly good life preserver or not—it might be a valuable old piece of furniture." Over time he refined this into a memorable aphorism: "If you are in a shipwreck, and all the boats are gone, a piano top buoyant enough to keep you afloat that comes along makes a fortuitous life preserver. But this is not to say that the best way to design a life preserver is in the form of a piano top."

The gathering was a success, and as Fuller prepared to implement his ideas on substitutions, he worked on other projects, including specifications for a steam engine to be manufactured in Brazil, tests for new fuels, and procedures for evaluating fibers for cordage. He did a capable job—Taub praised him as "an extraordinary engineer"—and he took advantage of his position to read secret intercepts on foreign industry, especially strategic patents.

Fuller was equally preoccupied with his *Life* article on the Dymaxion Map. To evaluate the projection, the magazine had convened a team of cartographic experts, including the geographer Samuel Whittemore Boggs and the science writer Lillian Rosanoff Lieber, but the panel's response was critical. Osborn Maitland Miller of the American Geographical Society thought that the map had been somehow "pasted together," and, according to Fuller, they concluded that it was "pure invention," which he thought would help with a patent application.

Despite the panel's reservations, the magazine proceeded with publication for March 1, 1943. The cartographer Richard Edes Harrison prepared a gorgeous version of the map that Fuller praised as "particularly fine," although he later criticized it for ignoring his grid of great circles. It was printed on two heavy sheets that could be folded into the polyhedron that Fuller was still calling the Dymaxion, which the article described as an "irregular solid first constructed by Archimedes."

He was rewarded with his greatest success to date. The *Life* piece opened with a photo of Fuller—sporting a mustache and wire-rimmed glasses—displaying his assembled map, which was colored to indicate different temperature zones. Various configurations showed the planet from the perspectives of the North Pole, the British Empire, Germany, and Japan, encouraging wartime readers to see new patterns of continuity in the oceans and continents.

Fuller saw children playing with the map on the street, and it made an impression on much higher levels. A set was prepared for President Roosevelt, while another was presented to Winston Churchill by Henry Luce. When two couriers, one of them Australian, came to take it for delivery to the British prime minister, Fuller put it together with Australia at the center, and he advised them to do the same with England when they gave it to Churchill.

In the meantime, Luce asked him to show off his map at a party, where Fuller spread it across the parlor floor. The event went unrecorded in his diary, but he loved to recount what Luce supposedly said afterward: "Bucky, in life every man must have his opposite; and, after long observation, I have come to the conclusion that you are my exact opposite." When Fuller asked what he meant, Luce explained, "I think man can change anything, whereas you think inexorable things are happening to man. I find this a very disturbing way of looking at things."

To prepare for a patent, Fuller created a more refined version of the three-way grid. On the advice of E. E. Norquist, he asked Benson Manufacturing of Missouri to make him two copper hemispheres, about twelve inches in diameter, with one sized to fit inside the other. Using the larger hemisphere as a great circle ruler, he inscribed a grid on the smaller one with a sharp tool, and to transfer the continents from a globe to a flat surface, he proposed employing a similar device made of transparent plastic—essentially a miniature dome.

The map's success encouraged Fuller to see himself as an educator at a time when the world was receptive to radical ideas, and he began to chafe against the constraints of a conventional bureaucracy. Taub was willing to find an outlet more suited to his strengths, and in March 1943 he told Fuller that he would be relieved of his regular duties in favor of "the research and development of an overall engineering and economic plan for a foreign country of your choosing."

To draft a report, Fuller was given the services of a secretary, and he naturally chose Lacey. It was his dream job, and it became even better two weeks later, when Taub said to him, "I'm going to make your hair stand on end." Taub revealed that he was expanding the project to investigate wider trends of postwar technology and economics, which Fuller explored in two essays.

In "Motion Economics," cowritten with Lacey, Fuller argued that society operated under the false premise that the universe was static, and even scientists still spoke of "sunset" as though the earth were flat. Americans had been forced to learn the truth "from their enemy rather than from their professed friends," but a lasting victory required a total reorientation of thought and language. Economics needed to redefine wealth as units of energy from the sun, while the words *up* and *down* ought to be replaced with *outward* and *inward*.

As a model, Fuller recommended the telephone industry, which had been transformed by dealing with energy and information into the "most financially successful private enterprise in the history of the civilized world." Echoing the technocrats, he argued that science and engineering should guide politics, which otherwise left power in the hands of the immobile "inferior citizenry," and advocated a system of daily electronic voting to shape public policy.

In "Contact Economy," he put these concepts into practice. A world commonwealth would allocate resources to nations, with businesses renting materials from the government and ideas from individuals, and industries would be socialized after they became too large to benefit from speculation. As a proving ground, Thornton Wilder suggested Greece, where these proposals could supposedly be tested without "charges of paternalism, meddling, or exploitation."

Fuller was only slightly disheartened by a letter informing him that Henry J. Kaiser's engineers had found major bugs in a prototype of his car, such as steering trouble and severe shocks at moderate speeds. The following month, Kaiser canceled the program. Fuller claimed later that Alex Taub had covertly assumed control of the project, and although he felt betrayed by Gerard Piel, he decided to let the matter drop, telling a business contact, "I would rather have Kaiser as a friend than an enemy. I wouldn't be surprised if someday he came to me again."

A favor for another friend was considerably more successful. The year before, Fuller had been contacted by Atherton Richards, the head of the Visual Presentation Unit of the recently formed US Office of Strategic Services—essentially the forerunner of the Central Intelligence Agency. Along with Merian C. Cooper, who conceived and produced the 1933 motion picture spectacle *King Kong*, Richards was developing a situation room for presidential briefings, which would include a huge globe to display weather and military data. He approached Walter Gropius and numerous other designers, and while Fuller was never involved, he kept in touch with the OSS.

Fuller had hoped to interest Richards in his polar warfare plan, but his connection to the intelligence agency paid off in a more surprising fashion. The painter Marion Greenwood, whom he had first met at Romany Marie's, had carried on a passionate romance with Isamu Noguchi in the late twenties. In 1939 she married the photographer and journalist Charles Fenn, who had traveled in China and Burma, and she asked Fuller to find a suitable position for her husband.

He agreed to talk to Fenn, whom he sounded out at a party at Greenwood's home, and they spoke again at his office the next morning. Fenn was looking for public relations work, and Fuller passed his name along to William H. Vanderbilt at the OSS. As a result of the reference, Fenn was dispatched to cultivate agents for the Americans in China. One of his recruits was a Vietnamese resistance leader whom Fenn provided with military support in exchange for cooperation in Indochina against the Japanese. His name was Ho Chi Minh.

By May 1943 Fuller had moved into a new apartment on Decatur Place, which he rented from the Yugoslav ambassador, and was researching his economic proposal for a foreign country. Instead of Greece, he settled on Brazil, seeing it as a nation where a modern economy could be constructed from scratch, with laboratories remotely connected by air and radio rather than by the "slowly erected extension of a languidly sprawling industry."

Decades later, he stated that the president of Brazil, Getúlio Vargas, had requested the report as part of a deal with the United States, but no trace of this arrangement appeared in Fuller's files. For two weeks, he interviewed executives who had worked on similar initiatives in Russia and Turkey. Peter Drucker, then a professor at Bennington, provided information on Turkish industry, and Fuller reached out to the architect Louis I. Kahn, a friend from his days at *Shelter*.

As Fuller reworked his report throughout the summer, he vigorously publicized himself elsewhere. He conducted another round of interviews about his unconventional nap schedule, now branded "Dymaxion Sleep," although he confessed that he no longer followed it himself. Fuller also pursued commercial applications of his work, especially his map, which was exhibited at the New York Public Library and the Museum of Modern Art.

The MOMA show provided two significant clues for the future. "Airways to Peace," directed by Monroe Wheeler and designed by Herbert Bayer, featured several innovative geographical displays, including a foldable icosahedral map by Irving Fisher that divided the earth into twenty triangular faces. Another was the "outside-in" globe, a wooden dome with the continents painted on its inner surface. Fuller wrote that a sailor would be "disgusted" by the concept, but both ideas anticipated designs that he would later claim as his own.

In the final draft of his Brazil report, Fuller used the country's lack of infrastructure to recommend that it skip the track stage and move directly to "trackless," bombing open the Amazon jungle to create "the leading skyport of the world." His plan assumed that the United States would take a leading role, and he advised Brazil to lease the development of its resources to American firms. The report was positively received—a contact later passed it through "unofficial channels" to the Vargas regime—and there was some casual talk of a similar project on Chinese industry, which Fuller described excitedly as "twenty times as big as the Brazilian job."

Instead, he landed an even more desirable assignment. His devotion to industrialized housing had become a running joke at meetings, but Fuller refused to give in, and in September he made a successful pitch to Herbert H. Lehman, the former governor of New York, at the Office of Foreign Relief and Rehabilitation Operations. Fuller was authorized to consult with the State Department on the housing industry, which he predicted would be the greatest activity of the postwar economy. To sustain employment, wartime production had to be converted to peacetime uses, and Fuller advocated the development of a "scientific" house to be manufactured at aircraft plants for Europe.

In private, Fuller had also been refining the specifics of his geometry. Like his theories of geography, his geometrical explorations came out of the Dymaxion Map, which was a fertile idea that evolved in response to the audience that he had in his sights. At the Cosmos Club, he had access to influential scientists, which encouraged his mathematical inclinations, and his apartment began to overflow with models that he built in his spare time.

He had continued to experiment with constructions based on great circles, as well as with the closest packing of spheres. In Washington, these approaches converged on the tetrahedron, which became one of the cornerstones of his work. A regular tetrahedron—a triangular pyramid with four equilateral faces, four vertices, and six edges—was both the simplest polyhedron and the minimum enclosure of three-dimensional space. No other structure, Fuller realized, could divide a system into an inside and an outside using fewer points.

Fuller had become interested in the tetrahedron through his map, since its triangular pieces could be seen as the exterior faces of eight tetrahedra that met at the earth's core, but he approached it primarily with sphere packing. Three closely packed spheres formed a triangle, and when a fourth was placed on top, it made a tetrahedron. If he kept going, he ended up with a cluster of twelve spheres around one in the center, or the equivalent of the polyhedron with fourteen faces—the vector equilibrium—that he had used in his cartographic projection.

Additional layers of spheres could be added to infinity, and if he connected their centers, it produced a uniform network that he saw as a vector diagram of the universe, which he defined as a "comprehensive system of energy processes." From there he developed a complicated series of calculations to express the relationships between different polyhedra. By using discrete vectors instead of infinite lines, he avoided the irrational numbers and geometric abstractions that he distrusted, and by taking an experimental approach based on whole units, he produced models that he thought would be accessible to laymen.

Fuller referred to his ideas as "energetic geometry," which he positioned as an alternative to the lines and planes of Euclid, and his inventory of its principles expanded enormously. By adding up the balls in his models, for example, he determined that there were $10R^2+2$ spheres in the outermost layer of a vector equilibrium, where R was the number of layers, not counting the sphere at the center. The third layer would have precisely ninety-two spheres, which he related to the list of naturally occurring chemical elements, as well as to the atom itself.

Thornton Wilder advised him to postpone publishing his results until he was sure that they were right, which Fuller claimed to have made "my life's policy." In fact, he began to share them immediately at the Cosmos Club, delivering lectures that amounted to a dry run for the rest of his career. His listeners in Washington tended to be skeptical of his ideas, however, and he felt that they treated his discoveries "as a joke, like numerology."

One of his few supporters was the classicist John Wolfenden, the senior officer of the British Commonwealth Scientific Office, who thought that his work might shed light on quantum mechanics. Another was John V. L. Hogan, the assistant to the engineer and administrator Vannevar Bush at the US Office of Scientific Research and Development. Hogan proposed "relating the discovery" to his superior, and Fuller subsequently said that the Manhattan Project, initiated by Bush, found his concepts "powerfully effective in their nuclear calculations," although no evidence survived for this claim.

In another talk, Fuller presented a timeline that he had drafted with Lacey of the dates of discovery of the elements, which he related to the accelerating curve of scientific progress. After listing the nine elements known in prehistory—carbon, lead, tin, mercury, silver, copper, sulfur, gold, and iron—he observed that only three more were cataloged before the eighteenth century. A burst of activity in the early modern period was followed by a nearly vertical ascent in the number of identified elements, with visible slowdowns during wartime. When he showed it at the Cosmos Club, he said, the response was "unanimous surprise."

Fuller warmly praised Lacey's contributions to their work. In a letter of recommendation, he wrote later, "She aids me in clarifying my own text, shows me how a shifting of phrases, a careful simile, or a pertinent example can more readily disclose my thoughts." After Anne arrived in the city with Allegra, who enrolled in the Madeira School in Virginia, he could no longer spend as much time with Lacey, but she remained an uncredited collaborator on all his essays. Their most ambitious effort was "Dymaxion Comprehensive System," a dense summary of his geometry that he privately circulated to Wilder and other members of the Cosmos Club, noting that he was more concerned with establishing priority than with any attempt at "lucidity."

His prospects at work were less favorable. Earlier that year, a dispute within the Roosevelt administration had transferred the mandate of the BEW to the newly created Office of Economic Warfare, which was reconstituted as the Foreign Economic Administration. In a staff meeting, Taub warned, "It may be that we will be spread out so thin we will not be able to continue the kind of independent research Bucky has been doing. After all, it is a kind of luxury."

In December Taub abruptly invited Fuller to submit his resignation: "The 'higher-ups' have decided that the kind of work which you have been doing here should not be continued. Frankly, I do not feel that there is any other work in the Engineering Service into which you would fit." Fuller looked into alternative positions, and in January 1944 he joined the staff of James L. McCamy as the special assistant to the executive director.

After a German magazine republished a *Life* article on his polar warfare plan, Fuller had become more convinced than ever that his work had strategic importance. His ultimate goal was to produce a house for the price of a car, with a chassis for utilities and a shell consisting of "an aluminum bell supported on the mechanical assembly mast, with a full-view plastic strip window." All the basic elements were in place for what he was already calling a dwelling machine.

Shortly after the academic journal *American Neptune* printed his article "Fluid Geography," which featured a revised version of his map, he contracted with Robert Ducas, a British businessman affiliated with the Reconstruction Finance Corporation, a US government agency, to license his housing and automotive designs overseas. In a note to Fuller, Lacey said, "I prayed so hard last night, I almost feel like claiming personal credit. All joking aside—I can't think of a more beautiful *time*—strategically—for you and I to quit."

For his next move, Fuller decided to approach Walter Reuther, the progressive vice president of the United Automobile Workers. His old roommate Ken Beirn had told him that Reuther was interested in converting aviation plants to housing, for which union support would confer a huge advantage. As mediators, Fuller chose Herman Wolf and Gregory Bardacke, two public relations officers from the US War Production Board whom he had met through Clare Boothe Luce. Both had strong labor connections, and they agreed to sound out aircraft companies as soon as possible.

Fuller wanted to outmaneuver a rival project at the Office of Economic Programs, which advised on foreign development, and he was mindful of his own conflicts of interest. He said that he was in "the uncomfortable position of knowing" that scientific housing might not happen without him, and since he would be unable to profit through the government, he shifted his focus to private industry. He presented his case in John Entenza's *Arts & Architecture* magazine, which called for the development of shelter with "all our experience as a nation organized for war production," and his involvement in the issue allowed him to work for the first time with the married designers Charles and Ray Eames.

On July 12, 1944, his forty-ninth birthday, Fuller conferred with Bardacke and two representatives of the International Machinists Union over lunch, where they decided to approach Beech Aircraft in Kansas. "All three agreed that Beech was by far the best of all airplane manufacturers so far as labor relations were concerned," Fuller wrote, and McCamy's response was enthusiastic: "I think this is the most important news I have heard in a great many years."

In August Fuller sketched a "round house with central chassis of mechanical system" for Reuther, who agreed that housing could sustain the postwar aircraft industry. Later that month, Fuller and Bardacke flew to Wichita to see Jack Gaty, Beech's general manager. Gaty was a staunch conservative—he left more than a million dollars in his will to "promote individual liberty and incentive as opposed to Socialism and Communism"—who had nothing in common with Bardacke, a dedicated Socialist who had been married to the Ukrainian-born experimental filmmaker Maya Deren.

Fuller had to work hard to meet them in the middle, and over the course of four hours, he argued that the dwelling machine could be produced only with the combined resources of labor and aviation. When he finished, Gaty reportedly asked, "Well, what are we waiting for?" Beech offered to cover overhead and provide drafting and labor services to develop a prototype, giving Fuller access to a full factory operation to realize his dreams.

They returned "triumphant" to Washington, although Fuller didn't inform Anne of his plans for another month. Ducas, who was disliked by the others, was pushed out, and as a lead investor, they settled on William Wasserman, a Philadelphia businessman with valuable contacts overseas. At one meeting, Wasserman declared, "Bucky, you may save the capitalist system."

In the fall, Wasserman pledged to advance $5,000 of his own money and raise an additional $50,000, while Herman Wolf secured loans from his wealthy wife, Emily, who had studied architecture at Columbia. They looked to small investors for the remaining funds, which would provide the capital base for the firm that they agreed to call Dymaxion Dwellings Inc.

Bardacke was originally set to serve as president, but when his connections with labor threatened to cause complications, Wasserman took the position instead. Fuller would be chairman of the board, although any specific title was less important than his public association with the house itself, and he intended to transition to pure research as soon as possible. His government contacts supported the move, and even Vice President Wallace asked to be kept up to date.

After resigning from the Foreign Economic Administration, Fuller prepared to oversee the project at Beech. The war had finally given birth to a dwelling industry, as he had predicted all along, and no questions had been raised so far about his difficulties working with others. Anne and Allegra would remain in Washington, and while Cynthia Lacey had originally planned to stay behind to finish up his projects, she and Fuller soon decided that it would be best if she joined him in Wichita.

On October 23, 1944, Lacey went to see Anne, who was loading boxes into a van at the apartment on Decatur Place. Anne was evidently unaware of her husband's affair, and the two women were on good terms, if not especially close. They spoke politely about their respective changes of situation—Allegra was staying to finish school in Virginia, while Anne would head for New York until Thanksgiving—and Lacey said that she hoped that they could get together soon.

The day before, Lacey had said good-bye to Fuller without knowing when they would meet again. She kept his industrial charts and a suitcase of his energetic geometry papers, and, at the office, she noticed that, in the haste of his departure, he had knocked over his coffee cup. Lacey set it tenderly beside hers, and when a secretary asked if his work would keep her busy, she replied curtly, "More than busy."

In a letter dated "First Day Alone," Lacey wrote that she longed to join him in Kansas: "I would like very much to make good here, to prove that your confidence in me was nothing more than exactly that, and was warranted, but other things are more important than satisfaction of my personal whims." The sight of his blue kimono at her home reminded her of their breakfasts together, and when they spoke by telephone, Lacey thought he sounded "wistful."

Their physical attraction was still strong. Fuller once called when Lacey's mother was there, and when he said that he wanted to make love to her, she pressed the phone tightly against her ear so that her mother wouldn't hear. As for her moving to Wichita, Lacey wrote, "Actually, our having two apartments is mostly for appearance's sake, since I can't imagine us being in one alone. . . . I think it's rather important to get a good bed and an easy chair, the latter for leisurely or playful progress to the former."

Fuller also heard from Anne, who was staying with her brother Arthur on Long Island. "Darling," she wrote, "I hope everything goes right along. Keep *rested* and *don't get agitated* and get your overcoat relined and cleaned." Once again he was navigating a complicated private situation, but his distance from the two women in his life allowed him to push it out of his mind.

In Wichita, the company finalized its contract for the prototype, although Wasserman embarrassed the others by asking about Walter Beech's reputation for drunkenness within earshot of the founder's chauffeur. Fuller was assigned a mechanic and a draftsman at Krehbiel Plastic Products, a Beech subcontractor on North Barwise Street. Mice and birds could be heard overhead, building nests out of aluminum and plastic scraps, and the company moved to a better space at the plant toward the end of the year.

The whole city seemed excited by the venture. Gaty said that it had halted an exodus of factory employees, who had been seeking work elsewhere in anticipation of a fall in production, and union officials eagerly publicized it. The press emphasized the company's ties to labor, and Fuller became a district union delegate in December— thereby allowing him to describe himself long afterward as "a card-carrying journeyman-machinist in the International Association of Machinists."

At the company itself, the situation was less rosy. Herman Wolf noted that they were unable to obtain life insurance for Fuller because of the underwriter's "fear of Bucky's suicide if we failed." Wolf regarded William Wasserman, who had provided only a fraction of the money that he had promised, as utterly incompetent, while Wasserman had his own doubts about Fuller's qualifications—he wanted an outside engineer to consult on the dwelling's design, complaining that the bathroom was "too cramped and too dreary."

In early 1945 Cynthia Lacey moved into Fuller's apartment on East Third Street. After she joined Wolf at the office, Fuller installed their only telephone on an overhead carousel, which allowed it to be passed between the desks. As company secretary, Lacey capably kept Fuller on task, and she came to love Wichita's "huge skies and quiet fields, brilliant suns and stark storms."

Later that month, Fuller heard from Noguchi, who had protested the internment of Japanese Americans by volunteering to enter a relocation center in Arizona. The other internees were suspicious of his intentions, however, and after six months, he secured a temporary leave and never returned. He and Fuller had seen each other only rarely in recent years, but they had once taken a train together to Washington, where Noguchi confronted the officials on board about the injustices done to the Nisei.

When Noguchi offered to work for the company, Fuller declined, writing, "There will be plenty of time later to consider what you might be able to do for us with lighting or surfaces." He also exchanged letters with Frank Lloyd Wright, who had designed a house in Wichita that Fuller saw on his drive to the office. In his review of *Nine Chains*, Wright had said that mass production in housing might happen, "but not if I win," and Fuller seemed to finally have the upper hand.

As they proceeded to the model stage, Fuller assumed without question that the house would hang from a mast, but he experimented with various designs for the walls. At one point, he considered a hemisphere with a triangulated framework—a feature evocative of the domes to come—but settled on four horizontal compression rings suspended by wires. His most intensive efforts were reserved for the ventilator, which had been central to his earliest housing plans.

At Beech, he conducted trials in a wind tunnel with a transparent model full of colored smoke, while another was hung upside down to test its performance under hurricane conditions. He finally arrived at a system in which air entered through inlet holes at the base. Some of it exited through floor openings, carrying away household sweepings, and the excess went to a roof vent, which circulated the air and reduced heat loss by equalizing the pressure with the outside.

The rotating vent also contributed to structural stability. Fuller saw houses as "little ships" that sailed with the wind, and a lightweight shelter needed special features to avoid being blown away. His solution depended on a combination of anchoring, streamlining, and the intake system. During a windstorm, the ventilator, which Fuller modeled after a ship's rudder, would rise like the valve on a pressure cooker to keep the house from exploding.

As he refined the design, Fuller expanded his staff to five draftsmen and two pattern makers, and their workload increased after the army requested a production schedule for the Airbarac, which was a simplified shelter consisting of the shell alone. It was a welcome sign of interest, but the added work severely compressed their timeline, and the situation became even more challenging after they decided to push ahead with a complete prototype of the house at full scale.

To fund the Dymaxion Dwelling Machine, a majority of stock was sold to small investors, raising $100,000, while the rest was allocated to Fuller, who retained ownership of more than one-third of the company. As a bulwark against Wasserman, he recruited new members to the board, including J. Arch Butts Jr., a local auto dealer. Butts agreed to serve as treasurer, and he also acquired Dymaxion Car #3, which had turned up unexpectedly in Brooklyn—Fuller drove it around town, but it was in bad shape, and it was later sold and cut up for scrap.

In April they finished the cost analysis for the Airbarac, which the US Army Air Forces ordered for a summer delivery. Work was unaffected as the war concluded in Europe, but the situation was changed dramatically by the events of August 6, 1945. After hearing about the atomic bombing of the Japanese city of Hiroshima, Fuller thought back sorrowfully to his meeting with Einstein, who had reportedly told him that he could envision no "practical application" for his ideas.

The following week, Anne and Allegra raised the flag at Bear Island as the radio announced that Japan had surrendered. In Wichita, the end of the war had an immediate impact. Following the standard practice for wartime contracts, the army suspended its order, indicating that it would let the company decide whether or not to proceed. Fuller and his partners accepted the cancelation, reasoning that it would be best to reposition the house as a family home for returning veterans.

Wasserman, still on bad terms with Fuller, finally resigned in August. "No matter how hard we tried or what pressure we used, we could never get a finalized plan from Bucky," he recalled. "Every time we thought we had one, he would come in with a whole new variation on the design. For instance, if there were rumors of unrest in the plumbers union, he would insist on his self-contained, shit-packaging john, which nobody had invented yet, not even Bucky."

By September a prototype of the house had been assembled indoors, prompting Lacey to write, "As Bucky so aptly puts it, we have completed a solo flight across the Atlantic, and all that remains is to make a clean landing in Paris." Their agreement with Beech covered only the prototype, though, and as the factory turned to postwar aircraft manufacturing, they prepared to subcontract elsewhere for mass production. They also

Cynthia Lacey on the mast of the Wichita House prototype at the Krehbiel building.

began the process of securing approval from the government's Federal Housing Administration, which would allow buyers to qualify for mortgages, and started a second prototype on a farm to test the shell outdoors.

Since they no longer had to depend solely on Fuller's powers of persuasion to generate interest, he consolidated his influence while he still could. On the advice of advertising counsel, he rebranded the firm as Fuller Houses, which was the name that he had wanted for close to twenty years. Lacey was promoted to vice president to serve as a "liaison between management and production," and he hired other allies, including his brother-in-law Arthur Hewlett and Leland Atwood. His most significant recruit was the head of personnel, Edgar J. Applewhite, a polished man of twenty-six whose sister had married a Hewlett cousin in the thirties.

Fuller had packed the company with people who would take his side in a dispute, and he was heartened further when he was granted a patent for the Dymaxion Map on January 29, 1946. He claimed later that it was the first projection to receive a patent since the turn of the

century, although he must have known that was untrue: his files included at least two recently patented polyhedral maps, including one by
Bernard J. S. Cahill, which Fuller had carefully studied.

It was still a major accomplishment, and it encouraged him to in-
dulge his theoretical side. As the company geared up for production,
engineers were hired to perform structural tests, and before they saw
the prototype, Fuller delivered a lecture that Applewhite edited into
the booklet *Designing a New Industry*. Fuller was aware that he was
addressing an intelligent crowd, and the result presented his vision of
housing more persuasively than he ever would again.

He opened by noting that shelter had traditionally been a defensive
measure, built using whatever materials were available, which had to
be reimagined for the modern production line. Fuller pointed out that
the use of a track for construction could be traced back to shipbuild-
ing. A British ship might have sailed from England to obtain riggings in
Maine and copper sheathing in South Carolina—the implication being
that the first assembly lines had been eight thousand miles long.

The next breakthrough had been due to Henry Ford, whose greatest contribution was the principle of mobile inventory control. Industrial
materials were in constant worldwide motion, and the only place where
they paused long enough to be seen was the factory floor, which allowed Ford to evaluate his needs at a glance. "The stacks of mudguards
or tires were his visible check on inventory," Fuller explained. "He used
no warehouses. He could really see the stacks getting low."

Since the manufacturing of industrialized housing demanded the
same kind of constant movement, the house's weight became hugely
relevant. Fuller said that every industry had a "mortality point" where
the scale of operations amortized its costs, which amounted in this case
to a hundred thousand houses per year. If each house weighed 150 tons,
it would overwhelm the national transportation system, so it had to be
as light as possible.

Once a shelter reached a practical weight—3 tons, according to
Fuller's calculations—it could be made available on demand, with
residents buying new homes as casually as they switched phones. In an
unconscious anticipation of the dome, he noted that the simplicity of the

design excited builders, who could see how it fit together: "You can sit here in this room and think of that house, scanning it with your mind's eye from the top down, and write out the parts list from memory."

Many practical matters remained unresolved. The firm recapitalized to continue operations until it began mass production, which would take at least another ten months, and prepared to bring its case to the public with media showings and a newsreel by Paramount Pictures. In March 1946 the company was toured by the Black contralto Marian Anderson, who praised the house's acoustics, and her visit was written up for the "Negro press," in an effort to access a vast potential market for affordable housing.

Their publicity push culminated on the East Coast, where Fuller and Wolf showed a model at the Waldorf-Astoria in New York, followed by a press conference at the Willard Hotel in Washington. At that point, no finished house existed: neither of the two prototypes was complete in itself, and most articles reproduced a picture of the model, which, at a glance, was indistinguishable from the real thing. All the

Outdoor prototype of the Wichita House.

same, it was an auspicious debut for Fuller's "stationary airplane," which would be known ever afterward as the Wichita House.

The house weighed 6,000 pounds and was 118 feet in circumference, with an area of 1,110 square feet. Its mast, which stood 22 feet high, was a hexagonal bundle of steel tubes planted in concrete. Tension spokes bore horizontal rings that supported the roof and the floor, which was anchored with cables at twelve points. The roof was made of aluminum cowling gores, or curved triangular pieces, flexible enough to be flipped into place, and the gaps were left unsecured, allowing the shell to expand or contract.

A standard floor plan consisted of an entryway, a living room with a fireplace and optional balcony, a kitchen, two bedrooms, and two copper bathrooms, which Fuller hoped to replace with aluminum and plastic. The furniture included a table designed by Fuller and "ovolving" conveyor shelves built into the walls; a single rectangular opening disclosed bins that rotated at a touch of a button. Interior partitions could be moved to adjust the rooms, which were divided by accordion doors, and indirect lighting was offered in a choice of colors.

Because of its unorthodox design, the Wichita House was ruled out by most city building codes, but Fuller believed that demand from outside areas would absorb the initial run. Delivery and assembly were equally advanced, with a single cylindrical shipping unit enclosing all the nested parts, each of which was light enough to pick up with one hand. After the roof had been assembled close to the ground and raised using an electric motor, the ventilator was hoisted up, and the ring of Plexiglas windows was installed. Its floor was constructed of plywood wedges over aluminum ribs, without any fastenings, which produced a distinctly springy sensation underfoot.

With all amenities, the total price of the house would be $6,500. It received a positive response from the media, including an article in *Fortune* by the modernist designer George Nelson, who said that it was "likely to produce greater social consequences than the introduction of the automobile." Although they had yet to obtain mortgage approval, they ultimately received 3,500 orders—a number that Fuller later inflated by a factor of ten—and estimated that they could soon produce

50,000 houses every year. Fuller stated modestly that he had predicted in 1927 that his house would be realized in twenty years, which meant that he was right on schedule.

It felt like a moment of triumph, but the heightened attention came at a cost. Tensions at the company had erupted over minor issues, including the hiring of Donald C. Miller of Chevrolet to handle sales and promotion. Fuller claimed to object to Miller's spending, but it was really a question of control. Now that his house was out in the world, he feared that he might no longer be needed.

Conflicts also arose over the house itself. Wolf wanted more conventional furniture—he especially disliked the shelves—and the placement of the tension rods forced the exterior doors to be impractically narrow. The flexible roof panels and springy floor were criticized, and a visitor recalled of the ventilator, "When the fin was in the wind, it damn near sucked out the inside of the house."

Fuller resisted these changes but pushed for other modifications before the design was locked, which would only delay production further. He was reluctant to step aside, and he suspected that his colleagues were talking behind his back. In a letter to the architect Edward Durrell Stone, Fuller imagined what they were saying: "Dear Bucky has worked so hard and done so well—now he is tired and is getting a little old and quarrelsome." Going on the offensive, he stamped his own drawings as obsolete, which he later described as an act of defiance during "the Ides of March."

At the time of the breach, Anne was living in a modest apartment at 6 Burns Street in Forest Hills, Queens, provided by Alrick Man, the gentleman friend—she still teasingly called him her "beau"—who had inspired Fuller's jealous walk from Lawrence to New York twenty years before. Allegra found the neighborhood less than fashionable, but Anne liked its proximity to her old social circle. In Washington, Allegra had encouraged her mother to return to art school, but she had attended classes for just one day, and now she was eager to settle back into the world that she had missed.

Her serenity was destroyed by the dispute at the company. "I just feel absolutely sunk about the whole thing," Anne wrote to Fuller, and she urged him to see matters from Herman Wolf's perspective, observing that he had more at stake financially. She reminded her husband that he was always accusing others of stealing credit, even though all of the publicity had been in his favor, and advised him to find an objective arbiter, apart from Lacey, who, she said emphatically, was "just a kid."

Anne's letter was accompanied by one from Allegra, and the eighteen-year-old addressed her father with unusual bluntness: "You have a tendency to think that you are always right, you usually are, but sometimes you're not, and business isn't one of your strong points, as you must know from the fact that so many of your ventures have been failures financially." This would have been difficult for Fuller to hear from anyone, let alone his own daughter, and it only hardened his resolve.

In New York, Wolf attended a show with Anne, who made a point of saying afterward that she trusted him. Wolf later told her over the phone that the trouble was Fuller's reluctance to give up control. After Anne acknowledged that her husband could be excitable, Wolf replied, "*I* know that. He's said things to me during the past few weeks that were pretty hard to take."

When Wolf said that he still hoped to make them millionaires, Anne noted that Fuller hated the idea that people were conspiring behind his back, even when it came to "giving a surprise party." After the call, she drafted a second letter to Fuller, stating definitively that he should let Wolf run the company.

Another version of the encounter in New York told a different story entirely. According to the account by Alden Hatch, which was reflected nowhere in their correspondence, Wolf came by the house one day to speak to Anne about a personal matter. "I think you should know," Wolf said, "that Bucky is having an affair."

"You have no right to say such things about Bucky to me," Anne allegedly replied. "And even if it were true, it's none of your business." Even if she knew about her husband's infidelity, Wolf may well have crossed an unspoken line, which evidently lingered in her memory. When Wolf's name was mentioned years later, Anne remarked, "I just hate that man."

Her real concern was that Lacey might push Fuller to resign, and when they spoke by phone, Anne said nothing about any other allegations. Fuller asked if William Parkhurst, their lawyer, had told her to send the previous letter, which she denied. Feeling reassured by their conversation, Anne wrote, "It was wonderful to talk to you tonight, darling, and I was so thrilled to hear things were going better."

On April 6, 1946, it all blew up. At a staff meeting, Wolf announced that Fuller, at his "express desire," would withdraw from the production side. The following night, Wolf and Bardacke came to Fuller's apartment with a written offer for him to remain as a research director and honorary chairman. For now, he accepted what he would dismiss later as a "false agreement"—which additionally called for Lacey's departure—while preparing secretly for a more definitive break.

In a handwritten note, Wolf expressed "love and admiration" for Fuller and Lacey, adding, "In the months ahead, I shall need your counsel and guidance." This affection was not reciprocated by Lacey. She despised Wolf, as well as Don Miller, who had insulted her with his attitude at the office: "He never once looked at me or acknowledged my presence in the room."

Later that month, Lacey told Charles Eames, who had visited in March, that she had left the firm to assist Fuller in his new role as a consultant. On April 27 Fuller formalized his departure from the board in a terse letter of resignation, and Applewhite left soon afterward. Only a few weeks had passed since the house was revealed to the public, and now the company lay in ruins.

Almost at once, Fuller began to disseminate his own version of events. He wrote to Safford Colby of the Aluminum Company of America, or Alcoa, that he had found himself "opposed as chief policy maker" in Wichita, and he placed the blame on his partners: "I am experienced in the abrupt shifts in attitudes of new developments and familiar with the ambitions suddenly engendered by initially successful demonstrations."

Over time this account was embellished even further. Fuller gave the most extensive version to Alden Hatch:

> What I was really developing was a good prototype which was fine for the air force, who could handle it. But it was not ready

for the general public. People wanted to buy; but there was no industry to distribute the houses. . . . There were a number of people who wanted to come in and take over the business and put up the ten million. The politicians in Washington were willing to back us to get the house into production. Herman Wolf wanted to get it going. And I kept vetoing. I said, "It is not ready. Here are all those orders, but who is going to install it?"

He concluded that since the company was in no position to enter manufacturing, it was "just not honest" to promise something that it couldn't deliver: "I was forced to do just what I had done before in Bridgeport: to shut down solvent. Nobody made any money."

This was a total fabrication. At the time of Fuller's resignation, there were no real orders, nothing even approaching government support, and no prospect of a fraction of the $10 million that he claimed had been offered. Investors would be even less willing to take a chance on a company in crisis, and Fuller's departure immediately after their first round of press coverage had guaranteed that it would draw negative attention at the worst possible time.

Despite his statements to the contrary, Fuller was far from believing that the firm was unprepared for production on a limited scale, which the group had been actively pursuing before his departure. To another correspondent, he asserted implausibly that he had determined that Fuller Houses would be better off without him: "Upon completion of the house, I withdrew from the active management of the company, as I believe that an idea can grow more rapidly when it is weaned from the inventor after completion of prototyping."

This was the opposite of what he had stated in *Fortune* and elsewhere about the role of inventors in business, and it was contradicted at once by his own actions. Fuller had resigned in anger, but it was also a tactical maneuver. Now that the house had been proven in principle, he believed that he could secure funding as soon as his partners were out of the picture, and instead of stepping back, he began to wage a protracted war for control.

"It may be that one must be tremendously interested in money to make it," Fuller wrote to a contact two days after he resigned. The following month, he told George Nelson that he had been ready to leave: "I welcome the opportunity to turn responsibility of growth over to others." In fact, Fuller had no intention of fading away. His actions would only repeat the tendencies that had derailed the Pierce Foundation bathroom and the Dymaxion Car, and in the authorized accounts of his life, the events of the ensuing year would disappear entirely.

Instead of returning to New York, Fuller remained in Wichita, keeping an eye on his estranged partners through back channels and waiting to make his move. His resignation had introduced a random element, but the company was still moving forward, with Wolf seeking government loan approval, resolving issues with city codes, and searching for new management.

Beech had agreed only to build a prototype, and as the demand for aircraft rose, it was disinclined to enter production, which required hard tooling rather than the soft tin dies—intended for short runs of components—that it had used so far. Given the uncertainty over mortgages, not to mention Fuller's resignation, the company was unlikely to convince banks to lend it millions of dollars, and Wolf decided on an alternative approach modeled on auto dealerships, with working capital secured by individual distributors.

Finding partners was easier said than done, and they lacked the money for a new prototype, which was urgently needed to address problems with the house. After laying off half the staff, Wolf hired an engineer named Paul Fisher, who delivered a plan that replaced the cable system with concrete, eliminated the conveyor shelves, and used curved plate glass in the windows instead of Plexiglas.

Fuller dismissed Fisher as "a stooge," and, indeed, the design, which increased the weight to twelve tons, was undeniably flawed. At the moment, however, the personal dispute was more pressing. William B. Harris, an editor at *Fortune*, offered to mediate, noting that the company's very survival was at stake: "People just aren't going to come in with you two at sword's points, because either one of you is in a position to upset the hell out of the applecart."

In a disingenuous response, Fuller declined. He said that he had been willing to leave, but Wolf, having already "arranged to do battle," had attacked him anyway. "There is no feud," Fuller contended. "I have no position in the company. I am just a stockholder. I have more to lose than anyone, both in shares and in name, and am therefore as anxious as anyone to have the company succeed. Herman wanted to take over and did not want my thinking." Fuller stated that he had offered Wolf his "blessing," but in private, he assured Lacey that they would "put Herman where he belongs."

Lacey returned to Virginia in May to care for her sick mother. Before her departure, she and Fuller had a rare argument, which she alluded to afterward in a letter: "[We] were making small headway towards peace until we started talking about world problems and their meaning and our possible small role in them." Both were feeling strained. Lacey feared that she was losing her hearing from otosclerosis, and her mother was suspicious of her relationship with Fuller, leading Lacey to propose to him that they write in code.

Because she was closer to the East Coast, Lacey was in a better position to meet potential contacts, but her departure separated them at a critical time. In her absence, Fuller worked with her younger sister Ginnie, who had arrived in Wichita the previous fall, and also made a show of exploring other options. Since he had no regular income, he looked into teaching and research positions, and he spoke at the University of Kansas at the invitation of engineering students in June 1946.

Fuller was reportedly offered a deanship, but he was unwilling to abandon the company just yet. As he waited for an opening, he occupied himself with plans for a rowing catamaran, which he designed as a boat with a pair of tubular hulls that could be mounted on an automobile rack. He also took flying lessons—he was certified by the end of the year—and considered buying a small plane.

All the while, he was escalating his efforts to regain control of the firm. Fuller still owned nearly 30 percent of its stock, but Wolf was trying to acquire a large enough stake to be able to approve the appointment of directors and officers. As an ally on the board, Fuller wanted to win over the real estate developer William Graham, his landlord, who had envisioned putting up hundreds of the houses on lots throughout Wichita. Another key figure in the balance of power was J. Arch Butts, who Fuller thought might be persuaded to push out Wolf in his favor.

In New York, William Parkhurst, who handled the firm's legal affairs, confirmed to Lacey that Wolf wanted to raise $250,000 to build a new prototype. It was this amount, not millions of dollars for tooling, that they needed to survive, and Fuller and Lacey decided to mount a rival bid. They expressed no concerns over the problem of mass production, and their real liability was Fuller. As Parkhurst noted, he was seen as "just a little too difficult and hard to get along with."

Since Wolf was chasing many small commitments, Fuller decided to seek one big investor. Before his resignation, he had dispatched three deputies to investigate aviation plants out west as alternatives to Beech. The search had yielded no results, but he revisited it now with a plan for a California prototype, with Fuller returning in any role "which they may deem necessary to the best interests of the company." After this failed as well, their only remaining option was in Washington.

Both sides were desperate to show progress, and the first group to bring in real money would win. As the rival factions pursued government funding, Fuller planned to use Ted Larson and Knud Lönberg-Holm, his colleagues from the Structural Study Associates, to approach the Department of Commerce or the National Housing Agency, but he needed a production contract first. Lacey cautioned that this might require hiring another engineer: "Remember that almost everyone you will have to deal with will be *lacking*," she said, "but must be made to serve."

Fuller thought that he still had narrow support among the shareholders, but his impulsive resignation had confirmed a widespread sense among outsiders that he was "impractical" and "unreasonable," and the Hughes Aircraft Company was advised to avoid him. Lacey told him to act in a manner that would silence the doubters, but, she wrote glumly, "The April debacle finally resolved to the personalities involved, including yours and mine, worse luck."

As Fuller considered his next move with Lacey, he deliberately excluded Anne, to the extent that she had been obliged to ask of the latest design, "Is it still round?" The only positive development came from David Noyes, a newspaper owner who had expressed interest in building a test house in Hollywood, California. After Fuller recruited him for an advisory role, Noyes clashed with Wolf, and there were signs that he would support Butts in a competing bid for the presidency.

Wolf had one last card to play. In early 1947 he connected with a group of businessmen who wanted to invest in GI housing, with the intent of leveraging their government connections into a loan, but his gambit backfired after they sent an engineer to Wichita. Upon reviewing the prototypes, he recommended returning to Fuller's design, which was lighter and more efficient.

After Fuller was reinstated as engineer in charge, they received a tooling quote from Beech, which they had previously written off as a prospect, stating that a minimum investment would require $150,000 for two hundred units using soft dies or $3.75 million for a run of fifty thousand. Fuller considered the first number a reasonable goal that would allow them to approach banks for additional support.

The deal was a complicated one, and, before long, Fuller began to suspect that he was being excluded from negotiations. In February he and Lacey flew to a speaking engagement in Cleveland in their new Luscombe monoplane. When bad weather stranded them in Greenfield, Indiana, he cabled urgent requests for updates to the investment group, but he received only limited information in response. On his return, he learned that the team was partnering with Alcoa, and when he protested that he hadn't been informed, Butts turned to another member and said, "You had better notify them that Fuller is backing out of the deal."

"You had better notify them of the precise wording of my coming into the deal," Fuller shot back. This infuriated Butts, one of his few reliable allies, and he walked out angrily. Once again an agreement had foundered over control, and a trip by Fuller and Lacey to find a California client came up empty. Wolf remained in charge, and after his pursuit of a government loan collapsed, Lacey complained that the management was chasing "one nebulous 'big deal' after another."

This disaster effectively marked the end of Fuller's interest in the company. After a year of unrelenting effort, with his fifty-second birthday approaching and most of his money gone, he had accomplished virtually nothing, and he reluctantly turned to "an alternate plan." While Lacey settled in Virginia, he would return to his family in New York to promote his work in other ways. He was still estranged from Wolf, who recalled later, "I'm a very patient guy, and very few people would have put in the two years I'd put in with this, quote, 'wonderful, loving' madman."

The year before, the Fuller Research Foundation had been incorporated as a nonprofit to develop ideas that were too radical for conventional industries, such as his map, the rowing catamaran, geometrical models, more books, and yet another car. Fuller asked Charles Eames and George Nelson to serve as board members, and he outlined his objectives in a small booklet titled *Earth, Inc.* In the introduction, Cynthia Lacey wrote that the world's shocked reaction to the development of the atomic bomb pointed to a lack of understanding of historical and scientific trends: "We may not be able to dispel the dangers which accompany the event, nor lessen the urgency of systems for control of its power, but we can regain our composure."

At the heart of the booklet was their timeline of the elements, which was described as a report card of mankind's progression from ignorance to knowledge. Fuller noted that war forced society to adjust to progress that science had already made, but it was better to advance in peacetime through the enlightened use of technology, or "philosophy resolved to the most cogent argument." He also discussed the closest packing of spheres, observing that when their centers were connected, each shared a vector with its neighbor. By implication, Fuller concluded, "Unity is inherently two."

As usual, when his more ambitious plans failed to gain traction, Fuller fell back on writing and speaking. In April 1947 he was invited by the College of the Ozarks to serve as a guest instructor for two weeks. Fuller flew his second plane, a Republic RC-3 Seabee, to Point Lookout, Missouri. The students received him warmly, and he was reminded that he enjoyed sharing his ideas with young people, who, he hoped, would "someday integrate them in their activities and lives."

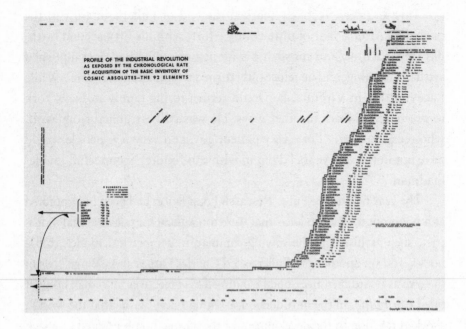

Fuller and Cynthia Lacey's timeline of the discovery of the chemical elements.

With his housing plans on hold, Fuller was more open to a teaching career, which would become a tremendously important avenue for the future. Fuller looked into buying property near the college to establish a research center, but Lacey pointed out that his finances were precarious. Despite his obsession with mobility, he continued to think in terms of something fixed and institutional, and he had yet to make the crucial leap to the ephemeralization of his own life.

The experience provided a timely reminder of his strengths, and it was underlined when he flew out for a talk at Dartmouth College. Upon introducing Fuller, his host mentioned that he had first appeared there in 1931. As he rose to speak, Fuller felt obliged to issue a correction. After a fire a decade earlier, Dartmouth Hall had been entirely rebuilt, and over that period of time, the flesh of his body had been replaced by new atoms. All that remained the same, he said, were his glasses.

His lecture impressed a student named Christian William Miller, an artist, model, and prominent member of the underground gay community in New York. In a letter to Monroe Wheeler, who had directed the

"Airways to Peace" exhibition at the Museum of Modern Art, Miller wrote, "This little man is an angel of the lord." It testified to Fuller's undiminished impact on younger men, although Miller also noted that some of the other students called him "R. Fuckminster Buller."

From there Fuller flew to a talk in Portland, Maine, followed by an unscheduled visit to Christopher Morley on Long Island. Fuller recalled that Morley was annoyed: "It was an amphibious plane so I could land on water. He was terribly upset and waved at me as if he wanted me to go away. I just thought of it as coming home, but he let me know he thought I was intruding." The encounter troubled him, and it cast a pall over his return to Forest Hills, where he reunited at last with Anne.

It was their first time living together for an extended period in more than two years, and his affair with Lacey hung between them. Allegra remembered that it was "probably the only time that I saw my father being very depressed," and he spent his first evening at home drawing alone at the dining table. The next morning, however, Fuller seemed excited, and Allegra came to believe that this was the moment that he came up with the idea of the geodesic dome.

There was no immediate evidence of any such breakthrough, though, and Fuller was still finding his footing. In May he was hired by the Mike Goldgar advertising agency to produce and market a simplified version of his house, without a mast, for a veterans community of one hundred units in Mount Kisco, New York. Goldgar was little more than an unscrupulous promoter—he later sold deeds for lots on the moon—but Fuller was ready to seize any proposal that might keep him in the public eye.

With its modest goals, the Mount Kisco project badly undermined his claim that he had refused to consider anything less than a complete housing industry, and after it failed to raise a paltry $3,000, Cynthia Lacey was shaken. She wrote to Fuller that she had always believed that their best move was to build the house in any form, but she was unsure what to do now: "I must rather count on you to tell me wherein our hope lies."

Fuller was doubtful of this himself, but he struck a deal with Goldgar to start a promotional design group, which he spun as a revolutionary fusion of marketing and science: "The advertising man has been asked to come in, after the product has already been manufactured, and find some way to sell it. But we feel that this procedure is working backwards. . . . We believe that the public needs certain things and that when the right things are submitted to them, they will ring the bell."

It still felt like a retreat, and the uncertainty affected his relationship with Lacey. He grew upset when she turned down an offer to travel with him, and she was saddened by their inability to communicate. "Your voice on the phone made my heart sink," Lacey wrote. "I thought our total experience was such that you would not ever again feel desperate or rejected. I have been sadly remiss, I guess, in my reporting—or we couldn't have arrived at such a state."

Out of concern for his feelings, Lacey had been keeping aspects of her personal life a secret. While taking flying lessons at Hybla Valley Airport in Alexandria, Virginia, she became close to a pilot and aircraft mechanic named David Floyd. In her letter of May 27, she reaffirmed her love for Fuller but finally told him of her new relationship. "His intentions are honorable," she wrote of Floyd, "and I very much hope that you will not be fearful for me or of him."

When Fuller went on a restorative trip to Bear Island, Lacey joined him, although not without reservations: "Please be a little more detailed than you have been lately, and tell me how Anne really feels, whether she would really not be uncomfortable to have me, or whether I'm just a stopgap between you and certain death." At the island, Rosamond—who had divorced her first husband and married a man named Alphonse Kenison—was taking paying guests. Anne gladly accepted a salary to help out, but when Fuller flew in with Lacey, the atmosphere was tense.

Fuller was happiest around his daughter, who was preparing for college. Allegra was drawn to dance—she had spent summers at the School of American Ballet, under the direction of George Balanchine, and organized collections at the Museum of Modern Art—but she had also worked with her father on the principles of his geometry. He had expected her to attend MIT, and after she chose Bennington instead, he bought her a boat to reassure her of his approval.

Like the planes that he had purchased, the gift was a questionable extravagance. Along with everyone else, Fuller had lost money on the Wichita House. Apart from a sight gag in the 1947 film comedy

The Secret Life of Walter Mitty, starring Danny Kaye and Virginia Mayo, its only visible manifestations were its two incomplete proto-types, which William Graham purchased for $1,000. He combined them into a house by an artificial lake, placing it on stone foundations and immobilizing the ventilator, since his children had feared that it would suck them outside. The modifications displeased Fuller, who lamented that Graham had "forever grounded this airplane."

As the architecture and design critic Martin Pawley wrote later, the Dymaxion Dwelling Machine had been "the most important pre-fabricated house design of the twentieth century, and certainly the greatest lost opportunity of the years of postwar building recovery." Fuller avoided discussing the real reasons for its failure, however, and his image as a detached developer of ideas was unable to sustain the fact that he had tried so hard to raise money for the house—which he never even managed to patent.

He was also drifting apart from Lacey. After he missed her departure from Bear Island, she joined the Food and Agriculture Organization of the recently founded United Nations. When she was scheduled to sail from New York to Geneva, Switzerland, for a conference, she went to the city early to say good-bye to Fuller, but they failed to see each other in time.

On August 19, 1947, Lacey wrote to him from the ship to confirm what they both knew already: "When I return, I am going to marry Dave Floyd. . . . This doesn't and couldn't change my feeling for you. It simply means that I have found someone with whom I would like to do the things a young girl does—marriage and babies and a home life—and Dave believes in me and takes away the loneliness."

The following week, Lacey sent him another letter from Europe, leaving no doubt of how important he had been to her:

So often while here, I have thought of our wonderful times to-gether when we explored Wichita and westward to California. No one makes it seem so real and exquisite as you, for which we must not mourn but only be grateful to have known at all. In the meantime, I try to look at things as you would see them and try to study and comprehend as you would do. . . .

I shall be anxious until I hear from you about my forthcoming marriage. Please search deep within your love and make it possible for me to do this without the aching heart I would have from your unhappiness.

Fuller eventually gave Lacey his blessing, but he never forgot her. At some point, he put a pair of photographs in his wallet. One was a picture of the two of them on a lawn with a dog, Fuller in his suit and tie, Lacey kneeling beside him in a ruffled blouse and striped skirt. Another snapshot showed her facing the camera, caught in the moment in her twenties when their lives had intersected. Fuller quietly tucked the photos away, and he kept them until the day he died.

Cynthia Lacey.

PART THREE

GREAT CIRCLES

1947–1967

One single picture—one frame—does not tell the story.
The single-frame picture of a caterpillar does not foretell
or imply the transformation of that creature, first, into the
chrysalis stage and, much later, into the butterfly phase of
its life. Nor does one picture of a butterfly tell the viewer
that the butterfly can fly.

—BUCKMINSTER FULLER, *SYNERGETICS*

SEVEN

GEODESICS

1947–1953

When we say "we think," our feedback has variable lags that may take overnight or months of time, for all we know. Because we want to understand—that is, to know the interrelationships of clusters of experiences—our first great discovery is dismissing irrelevancies, the macro-micro characteristics. Add: forgotten questions; different rates of feedback; persons' names; random questionings; the challenging set you would like to understand; our friend intuition.

—BUCKMINSTER FULLER, *SYNERGETICS*

In the fall of 1947 Fuller visited the industrial designer George Nelson in Quogue, New York. They had first met in the thirties, when Nelson was working as an associate editor at *Architectural Forum* and *Fortune*, and had remained in touch ever since. At his Manhattan studio, Nelson had emerged as a powerful figure in American modernism, and as the design director for the furniture company Herman Miller, he had recruited Charles and Ray Eames.

Fuller spent a pleasant weekend in Quogue with Nelson and the Eameses, whose lives had clear parallels to his own. At forty, Charles Eames was attractive to women and capable of drawing devoted followers, and he had left his first wife to marry Ray Kaiser, his ideal partner. Working together, they had perfected the famed plywood chair that bore their name, drawing on advanced materials and methods to solve problems that had frustrated Fuller in housing.

At Nelson's home, Fuller showed them his timeline of the elements and his drawings of the Dymaxion Car, which moved Charles to comment, "How blind people can be." Charles told the former Cynthia Lacey, now married to Dave Floyd, that the encounter had been "like lungs full of fresh air, and cool drinks of water, and a Finnish bath all at one time," and he thanked Fuller for his discourses on geometry: "Ray of course wanted to hear every word, but I am afraid that I could do little more than whip up a few additional tetrahedrons."

As Charles took snapshots of Fuller, they spoke of more practical matters. Fuller was out of money: he was selling his two small planes, and he would soon be forced to ask Allegra to cover her tuition at Bennington. He was trying to overcome his reputation for erratic behavior, but his only source of income was his doubtful arrangement with Mike Goldgar, and when Nelson offered him additional work as a scientific consultant, Fuller accepted gladly.

Fuller and Charles also discussed an invitation from the Institute of Design in Chicago, formed in 1937 to perpetuate the principles of the Bauhaus. It had obvious affinities with Fuller, who had once been considered as a candidate to succeed its founder, the late László Moholy-Nagy, with whom he had corresponded about the Dymaxion Bathroom. The current director, Serge Chermayeff, had assisted in exhibiting the bathroom at the Museum of Modern Art, and in July he had written to Fuller, "I had a long talk the other day with our mutual friend, Charlie Eames, here in Chicago, who tells me that there might be a possibility of your coming to lecture to us here at the institute."

Fuller's asking price of $500 was more than the school could afford, and although he might have compromised for another institution, his competitive streak made him reluctant to show vulnerability. In Quogue, Charles advised him to reconsider, which he eventually did, telling Chermayeff that he would participate for a lesser amount in "a public appearance through joint auspices." It was a fine distinction that allowed him to save face, and Chermayeff offered $100 for a talk in January 1948.

In his standard narrative of this period, Fuller portrayed himself as cut off from the world, as he concentrated on mastering the geometry behind the geodesic dome. His connections with design circles remained strong, however, and he was visibly active on numerous fronts. One involved the mathematician Ernst Straus, Albert Einstein's assistant at the Institute for Advanced Study in Princeton. Straus hinted that Einstein, who remembered Fuller, might agree to a meeting if he received a written synopsis to distinguish the request "from the hundreds he is approached by every day."

Fuller was excited enough to prepare six drafts of an introductory letter describing his ideas. He sent "a woeful bit" for mailing to Cynthia Floyd, who was handling his administrative work, but the result, finished shortly before he flew to Chicago, did little to separate him from the legions of cranks. If it ever reached Einstein, Fuller would have lost him by the opening of the second paragraph: "In all humility, I state that I seem to have articulated aright the 'open-sesame' to *a comprehensive system of sublime commensurability*."

At the Institute of Design, the reaction was more encouraging. On January 22, 1948, Fuller spoke from nine in the morning until ten at night. After a standing ovation, Serge Chermayeff took the podium to announce that he had extended Fuller an invitation to teach in the fall. In his upcoming class, Fuller would present a course on Comprehensive Design Technique, moving from "an unsolved social problem to demonstrated mechanical product in full scale operation."

Fuller's hopes of doing similar work for Mike Goldgar had generated just a handful of uninspiring assignments, including one for leather bands for men's hats, and he quit to join Nelson's office, which was far more promising. For a modest retainer, he focused on designing furniture, including an aluminum "seating tool" that turned out to be less than comfortable. After a sketching session late one evening with Fuller and Noguchi, Nelson returned to find a drawing of what became the firm's iconic ball clock, recalling years later, "I don't know to this day who cooked it up."

Nelson told Fuller that he didn't want to distract from his geometrical investigations, which became a private refuge as he struggled for professional traction. Fuller tended to fall back on less costly activities after his larger plans faltered, and for geometry, he could use materials from a hardware store. He built small models and performed calculations by hand, and the operational nature of the work was well suited to an autodidact who distrusted his own abilities at higher mathematics.

Fuller often worked at home until three in the morning, prompting his downstairs neighbors to pound on the ceiling to complain. His efforts climaxed in a feverish period of activity in the first four months of 1948. One thread focused on the packing of spheres within polyhedra, especially the cuboctahedron, which Fuller later called the vector equilibrium. In a series of detailed diagrams, he related its patterns to the structure of the atom, excitedly proclaiming that it was "what Archimedes sought and Pythagoreans and Kepler and Newton."

He also explored the transformations of one polyhedron into another. Years earlier, he had described the vector equilibrium as "a

The jitterbug transformation moves through the vector equilibrium, icosahedron, and octahedron as it rotates in either a left- or a right-hand contraction. Additional folding of the final stage produces the tetrahedron.

railroad station to the different units," and now he experimented with this concept in a tangible form. Building a vector equilibrium with rigid edges and rubber joints, Fuller found that if he pressed down on it, the entire construction twisted and contracted. As the six square faces were compressed, each became the equivalent of two triangles, which made twenty in all when combined with its eight triangular faces.

In other words, the vector equilibrium became an icosahedron. When he pushed the model down farther, the triangles nested into one another to produce an octahedron, and from there it collapsed into a flat structure—consisting of four smaller triangles—that could be folded into a tetrahedron. This sequence of polyhedra was an original discovery that delighted Fuller, who named it after a dance that he would have seen at the Savoy Ballroom in Harlem: the jitterbug.

Fuller's work on the jitterbug transformation and the arrangement of spheres in closest packing, which he related to the "identical axial rotations" of the vector equilibrium, led him to take a closer look at the patterns made by rotating polyhedra. When a polyhedron was spun

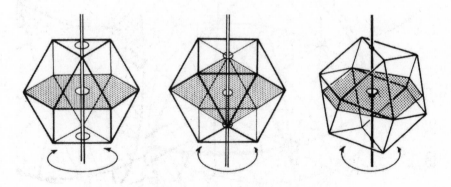

Rotation of the vector equilibrium around its faces, vertices, and edges.

on an axis, as a gambler might twirl a die between a thumb and fore-finger, its equator traced a circle perpendicular to its line of rotation. An axis could be formed by any two opposing points, and he focused on the circles that were generated by spinning a polyhedron around its faces, vertices, and edges.

The vector equilibrium, for instance, had fourteen faces, twelve vertices, and twenty-four edges, each of which could be paired with the opposite side to form an axis. Rotating the vector equilibrium around each axis made twenty-five different circles. Each polyhedron produced a unique rotational set—thirty-one circles for the icosahedron, thirteen for the octahedron—that could be mapped onto a sphere, where they intersected in a network of triangles.

As Fuller sketched diagrams of the great circles that resulted from rotating different polyhedra, he noticed that they resembled an architec-tural framework. On March 1, 1948, he did a rough drawing of what he called the "Atomic Buckalow": a curved lattice shell enclosure made out of a tensed system of flexible aluminum or plastic strips, which he

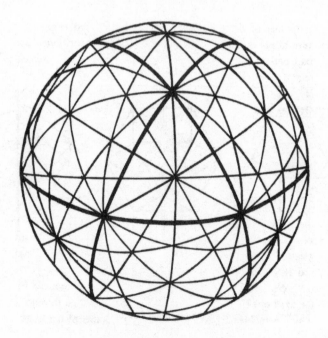

The twenty-five great circles produced by rotation of the vector equilibrium.

based on the "triangular intertension" of the twenty-five great circles of the vector equilibrium.

In this sketch lay the origins of the geodesic dome. It had arisen almost inadvertently from his geometric explorations, and he based it on a network of great circles for reasons that had less to do with engineering than with the aesthetics of his geometry. The approach yielded an overly complicated pattern of different triangles, but it had many points in its favor. As a hemisphere, it enclosed the maximum space within a given surface area; it was a clearspan structure that required no internal supports; and, best of all, it could be prototyped at a price that he could afford.

Fuller claimed retroactively that the dome had emerged from the mast and metal stays of the Wichita House, which he expanded into a triangulated "spherical mast" that doubled as both shell and support, and he never mentioned that he had kept what amounted to a miniature dome in his office for years. In Fuller's patent for the Dymaxion Map, he had recommended using a transparent plastic hemisphere to transfer

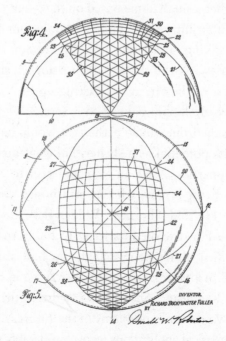

Illustration of a hemisphere with a three-way grid from the patent for the Dymaxion Map (1944).

information from a globe, and its grid of triangular crossings had been staring him in the face for most of the forties.

For the rest of his life, Fuller retold the story of the dome repeatedly, but he never revealed the pivotal decision that followed. He lacked the resources to start another business himself, and the dome might have remained another unrealized concept if he hadn't been in a position to show it to someone ideally placed to make it happen. Within a month of his first sketch, he brought the idea to George Nelson, who saw its value at once.

As a first approach, Nelson contacted Henry Sonnenberg, the founder of the window treatment company Hunter Douglas, which would be a useful source for a project that needed large quantities of flexible tension members. On April 6 Nelson confirmed to Fuller that Sonnenberg was interested in developing a structural system that utilized thin strips of metal, and he advised him to build a model in time for a meeting around the end of the month.

Working in Nelson's studio, Fuller used venetian blind strips to quickly put together a small dome based on the twenty-five great circles of the vector equilibrium. Although it was only four feet in diameter, it was visually striking, and Fuller proudly brought it to show off to Christopher Morley. At the threshold, he tripped, shattering it, and they spent the rest of the afternoon picking up the pieces. Fuller, undaunted, was already looking ahead to a version with the thirty-one circles of the icosahedron, which he thought would be stronger.

Apart from its potential as a shelter, Fuller wanted to promote its mathematical properties. After an informal visit to MIT, he went to Princeton to see Ernst Straus, who he still hoped would reintroduce him to Einstein. He told Thornton Wilder that the encounter went well, but Cynthia Floyd wrote in a letter to Fuller that she had heard otherwise: "[Straus] didn't believe that you had anything new and startling. He said that your lack of formal training in mathematics was probably a great handicap." She also warned him that his inability to accept criticism would cause problems in the future.

The dome's practical applications were considerably more compelling. When Serge Chermayeff proposed that the class at the Institute of Design develop "a greenhouse enclosure, a living garden," Fuller responded with an uncredited reworking of the "outside-in" dome from the Museum of Modern Art. It was a transparent dome with a world map on its inside surface, positioned to allow its occupants to see the stars at the correct zenith points, and it could even be placed above a hemispherical pool to show the southern constellations.

His work on his "private sky" was interrupted by a fateful call from the architect Bertrand Goldberg, asking if Fuller would be interested in teaching at the summer session of the experimental Black Mountain College outside Asheville, North Carolina. Goldberg's schedule had forced him to decline, and Leland Atwood, one of the architect's former associates, had suggested Fuller as a replacement. The call was soon followed by another from Josef Albers, the college's head of arts, who said they could pay him twenty-five dollars a week.

It was less than what Fuller supposedly charged, and it would take him out of New York at a crucial time, but Allegra urged him to accept. When he agreed, it was partly out of financial necessity, but he also sensed that it could introduce the same random element in his life that he had discovered at Romany Marie's. The college's founders, John Andrew Rice and Theodore Dreier, had conceived of Black Mountain as a program to build complete individuals, with students granted equal voices with teachers as they engaged in meaningful labor. In practice, a lack of structure led to cults forming around figures such as Rice, who compared its dynamics to "the Oedipus complex."

As a result, Black Mountain was one of the few places where Fuller fell into an existing template. In his 1973 history of the college, the author Martin Duberman noted that the summer institute, which was designed to raise funds, differed in fundamental respects from the academic year: "The summer artists generally viewed Black Mountain simply as a nice spot in the country, a pleasant change of pace, an agreeable refuge. . . . The summer people weren't trying to make a life at Black Mountain; they were trying to put together a concert or an art show."

It was an ephemeral program in every sense of the word, and it gave Fuller what he needed at exactly the right moment. Leaving his family behind yet again, he loaded his models into a trailer and arrived on July 12, 1948, his fifty-third birthday. In a remarkable coincidence, it was the exact date that he had specified a decade earlier for the fulfillment of his list of predictions in *Nine Chains to the Moon*.

When Fuller showed up, the summer session had been under way for two weeks, and the campus in the meadowlands of the Blue Ridge Mountains was filled with the shrill buzzing of cicadas. Emerging from his car, Fuller introduced himself to the onlookers with his full name, but he added reassuringly, "Call me Bucky."

The students helped the new instructor to unload his equipment. One of them, an aspiring twenty-year-old artist named Kenneth Snelson, was studying painting with Albers, who had asked him to help Fuller prepare for a talk that night. Snelson had expected little more than a few miniature house models, and upon entering the trailer, he was astonished to find "cardboard polyhedra of all shapes and sizes, spheres made out of great circles, metal-band constructions, plastic triangular items, and fragile globs of marbles glued together."

Fuller gave his first lecture after dinner. At first, he stood before his listeners for a long time with his eyes closed, and Snelson thought that he seemed "unknown, humble, even a bit pathetic." The instructors in the audience included the abstract expressionist painter Willem de Kooning and his wife and creative partner, Elaine, who murmured, "He looks stuffy."

"Wait until he opens his mouth," Willem replied. At first, Fuller stammered slightly, but he rapidly built momentum. In his talk, he said that his listeners—there were around seventy students that summer— had the power to save mankind, but his most convincing arguments were visual. Taking a small dome of venetian blinds, which was bound into a tight package, Fuller removed its rubber bands and tossed it gently into the air, demonstrating how it sprang into shape at once.

By the time he finished, it was after midnight, and they could hear the tree frogs croaking outside. Elaine was entranced by the jitterbug, his praise of "the numbers nine and three, the circle, the triangle, the tetrahedron, and the sphere," and "his complex theories of ecology, engineering, and technology." When he sketched a diagram on the blackboard of "our old friend, the hypotenuse," she was won over for good, whispering to her husband that she'd decided to attend Fuller's classes. Willem responded, "I knew you would."

Fuller struck the German mathematician Max Dehn as a charlatan, but the other teachers were impressed. One was Albers, a leading member of the Bauhaus with an interest in efficient materials and the properties of folded paper, whom Philip Johnson had recruited for the college along with his wife, Anni, the groundbreaking textile artist. Fuller admired Theodore Dreier, whose aunt Katherine had been one of his art world friends, as "a great idealist," and he discussed geometry with the Moscow-born Natasha Goldowski, a former ballerina who taught physics and chemistry. Their letters afterward hinted at more than casual affection, with Goldowski telling Fuller, "I love you very much."

The de Koonings found Fuller fascinating. Willem rode with him on supply runs into town, and the artist instantly solved a cube puzzle that had stumped Fuller's students for more than an hour by looking "for the least logical way." Elaine thought that Fuller had "the eyes of a visionary, a saint," while the American sculptor Richard Lippold compared his talks to hearing "Zoroaster speaking Islamic."

His other colleagues included stage director Arthur Penn, there to teach method acting; composer John Cage; and choreographer Merce Cunningham. The latter two men were romantic partners and collaborators—Cage was in his thirties, Cunningham several years younger—and both were instantly drawn to Fuller. Cage, who had briefly had an affair with Philip Johnson, was best known at the college for playing the piano with the window open, and he was years away from the experiments with silence and chance that would make him famous.

The three of them often met for breakfast under the trees, where they joked about starting a caravan school. Cage loved Fuller's "liveliness and optimism and generosity," and, he recalled, "From the beginning of my knowing him, I had, as he did, confidence in his plan to make life on earth a success for everyone." Cunningham, who had worked with Noguchi, was struck by Fuller's observations on space: "Oh, isn't that marvelous—that's what I think of dance." His lectures on geometry, during which he dramatically opened a curtain to reveal his polyhedra, reminded Cunningham of the Wizard of Oz, and he spoke fondly of "Bucky Fuller and his magic show."

In his attic room, where he set up his models, Fuller prepared for the construction of a real dome. Before he departed for North Carolina, Henry Sonnenberg's partner at Hunter Douglas, Joe Hunter, had sent Fuller six rolls of a new kind of aluminum blind strip, which George Nelson thought would be enough "to rehouse Black Mountain." With thousands of yards of material at his disposal, Fuller decided on a dome based on the great circles of the icosahedron, measuring forty-eight feet in diameter and weighing less than fifty pounds. The complicated structure had to be measured precisely, and he worked on the calculations late into the night, as students undertook the tedious job of punching holes for aircraft bolts.

At last, they assembled the dome in a field on a slightly rainy day, with observers under umbrellas watching from a nearby bluff. As Fuller's assistants, including Penn and Elaine de Kooning, bolted together the color-coded segments, the others waited for the structure to ascend.

The "supine dome" under construction at Black Mountain by Fuller (far right) in the summer of 1948. A model of the thirty-one great circles of the icosahedron is visible on the ground.

Instead, it sat on the ground in what the student Ruth Asawa later compared to "a giant's plate of spaghetti." Going up on a roof to study it from above, Fuller claimed that it formed the exact pattern of the chromosomes of a fruit fly as seen under a microscope, and everyone politely agreed.

Fuller said later that he had meant to fail in order to discover the "critical point" at which a dome would collapse. Penn backed him up, saying, "It was predicted to fall down." In reality, it was so much larger than any size that the aluminum could plausibly support that it provided no useful information. According to Elaine de Kooning, who jokingly dubbed it the "supine dome," they lacked the material to double the strips, which would have increased their strength, but Fuller felt that he had to proceed. "Let's put it together anyhow," he said. "One never knows the ways of the Almighty."

Cage was reminded of his father, who was also an inventor, and Fuller told him that he was delighted by the outcome, explaining, "I only learn what to do when I have failures." In subsequent accounts, Fuller spun it into an outright success, saying that he managed to get a section to stand by taping wooden struts to the blinds but ran out of supplies before it could be salvaged. Elsewhere, he indicated that when the dome "gently collapsed" as he was about to complete it, he miraculously restored it in full view of the crowd. No evidence of either solution survived.

Despite this setback, Fuller remained popular on campus. Accompanying Elaine de Kooning to a local watering hole called Peek's Tavern, he diagrammed the steps of the folk dancers there, using the icosahedral notation of the Austrian choreographer Rudolf Laban, which Cage and Cunningham encouraged him to explore for mapping patterns of forces. Another lead resulted from a visit by James Fitzgibbon, an architectural professor at North Carolina State College in Raleigh, who raised the possibility of Fuller's presenting a guest lectureship next year.

Fuller captivated most of the students, with many adopting his experimental sleep schedule and joining him for hikes in the mountains. His admirers included Paul Williams and Albert Lanier, both future architects, and the twenty-two-year-old Ruth Asawa, whose coiled wire sculptures would one day be hailed as masterpieces. As the campus barber, Asawa gave Fuller a haircut—he held his thumb and finger apart to show that he wanted it trimmed to a quarter inch—and was rewarded with a pole made of red and yellow blinds.

His most devoted fan was Kenneth Snelson, who reportedly said of the jitterbug, "I wish you hadn't discovered this first," although he also noticed how Fuller would chuckle nervously "to skip over a discrepancy or a claim short on evidence." The more skeptical students joked that Fuller had invented the tetrahedron, and one confessed, "When I listen to Bucky talk, I feel I've got to go out and save the world. Then when I go outside, I realize I don't know how."

Apart from the supine dome, the other high point of the summer came through Cage, who had been performing the music of the French minimalist composer Erik Satie. Cage followed Satie in disparaging Beethoven, which led to a mock duel with the harpsichordist Erwin Bodky, with Fuller serving as the referee as students and teachers fought using crepes and sausages.

Cage's performances culminated in Satie's 1913 absurdist play *The Ruse of Medusa*, in which musical interludes alternated with short scenes translated for the occasion by the poet Mary Caroline Richards. Lippold worked on the costumes, the de Koonings on sets, and Asawa and Lanier on a throne of venetian blinds, while the cast featured Fuller as the elderly Baron Medusa, Elaine as his daughter, and Cunningham as his mechanical monkey.

Arthur Penn, recruited as a play doctor, staged the action so that it spilled out into the audience, and he encouraged the actors to depart from the script. To his surprise, he found that Fuller—who provided the Baron's magnifying glass and thermometer from his personal possessions—had trouble acting in public. The rehearsals were open to the entire campus, and Fuller confessed to the director, "I'm afraid of making a damn fool of myself."

Penn, who was just twenty-five years old, gave him exercises to break down his mental barriers: "We skipped around, did giddy things, laughed artificially, and rolled on the floor." Fuller enjoyed the finished production, in which he wore striped pants and a top hat, and in the second performance, he allowed himself to improvise. He said later that the lessons with Penn shaped his style of public speaking, encouraging him to embrace a more theatrical approach. The experience, he concluded, "let me learn to be myself on the stage."

When Fuller left at the end of August—he narrowly missed the arrival of an art student named Robert Rauschenberg—he had made friends whose careers would become as legendary as his own. Cage proposed that he design a studio in New York for a performance of *The Cantos* by the poet Ezra Pound, while Snelson wrote that he was using energetic geometry in his sculptures, including a "tinker toy" model and a cube made out of wire and thread.

Fuller replied with what he called the most important letter that he had ever written. He told Snelson that the jitterbug was proof of the unified field theory for which "Einstein has been searching the last quarter century," and he closed with his conviction "that this communication must eventually bear important fruit." He forwarded part of their correspondence to Cynthia Floyd, who thought that he had found "a wonderful young protégé," while Snelson was left dreaming that he "might turn out to be a reincarnated, young Buckminster Fuller."

In October Fuller headed to Chicago to teach at the Institute of Design for the 1948–49 academic year. Compared with the myth of Black Mountain, his time there would fade from his standard biography, in part because it complicated the question of credit. If Black Mountain was a story about teachers, with John Cage and the de Koonings as his peers, the Institute of Design was about his debt to his students. To recruit a permanent group of followers, he put the summer's lessons to use, plunging into his role as eagerly as though he had internalized Baron Medusa.

He wound up living in an Airstream trailer—a true example of mobility—in a parking lot in the Chicago Loop. Before Fuller's arrival, Serge Chermayeff had asked a carpenter to build drawing boards for a classroom in the armory basement, only to have the new instructor immediately push them against the walls to use as shelves. Occasionally, he used them in other ways, as the critic Peter Blake, a protégé of Philip Johnson, learned when he asked to meet Fuller. Advising him to try after lunch, Chermayeff said enigmatically, "He seems to be conducting some kind of experiment on himself."

Blake waited in the basement until Fuller appeared. On seeing the young stranger, Fuller announced without any preamble, "What I have just discovered is that bebop has the same beat as the new mathematical shorthand I have been working on." Leaping up onto one of the drafting tables, Fuller started to tap his feet and snap his fingers in time, calling out what Blake remembered as "an incomprehensible sequence of numbers." Fuller, still dancing, called down to the bewildered critic, "You see what I mean, don't you, dear boy?"

His triumphant return to Chicago was scored by bebop. Like many of the instructors, Fuller became obsessed by this novel form of jazz, which was based on breakneck tempos and complicated syncopation. He referred to himself to his followers as "your bebop representative," and he made the rounds of jazz clubs with Chermayeff and designer Robert Brownjohn, the director's talented assistant, who later became famous for his title sequences for the James Bond movies.

Fuller often assumed the role of a harmless eccentric, and Blake was amused when he flung himself downstairs to practice his football moves: "The next time I saw Bucky, he was lightly bandaged, and beaming." His colleagues spoke about him with an affection that bordered on condescension, but he was an overwhelming presence for impressionable admirers such as Brownjohn, whose insecurities led him to become addicted to heroin, a friend said, "because he was mixing with these gods at the Institute of Design—people like Buckminster Fuller."

In general, Fuller was too busy to notice. He gave six lectures a week to his students, whose models filled the basement studio that Chermayeff compared to "a Merlin's cave." Ludwig Mies van der Rohe, then the architectural chair of the Illinois Institute of Technology, attended his talks, and Fuller found a kindred spirit in the architect Konrad Wachsmann, another German modernist, who had designed a "packaged house" with Walter Gropius, although Fuller was careful to downplay their similarities.

Like many magicians, Fuller was only as good as his assistants, and at the Institute of Design, he found protégés who possessed the skills and imagination that he needed, notably Jeffrey Lindsay, Donald Richter, and a Black student named Harold Young. He promptly commenced a technical program that few at Black Mountain would have been capable of pursuing, based on the separation of the house into a shell and a utility core, which Fuller still treated as a pair.

In the classroom, he placed equal emphasis on the dome and on the utilities of what he called the Autonomous Living Package, drawing on the Mechanical Wing trailer that he had published in *Architectural Forum*. Instead of a camping trip, the exercise now took place against the backdrop of a wartime evacuation, with Fuller challenging his students to fit everything that a household needed into one portable unit, including a "kenning space" with television and maps.

Another element, a bathing device known as the fog gun, had appeared in his earliest housing plans. Fuller said that his idea of washing with atomized water had been inspired by his patrols in the navy, in which he had noticed that the fog would wipe the grease from his face. Taking photos to show how dirt clung to pores, he sent students to interview dermatologists, who allegedly agreed that "the worst thing you could have for your skin is soap." A prototype made from an air compressor delivered a pressure bath for an hour using a pint of water, and Fuller hoped to combine this with a packaging toilet to minimize plumbing.

Although the utility core took up much of his attention, the most important research was done on the shell, which his students developed in a workshop on Kedzie Avenue that they shared with a gambling den. One attempt consisted of a small dome made of venetian blinds covered in cement, while another was a structure of aluminum pans that were shaped into wedges, assembled into a hemisphere, and cinched with a strap around the equator.

The most successful effort was the Necklace Dome. Instead of strips that followed the arcs of great circles, it used short chords of aluminum tubes threaded with aircraft cable, which were connected with hubs like Tinkertoys. After all the components were strung loosely together, the turnbuckles at the base were tightened with a wrench, raising it to its full height. At ten feet across, it was smaller than its Black Mountain counterpart—and thus considerably more practical—but Fuller was already imagining a version that covered eight thousand square feet.

In January 1949 he gave a talk at the Illinois Institute of Technology that linked synergy—the behavior of a whole unpredicted by its parts—with industrialization, which had benefited from the unexpected properties of alloys. He also related his work to the Cold War. Since America and the Soviet Union would be evenly matched in combat, he advocated a strategy of decentralization based on the dome, predicting, "The winner will be that side which has the most effective defense."

Instead of privately funding a complete prototype for the military, Fuller decided to present a rudimentary concept to secure the green light for additional development, allowing him to avoid the outside investments that had produced so many disputes in the past. Early in 1949 he reached out to an old friend, Reginald E. Gillmor of the Sperry Corporation, for an introduction to the Army Air Forces, and they arranged for a demonstration in Washington.

In advance of the meeting, Fuller drove with Anne to North Carolina State College, where James Fitzgibbon had invited him to conduct a seminar in March. His ideas tied into existing dispersal studies at Raleigh, but other faculty members thought that his lectures were difficult to follow, and even the artist Manuel Bromberg, who later became a colleague, remembered of Fuller, "He seemed to be a thinker and con artist at the same time."

From there Fuller headed for the Pentagon. With Dave and Cynthia Floyd in attendance, he installed a fourteen-foot Necklace Dome in the garden, assisted only by "a civilian stranger and a GI passerby." If he had known that it would be outdoors, he added, he could have built one that was fifty feet in diameter. Officers whose windows overlooked the site sent subordinates to investigate, and an engineer reportedly remarked that it was "as good as a tent."

Fuller had also brought a model by student Harold Young of the Garden of Eden, also known as the Skybreak House, with greenery enclosed by a transparent skin, based on Serge Chermayeff's idea of connecting a shelter with nature. His military contacts were more interested in the Necklace Dome, which they discussed testing in a cold hangar to evaluate it for deployment in polar regions. Fuller had even loftier goals in mind, and on returning to Chicago, he asked his students to work on a hydraulic dome that could be "shot to the moon" to open under its own power.

In April he returned to the theme of dispersal at a lecture at the University of Michigan, arguing that America's defensive advantage resided in its ability "to dodge widely and without loss of poise." For most of history, Fuller said, mankind had limited itself to linear solutions, such as a highway or railroad, which he compared to a pipeline for the commuter: "His car or his train is a section of pipe surrounding him." The future demanded an omnidirectional perspective that would usher humanity into the trackless phase.

Fuller's answer was the dome, "a super-camping structure" designed to eliminate the outdated notion of the house itself by addressing shelter as a purely mechanical problem. If it seemed uncomfortably radical, he advised his listeners to see the accelerating rate of change as a return to fundamentals: "We need only revolve our charts to ninety degrees of angle, so that we may see the curves descending precipitously from the old heights of ignorance and abnormality."

His tone had grown more prophetic, but he maintained casually affectionate relationships with acquaintances from Black Mountain. In New York, he reunited with Cage and Cunningham for an evening that included Le Corbusier and Maya Deren, as well as the mythologist Joseph Campbell and his wife, the dancer and choreographer Jean Erdman. Ruth Asawa wrote to him of a visit to the Wichita House: "I felt as though I was spreading the wings of a dragonfly to see what made it move." When she became engaged to Albert Lanier, Fuller designed her a silver ring modeled on the vector equilibrium.

Fuller was closest with Kenneth Snelson, who saw him in Forest Hills and Chicago. Snelson had moved back home to Oregon, where he built models out of drinking straws in his basement, and he mailed two small constructions of cardboard and thread to Fuller. He clearly looked up to his former instructor, and he even asked Fuller for advice on sex: "How did you solve this problem of the ever-present demon of desire?"

It was a question that Fuller had yet to answer to his own satisfaction, and he was distracted by word of some drama back at Black Mountain. In September the science lab had burned down, and tensions were rising over dwindling funds, leading Natasha Goldowski to describe the college to Fuller as "a concentration camp." Albers and Dreier both resigned, and Goldowski, remaining as secretary, was tasked with bringing the financial situation under control.

The summer session had always been a reliable way of raising money, and Goldowski asked Fuller, who she hoped would build them a new lab, to serve as dean. Fuller agreed to return for six weeks, and he recruited instructors from the Institute of Design, along with fourteen students, including Richter, Lindsay, Young, and Masato Nakagawa, who had won a Silver Star for heroism during the war. Snelson wrote gladly, "I shall be there if it means riding my bicycle all the way."

Fuller's followers in Chicago, who would be in charge of building a dome at Black Mountain, had formed a community of their own. Its members called themselves Spheres Inc., and Fuller charged one dollar for a license that allowed the beneficiary to use the title "Student Dymaxion Designer." The team headed to Asheville, where they soon became aware of a cultural divide between their rigorous approach to design and the "somewhat escapist tradition of a mountain school."

At the end of the spring 1949 semester, Fuller and Anne drove down to North Carolina with a trailer full of models. Fuller would be the dean for around fifty students, with a curriculum that he outlined in an essay titled "The Comprehensive Designer." In the tradition of Black Mountain, he focused on the creation of an exceptional individual, "the comprehensive harvester of the potentials of the realm," who would play the same role in the future as the architect did in feudal times.

"Man has now completed the plumbing and has installed all the valves to turn on infinite cosmic wealth," Fuller said, and its realization called for generalists who combined the strengths of artists, inventors, and economists. In the past, such thinkers had clashed with capitalists, requiring them to achieve their goals "by indirection and progressive disassociations," but politicians would soon be forced to empower designers to increase the standard of living. Fuller's own example was more inspiring than any one project, but it required physical artifacts, and it was no coincidence that his myth appeared in its mature form at the same time as its most potent symbol.

On July 15, 1949, Fuller supervised the construction of the Necklace Dome at the north end of the studies building, below his studio window. Its latest feature, a transparent cover of inflatable plastic, transformed it into a gossamer inverted bowl. As Nakagawa took pictures, the Chicago group unfolded the frame, raised it using a temporary mast, and tightened the turnbuckles that slowly expanded it into a complete hemisphere.

After three students lifted it on their fingers to demonstrate its lightness, it was fixed to a ring of posts. Fuller and his team hung from it like children on the monkey bars, testing its strength, until it held

Students lifting the Necklace Dome at Black Mountain (1949). Photograph by Masato Nakagawa.

fourteen of them at once. A wooden platform was suspended inside, allowing an observer to stand upright to look through the opening in the center. Fuller lay down on it, gazing up at the sky, and switched to a chair as the plastic skin was filled with air.

Once the covering was ready, three women, including Anne, stood on the inner platform as the dome was hoisted upward. Perching on a student's shoulders, Fuller emerged up to his waist from the central aperture, beaming upward for several photos taken from a neighboring roof. Just as they finished, a thunderstorm broke, and they sheltered in the dome, looking out at the mountains in the rain.

At dinner that night, Fuller's achievement received a standing ovation, and the dome stood in place until September. Students tossed stones at the plastic to simulate hail, and Fuller found that its interior was cooler than the outside, in an apparent example of the circulation effect that he had noted in the Dymaxion Deployment Unit. The only setback was an unsuccessful effort by Jeffrey Lindsay to make fiberglass panels, which failed to harden properly and were thrown into a ravine.

Early X Piece by Kenneth Snelson. This version of the lost original was re-created in 1959.

It was a triumph of engineering, but the most significant development that summer came from another direction entirely. Kenneth Snelson had traveled separately from the others, and he arrived at Black Mountain one sweltering day, encountering Fuller on the dirt path between the dining hall and the studies building. They shook hands, and Snelson asked if they could discuss his latest construction, which he was carrying in a cardboard box.

Inspired by a children's toy and the mobiles of Alexander Calder, Snelson had been building sculptures with wire elements balanced on a vertical framework. He wanted to eliminate the balancing members entirely, keeping it upright through tension alone, and came up with a pair of plywood pieces, each shaped like a cross, that he connected with string. When the tensile components were uniformly tightened, the compression units seemed to float without visible support.

Snelson had described it in letters to Fuller, but nothing compared to seeing the result in person. "When I showed him the sculpture, it was clear from his reaction that he hadn't understood it from the photos I had sent," Snelson recalled. "He was quite struck with it, holding it in his hands, turning it over, studying it for a very long moment. He then asked if I might allow him to keep it."

Although Snelson had intended only to show it to Fuller, he agreed, relieved that his teacher wasn't upset that he had used energetic geometry for a mere work of art. Fuller gave no sign that anything important had occurred, but the next day, he informed Snelson that the piece should have been based on a column of stacked tetrahedra. Snelson had already tried this approach with an earlier sculpture, and he had even sent Fuller pictures, but he was willing to do it again.

At a Woolworth's five-and-dime store in Asheville, Snelson bought a dozen telescoping curtain rods to build a mast out of tetrahedra, which he finished the following day. Fuller loved the tetrahedral version, which stood five feet high, and asked if he could stand beside it for a picture. "As I photographed him with it, I felt a numbing inside from what was happening, but I was not yet distrustful," Snelson remembered. "After all, Bucky knew as well as I did whose idea it was—and, besides, teachers don't go around stealing students' ideas and claiming them as their own."

Fuller's account was very different. During his lectures that summer, he presented the mast as his own design, and he subsequently insisted that Snelson had found only "a special case demonstration of a generalized principle for which I had been seeking." In fact, even if Fuller had inspired the sculpture, which would later be known as *Early X Piece*, it was a novel development that had been enabled by distance, and it had been influenced to a considerable extent by Calder.

Snelson always thought that it had more to do with art than engineering, while Fuller would make it central to his life's work. "No one else in the world but I could have seen the significance I saw in what you showed me," Fuller wrote long afterward to Snelson, but it was equally true that no one else could have made it, and although the original was soon lost—Fuller said that someone stole it from his room—its reverberations would be felt for decades to come.

The rest of the summer passed quickly. Fuller lectured on geometry and prototyping, tinkered with polyhedral models based on Snelson's discovery, and asked a guest to read aloud one of his poems at a party—it turned out to be two hundred pages long, and several attendees dozed off. Anne had tea with the students, whom she delighted in telling that she had designed a cottage in Connecticut long before her husband had ever built a house, while Snelson busied himself with a structure of rotating model airplane wheels, which Fuller quietly copied years later.

Tensions arose between the Black Mountain circle and the group dismissively known as "Christ's Dymaxion disciples," and by the summer's end, Natasha Goldowski had stopped speaking to Fuller. Anne went home early, while Fuller lingered to make his farewells to a school that would play an outsized role in his legend. One student recalled of his departure, "I can see him still, in the back of that old, open convertible, waving good-bye to us with one bare foot."

Fuller's clownish behavior, with its shades of Baron Medusa, concealed his true intentions toward his students. In the essay "Total Thinking," which he finished at Black Mountain, he evoked the philosopher Alfred Korzybski, writing that man was the only animal to participate "in the selective mutations and accelerations of his own evolution." Actively directing this process demanded an awareness of historical trends, and he advised designers to learn to use statistics to plot "not informative but provocative curves."

Once a trend had been identified, it could be guided by the comprehensive designer, or synergist. In practice, large projects required more than one person, along with independence from fixed institutions, and Fuller achieved this by combining two developments. One was a postwar rise in college enrollment that offered an abundance of students, many of whom were veterans with practical experience. The other was the dome, which embodied his ideas in a form that he could prototype on his own. Fuller would fail again, but there would never be any question of control.

At that point, no one had expressed any concerns over who would receive credit for concepts originated by students or how the effort might change Fuller himself. The only person to sense these issues was the poet Charles Olson, the other dominant figure at Black Mountain that year. Prior to the start of the summer session, he had written to Fuller to ask him to take the lead in a play that he hoped to stage. It was titled *Kyklops II*, and Olson wanted Fuller to play Odysseus.

Olson's darker counterpart to *The Ruse of Medusa* was never performed, but its identification of Fuller with Odysseus was remarkably shrewd, and the part was manifestly shaped with him in mind. In the opening scene, Eurylochus, the ship's second-in-command, describes his captain: "Every night he sleeps encyclopedic dreams." When Odysseus appears, he tells his crew, "This is not a trip to gather beauty. I take back only strength, bold form, lines that boldly form a sphere."

Not surprisingly, Olson failed to get along with Fuller, whom he dismissed in a letter to the poet Robert Creeley as "that filthiest of all the modern design filthiers." Some observers thought that they were contending for the soul of Black Mountain, although Fuller was already turning his attention elsewhere, while Olson would long be associated with the college. Nevertheless, the clash between their philosophies was undeniable. After they argued over industrialization one day, Olson threw Fuller out of his house with an unanswerable question: "In what sense does any extrapolation of me beyond my fingernails add a fucking thing to me as a man?"

On August 16, 1949, Myron Goldsmith, an architect in the Chicago office of Ludwig Mies van der Rohe, sent a letter to Fuller to follow up on a recent conversation. In a description of the photograph that he had enclosed, Goldsmith wrote, "As nearly as I can tell, it is a photo of the dome of a German planetarium. I was able to find another photo which clarified the pattern somewhat." It was evidently the Zeiss Planetarium in Berlin, which featured a domed roof derived from a design that the engineer Walther Bauersfeld had demonstrated in the German city of Jena in 1922.

Bauersfeld had developed a patented method of geodesic construction that predated Fuller's work by more than a quarter century. The Jena planetarium was twenty-five meters across, with a thin concrete shell over an iron framework, using a triangulated pattern based on the icosahedron. Bauersfeld had been more interested in the projection system that displayed the stars inside the hemisphere than in the geometry of the dome itself, which Walter Gropius and László Moholy-Nagy had studied while it was only partially complete.

Whether Fuller had known about Bauersfeld's work was a question that would persist for decades, but the evidence indicated that he was unaware of it until he received Goldsmith's letter. Fuller's dome had arisen independently from his interest in great circles, and it had yet to occur to him to utilize the icosahedron, which he had already rejected as a cartographic projection. It would have been simpler and sturdier than the designs that he had used so far, and he would have been unlikely to deliberately ignore such an obvious solution.

All the same, it complicated his claim of discovery. Fuller wrote much later to Snelson, "If the Zeiss engineer had in 1922 anticipated geodesic domes' unlimited spanning capability, [Hermann] Göring would have used geodesic domes for his Luftwaffe hangars during World War II. . . . No one was the inventor of geodesic structure. I was the conceiver of the engineering theory which showed that they had no limit of clear span enclosing capability and of their practical and economic realizability."

Ultimately, the dome was less innovative as a structure than as the organizing principle that Fuller had sought since Wichita. In the privately circulated essay "Universal Requirements for a Dwelling Advantage," in which he revised a list of the basic functions of shelter that he had been updating for two decades, he furnished an entire manual for a lean start-up. An invention, Fuller wrote, began with one person, whom he advised to keep a detailed journal, while the next stage called for associates to build models and study production curves. He described his hypothetical company all the way to public relations, which was guided by rules that he would follow only erratically:

RULE I: NEVER SHOW HALF-FINISHED WORK.

- General magnitude of product, production, distribution. But no particulars that will compromise latitude of scientific design. . . .
- Publicize the "facts," i.e., the number of steps before "consumer realization."
- Understate all advantage.
- Never seek publicity.
- Have prepared releases for publisher requests when "facts" ripe.

His most lasting invention was a new kind of company, and while it might not have been suited for every enterprise, it was perfect for geodesics. Fuller stated later that a dwelling industry would have cost $1 billion in 1927; by 1946, that figure would have fallen to $10 million. Now it took the form of a business that lacked "any tool-up expense whatsoever," with components that could be purchased "spontaneously" at a five-and-dime store.

It still required a minimum amount of funding, which Fuller found close to home. As one story had it, he had discussed his needs with Anne as they drove to North Carolina State in the spring. "I am absolutely confident about the geodesic thing now," Fuller said. "But I have exhausted all our money, and we are not getting enough from this lecturing to move very far."

When Anne asked how much he needed, Fuller said that it was around $30,000. Anne allegedly replied at once, "Then go to me. I can lend it to you." According to Fuller, she had recently inherited a block of IBM shares, which she cashed out to give her husband a loan. Years later, Fuller told Alden Hatch proudly, "Had she kept it, she would have been a millionairess now."

Elsewhere, Fuller said that the start-up cost was closer to $12,000, including the "overhead" of giving up other income, and that he secured it "by borrowing on life insurance, and investing lecture and seminar fees." More to the point, neither version revealed that most of the funding was provided by his younger associates. Jeffrey Lindsay, for instance, put up a stunning $32,000.

Like the housing industry in *4D Time Lock*, his operation could remain decentralized as long as he could remotely control his followers, who sorted themselves into two divisions of the Fuller Research Foundation. The teams were encouraged to compete, and he kept them in a state of continuous tension, which the critic Mark Wigley later called "an interlaced network to study interlaced networks."

One was located in a basement workshop on East Ontario Street in Chicago, where the group concentrated on a dome with a dimpled "involute" shell. Snelson left after growing upset over Fuller's appropriation of his concepts, and its core membership came to consist of twelve enthusiasts from the Institute of Design, including Harold Young, Donald Richter, and a woman named Lee A. Hukar, who had been Serge Chermayeff's secretary. Another former student, Harold Cohen, declined to participate. "I knew that if I went to work for him," Cohen said decades afterward, "he would crucify me."

A second branch was established by Lindsay in his family's row house on McTavish Street in Montreal. After the back room overflowed with models, he expanded into the coach house, which served the same function as the garages of a later generation of start-ups. Public relations was handled by an advertising man named Ted Pope, but the engineering was driven by Lindsay alone.

A rivalry soon developed between the two groups. For example, in September 1949 Lindsay shared with Hukar an insightful strategy for rapid prototyping: "I have determined that the thing to do is produce 'structure' as quickly as possible, washing out obsolete models quickly, and arrive at a structure for the public by 'make and break' rather than by trying to solve it all on paper." But when Hukar subsequently wrote to Fuller, she was dismissive: "You will doubtless understand our complete lack of comprehension of what the hell he's getting at."

Like many founders, Fuller cultivated an emotional bond with his supporters, but he could turn on them without warning. After his car was damaged by a student named Ysidore Martinez, Fuller shut him out, and he admonished Hukar after hearing that she had aired some of her concerns about the group's progress with Young: "Have you lost faith in my understanding of any of your doubts or hopes?"

Hukar replied that she was "thoroughly chastened," and Fuller would resort to emotional manipulation again. When the Chicago branch innocently prepared to raise money with an auction, he objected that it fed into his reputation as a "promoter," and he scolded them about their financial misgivings:

> It is unwise to present "hooded Klan voices," which say to me "We will pay your goddam bills with our money," and it is incongruous to furnish such voices with proffered love. I do not believe you bear me ill will. If any do, they have been acting as hypocrites. . . . It has not been any of your experience that I have been parasitic upon you. On the contrary, you came to me to ask that I share with you. I have given you all there is of me and gladly, not for sentimental reasons but because I have received the impression that each one has been genuine in voluntarily joining up. . . . Please enlighten me as to whether this belief is true.

For now, the change in tone went unrecognized. To retain credit and control, Fuller was regressing to the tendencies that had caused conflicts with his previous business partners, but now they were directed toward his younger admirers, who wanted nothing more than to earn his approval.

Like the first round of hires at future start-ups, they took consolation in being part of a movement that they truly believed could change the world, even as Fuller turned more of his attention to students at Harvard and MIT. In December the Canadian group attended a crowded event for Fuller at the University of Toronto, filling Lindsay with a heightened sense of devotion: "Bucky knows each one of us like we do not even know ourselves."

If Don Richter was Fuller's most pragmatic disciple, Lindsay was perhaps the most gifted. At twenty-five, he was a skilled sculptor and photographer who had trained as a navigator during the war. His family's wealth had allowed him to underwrite the Necklace Dome, and he had arranged a meeting that year between Fuller and the British metallurgist Cyril Stanley Smith, a former member of the Manhattan Project, who became an enthusiastic supporter of energetic geometry.

In Canada, Lindsay had access to aluminum, not plastics, forcing him to be strategic about his designs. Like Fuller, he had wavered between approaching the military or "peaceful types," and he found a client in the Scottish glaciologist Patrick Baird, who needed a structure to protect equipment on Baffin Island. Lindsay developed a metal "firecracker umbrella" with a nylon shell, which he and Fuller installed for testing on Mount Washington in New Hampshire.

Larger domes presented a challenge, partly because their structure was still based on great circles, requiring numerous hub and spoke types. Although Fuller was reluctant to change it, he finally switched to an icosahedral design. To be sturdy enough to stand, the framework of any substantial dome had to be subdivided into smaller triangles, and he decided to divide the faces of the icosahedron into the three-way grid of the Dymaxion Map, even as he continued to overlook the invisible gaps, or "windows," where the lines failed to cross exactly.

For now, the pattern that became known as the regular grid—distinguished by hexagonal clusters of triangles interspersed with pentagonal groups at the vertices of the original icosahedron—was a marked improvement. Because the struts were slightly longer or shorter based on their placement, the plan could be reduced to a table of lengths and a simple diagram. The specifications became known as the "chord factors" for a particular dome, and they were treated as a trade secret.

Much of the new grid's appeal lay in what Fuller called "the psychological factors regarding simplicity of coding," with a straightforward design that could be tooled for various materials and even printed on cardboard. As long as the number of divisions of each face, or frequency, was high enough to maintain its strength, Fuller thought that it would grow even more efficient with size, and he did calculations for a clear-span to cover much of Manhattan. "Beyond certain magnitudes of structuring," Fuller wrote, "only domes are possible."

As a familiar face in the media, he was perfectly positioned to sell it. The astronomer Harlow Shapley encouraged the magazine *American Scholar* to publish an article by Fuller, whom he praised as "the brightest man alive." Although the editorial board declined, Fuller received positive coverage elsewhere. An item in the November 7, 1949, issue of *Time* noted that architects tended to "chuckle indulgently" at his ideas, but it hinted that the dome might be different: "If this one turns out to be practical, no one will ever again chuckle at Bucky's dreams."

The Wichita House had received the same kind of attention, leading at once to an internal struggle, but Fuller had learned from his mistakes. By working with his former students, he would retain unquestioned control, and he reinforced his claims by making the rounds of influencers. His ideas interested Walter Gropius, who hosted him in Massachusetts in January 1950, and he spoke at the MIT Faculty Club on "Continuous Tension, Discontinuous Compression Structural Systems," which he had encountered for the first time in Snelson's sculptures.

Some of his contacts during this period were more questionable. In May Fuller heard from John Moehlman, a design consultant who had worked on his models at North Carolina State, about the recently published book *Dianetics: The Modern Science of Mental Health* by L. Ron Hubbard. Two weeks later, he received a form letter regarding "an intensive professional course" on Hubbard's method of psychotherapy in Elizabeth, New Jersey, from the physicist W. Bradford Shank. In a handwritten addendum, Shank wrote, "Don't miss this—it's the thing for which we have all been seeking."

Fuller was receptive, and he gave a talk at the Hubbard Dianetic Research Foundation in Elizabeth on August 1, 1950. No record of any meeting survived, but he almost certainly saw Hubbard, who delivered a lecture the next day. They also had striking similarities. Both drew inspiration from the navy; both named foundations after themselves to conduct research outside conventional institutions; both envisioned a new type of human, variously called the comprehensive designer or the "clear"; and their stories would intersect again in the future.

In the meantime, neither division of the Fuller Research Foundation could point to a major commission, and Richter warned that they were short of money: "I cannot continue living on the understanding good will of my folks." Eleanor and Victor Cannon, a married couple in Woodstock, New York, were interested in a dome, which Fuller described as a valve to regulate the exchange of energy with the environment, but it was beyond his capabilities, and no large prototype had been tested in the field.

Over the summer, a relatively sophisticated tent dome was erected by Lindsay, whom Fuller and Anne saw in Montreal. He apologized to

The Weatherbreak Dome on West Island, Quebec (December 1950).

Fuller afterward for the stressful tone of the visit: "If sometimes I seem to be losing confidence, it has nothing to do with the philosophy but rather of the insecurity which I naturally developed due to my industrial naïveté."

He was playing down his abilities, and in the absence of a client, Lindsay decided to personally fund an ambitious proof of concept, which rose in a snowy field on West Island, Quebec, in December. With the help of the McGill University ski team, Lindsay oversaw the dome's construction over two days. Its aluminum frame was reinforced by sprits, or short tubes extending from its hubs, which were strung with steel wire that sang as it vibrated in the wind.

At forty-nine feet across, the Weatherbreak Dome was the largest yet, and Lindsay informed Fuller later that it was "the first convincing proof of your theory." In the spring of 1951, it would be covered in nylon, but even the bare framework yielded beautiful photos, and it generated more interest in the dome than any previous effort. Its pictures entranced Arthur Drexler, the recently appointed curator of architecture and design at the Museum of Modern Art, and Anne wrote to Lindsay that it was "a dream come true."

Fuller, on the other hand, was more ambivalent. He treated all of the ideas that originated in his circle as his personal property, but Lindsay had clearly acted on his own. Although Fuller had played no part in its design or execution, within a few years, he was speaking as though he had been there himself: "When the structure was completed, we looked up at the blue sky through this thing and began to realize that something very pleasantly exciting was happening to us."

Lindsay's ingenuity concealed the issue of the windows in the regular grid, which would cause difficulties as components were manufactured with greater precision. At MIT, Fuller asked students to check his math using trigonometric functions that went to fifteen decimal places, and when they refused to come out right, he interpreted it as "a message from God" that the tables themselves were wrong.

Fuller was still annoyed, and he hoped that a new grid would avoid the problem entirely. He had formed a valuable partnership with James Fitzgibbon and his colleague Duncan Stuart at North Carolina State

College, who were more experienced than most of his other associates. To eliminate windows and reduce the number of different parts, Stuart explored a design of lesser circles, which were produced by planes that bypassed the center of the sphere. Although it violated the aesthetics of Fuller's geometry, it led to an elegant solution that became known as the triacon grid.

After a branch of the Fuller Research Foundation was incorporated in Raleigh, Fitzgibbon struck a deal with Kidd Brewer, a real estate investor who wanted a hilltop dome for publicity. They intended to use a plywood frame with a plastic skin that Lindsay had developed, but Fuller killed the plan. He had concerns about Lindsay operating outside of Canada, as well as a loss of influence if the separate groups started to work together, and he was determined that the first major commission be under his control.

Instead of landing clients, they built more prototypes, including a wooden dome by an MIT sophomore named Zane Yost. Fuller and Richter assembled a three-quarter sphere on Long Island, while Lindsay finally sent a dome to Patrick Baird. It resembled the dome tents that would appear decades later, but its struts were lost, and so the skin was hung on a makeshift system of guylines. Undaunted, Lindsay pivoted to a wooden dome lined with foil, which became the first to be occupied as a shelter after his brother purchased it to use as a ski lodge.

Fuller was still working with George Nelson, who privately told Charles Eames that he was "totally skeptical" of the dome as a business: "I have anticipated without pleasure a repeat of former disasters." As an incubator space, Nelson provided his office—Fuller lectured on his design principles while showing thousands of slides at parties—and they collaborated on another Garden of Eden dome, the Bubble House, which Eames featured in *Interiors* magazine. They hoped to install it at the Eameses' house in Pacific Palisades, California, or as a model home at the Museum of Modern Art, but it was never produced.

Charles and Ray Eames remained fond of Fuller, whose use of fiberglass was to shape their approach to furniture. "Every problem we work on is touched by your influence in some way," Charles told him admiringly, and he said later that he and his wife asked themselves the same question about every project: "What would Bucky Fuller say about it?" At the moment, however, Fuller's attention seemed to be wandering. He designed another Dymaxion Car, with separate engines for three pairs of wheels, and arranged for a book on his work by Professor Richard Hamilton of MIT.

Fuller was also dealing with changes in his family life. Two years earlier, Allegra had landed a job at the International Film Foundation in New York through Martha Hill, who taught dance at Bennington before moving to Julliard. Allegra spent her winter recess answering phones and on location shoots, and she became friends with Kenneth Snelson, whom she brought along as a cameraman.

A mutual acquaintance introduced her to Robert Snyder, a documentarian who specialized in educational films. "We sparked right away," Allegra remembered of Snyder, who was thirty-five, and despite their age difference, they entered into a romantic relationship that led to an awkward encounter with Anne. Fuller thought his wife was offended by the pragmatic tone that Snyder took in discussing a possible marriage: "He went to her, not to me, and he talked to her in a way that the Old World does talk about what parents might be able to do for a young couple when a good man takes their daughter as his wife."

Anti-Semitism was probably a factor—Anne's letters sometimes contained casually prejudiced remarks about Jews—but Snyder struck her as a risk for other reasons. He was colorful, creative, and working in an unreliable industry, although he reminded her that this had been equally true of her husband when they were first married: "Mr. Fuller was not a man of any substance." Snyder achieved validation in his own right that year for *The Titan*, a film on Michelangelo that won the Academy Award for Best Documentary Feature, and within a month, he and Allegra announced their engagement.

Allegra was married on the afternoon of her college graduation in June 1951. Rock Hall, which had hosted so many family weddings, had become too costly to maintain, and a ceremony was held instead at Bennington, where Fuller gave away the bride. Once the newlyweds settled in New York, Allegra returned to the International Film Foundation, while Snyder worked on a documentary about the dangers of totalitarianism. Allegra's mother eventually warmed to her son-in-law, but Fuller noted that it happened slowly: "Anne gave Bob a tough time for several years."

It was an emotional period for Fuller, who had conceived of his life's work more than two decades earlier with his daughter in mind. Now that he no longer felt required to provide for Allegra directly, his paternal impulses expanded outward—first to his younger followers and ultimately to the entire human family. He began to speak about a future in which 50 percent of mankind would reap the benefits of industrialization, which were currently enjoyed by just a quarter of the world's population. The turning point, he predicted, would occur around 1970.

On October 13 Fuller delivered a talk at the University of Michigan School of Architecture and Design in Ann Arbor. No complete record of his remarks survived, but he had recently told another audience that it was possible to feel the earth turning by looking at the North Star and "imagining that one is on a huge space ship sailing through the universe." Compared with the expansive metaphor that it later became, it was a modest image, but Fuller recalled that the Michigan lecture was the first time that he spoke of Spaceship Earth.

Fuller's widening perspective coincided with an increase in the complexity of his affiliations, which demanded a focal point to keep him at the center. The obvious solution was a patent, but because he presented the dome as a logical extension of geometrical principles, it raised the question of whether he could own the rights to its design. His attorney, Donald W. Robertson, advised him "to assert claims to the pattern of the geodesic structure," and they spent much of 1951 drafting an application that was filed toward the end of the year.

In a summary, Fuller wrote that the structure, far lighter per square foot than a conventional building, resulted in "substantially uniform stressing of all members, and the framework itself acts almost as a membrane in absorbing and distributing loads." The description closely resembled Jeffrey Lindsay's Weatherbreak Dome, which suffered from the windows problem—a "patented error" that evidently went unnoticed by Robertson. Fuller was well aware of it, of course, but he advocated the geometry of the three-way grid for years without mentioning its flaws.

His obsession with patents, particularly for foreign countries, was questioned by some of his collaborators, including Duncan Stuart, who thought that they produced "a paranoia that seemed to govern him." Fuller acknowledged that they cost more than they were worth, but they made sense for the dome, which needed a defining statement. As he noted later, with some truth, "If I had not taken out patents, you would probably never have heard of me."

Fuller's ephemeralized operation was taking root in colleges. With its clear sequence of design challenges, the dome had real pedagogical value, and it was adaptable to any budget. These factors appealed to administrators, while the students were just as drawn to Fuller himself. As one glowing article noted in *Architectural Forum*, "The expense in this one is in the thinking, and Fuller has always been a profligate spender of thought. In addition, he has today a crew of smart, hard-working young associates to pounce on his geometry, calculate it, and build it."

The dome was going viral, and, like many social epidemics, including dianetics, it had spread through an existing network by word of mouth. Fuller was somewhat like a virus himself, with a unique talent for using a host to reproduce, and he carried the discoveries of each group of students to the next school on his list, burnishing his image as a genius by assimilating the work of many others. When credited to him alone, it made him seem staggeringly versatile, and it was no accident that he began to be widely compared around this time to Leonardo, even by Charles Eames.

Fuller also used students for unpaid research that benefited him personally. All of his domes so far had been crafted by hand, and he badly wanted a contract for mass production. When he was approached by the US Navy's Bureau of Aeronautics about a program "to develop a portable cold weather aircraft maintenance shelter for use by Marine Corps Aviation Units at advance bases," he brought it to a visiting lectureship at MIT in the fall of 1951. The class came up with a model of a fiberglass dome for deployment with aircraft carriers in the Arctic, and although it was rejected as overly complicated, he would soon put a similar idea to good use.

As Fuller looked to manufacture the dome in the quantities that its design principles made possible, he was asked to lead a class at North Carolina State in designing a cotton mill—a structure with clear industrial potential. In January 1952 he arrived in Raleigh, where his students drove the project, although he later framed it as an homage to his apprenticeship in Sherbrooke forty years before. Early mills had been built vertically near rivers, allowing them to be run by gravity and the waterwheel, but the availability of electric power had led to more horizontal layouts. The student plan returned to the setup that he had seen in Quebec, with a vertical mill like a fountain.

On paper, the geodesic mill was eight stories high, with cotton blown to the top and directed down through a network of tubes, like the fibers in the machinery that Fuller had designed at Stockade. Its shell, a three-quarter sphere, could be replaced by successively larger versions as the mill itself expanded. To allow for the vertical movement of materials, each floor would be an open triangulated scaffold, in which every angle was the same and no strut duplicated the load of any other, like a vector equilibrium extended into a flat surface.

His students in Raleigh put together a test section of the floor, which became known as the octet truss. Like the dome, it raised questions of originality. Alexander Graham Bell, the inventor of the telephone, had constructed an identical framework at the turn of the century for a kite made of tetrahedral cells, although Fuller was unlikely to have copied it directly. Since the forties, Fuller had seen the universe as an omnitriangulated system of energy transactions—he called it the isotropic vector matrix—and the octet truss was essentially a single slice.

Fuller remained worried enough by the similarities to revise his famous anecdote from kindergarten, in which he had supposedly impressed his teacher with a triangulated structure of peas and toothpicks. In a curious attempt to establish priority over Bell, he claimed eventually that he had assembled a full octet truss, but he never explained why it took him a half century to pursue an idea that allegedly occurred to him when he was a small child.

In the meantime, Professor James Fitzgibbon hired Thomas C. Howard, a gifted student from North Carolina State who had built an exquisite model of the cotton mill, to construct a plywood dome in his backyard. Around the same time, Fitzgibbon, Stuart, and the artist Manuel Bromberg established the Skybreak Carolina Corporation in order to sell domes in Raleigh. They lined up several potential deals but were unable to proceed without express approval from Fuller, who wanted to begin with something truly impressive.

Back in Montreal, Jeffrey Lindsay requested a clean break, with his team retaining the Canadian rights to Fuller's patent and severing its ties to the foundation. Fuller countered with a proposed trial period of five years, and he didn't shy away from invoking their personal relationship to his advantage, writing to Lindsay, "Anne and I feel as though you were our own." Lindsay consented, although he felt that they were only postponing the inevitable, and told Fuller, "You achieved a far greater effort and investment from me through your general process of indoctrination than would have been possible through a contract."

In April 1952 Fuller headed to Cornell University, where he was about to attract the most loyal collaborator he would ever have. For his first class, held on a warm afternoon, he invited students to sit under an elm in the quadrangle. In a talk that lasted for four hours, followed by dinner at an Italian restaurant, he described their project, which would consist of a twenty-foot wooden sphere covered in wire mesh. Continental outlines would be fastened to the outside, with a ladder allowing observers to look out from the center of their private sky.

One of his students was twenty-five-year-old Shoji Sadao, who was born to Japanese immigrants in Los Angeles. During World War II, Sadao had been interned at the Gila River Indian Reservation in Arizona, where he developed an interest in architecture while working with a Quaker staff member on the grounds. Drafted into the army, Sadao served in a topographic battalion in Germany, and upon his return, he enrolled in the architectural program at Cornell, in which his background in mapmaking and technical drawing gave him an advantage over the other students.

Sadao was fascinated by Fuller's vision and range of knowledge. Students in the class were deployed to hardware stores and lumberyards to solicit donations of material, while Sadao drew on his training in cartography to make the continents, carefully cutting them from sheets of copper screening. Mounted on a rooftop, the sphere became a campus attraction. Unfortunately, it suffered an ignominious fate on Halloween, when some students, trying to move it as a prank, cut through the wires securing it to the roof and sent the globe crashing to the ground.

Fuller was impressed by Sadao, whom he recruited for a revision of the Dymaxion Map that was intended for the United Nations. Instead of a vector equilibrium, it used the unfolded icosahedron, which offered many advantages: it had twenty identical pieces, instead of a mixture of triangles and squares, and provided a closer fit to a sphere. Fuller had rejected the icosahedron before, but now that it was associated with the dome, he could embrace it as part of his philosophy.

Working from a layout by Fuller, who covered a globe in a wire overlay to find the best arrangement of the continents, Sadao used a plastic hemisphere as a ruler to transfer the information to a flat surface. This version of the map, printed by North Carolina State, became its most successful incarnation. What Sadao described as his "self-effacing" personality, which he attributed to his Japanese background, enabled him to work well with Fuller, and they grew increasingly close.

By then, Fuller had found the large project that he had been seeking, which he owed once again to his connections. For the Chicago World's Fair in 1933, the German-born industrial architect Albert Kahn had built the Ford Rotunda, an open pavilion with walls like a tower of nested gears. After the fair ended the following year, it had been dismantled and reassembled at the Ford plant in Dearborn, Michigan. For the company's fiftieth anniversary, the 130-foot-tall building's circular opening would be covered to allow exhibits to be placed in the central courtyard, but it lacked the strength to support a conventional roof.

Niels Diffrient, a young freelancer for the Detroit office of the industrial designer Walter B. Ford, suggested contacting Fuller, whom he had met through the architect Eero Saarinen. Fuller recognized it at once as the showpiece that he wanted, and it had obvious symbolic significance. Although Henry Ford had died in 1947, Fuller still saw the request as a sign of approval from "Mr. Industry himself."

In Dearborn, Fuller was taken in "a very fancy automobile" to view the courtyard of the Ford Rotunda, which measured ninety-two feet across. After proposing a geodesic roof that would be significantly lighter than steel, he agreed to provide the calculations, claiming later that he worked them out on an envelope. In fact, he followed his standard procedure by developing it with students, initially at Yale and MIT, before moving it to New York.

Over the summer, Fuller had installed a student team in the office of George Nelson, who paid them to work on design projects by day, while Fuller drew on their services for free at night. The members had originally been asked to build a model of the Bubble House for the Museum of Modern Art, and Nelson grew furious when Fuller abruptly reassigned them to the Ford Rotunda. Their most prominent member was T. C. Howard, who recalled of Fuller, "He may have been a machinist, but he was scary around the equipment."

Fuller's growing influence at colleges was causing consternation elsewhere. In the fall of 1952 he spent six weeks at Yale, where a paperboard dome was installed on the roof of Weir Hall. According to the author Tom Wolfe, after it was finished, "[The students] as much as dared the dean of architecture to do anything about it. He didn't, and the dome slowly rotted in its eminence."

The dean, George Howe, formerly of *Shelter*, was skeptical of the dome's value, saying bluntly in a letter to Fuller, "Personally, I think it is a waste of time." Fuller still inspired devotion on campus, however, and after learning that four of his students were in the glee club, he wrote them a song, "Roam Home to a Dome," which he sang with gusto to the tune of "Home on the Range." Its first chorus ran:

Roam home to a dome,
Where Georgian and Gothic once stood:
Now chemical bonds alone guard our blondes,
And even the plumbing looks good.

Fuller often rode the train to Yale with architect Louis I. Kahn, whom he had met years earlier through *Shelter*. Kahn, a professor at the university, had recently spent time in Rome, inspiring him to explore monumental forms, and although he was more interested in compression than tension, Fuller felt that he had impressed Kahn deeply: "I think this was very much of a turning point in his life."

He took credit for inspiring Kahn's tetrahedral ceiling for the Yale Art Gallery, which Fuller later dismissed as "aesthetic nonsense," and the triangulated frame of the unrealized City Tower for Philadelphia. In reality, the ceiling design had already been finalized, and any influence was more likely due to Kahn's lover and collaborator, Anne Tyng. According to Tyng, "Lou and Bucky did not really communicate, since they spoke such different creative languages."

For students whose thoughts on design were still evolving, no one was more exciting than Fuller, who assigned outlandish problems on the remote chance that they could pull them off. At the University of Michigan, he asked a class to design a dome that rotated in the wind to shed rain, and the students made a model with aluminum blades that spun on a horizontal bicycle wheel. It failed in testing, but one of the participants, Charles Correa, who later became a prominent architect and urban planner in his native India, recalled that Fuller predicted that it would spin faster "until the whole dome itself became invisible."

Fuller took greater interest in a dome greenhouse at North Carolina State, along with a more ambitious concept for agriculture, the Growth House, which borrowed the vertical layout of the geodesic cotton mill. He was learning to organize groups in tandem, and the Raleigh team helped with the skin for a wooden truss dome at the University of Minnesota. Fuller conceived of the project as a house for himself, complete with the Autonomous Living Package, although he eventually dropped the internal utilities as the more practicable shell took on a life of its own. It was more exciting to conduct tests on the skin, which included a dome covered in transparent Mylar at the University of Oregon.

A more important lead came through the Lincoln Laboratory, funded by the US Department of Defense, in Lexington, Massachusetts. The architect William Ahern, a former student, contacted Fuller about a program there for enclosures for air force radar antennas. Jeffrey Lindsay had once proposed a similar idea for the Signal Corps, and Fuller asked two MIT students, Bill Wainwright and John Williams, to work on a test section of what became known as the radome.

Fuller devoted more of his own time to the Ford Rotunda. Alongside a tent for a Ford station wagon, he worked with Don Richter on a new pattern for the dome. The alternate grid, as it came to be known, divided the faces into rows of triangles, eliminating the windows problem. He asked his team in Nelson's office to build a miniature version, which they finished a mere fifteen minutes before he had to leave for Detroit. Fuller and his students piled into a taxi to Grand Central Station, ran across the concourse, and barely got him on his train—where he found that the model was too large to fit in his compartment. But the pitch succeeded, and Fuller was officially granted the commission in November.

After establishing a Detroit office, he flew out every weekend for the first half of 1953. He performed systematic structural tests on the dome for the first time at the University of Michigan, where a model was used to display strain patterns in polarizing plastic. Fuller had initially wanted a single layer of struts, but when they threatened to buckle, he was persuaded to switch to the octet truss. A sample section supported more than a half ton per square foot, or more than a standard factory floor, but the challenges of manufacturing it meant that its first use in a dome would also be its last.

On March 23 they began construction. To circumvent building codes, Fuller was advised to call it a skylight, and he left the details to T. C. Howard, who, shortly after finishing the design, was drafted into the military. There had been some uncertainty over how to raise the dome—Fuller suggested floating it up on a balloon—and they eventually decided on a rotating hydraulic crane. On a deck suspended across the opening of the rotunda, workers could build the framework from the top down, lifting one level at a time before lowering it onto the roof in one piece.

By modern standards, it was overengineered, but Fuller took pride in the precision of the nearly twenty thousand aluminum

Interior view of the dome of the Ford Rotunda.

struts, manufactured with a tolerance of five hundredths of an inch. At workstations around the rim, the pieces were assembled into octahedra, which were then riveted into large triangles that were light enough for a worker to hold one in each hand. Colored tape was used to code the parts, and only two kinds of bolt were required.

Once they were set into the overall framework, the trusses were covered with fiberglass that had been tested by bombarding the panes with small objects. The inside of the entire dome was supposed to be sheathed with acrylic; it turned out to be unavailable, however, forcing them to seal the seams with tape. Work proceeded quickly, and the roof, weighing eight and a half tons, or a twentieth of a conventional steel frame, was finished in forty-two days.

A preview in May was followed by the official opening on June 16. Now that the Fullers' home life had grown more stable, Anne had assumed a greater role in supporting her husband's work, and she joined him for the trip to Michigan. In the weeks before the event, Anne had closely followed the coronation of Queen Elizabeth II on the radio, and it deepened her lifelong fascination with the British royals, whose lives offered a counterpoint to her own complicated marriage.

After picking up Shoji Sadao, they made the long drive in a green Lincoln Continental convertible, which Fuller had bought with his fee, and arrived for the unveiling of the roof in Dearborn. It had a temporary feel—the colored tapes were clearly visible—but its design was hailed as a triumph. With fifty candles flickering around the rim, the sunlight through the fiberglass produced an iridescent effect, and the automotive exhibits displayed in the rotunda evoked the Ford Industrial Exposition that had accompanied Fuller's visionary experience in 1928.

Fuller claimed that his clients had arranged for a construction firm to haul away the pieces if it collapsed, but if he thought his success would yield an ongoing arrangement with the Ford Motor Company, he was disappointed. He presented a copy of *Nine Chains* to Henry Ford II, who was unable to grasp what had made his grandfather a great artist. "Ford's son and grandson failed to understand old Henry's inspirational philosophy of real-wealth producing," Fuller wrote, "and learned to play only the game of moneymaking with the money they inherited."

In any event, as his first commission for a dome, it was a landmark. *Life* magazine mentioned its alleged resistance to atomic bombs, while *Architectural Forum* speculated that it might be used for airports or amphitheaters, although one engineer expressed concerns that observers would wrongly assume that it was the equivalent of a permanent roof. After promoting himself tirelessly for a quarter century, Fuller could at last justifiably say that clients came to him rather than the other way around—a distinction that he would emphasize for the rest of his life.

To others, it felt more like a conclusion. Four years had passed since Black Mountain, and something intangible would be lost as Fuller moved to a larger stage. His most talented collaborator so far was undoubtedly Jeffrey Lindsay, who traveled to congratulate him at the Ford Rotunda. Lindsay had hoped for encouragement from Fuller; instead, they had a bitter confrontation. Fuller questioned his motives over asking for Canadian rights, saying that Lindsay "should be prepared to compete on merit rather than hide behind patents," and he implied that he felt deceived.

In the end, Lindsay recalled, he assigned the rights back to Fuller, who promptly reversed himself and declared that they "were now friends with a great, prosperous, and trusting future together." His former student had reason to feel uneasy. Now that the dome had been proven in public, the stakes had risen for both success and failure, and Fuller would react accordingly. At the Ford Rotunda, it had rained, and because of the lack of a plastic skin, leaks in the taped seams had left puddles on the floor. Decades later, what Sadao remembered most clearly about that triumphant day was Fuller's anger as he ordered an official to find a mop to clean up the mess, which he said was "a goddamned shame."

EIGHT

CONTINUOUS TENSION

1953–1959

The most economical relationships are geodesic, which means most economical relationships. Ergo, we have events and novents: geodesics and irrelevance.

—BUCKMINSTER FULLER, *SYNERGETICS*

After his confrontation with Jeffrey Lindsay in Dearborn, Fuller drove west with Anne, Sadao, and George Welch, a University of Michigan student who had served as his representative with Ford. To pass the time on the road, they played a game that Fuller claimed to have invented in the thirties. It concerned an imaginary, unscrupulous corporation called Obnoxico, and the object of the game was for each player to devise a product or service that would generate "vast amounts of dollars," such as a business where parents could have their baby's last pair of diapers plated in gold. The travelers outdid one another with various schemes— peddling plastic rocks for gardens, for example—which Fuller used to underline the point that he personally had no interest in money.

In reality, he was more than willing to reap the financial rewards of his work, and when he arrived at the International Design Conference in Aspen, Colorado, on June 21, 1953, he knew that it would be a significant stage. Sadao viewed it as his mentor's debut in the design world, and although Fuller had been part of that community for decades, the resort made for a singularly magnificent backdrop for the University of Minnesota dome, which was assembled before four hundred attendees.

His magic show required occasional large effects, and his most spectacular achievement so far was a shelter that could seemingly be conjured out of nothing. The dome was erected in one and a half hours—the wind threatened to blow it away as the skin was installed—and served after its completion as a student dorm. It delighted Charles and Ray Eames, but Fuller had another prospect in mind. Walter Paepcke, the founder of the Container Corporation of America, was the festival's benefactor, and his patronage of modernist design would make him a useful client.

After the conference, the dome was transported to Massachusetts to provide temporary housing on another project for twenty students. Earlier that year, Fuller had been approached by an architect, Gunnar Peterson, who had bought an undeveloped lot in Woods Hole for a

Fuller with the dome constructed by students from the University of Minnesota in Aspen, Colorado, in June 1953.

motor lodge. Given the village's conservative Cape Cod atmosphere, it was an unlikely location for the first permanent freestanding dome, which Peterson planned to turn into a restaurant.

Fuller saw it as a trial run for a bomb shelter that he had designed at Princeton, based on the assumption that a curved surface would resist the shock wave of an atomic blast. After a survey in the spring, he arranged for the pine struts to be produced at the woodworking shop at MIT. A volunteer crew arrived at the beginning of August, and Fuller was there for most of the following month, although the construction was overseen by a student named Peter Floyd.

The dome went up in a festive atmosphere. A rope was hung from the top of the framework, allowing Fuller and the workers to swing around it, he recalled, "like Douglas Fairbanks," followed closely behind by a bat that had taken up residence inside. The dome also drew insects, especially after a powerful lightbulb was installed. One student remembered, "It cast shadows of the structure into the fog. We climbed up on it, and our bodies were cast like great gods in the sky."

Before it opened, the dome was skinned in Mylar, but persistent leaks forced diners to eat under umbrellas. The shell was transparent, revealing the stars, but it became uncomfortably hot in the summer, and, to Fuller's lasting regret, it was covered in opaque fiberglass. During dining hours, a woman played the zither, and because the dome amplified the music like an enormous loudspeaker, the neighbors complained about the noise. Despite the controversy, the Nautilus Motor Inn and Dome Restaurant was active for years, and it became the oldest surviving dome in America.

After a visit to Bear Island, Fuller prepared to build a sphere at Princeton based on "discontinuous compression, continuous tension." The principle called for assemblages of struts, linked by wire or string, in a uniformly stressed network in which none of the stiffening members touched. When its tension elements pulled tightly on one another, like a diagram of its own system of forces, it stood as if by magic, and versions based on different polyhedra were built by Fuller's collaborators, including a spherical model by Lee Hogden at North Carolina State.

Discontinuous compression, continuous tension sphere at Princeton University (1953).

The Princeton project was considerably more ambitious. In the fall, Fuller supervised fifteen students in assembling a sphere forty feet in diameter, with steel aircraft cable holding together ninety aluminum struts. After it was installed outside the architectural lab, Fuller found that tightening it in one spot tensed the entire system, which resonated at a higher pitch when one of the wires was plucked. When it was accidentally struck by a snowplow, a member bent on the opposite side, indicating that each part was affected by every other.

In press coverage, the Princeton sphere was compared to the experiments that had led to the atomic bomb, but Fuller seemed hesitant to discuss it with the one man who might have had some thoughts on the subject. Although he occasionally saw Albert Einstein from a distance on campus, he made no effort to approach him. Fuller was told that Einstein had inspected the sphere, however, and he took pleasure in the physicist's reported reaction: "I wasn't there at the time, but was told by the architectural students and faculty who were there that he was extraordinarily moved by it."

He returned in November to the University of Minnesota, where his students spent a month on a huge discontinuous compression dome made of 275 hollow fiberglass struts shaped like cigars. Fuller said that if it were expanded into a complete sphere, it would rise into the air, but the structure was so unstable that it had to be hung from the ceiling, and it amounted to yet another gigantic prototype for a nonexistent industry.

The military side had been dormant since his project for the Bureau of Aeronautics, but a break came in the form of Major W. L. Woodruff, who contacted Fuller on behalf of Colonel Henry Lane, the head of aviation logistics for the US Marines. In the aftermath of the Korean War, Lane wanted shelters for advance bases that could be airlifted and erected more rapidly than tents, and after reading about the Ford Rotunda in *Architectural Forum*, he saw the dome as a potential solution.

Early the following year, Fuller met Major Woodruff in Washington to discuss Colonel Lane's plans. In theory, the helicopter, which Fuller recognized as the trackless mode of transportation that he had envisioned since the twenties, could enable a more mobile military, but only if shelters could be deployed quickly. Because of the program's limited budget, Fuller offered to fold the design work into his classes. His personal operation had minimal overhead, and his administrative needs were met by a capable twenty-six-year-old named John Dixon.

James Fitzgibbon, who was also at the meeting, proposed a theatrical gesture to raise awareness of the project. Using a photo collage, he pitched lifting a dome with a helicopter, in order to show off its possibilities in a spectacular fashion. On January 28, 1954, they staged an airlift at Orphan's Hill in Raleigh, using the wooden dome that T. C. Howard had built in Fitzgibbon's backyard. After fifty students carried it into a field, it was hooked up to a Sikorsky copter piloted by Major G. W. Cox, who recalled of his historic flight, "It was no troubles at all."

Cox flew the dome a hundred feet in the air for a half mile, at a top speed of fifty knots. The other members of the Raleigh team went along for the ride, but Fuller stayed on the ground, positioning himself for an

Fuller witnessing the helicopter airlift of a wooden dome in Raleigh,
North Carolina, on January 28, 1954.

iconic photograph, while another picture ran on the front page of the
New York Times. Fuller hailed it as the fulfillment of his prediction of
"air-deliverable housing," but it was less a practical demonstration—
the wooden dome was hardly suited for military use—than a masterful
piece of guerrilla publicity for a project that had yet to be approved.

If the dome program had possessed genuine strategic importance,
it might have been classified, but, instead, it unfolded in full view of
the world. Lane wanted the marines to appear technologically ahead
of foreign countries and rival branches of the armed services, and the
dome did more than any other comparable design to convey the impression of advanced engineering at a low cost. On the level of public
relations, it was an exemplary instance of ephemeralization, and Lane
secured a contract from the Bureau of Aeronautics for a paperboard
dome.

Using a design that had been demonstrated in Raleigh, Fuller oversaw the project at Tulane University in New Orleans. Corrugated
paperboard was waterproofed with resin, folded into triangles for

stiffness, and assembled into a disposable hemisphere that was nicknamed the Kleenex House. Fuller assured Walter Paepcke that he was in "a position of authority governing the license of manufactures," but a version at Cornell wilted after a rainstorm.

He was encouraged to explore alternatives, and a wooden prototype for a metal dome at Virginia Polytechnic Institute was followed by a magnesium design that was airlifted at the Marine Corps base in Quantico, Virginia. Lane was impressed by his network: "You and your university projects with only the blue sky as the limit on ideas. Then your Fuller Enterprises to pick them up and put those applicable to production for us." As work proceeded, Fuller became notorious for speaking for hours on the phone via long distance, and his military contacts were reportedly ordered to fly out to him to save money.

According to another anecdote, Fuller objected when he was told that he needed security clearance, saying, "I already know my own top secrets." In fact, his participation was widely publicized, and it gave him legitimacy elsewhere. At Tulane, he worked on a concept by Charles Eames for a geodesic radio telescope that would float on a pool to dampen vibrations. The Sky Eye, intended for Australia, never made it past the model phase, but it was a sign of his growing confidence, which increased after his dome patent was granted on June 29, 1954.

Fuller was preparing for an even grander statement on the international stage. He had been contacted by Olga Gueft, the editor of *Interiors* magazine, about the Milan Triennale, the prestigious art and design exhibition held every three years in Milan, Italy. Gueft wanted it to feature a piece by an American, and she praised the paperboard dome—the only version that could be tooled for manufacturing—as "the most dramatic symbol of industrial design that any nation in the world can place on such an exposition."

The Triennale committee agreed to display two domes: one as a greenhouse, another as a model home featuring furniture by the architect Roberto Mango. Paepcke provided materials and manufacturing, while additional support came from Clare Boothe Luce, the object of Fuller's unrequited love twenty years earlier, who was now the United States ambassador to Italy. To erect the domes at the Sforza Palace,

Interior of one of the two paperboard domes at the Milan Triennale (1954).

Shoji Sadao was dispatched from the Raleigh office in August. The paperboard had to be covered in a translucent "bathing cap" for waterproofing, but despite this conspicuous compromise, the submission was awarded a Gran Premio.

As the domes were being displayed in Milan, Fuller was planning to conduct wind tests for the US Marines at Princeton, and his designs were spreading across the country. A magnesium dome went over a pool in Aspen, and at the University of Minnesota, he put together specifications for the Minni-Earth, a more advanced version of the globe that he had built at Cornell: it would be fifty feet in diameter, made of metal and plastic, and wired to display geographical and strategic information. He followed up at Washington University in St. Louis with a prototype for the Flying Seed Pod, which would release compressed nitrogen at its vertices to unfurl in an airdrop.

To handle commissions that exceeded the capabilities of his students, Fuller formally founded Geodesics Inc. Naturally, he was president, while Anne, who covered the office's expenses with thousands of dollars in personal loans, assumed the dual titles of vice president and treasurer. The new company would concentrate on military projects, with the marine shelters based in Raleigh and the radome group working out of Cambridge, where Bill Wainwright, who was funding research out of his own pocket, was joined by Peter Floyd and a twenty-year-old Institute of Design graduate named Bernie Kirschenbaum.

Operating with negligible oversight from Fuller, the Cambridge office had turned out impressive work. Radomes were urgently needed for the military's Distant Early Warning Line, a network of radar stations in the Arctic Circle that watched for signs of a Soviet air attack. To shield the antennas from ice, the enclosures had to be weatherproof and permeable to radar signals. For once, the dome, with its minimal surface area, was applied to a problem for which it was ideally suited, linking it to the invisible spectrum that Fuller saw as the next frontier.

Although Fuller fumed over a lack of control—he went so far as to accuse the team of "opening [his] mail"—the results spoke for themselves. John Williams perfected a structure of bolted fiberglass sheets, which were cured in a cardboard oven and used to build a thirty-foot prototype in Acton, Massachusetts. Installed on the roof of Lincoln Labs, the radome survived 1954's Hurricane Carol—a storm so violent that the National Oceanic and Atmospheric Administration took the name out of circulation for a decade—and was moved to Mount Washington in New Hampshire for winter testing.

Fuller wanted visitors in Raleigh to sign a guest book that assigned any ideas back to him, but he seemed reluctant to see projects through to completion. In January 1955 Geodesics closed a deal with the Western Electric Company, which would serve as the prime contractor for twenty-nine radomes. It would be the first time that domes were scaled up for mass production, but Kirschenbaum had to sit with Fuller until early morning in Forest Hills to "beat him down" over his fear that the venture would fail.

He also had a final break with Jeffrey Lindsay, who had consulted on the Sky Eye but wanted to work independently. When Fuller responded with a fresh string of accusations, Lindsay wrote back, "Your reaction was such as to crystalize for me the void between your teaching and your personal emotional constitution. . . . You should search yourself for the cause of the anxiety which you attribute to me." Lindsay informed his former instructor that he had made no efforts toward "planning your destruction," and, after he opened an office in California, the two never reconciled.

While Fuller claimed to express his ideas best in the form of artifacts, he was more comfortable building his reputation on paper. When the British artist John McHale, who coined the term *Pop Art*, asked if he had been influenced by the Bauhaus, Fuller denied it vehemently, saying, "The Bauhaus international school never went back of the wall surface to look at the plumbing." On reading this declaration afterward, Philip Johnson agreed that architecture was more about design: "Let Bucky Fuller put together the Dymaxion dwellings of the people so long as we architects can design their tombs and their monuments."

In a lengthy autobiographical letter, Fuller told McHale that his firstborn daughter's death had been "the negative stimulus or vacuum" into which his repressed energies had exploded. The name of his first grandchild, Alexandra, born in 1953, had obvious emotional significance, but he saw her less frequently after Allegra and her husband moved to Chapel Hill, where Robert Snyder directed the motion picture program at the University of North Carolina. A grandson, Jaime, was born in April 1955, three weeks before Fuller's older sister died in New York at the age of sixty-two.

As he neared sixty, Fuller was comforted by the thought that Leslie had lived to see his rise to fame. The year before, he had received an honorary degree at North Carolina State, which hailed him as "one of the most influential and controversial personalities of the machine age," and he was awarded a second doctorate at the University of Michigan. Another honoree was Dr. Jonas Salk, the developer of the polio vaccine, who asked over dinner, "I've always felt that those Dymaxion gadgets were only incidental to what you really are interested in. Could you tell me what your work is?"

According to his own account, Fuller responded, "Yes, I've been thinking about that definition for a long time. I've been engaged in what I call comprehensive anticipatory design science."

"That's very interesting," Salk said. "Because that's a description of my work, too."

Fuller began to correspond with Salk, for whom he wrote a long poem the following year. Using the image of two ropes spliced together, one manila and the other nylon, he noted that a knot passed from one end to another would change materials halfway. By analogy, a person was a kind of knot moving through time, a "pattern integrity" independent of the tons of food, water, and air that were progressively assimilated by the body. He counted himself and Salk among the rare individuals who felt validated by the universe, for which the best response was "grateful silence."

Although Fuller himself was rarely silent, he had reason to feel gratitude for the reception of his ideas. Geodesics Inc. had constructed a fifty-five-foot radome, the largest rigid plastic structure ever built, in Huntington, New York. At one test, it was placed on a concrete block and hooked up to gauges to determine its failure point. When a load was applied, the radome simply moved the whole slab, and the entire dome could be seen to contract simultaneously, confirming the principle of uniform geodesic stressing.

The Huntington dome also led to a coda to an important chapter in Fuller's past. Helen Morley Woodruff, Christopher Morley's daughter, was driving on Long Island one day and spotted the dome outlined on the horizon, which inspired her to reach out to Fuller. When Fuller visited his old friend, who was bedridden after a series of strokes, Morley encircled him with his good arm and sighed, "Oh, Buckling." Two years later, Fuller and Anne were passing through Illinois, where he had first seen the Butler grain bins, when he switched on the radio to hear the news that Morley had died.

In the summer of 1955 Colonel Henry Lane issued a final report on domes for the marines. It was drafted with Fuller's input—including a section that related the project's goals to the structure of a tetrahedron—

but still managed to be a model of clarity. Noting that only $80,000 in government funds had been spent on the dome program, Lane estimated that millions of dollars could be saved by what he called "the first major basic improvement in mobile military shelters in the past 2,600 years."

The marine shelter program had been a triumph of public relations, not military planning, and its high point was the United States National Air Show in Philadelphia in September, where a nylon-skinned dome was airlifted before a crowd that included Fuller and Noguchi. With America immersed in a Cold War dominated by the threat of nuclear Armageddon, the dome served primarily as propaganda for the role of traditional defense. The marines ordered several hundred, and domes were put to intermittent use by the military in Korea and Antarctica, but most were stored in underground bunkers in the event of a nuclear war.

While he occasionally disputed with the marines over royalties and the rights to his ideas, Fuller was happy to take advantage of the publicity. Despite his stated devotion to practical inventions, the geodesic dome was closer to a shorthand symbol for the future, and it could be turned to any ideology, including the display of American strength. After the air show, Fuller told Major George J. King that it demonstrated the nation's "advantage over our competitors in the historical struggle for world man's highest credit as his most effective leader."

This message was at odds with his criticism of the world's division into competing nations, which became more pronounced in his talks over time, and although his military work frustrated some of his followers, he knew how to make his case to King: "The cool barrel of the geodesic structures weapon—inadvertently adopted by the Marine Corps—is the barrel which can now hit directly, instantly, and effectively at the heart of every peacetime economic pattern the world around without unleashing hot war. And if we win the cool war first, then there will be no hot war."

⚬⚬

Radome on the Distant Early Warning Line.

For much of his career, Fuller said that he had never been interested in money, while at the same time noting casually that he had managed to make quite a lot of it. He treated financial success as a form of validation, but like many prophets of human potential, he earned more from talking about his ideas than through their realization. Fuller's implication that the dome gave him the revenue for his research was true for just a few years in the fifties, and it was due to a project in which he was hardly involved.

After the radome went into production at Western Electric in 1956, radar enclosures were installed across the Distant Early Warning Line. Fuller drew a contrast between the unionized workers who needed a month to erect a radome and the "Eskimos" who did it in fourteen hours: "Luckily, in 1927 I had cast my lot with proving the economic feasibility of comprehensive anticipatory design science, as applied to the building arts, by priority of experiment in the Arctic and in remote foreign countries. I detoured so-called civilization."

This was both a blatant revision of the map of his life—the Wichita House had been built close to the geographical center of the United States—and a misrepresentation of the radome, which was entwined with practical concerns back home that Fuller had difficulty managing. Geodesics Inc. subcontracted construction to outside firms, leading to a complicated tangle of deals, and many affiliates, including Western Electric, tried to circumvent his claims completely.

Fuller boasted that his "superbly drawn" patent was impossible to evade, but his haphazard business skills were unequal to the challenge—he often spent more on policing his rights than he earned in royalties—and the Cambridge group asked him for a loan to start a separate company. As Geometrics Inc., they licensed his patent and made improvements, including a "truncatable" grid by Bill Wainwright. The decision hurt Fuller, but the benefits were clear, as Wainwright recalled: "We generated enough money so that he could go on being Bucky Fuller."

The radome program became his only truly lucrative enterprise. More than five hundred were built over six years, with Fuller earning a royalty of 5 percent of fabrication costs. It yielded him close to $2 million, leading to a flurry of spending on new cars, a racing sloop, and renovations on Bear Island. The radomes were widely publicized, and later models were erected in prominent locations, including a listening station that loomed ominously over Berlin for the US National Security Agency.

As the radome became a symbol of American technology overseas, Fuller wanted to make a greater impression at home. In 1955 he founded Synergetics Inc. to pursue civilian clients, with himself as president, James Fitzgibbon as vice president, and a design staff based in Raleigh. The company would conduct research and offer consulting, while Fuller continued to search for suitable projects.

Almost immediately, he stumbled onto one of the most promising leads he would ever have. Walter O'Malley, the owner of the Brooklyn Dodgers baseball team, wanted to replace aging Ebbets Field with a new stadium on Flatbush and Atlantic Avenues. Norman Bel Geddes had worked on a design with a retractable roof that would allow for events throughout the year, but it was unworkably expensive. After seeing coverage of radomes and the geodesic cotton mill, O'Malley contacted Fuller about the new stadium that he had in mind, which he conceived as an architectural marvel.

Fuller recognized its public relations value, and he wrote back that he was thrilled by the prospect: "By coincidence, I have already thought a great deal about this subject and have made many calculations in respect to it." He noted that he had previously been approached about stadiums in Denver and Minneapolis, and he conveniently arrived at the exact dimensions and budget as O'Malley, to whom he proposed an octet truss dome "skinned with translucent fiberglass petals opening and closing to the sky."

O'Malley was intrigued enough to visit Huntington, where his friends tested the radome by pelting it with rocks. After Fuller offered to conduct a study at Princeton, they announced their plans to the media. The aluminum dome was evoked in glowing terms, with a diameter of 750 feet and a roof 300 feet high, which they believed was the greatest height to which a ball could plausibly be hit. With no columns to block the view of spectators, it would be the largest clearspan in history.

Over the fall, Fuller developed a model with his Princeton students. During an interview with a reporter for *Sports Illustrated*, who described him as "built along the lines of a jar of yogurt," he inverted a sugar bowl to explain his design, which he presented at a press event on a snowy day in November. O'Malley, puffing on a cigar, expressed delight with Fuller's stadium and said he hoped it would be the first landmark that ships saw when approaching the city.

The coverage implied that O'Malley had hired Synergetics for what he called "one of the wonders of the world," but Fuller lacked the experience for such a massive undertaking, which required someone who commanded more power than both of them combined. O'Malley wanted Robert Moses, the parks commissioner, to acquire the necessary land in Brooklyn through a public agency. Moses, the autocratic official whose projects and priorities permanently shaped New York City, countered with the offer of a ballpark in Queens, but O'Malley declined, and the Dodgers eventually announced that they were moving to Los Angeles following the 1957 baseball season.

Fuller continued to correspond with O'Malley, but his involvement ended with the model, which succeeded in the limited objective of building enthusiasm for a roofed stadium. Both men were just as interested in the publicity value of geodesics as in its suitability for this particular plan, and Fuller came away with little more than an appreciation of the value of a major showpiece.

In the meantime, Fuller focused on theoretical endeavors. At North Carolina State, Duncan Stuart worked with students on a booklet about "energetic-synergetic geometry." It included a section on the principle of discontinuous compression, continuous tension, which Fuller would soon rebrand as tensional integrity, or "tensegrity." Stuart pitched the US Air Force on using geometrical concepts to analyze defense systems, but Fuller was reluctant to publish this material, in part because he wanted to reserve it for the classes that generated much of his income.

At a typical seminar, Fuller assigned students a dome problem, dividing them into departments based on the aircraft industry, while retaining all the rights to their ideas. He was still working on the Minni-Earth, which he expanded into a plan for a transparent computerized sphere—165 feet in diameter—that would remind viewers that the planet was a huge "Space Ship." Fuller suggested that it be installed within view of the United Nations, where it would reveal patterns in resources and migration that were normally too slow to be seen.

He also reconnected with T. C. Howard, who had finished his stint in the military. At the invitation of Anne, with whom he had become good friends, Howard joined Synergetics Inc., where he partnered with Fitzgibbon on a dome for a model-home show in Grand Rapids, Michigan. Although the event never took place, the Raleigh group designed a "catenary" dome with an interior skin attached by short chains, which they contracted to build for the Mid-America Jubilee in St. Louis.

As it turned out, the real debut of the catenary dome occurred more than seven thousand miles away. In the fifties, trade fairs were important sites of cultural exchange, as the United States responded to Communist exhibitions with an extensive program of its own. The latest battleground in the charm war was the Jeshyn International Fair in Kabul, Afghanistan, a country that the Soviet Union was seeking

The catenary dome of the United States Pavilion in Kabul, Afghanistan (1956).

to influence. Nelson Rockefeller, a special assistant for foreign affairs to President Dwight Eisenhower, was chagrined to discover that there was no American submission, and he ordered one to be undertaken at once.

This unenviable assignment landed in the lap of Jack Masey, a designer for the United States Information Agency (USIA), which was charged with promoting public diplomacy. Masey, who was based in New Delhi, India, faced a daunting task: the usual pavilion was out for repairs, and he had less than four months to find a substitute. Hoping to outshine the Soviets on a tight deadline, Masey thought of Fuller, whom he had met at Yale.

Over breakfast with Fuller in New York, Masey became "excited as hell" about using a dome as a pavilion, which was an idea that had been previously explored. The paperboard dome had appeared at international expositions, and Fuller had hoped that an exhibition in Sweden would feature the Flying Seed Pod, although it wound up sidelined by design issues. Jeffrey Lindsay had been approached separately about furnishing a dome for a Canadian pavilion in Ceylon, and the proto-type was already in the works when Masey spoke with Fuller.

The Raleigh team spent a month on the catenary dome, which was loaded in pieces onto a Douglas DC-4 transport plane on July 15, 1956. Four days later, it landed in the Afghanistan capital. Fuller had a prior commitment at Southern Illinois University in Carbondale, so he sent John Dixon to Kabul to advise on the construction, alongside Jefferson Davis Brooks III of Synergetics. Over two days, four local workers assembled the color-coded struts, and the Afghan minister of mines and industries drove in the last golden pin.

At first, the response was muted. One businessman complained that it seemed small, while another called it "a very poor show," and the exhibits—consisting of "talking chickens, talking cows, bouncing ball bearings" that Masey had scrounged from other fairs—were less than inspiring. Before long, however, the dome emerged as the high-light of the exhibition. It glowed beautifully at night, and during the day, it attracted larger crowds than any other pavilion. The Soviets took notes as attendees climbed over the framework, and Afghanistan's longtime king, Mohammed Zahir Shah, asked to keep it as a gift. He was turned down.

Many descriptions of the dome were culturally charged, with Fuller's friend David Cort writing in *The Nation* that the visitors "came in-side, fell on their knees, and prayed." Fuller himself compared it to a yurt, and he was unable to avoid a note of condescension: "The natives often thought they were building a cubical building and were amazed to find they had contrived a spherical structure." He also said that it was praised as a vindication of their "primitive" techniques by Mohammed Kabir Ludin, the Afghan ambassador to the United States, who had studied engineering at Cornell.

Neither Fuller nor Cort had been there in person, but both were conscious of the site's historical resonance. Fuller was an avid reader of Sir Halford Mackinder, the British geographer who originated the heartland theory of the strategic value of Central Asia, and the Cold War contest for regional influence through infrastructure was a display of soft power with shades of imperial conquest. The inadvertent discovery that a pavilion could send as much of a message as the exhibits inside delighted Fuller, who hailed the dome's appearance at the fair as "a logistical Iwo Jima."

In the aftermath, the catenary dome, which owed the specifics of its structure to Fitzgibbon and Howard, became an emblem of American innovation, as well as a symbol of Fuller's growing reputation. Its appearances at trade fairs throughout Asia led him to describe it as the first building to circle the globe, and the Department of Commerce ordered eight more, including one for the next Milan Triennale. Along the way, Fuller established a rewarding relationship with Masey, who became as responsible as anyone else for his rise to fame later in life.

The USIA mission was uniquely aligned with what Fuller represented, and, at home, he benefited from his newfound stature. He had skipped Kabul to teach at Southern Illinois University at the request of his former student Harold Cohen, the head of the design department. Fuller lectured in a scorching studio, slept on the floor during his lunch break, and proposed building a dome on campus that would be the largest in the world. It never went past the calculation stage, but the connection to Cohen would pay off tremendously.

His contacts in industry were equally valuable. During the war, Fuller had been friendly with Edwin Locke, the president of the Union Tank Car Company, which leased railcars to the petroleum industry. Locke was developing a maintenance facility in Baton Rouge, Louisiana, based on a proposal by company manager Richard A. Lehr to service multiple cars on a massive turntable, requiring a huge open floor. Locke decided to cover it with a dome, and Fuller accepted a royalty payment to bring the project to Synergetics, where it would be overseen by Howard.

Every dome doubled as the prototype for an industry—Locke hoped to launch an entire side business in geodesics—and Fuller was about to land the most promising partner of all. Earlier that year, he had reunited with Don Richter, who was working on product development at Kaiser Aluminum. As a test run for an aluminum dome, Richter arranged for students from the University of California at Berkeley and Los Angeles to design a cage for migratory birds at Lake Merritt

The grand opening of Kaiser's geodesic auditorium at the Hawaiian
Village Hotel in Honolulu on February 17, 1957.

in Oakland, using materials furnished by Kaiser, in the first permanent
geodesic structure on the West Coast.

It soon led to much more. As one story had it, Henry J. Kaiser, who
had fallen out of touch with Fuller since their abortive car project, was
passing Richter's desk when he saw a model of a dome. After Richter
explained what it was, Kaiser became enthusiastic. Fuller's efforts to
lure the magnate into the shelter business had once foundered, but
Kaiser had invested heavily in aluminum, for which a metal dome had
obvious commercial potential.

Kaiser commissioned a dome as an auditorium for his Hawaiian
Village Hotel, which had opened the year before as part of his program
to reshape the city of Honolulu. Synergetics contracted to draft plans
and conduct tests on a truss of diamond panels, developed by Richter,
that combined the framework and skin into a single structural element.
The dome, which was to be built from the top down with a temporary
mast, would be 145 feet in diameter and stand 49 feet high.

After a test assembly in California, construction in Honolulu began in January 1957. According to another famous legend, Kaiser wanted to watch it go up, but when his plane landed, he was astounded to learn that it had already been finished—in just twenty-two hours—and was ready to host a symphony concert that night. In reality, the shell was supposed to be built in five days, but when supplies were unloaded ahead of schedule, Kaiser himself gave the nod to proceed. It took thirty-six hours, and it was Richter, not Kaiser, who flew out too late for an inspection.

At that point, the dome was still only a shell, not a functional auditorium, and it took several more weeks to install the seats and stage required for performances. On February 17 it opened officially with a concert by the Honolulu Symphony and a popular Hawaiian baritone, Alfred Apaka, which was watched on television by millions. The acoustics were applauded, and many records of the exotica genre—a kitschy form of lounge music featuring sounds inspired by the South Pacific—would be recorded there in years to come.

It was the first great permanent dome, and it was meant to be just the beginning. Kaiser announced a building initiative that could provide a roof for Ebbets Field, an arena for New Orleans, and community centers for every town. For exclusive rights, Kaiser paid Fuller a minimum annual royalty of $25,000, and more than a dozen domes were built over the following decade. A housing industry seemed close enough to touch, but it was never as lucrative as Kaiser had hoped, and the dome's stated principles of efficiency were inconsistent with his objective of selling aluminum.

Fuller spent more of his own time on domes made from uncut plywood, a building material that had interested him since his days at *Fortune* magazine. When a prototype was built at Washington University in St. Louis, his patent attorney, Donald W. Robertson, thought that the sheets fell into place so elegantly that they seemed to say, "We want to be geodesic." Fuller proposed it for the Union Tank Car Dome, but after a test section buckled, Locke concluded, "I don't want to mess with wood. We are not wood people, we are steel people."

To market the plydomes, Fuller pivoted to an existing partnership. Two years earlier, the magazine *Better Homes and Gardens* had approached him for a piece about "dynamic developments" in design. Instead of a mere article, he had sold the assistant editors Alvin Miller and Ken Olson on his paperboard dome, and the magazine decided to promote it to readers. Fuller dreamed of a civilian business on the same scale as the radomes, although Fitzgibbon feared that it would require establishing an entire service division.

The paperboard dome was never effectively waterproofed, and his partners switched to the plydome, with Miller in charge of development at Fuller's latest company, Plydomes Inc. After a prototype was built by a teacher named Fletcher Jennings in Iowa, Fuller noted on a visit there with Robertson that its interior was comfortable in frigid temperatures—even without a door—and credited this to the curtain of warm air descending from the top. In fact, Jennings had taken pains to seal and insulate the dome, which was the first to be permanently occupied as a house.

For now, though, few of its kind were available. Fuller promoted the plydome on television and worked on a "pinecone" version at Cornell of overlapping plywood sheets, but issues persisted with building codes—there were concerns over heating and fire safety—and it was usually positioned as a ski hut or storage shack. Fuller eventually closed Plydomes Inc. to reduce his overhead, and only a handful were constructed, including chapels in Korea and the Philippines.

A more successful variation was the playdome. The first examples were built by Matrix Structures of Cambridge, founded by a group of engineers who realized that a miniature radome would make a wonderful climbing toy. Their original playdome stood five feet high, with a wood frame and a removable vinyl cover. Although it felt like a humble application of Fuller's theories, it was an ideal use of the dome, especially when reduced to a metal framework, and its successors would be seen on playgrounds well into the following century.

In June 1957 the Poznań International Trade Fair in Poland featured the dome's first appearance behind the Iron Curtain. *Life* printed photos of attendees admiring dishwashers and air conditioners, which annoyed Dave Cort: "Am I subversive," he wrote in *The Nation*, "if I say that America stands for something very different? We were once a masculine, ascetic, roving, adventurous people who created new solutions as fast as new problems arose." In Cort's opinion, the dome exemplified that pioneering spirit because it took "the simplest and the best way to a difficult thing."

This was how Fuller saw himself, and the world seemed to be catching up at last. On July 12 he celebrated his sixty-second birthday by witnessing the airlift of a catenary dome from the aircraft carrier USS *Leyte* to Hampton Roads in Virginia. The helicopter carried the dome over the water, its skin fluttering in the wind, and Fuller was filmed standing in the rectangle of its doorway on shore, as if marveling at what he had done.

Anne, who handled much of his correspondence, wrote to Peter Floyd in September that the dome was everywhere on the front lines:

> The first line of defense and also the bridgehead operation of the swift retaliatory offensive with the Marines. The front line of this cold war. . . . First line of agricultural offense with the plydomes . . . for agricultural stockpiling and machine protection. First line of cultural offense with the municipal auditorium at Hawaii, Virginia Beach, and Borger, Texas. . . . On the infantile frontier, cradle-to-grave coverage "playdomes." On the educational frontiers in developing new engineers and scientists.

And the most remarkable phase was still to come. Fuller was on the verge of taking the ephemeralization of his career to its logical conclusion, in which he would do seemingly everything with almost nothing at all.

On November 1, 1957, the Kaiser auditorium at the Hawaiian Village Hotel hosted a gala showing of *Around the World in 80 Days*. As searchlights swept the sky, and the movie's title lit up the dome, photographers swarmed around the film's producer, Mike Todd, and his glamorous wife, the actress Elizabeth Taylor. The event coincided with the announcement of a grand collaboration between Todd and Kaiser, who planned to open a chain of aluminum-domed theaters to present widescreen epics across the country.

For the project's "master architect," they arrived at the obvious choice. It was Frank Lloyd Wright, whom Todd and Taylor met the following month at Taliesin West in Scottsdale, Arizona. Although Todd complained that Wright was "the biggest ham actor in the world," they reached an agreement. Wright delivered a design that modified the dome's profile to produce a "gentler line of curvature," but the enterprise was cut short when Todd died in a plane crash.

In the fall, Fuller had visited Wright in Scottsdale to present a lecture to the Taliesin Fellowship, but he was never asked to join the team. Instead, he threw himself into an international adventure of his own, one that would take him around the world in just over a hundred days. In his talks, he often invited audiences to imagine a globe, twenty feet across, that represented how long it would take to circumnavigate the planet on foot. A horse reduced the diameter to two yards; a clipper ship, to the size of a basketball; and a jet plane, to that of a marble.

It was a cornerstone of his message of mobility, but Fuller himself hadn't been overseas since 1919. In his sixties, he became a model of worldwide motion, but only after the dome had made the trip ahead of him. The "first of his subsequently multiannual circuits" of the world was due to the United States Information Agency, which flew him to countries where the dome had already appeared. It was easier to transport Fuller himself, and he embraced it with more energy than anyone might have guessed for a man approaching retirement age.

His domestic affairs remained increasingly disorganized. Fuller's scattered tendencies led to more than sixty licensees for domes, creating a web of arrangements that frustrated his associates. Spitz Laboratories built a planetarium for the US Air Force Academy in Colorado Springs; Berger Brothers of New Haven, Connecticut, designed an inflatable "moon hut" to join four navy domes in Antarctica; and Fitzgibbon worked on an unrealized proposal for a domed shopping district in Montreal.

Fuller relied on others to take care of the details, but he knew how to get the ball rolling, and he landed an especially enticing prospect just before he left. He spent the week of February 24, 1958, leading students at the University of Houston in an attempt to construct a dome out of Plexiglas, which was reinforced with steel after the panels kept shattering. The trip was more noteworthy for an informal meeting with Roy "Judge" Hofheinz, the colorful businessman and former mayor, who wanted a covered shopping center for his own city.

Soon afterward, Hofheinz met with a local architect, Si Morris, who was conducting site studies for a ballpark. "Now, I've been talking about this thing to Buckminster Fuller," Hofheinz said. "He thinks we can build a stadium that will seat forty-five thousand for baseball and sixty-five thousand for football and have a roof over it, and we can grow grass, and it'll be completely air-conditioned. What do you think about that?"

Morris pretended not to see what Hofheinz was implying. "Well, why do you ask?"

Hofheinz cut to the chase: "If you don't think you can do it, I'm going to hire Buckminster Fuller."

"Hell, we'll do it," Morris said. Later that year, Fitzgibbon flew out to consult on a covered stadium—the first in the world—for twenty thousand spectators. It was the beginning of what would become the Houston Astrodome, the most important commission that Fuller and his colleagues would ever have the chance to pursue.

Fitzgibbon was nominally there on behalf of the Union Tank Car Company, which had begun construction of the dome in Baton Rouge in October 1957. Instead of calculating the angles by hand, the Raleigh team generated the specifications for the hundreds of panels using a program in the Fortran computer language by Richard Lewontin, a North Carolina State professor and population geneticist, in one of the first applications of computing in architectural design.

The outer dome was finished in the spring. At 384 feet across by 120 feet high, it was the largest clearspan structure of all time. A second interior dome would serve as a control tower, with the railcars rotating between repair stations arranged around it like spokes. Heat

The interior of the Union Tank Car Dome in Baton Rouge, Louisiana (1958).
A smaller internal dome served as a central hub and control tower.

and humidity caused operational issues—the temperature rose to more than a hundred degrees—but it was undeniably impressive, and the company's Graver Tank division planned to build a sister dome as a maintenance plant in Wood River, Illinois.

Its completion overlapped with the announcement of a dome by Synergetics Inc. for the American Society for Metals, an organization for materials scientists located near Cleveland. At the unveiling of the project, Fuller noted that stone became eight times heavier when its linear dimensions were doubled, which was why cathedrals required massive supports. Using geodesic principles, structures could be designed that were more efficient the larger they grew. As his friend Knud Lönberg-Holm had been the first to observe, Fuller said, "buildings are trending towards invisibility."

The commission from the American Society for Metals was the ultimate expression of this idea. Instead of a closed shell, it would cover the headquarters and garden with a decorative aluminum framework, like a huge playground dome. A design by T. C. Howard would be built by North American Aviation, which had become interested in the dome business as well, and the dream of a dwelling industry based on aircraft manufacturing seemed closer than ever before.

On March 19, 1958, Fuller left from Los Angeles on his world tour, as Anne stayed behind to meet him later in London. His first stop was Honolulu, where he waited three days in hopes of pitching Henry Kaiser on a dome rental business. Kaiser, who had cut back his dome program, managed to avoid seeing him, and Fuller left his materials with a subordinate. He always advised his students to approach the head of a company first, but Kaiser had eluded him yet again.

Any negative feelings were swept away by his warm reception in Japan, which was a promising market for his views on economic planning and industrialization. Fuller made a flurry of appearances to coincide with a dome at a trade fair in Osaka, and at the end of March, he left for Hong Kong and Bangkok, where he marveled at the accuracy of the shipbuilders on the Chao Phraya River. In Burma, he acquired a toy ball made of bamboo, in which he recognized his three-way grid of great circles, and then he headed for India.

On April 11 Fuller delivered a speech in New Delhi to an audience that included Indira Gandhi, who was serving as the personal assistant to her father, Prime Minister Jawaharlal Nehru. Afterward, Fuller presented a tensegrity model to Gandhi, who invited him to meet the prime minister. Fuller was unclear about their relationship—he thought that Nehru was her husband—but promised that he would cancel his visit to the Taj Mahal to make the appointment. Gandhi jokingly warned him not to tell Nehru: "He doesn't like people to come to India and not see the Taj."

The following morning, Fuller arrived at the prime minister's residence, where he laid out the Dymaxion Map on the parlor floor. Gandhi, then forty years old, recalled that Fuller gave her "nearly five hours of higher mathematics," which was "quite exhausting but so stimulating." As they were talking, Nehru appeared in the doorway, dressed in a white kurta. At Gandhi's urging, Fuller approached her father, who remained on the threshold, and launched into a description of his philosophy.

Fuller was unable to read the prime minister's expression, and after twenty minutes, Nehru left without a word. He was suffering from exhaustion, which may have played a role in his silence, but Gandhi reassured Fuller that he had listened carefully. Fuller was excited by his first meeting with a head of state—later accounts inflated it to an hour and a half—and he began to think seriously about the benefits of access to a developing nation. Nehru was embarking on an ambitious program of five-year plans, and his daughter told Fuller to call again whenever he was in the country.

From India, Fuller proceeded to the Kenyan capital of Nairobi, followed by a series of events in South Africa, including a seminar at the University of Natal in Durban. After two days of lectures to students, who wondered what they would be asked to build, he revealed that the answer was right in front of their eyes. It was the woven ball that he had picked up in Burma, which he wanted to use as the basis of an aluminum hut to replace the traditional grass shelter, or indlu, employed by the Zulu.

Fuller claimed that the thatch was constantly being eaten by cattle, giving him an excuse to experiment with a new construction method. The students built a prototype of corrugated sheets, laboring around the clock, and Fuller said later that they were sleeping in the dome when they heard workmen talking outside: "They were all Zulu boys on the campus, and as they came to this, they said, 'This is the way houses . . . should be built.' They were terribly enthusiastic about it."

The anecdote inadvertently underlined his failure to consult the people for whom the dome was intended, and aluminum was ludicrously expensive in comparison with local materials. A student report concluded uncritically, "If the technician could supply the answer to a truly economic native housing unit, then he would have contributed more to the welfare of this state and continent than any number of political measures." This would have been a questionable statement in any context, but it was especially doubtful during the era of apartheid.

Fuller flew to London, where he was joined by Anne in June. At a press conference at the US embassy, he said that the purpose of his trip was to show that design was "demonstrably effective in treatment of accelerating world evolution problems." He expanded on this point at a dinner of the Royal Institute of British Architects, with a crowd that included John McHale and the critic Reyner Banham, both of whom became enthusiastic advocates for his work.

"We find that industrialization is inevitably headed towards automation," Fuller said, "that is, towards disenfranchisement of man as a physical machine." Rather than framing it as a threat to employment, which was an issue of growing political importance, he saw it as a form of liberation, as workers moved from craft labor to the intellectual challenge of optimizing global industry.

Since patrons were unable to articulate what they needed, Fuller advised designers to look to the example of ships and planes. Houses could get away with poor design because catastrophes were rare, allowing architects "to build in between the earthquakes," but at sea and in the air, the problems presented by huge displacements of energy were obvious: "You cannot stay out there on a myth."

Describing his insight of 1927, Fuller reflected, "The minute you were not concerned with earning a living and really tackled problem after problem that the other fellow was not tackling, there proved to be a wealth of solvable problems. In fact, the whole mass of problems that are worth tackling is so great that any average individual who goes into that kind of a paradise wilderness garden ought to make very good progress." He added, "I am not afraid to suggest to a young man today that it is possible to forget altogether about the priority concept of earning a living."

The speech was a success, and Fuller opened the last stretch of his tour with appearances at the Hague in the Netherlands and in Germany. In Paris, he was introduced to famed film director John Huston by Isamu Noguchi, who was experimenting with stone—the opposite of tensegrity—at his landmark garden for UNESCO. By August, Fuller was back in Maine. He wrote of his voyage, "I hurry to Bear Island to see how it compared and found it to be the most wonderful of all."

Fuller had been granted a glimpse of his future, but his odyssey had withdrawn him from his business affairs. In Wood River, the dome by Graver Tank was in trouble. It had been conceived as the pilot for an expansive building initiative, but Fitzgibbon, who dismissed the company's approach as "refined blacksmithing," delivered a plan to accommodate their unskilled techniques. Edwin Locke was offended, and Synergetics was barely consulted at all.

After construction began, the site was found to lack a solid bottom, forcing them to sink concrete piles at huge cost, and local union rules allowed only one group to work at a time. Unlike the Baton Rouge plant, which had been built from the bottom up, the dome was assembled from the top down, using a combination of hydraulic jacks and a giant inflatable bag. It ran into difficulties, and Dick Lehr, the manager of the dome program, was fired.

Out of deference to Locke, Synergetics had backed away from steel structures. The company expected to consult on the engineering side, but Fuller circumvented it by working directly with Locke on a proposal for William Zeckendorf. The prolific real estate developer wanted a giant dome—a stunning 1,500 feet in diameter—for his proposed Freedomland theme park in the Bronx, but the plans fell through, and

Rendering of floating cloud structures (Cloud Nines) by Fuller and Shoji Sadao.
This photomontage dates from around 1960.

the Raleigh team became exasperated by Fuller's fixation on unlikely projects.

All the while, the firm was operating at a loss. Synergetics had received just one minor contract since the American Society for Metals, and Fuller had trouble collecting royalties on radomes, leaving him unable to cover their salaries. By the time of his return, he owed the office tens of thousands of dollars, forcing Fitzgibbon to lay off a third of the technical staff and all the part-time workers, including Wolcott Fuller's daughter Lucilla and her husband, Tom Marvel.

Their only hope was a large commission, which Fitzgibbon worked furiously to land without Fuller's participation. One possibility lay in Frobisher Bay, a community on Baffin Island that Fitzgibbon proposed enclosing in a single enormous dome. A more plausible prospect was the domed greenhouse for the Missouri Botanical Garden in St. Louis, which became known as the Climatron. Synergetics handled the engineering, with Howard primarily in charge, and although Fuller subsequently took credit for it, he wagered privately that it would collapse.

As a practical matter, Fuller lacked the background to oversee complicated projects, and Howard recalled that his fears nearly gave him "a nervous breakdown" over the Union Tank Car Dome. For all his grand visions, Fuller was more at ease with smaller domes intended for mass production, such as a plan for houses by an Ohio building company, Pease Woodwork, and he clearly longed for a return to unconstrained speculation—allowing him to avoid building anything at all.

Fuller was still fascinated by the unproven potential of tensegrity, claiming that it could be used to cover entire cities. A tensegrity sphere of more than a mile in diameter, which he compared to a "hollowed-out balloon," would supposedly float into the sky, since the weight of its shell was a tiny fraction of the mass of air inside. A slight reduction of pressure—perhaps from the heat of the sun—would cause it to rise, and he later envisioned "Cloud Nine" structures as colonies for thousands of passengers or even as "launching platforms for missiles."

As Fuller prepared to file a patent on tensegrity, its real function was to excite the students who had been enthralled by the dome before it had been reduced to a business. In January 1959 he oversaw a tensegrity sphere of identical components at the University of Oregon, where he gave a lecture inspired by the British philosopher and mathematician Alfred North Whitehead. Because the most competent professionals were encouraged to specialize, Whitehead observed, politics had been left "to those who lack either the force or the character to succeed in some definite career."

Fuller's solution lay in "the self-commissioned architect." Following his own advice, he commissioned himself to focus on his theories of pattern and geometry, which he outlined in the manuscript "Omni-Directional Halo." To understand the "omnidirectional universal maelstrom of nonsimultaneous near and far explosions" that made up the universe, he called for a new kind of thinking that he defined as the tuning out of irrelevancies. Most phenomena were too large or small to be seen, with twilight areas on both sides that were tantalizingly close to comprehension.

To expand the "zone of lucidly tuneable relevancy," he advocated refining one's intuition—a valuable shortcut for individuals without access to conventional resources—with a geometrical system based on the tetrahedron, which was the most basic system of any kind. As he moved into what would later be known as systems theory, the risk was low, while the upside was incalculable, and it required no clients, funding, or center of operations. Over time his main product line became the weightless principles of his geometry, which Fuller saw as a form of wealth, and within a few years, he had rebranded it as "synergetics."

As Fuller pursued these investigations in private, his public talks grew into a form of invisible architecture that seemed more monumental than ever. On January 31, 1959, he spoke at San Quentin State Prison in Northern California, having arranged to appear through a class on general semantics, the cognitive discipline developed by Alfred Korzybski. He arrived at seven in the morning, and when he went onstage, he saw the inmates entering with their heads down. Most were men of color, and few were out of their twenties.

Fuller never prepared his talks in advance, and as he looked out at the audience, he found himself reflecting on his life: "My mother was really very scared that I was on the way to San Quentin," he confided. "She told me so many times." He grew so caught up in his words that he closed his eyes, and after an hour, he realized that his listeners were staring "because I was almost at the point of falling off the stage."

After lunch, Fuller kept going. He was scheduled for five hours, but he talked for much longer, and when he was done, he claimed, the inmates were so elated that they rushed up to shake his hand. Fuller was touched, and he never questioned the sacrifice that they had allegedly been willing to make: "They were just going to sit there and listen, even though it would have meant missing their head count. And if anybody's late for a head count, it means solitary, and here these men were agreeing to take a week or a month of solitary to hear me talk for another minute."

Shortly after returning from San Quentin, Fuller visited the office in North Carolina. The big contract that Synergetics needed to survive was still elusive, and Fuller alluded at a meeting to a "turning point in our affairs." After discussing the predicament with the others, Fitzgibbon wrote to him on March 12, 1959, to propose that "some of us in Raleigh purchase Synergetics Inc. from you and operate it as the geodesic, architectural engineering consulting firm it is tending to be."

Fitzgibbon advocated a relationship that was similar to Fuller's arrangement with Bill Wainwright and Geometrics, although he hoped that Synergetics would remain more involved with his research. He offered to use the staff's unpaid salary accounts as money for a buyout, which would essentially forgive Fuller's debts in exchange for ownership. As its founder, he would be listed as president and kept up to date on its activities, but their financial ties were cut by June, as another group of associates decided that a formal connection with Fuller was no longer worthwhile.

At exactly the same time, a landmark project was taking shape that would cement his global reputation. The American National Exhibition

Fuller and Frank Lloyd Wright at Taliesin West (March 1959).

in Moscow, scheduled for the summer, had been taken over by the government after its private backers withdrew. Given the six-week event's diplomatic significance, it was granted a substantial budget, which was placed in the hands of Fuller's friends. Jack Masey was responsible for the overall planning, and he hired George Nelson to oversee the exhibits for "the big blockbuster" of the trade show era.

The exhibition would carry the charm war into the heart of the Soviet Union, and a dome by Kaiser, conceived by the organizers as an "information machine," was their first choice for the pavilion. As the centerpiece of a network of temporary structures in Sokolniki Park, including a smaller dome for screenings of a Circarama film by Walt Disney Productions, it would be two hundred feet in diameter, with its aluminum shell anodized to a golden sheen.

In the spring, the unfinished dome was toured by Nikita Khrushchev, who joked with workers and playfully pretended to slip a rivet into his pocket. The rotund Soviet premier approved of what he saw, saying, "I am thinking of authorizing the same thing here." According to

Khrushchev, the Russian architect and painter Konstantin Melnikov had built something much like it decades earlier, but he concluded, "This dome is very interesting, very good, worth copying many times over in the Soviet Union. J. Buckingham Fuller [*sic*] should come here to lecture our engineers on his invention."

Fuller's actual involvement was nonexistent, and he devoted the first half of the year to extending his vision of global transportation networks into the ocean. His latest invention, the Submarisle, was a floating undersea island—anchored by cables that formed a pair of "counter-torquing tetrahedra"—that could serve as a drilling rig or a refueling station for submarines. Fuller filed for a patent, but it would never exist outside his mind's eye.

In March he visited Frank Lloyd Wright at Taliesin West, where they posed outdoors for pictures and watched as a chartered helicopter took aerial photographs of the grounds. "I am an architect interested in science," Wright informed the Taliesin fellows. "Buckminster is a scientist interested in architecture." The passage of time had turned them into what felt effectively like contemporaries, but they were still a generation apart, and Fuller was saddened a month later by the news that Wright had passed away at the age of ninety-one.

For nearly four decades, they had regarded each other with alternating wariness and respect. At the beginning of his career, Fuller had predicted that his ideas would be appropriated by Wright, and he dismissed the renowned architect's conception of building with nature as "having a stream in your living room." More generously, Fuller saw Wright as "the last of that era of great composers" who combined art with science, while he was merely "the instrument man."

Wright, in turn, had called Fuller a person "with more absolute integrity than any other man I have ever known." Fuller's influence was visible in the central core that supported the Johnson Wax Research Tower in Wisconsin, which Wright compared to a tree, while Fuller thought that the Price Tower in Oklahoma had been inspired by his designs from the twenties. Fuller claimed that he was often consulted on engineering by Wright, who had granted his own form of praise two years before his death: "I am one of Bucky's patron saints."

On June 8, 1959, Fuller was one of four featured speakers at a memorial dinner for Wright at Taliesin East in Wisconsin. In a speech that he delivered before the closing remarks by Democratic politician Adlai Stevenson, Fuller said, "Of all the befriendings I have ever had, there is none that has given me such courage and pride." He stated that Wright had encouraged him to focus on structure, which was the gift that he had been given by God, and ended, "Frank Wright, as the ages go by, will be seen as the great architect who was responsible for carrying the Old forward into the New; and thereby really gave us what I, as a student of pattern, might not have been given otherwise."

In Wright's absence, Fuller had arguably become the most visible prophet of architecture in America. In an interview with Calvin Tomkins of *Newsweek*, he returned to the provocative concept of a dome over Manhattan—he claimed that it would pay for itself through savings from snow removal and climate control—and his assertions for the scale of tensegrity structures were as overwhelming as ever: "There's no inherent limit to the size you could make them."

Rendering of the Dome over Manhattan by T. C. Howard.
This photomontage dates from around 1960.

He enjoyed describing himself as "a research laboratory for architecture," but his routine was upended in June. While Fuller and Anne were traveling, a fire broke out at their home in Forest Hills, where Sadao had been staying while they were gone. In anticipation of their return, their houseguest activated their aging refrigerator, which ignited. The flames spread from the kitchen into the living room, and the refrigerator plummeted through the floor to the apartment below.

Fortunately, before they left, Anne had covered the furniture and bookshelves so that the walls could be painted, saving most of Fuller's papers from damage. The blaze still left him with no fixed address, aside from an office above "a ramshackle store," and while Anne moved in with her aunt in Lawrence, he took off for a European trip that would culminate in a visit to Moscow.

His first stop was England, where his most significant encounter occurred in private. Earlier that year, John McHale had seen an article in the British newspaper *The Observer* on the structure of the polio virus. The biophysicist Aaron Klug and the crystallographer John Finch had determined that its shell had icosahedral symmetry, which appeared to be "the most economical way of 'packing' the small protein units around a central core." McHale informed Fuller, who saw Klug and Finch in London, and the meeting set a series of events in motion that would result in a major breakthrough.

On July 25 the American National Exhibition opened to the public. Khrushchev and Vice President Richard Nixon cut the ribbon at the dome, which they entered to watch an innovative film by the Eameses, *Glimpses of the U.S.A.* At a model home across the park, they engaged in the famed "Kitchen Debate," where Khrushchev objected to Nixon's claim that Americans liked to upgrade their houses after twenty years: "Some things never get out of date. Furnishings, perhaps. But not houses."

Two days later, Fuller arrived for his triumphant week in Moscow, serving as one of twelve cultural representatives, along with poet Carl Sandburg and photographer Edward Steichen. Fuller took what Don Richter described as "a lot of bows" for the dome, and when Peter Blake, who had curated the exhibit on modern architecture, escorted

Fuller at the dome built by Kaiser Aluminum in Moscow (1959).

him to Red Square, they drew curious looks. Fuller remarked, "You see, dear boy, these people know all about my work."

Blake suspected that the passersby were more interested in Fuller's glasses, "a pair of enormous, thick lenses held in place by a wide, black rubber band that circled his shiny, bald head." Yet his designs had unquestionably received positive attention from the Soviets, who had agreed the year before to purchase the dome as a venue for sporting events and trade fairs. "We are going to have your dome forever," an interpreter told Fuller, who suspected that they wanted to study its underlying structure before entering geodesics "in a big way."

Fuller related this development to the launch of *Sputnik 1*. When the first artificial satellite went up on October 4, 1957, stunning—and terrifying—much of the world, he had been lecturing at Southern Illinois University, where he entered a classroom the next day brandishing the newspaper "like a kid who had just gotten a long-awaited birthday present." While most Americans regarded this momentous develop-ment with dread, Fuller saw it as a sign that the world was falling into line with his predictions, and he prepared to assume a leading role. On a tour with Soviet engineers, he basked in their praise: "They said that I was recognized in Russia as one of the pioneers, if not *the* pioneer, in applying airplane technology to architecture."

His personal impressions of Moscow were largely uncritical. As he wrote later, "Many a US worker citizen, participating in the installation of the American Exhibit dome in Moscow, looked with envious reminiscence upon the unstinted, unlimited hours of enthusiastic work dedication universally exhibited by the Russian laborers. . . . The more work we gave them, the more friendly they became." Along with the implication that he had participated in the dome's construction, Fuller was characteristically eager to contrast unionized labor with the industrious spirit of other countries, regardless of the political situation.

In many ways, his sympathies lay less with Nixon than with Khrushchev, to whom he bore a certain physical resemblance. Fuller declared confidently that the Soviets wanted to reallocate military spending to industrialized housing, which could deliver towns by air, and he was willing to give the premier the benefit of the doubt: "Khrushchev is glad to have his adversaries think of his desire for disarmament as constituting only propaganda, for this will give him just so much of an advance start in capturing consumers for his conversion of weaponry to livingry."

The trip itself was a triumph of "livingry," but Fuller was still without a home of his own, for which his academic connections provided a possible solution. During the summer, he had received another honorary doctorate—his fourth—at Southern Illinois University. Delyte Morris, its president, wanted a marquee name for the design department, and Harold Cohen proposed hiring Fuller as a research professor, although, he warned, "He's not a teacher—he just talks." It came with access to an archivist and storage for his files, and he would spend only part of the year in Carbondale, which Cohen saw as a successor to the Institute of Design.

Anne was less than excited by a move that would take her far from her family and friends, but the advantages for her husband were clear: it would provide a headquarters from which he could rove safely, and he would report directly to Morris, who was unlikely to burden him with responsibilities. On September 18 the SIU board approved Fuller's appointment to the faculty for the fall semester. A quarter of his time would be spent teaching, with the rest devoted to research and directing a program in product and shelter design, at an annual salary of $12,000.

It was a welcome source of stability, and it came just as Fuller was making strides elsewhere in the establishment. Arthur Drexler, the director of the Department of Architecture and Design at the Museum of Modern Art, had been intrigued by Fuller ever since the Weatherbreak Dome. For the museum's thirtieth anniversary, he wanted an exhibition that would reject the modernist emphasis on function to celebrate the forms of technology for its own sake, and Fuller was the natural choice.

Fuller played only a nominal role in the show, which would feature a space frame, dome, and tensegrity mast in the museum garden. A huge golden octet truss was designed by Howard, who placed its base strikingly off center, and manufactured by Aluminum Limited of Canada. The tight schedule meant that a radome had to be borrowed from Lincoln Labs, but its designers went uncredited, deeply offending Bernie Kirschenbaum.

For an updated version of the tetrahedral mast from Black Mountain, Fuller turned to Sadao, who had landed a job through Noguchi with Edison Price, a lighting consultant based in Midtown Manhattan. Sadao and Price developed a revised design with aluminum struts that they learned to balance by ear, tightening the rods until they all resonated at the same pitch. When they lowered it from their studio window and carried it down the street, onlookers gathered to stare.

With the help of various hands, including Noguchi, the exhibits quietly rose into view above the MOMA garden wall. As the opening day drew near, Fuller made an equally significant impression in the *New York Times*, which published a laudatory profile by the freelance writer Robert W. Marks. One admiring quote was attributed to an anonymous engineer: "The domes do so much at such low cost that they are bound to make Fuller one of the richest and most powerful men of our time."

When it opened on September 22, 1959, "Three Structures by Buckminster Fuller" did for its subject what a similar retrospective had done a decade earlier for Ludwig Mies van der Rohe. Drexler said that domes might be used on the moon, that the octet truss could make it possible to "weave" entire buildings, and that tensegrity called for a new sense of scale: "In effect, the city would be one building. . . . We could climate-control and reclaim whole areas of the Sahara, or of the Arctic. We would have to rethink architecture as we know it now."

The *Times* was similarly effusive. Ada Louise Huxtable praised the exhibit as a return to the museum's grand tradition of showmanship, hailing Fuller's designs as "the greatest advance in building since the invention of the arch." Calling him the originator of "the newest look in architecture since the Bauhaus," John Canaday singled out the tensegrity mast: "It could remain exactly where it is as a piece of extraordinarily ingenious abstract sculpture of the mechanistic school."

Canaday was closer to the truth than he knew. For years, Kenneth Snelson had been demoralized by his mentor's appropriation of his ideas, and he had been especially pained by an unattributed photo of the mast in *Architectural Forum*. "I had been successfully swallowed up," Snelson recalled years later. "The damage was already done. He had access to publicity and, lesson of all lessons, the power resides in the press." At the time, though, Snelson never mentioned these feelings to Fuller, whom he saw occasionally in New York, although he daydreamed sometimes about "cutting him to pieces."

Shortly before the exhibit opened, Snelson received a call from John Dixon, Fuller's assistant, who invited him to the museum. Upon his arrival, Snelson saw the mast lying on the ground. "I was suddenly gripped with fear that I would, in this instant, be thrust into an ugly showdown with Bucky over my long grievance which had festered in silence," Snelson remembered. "The truth was, I was afraid of the man; that enormous father figure. Bucky was not known to abide challenges, open or otherwise, and those around him were familiar with his tantrums when he was crossed."

For now, instead of saying anything, Snelson accompanied Fuller and the others into the radome, where their voices echoed against the walls. Over lunch in the garden, Fuller spoke glowingly of the mast that Sadao and Price had made: "Pure jewelry, Ken, pure jewelry."

Snelson heard himself speak. "I hope my name is going to be on it this time, Bucky."

Fuller stared at him for a moment. At last, he stammered, "Oh, I know, Ken. Yes, I'm sure I've told Arthur Drexler all about you."

Dixon, who had been listening in silence, spoke up. "Gee, Bucky, I don't think Arthur knows about Ken."

Fuller tried to dodge the issue. "Oh, I'm certain I've spoken about Ken to Arthur . . ."

"Well, we could ask." Dixon rose to check with Drexler. Snelson followed, sensing that Fuller's assistant had few illusions about his employer, and when Dixon introduced him with a description of his contributions to tensegrity, it was clear that the museum director knew nothing about it.

Going to Fuller, Drexler said that he would amend the catalog and text to credit Snelson. Fuller only nodded, knowing that he had to put the best face on it that he could. "Yes, wonderful."

At what should have been an hour of triumph, Snelson had complicated the narrative. It was too late to revise the press release, but Drexler arranged for a second exhibit, which made an effort to correct the record. The show included a collage by T. C. Howard of the Dome over Manhattan; a design for an athletic center by Synergetics; and a display case devoted to Snelson, including a replica of the sculpture that he had given to Fuller at Black Mountain.

This smaller exhibit closed quickly, while the outdoor exhibition ran well into the following year, with the radome serving as the workspace for the Swiss artist Jean Tinguely's sculpture *Homage to New York*, a huge kinetic assemblage that destroyed itself in full view of the crowd. The belated recognition was less than perfect, but it inspired Snelson to show his art "with the feeling that now I was free and on my own."

As for Fuller, he responded with even more outlandish claims for tensegrity, seeing it in everything from the atom to the solar system, as well as at the heart of his career: "All of my structures from the

Dymaxion House of 1927 and all the geodesic domes are tensegrity structures." Neither statement was true. There was a vast difference between the segregated compression members of tensegrity and the tension elements of his earlier work, which had more in common with the bicycle wheel and suspension bridge, and the dome was a separate line of development entirely.

His attempt to link them all together was a purely rhetorical gesture, but Fuller may well have believed it, much as he felt that his followers owed all their insights to his influence. As Snelson observed, "They became his ideas the moment he recognized them." His reputation as an innovator was his greatest asset, and the question of credit outweighed any financial reward. As a result, while he sometimes acknowledged Snelson in his lectures, he rarely did so in print. When Snelson once asked him why, Fuller's response revealed how conscious he was of the limited time that he had left: "Ken, old man, you can afford to remain anonymous for a while."

NINE

INVISIBLE ARCHITECTURE

1960–1967

Only humans play "Deceive yourself and you can fool the world"; or "I know what it's all about"; or "Life is just chemistry"; and "We humans invented and are running the world." Dogs play "Fetch it" to please their masters, not to deceive themselves. The most affectionate of dogs do not play "Burial of our dead"—"Chemistry is for real." Only humans play the game of masks and monuments.

—BUCKMINSTER FULLER, *SYNERGETICS*

As Fuller established himself at Southern Illinois University, he still spent most of his time on the road. In March 1960 he lectured for two weeks at Princeton, where students built a plastic globe six and a half feet in diameter. With its continents aligned with the stars, it was hyperbolically described as a map "believed to be the largest and most accurate in the world." Using the standard scale of airborne photography, Fuller said that it would be possible to build a computerized version big enough to show individual houses, which he had recently rechristened the Geoscope.

If the earth's geography had once been theoretical for Fuller, who had spent most of his life in America, it had since become very real. As he moved into his more expansive future, however, a key connection to his past was severed. His brother, Wolcott, had died in November in Pittsfield, Massachusetts, at the age of sixty-one, after a career in sales at General Electric. Rosamond was now Fuller's last surviving sibling, and as he settled into Carbondale, mortality weighed on his mind.

It soon became clear that Fuller would be traveling as often as before. Students liked him, but his lectures could be hard to endure—attendees were quietly provided with caffeine pills—and some complained of a lack of practical instruction. He seemed detached from the rest of the faculty, and he maintained his transient lifestyle by staying as a houseguest with Harold Cohen, whose wife, Mary, patiently edited his manuscripts. The arrangement lasted six long months, but he got along well with their children, entertaining them with stories about "the gnomes who lived in domes."

For his permanent residence, Fuller had envisioned a dome by Kaiser, but he decided on the standard model from the Pease Woodworking Company instead. It would be the first house that he had ever owned, as well as a promotional tool: Pease's domes had been available for over a year, but despite the efforts of its salesmen, who carried models in custom sample cases, not a single one had been sold.

The house would be located on an undeveloped lot at the corner of Cherry Street and Forest Avenue, less than a mile from campus. To build it, Fuller asked for student volunteers, insisting, "We can do it in a day." In fact, nearly all the labor was handled by five local construction workers, who erected most of the exterior on April 19, 1960. Alvin Miller, its designer, oversaw the work, while Fuller lectured to onlookers. The shell, consisting of sixty plywood sections bolted to wood frames and sealed with tape, was largely finished by dinnertime, but the rest took months, as electricians and plumbers struggled to make sense of a house that lacked conventional angles.

As a typical Pease model, the Carbondale house was 39 feet in diameter and 16 feet high, with 1,400 square feet of space, including a balcony level. The large windows let in plenty of light, but they also forced Fuller to put up a redwood fence for privacy, closing him off from the neighborhood. Separate bathrooms were installed for him and

Anne, who designed the interior to unfold as a continuous space that Fuller compared to "coming out of a conch shell."

When Anne joined Fuller in what he referred to as his "private motel," she found that the pictures she hung "would be just sort of dangling out from the curve." She furnished it with antiques from China and Italy, a sculpture by Ruth Asawa, and a chair that had belonged to Margaret Fuller. Housekeeping was fairly easy, but when the seams in the roof failed to expand properly, leaks were a constant issue. Fuller was seen reading at home with a book in one hand, an umbrella in the other, and a flashlight clutched between his teeth, and after he was unable to fix the problem with plasticized paint and foam, he finally shingled the outside.

Apart from the minor nuisance of children climbing on the roof, whom Fuller drove away with pebbles, they settled in comfortably. Carbondale had taken Anne far from her family, but she became more popular there than her husband, serving tea to students, cooking up shrimp creole for faculty parties, and handing out candy on Halloween.

Fuller and Anne in the living room of their dome house in Carbondale.

Fuller was harder to like, even for Harold Cohen, who had played a major role in recruiting him. "Bucky was brilliant but somehow devoid of emotion for the plight of people, except in the abstract sense," Cohen observed. "He wouldn't give money to a beggar in the street, but he would design domes for 'the poor.'"

Cohen thought that Fuller's energy came with "an ego to match," and he became even more overbearing after the 1960 publication of *The Dymaxion World of Buckminster Fuller* by Robert W. Marks, the most comprehensive book on his work so far. Fuller's hand was visible in the picture section's lengthy captions, which established a canonical sequence of his ideas—tensegrity came before the domes, implying that the latter were based on the former, and it featured many structures that had barely involved him at all.

According to Marks, Fuller was interested in domes primarily as an expression of his geometry, and he considered no single building to be of "generic importance." To anyone who was paying attention, this statement was contradicted by his eager pursuit of showpiece projects that seemed capable of securing his reputation. Although he no longer controlled Synergetics Inc., Fuller had access to powerful deal makers, and he benefited from the publicity in the meantime.

Fuller had kept in touch with Judge Hofheinz, who was still advocating for a domed ballpark in Houston. The city was about to get its first professional baseball team, the Houston Colt .45s of the National League, one of the new expansion clubs being formed for the 1962 season. A contract to build the first covered stadium in the world would be a huge prize, and Fuller's patience paid off when the Harris County Domed Stadium—later known as the Astrodome—was officially announced in August 1960. It would be 720 feet in diameter, with room for forty-five thousand spectators, and construction was scheduled to begin the following year.

As Fuller had intended, he was featured prominently in the press coverage, which indicated that he would "collaborate with local design and consulting architects." His name was really only a form of branding for the concept of a dome, just as it had been for Walter O'Malley— there was no deal or plan, and he was hardly in a position to provide the engineering himself. Synergetics stepped into the gap yet again, as T. C. Howard submitted a quote that ended up competing with several rival bids, including one from Jeffrey Lindsay.

In the fall, another enormous prospect appeared in the person of the Japanese newspaper and television magnate Matsutarō Shōriki, the seventy-five-year-old owner of the mighty Yomiuri Giants baseball team. To explore the possibility of building his own covered ballpark, Shōriki dispatched sports columnist Sotaro Suzuki to Los Angeles to consult O'Malley, who then sent him to Fuller in New York. Suzuki submitted a favorable report, and Fuller soon received a telegram inviting him to Tokyo.

Fuller welcomed the overture from Shōriki, who was the kind of patron with whom he could deal directly. The prudent approach would have been to hire Synergetics, which had worked for years on the stadium problem, but Fuller came up with a new design that he could call his own. It was an aspension dome, for "ascending suspension," made of a series of concentric circles that evoked a Japanese lantern. With each ring supported by tension members attached to the level below, it could be assembled on the ground and raised one story at a time.

It existed only in his mind, of course, and wasn't even geodesic, but Fuller asked Sadao to begin work immediately. He wanted to announce it a month later in Tokyo, but any public disclosure would make patenting it there impossible unless an application had been filed already in the United States. Early the following year, Sadao completed the sketches, which were used as the basis for renderings by American sculptor Richard Stankiewicz, and the paperwork—with some illustrations still in pencil—was handed to Fuller to sign during an airport layover on his way to Japan.

In February 1961 Fuller arrived in Tokyo with Anne and Sadao. He brought a beautiful model of the aspension dome, but Shōriki was unconvinced. Its tensegrity hemisphere lacked the stability of a complete sphere, and there were questions about its performance in snow, which Fuller proposed melting with infrared heaters. By departing from his most famous design, he had made his task even harder, but they agreed that he would conduct additional research on a stadium, along with a relatively modest dome as a golf clubhouse for the Yomiuri Country Club.

Fuller was more at home on a smaller scale. To consolidate his portfolio of dome designs, he applied for a patent on what he called the laminar dome, constructed of diamond pans similar to those that the Chicago branch of the Fuller Research Foundation had developed decades earlier. The Monsanto Chemical Company shipped two hundred foamcore domes to Puerto Rico for the Peace Corps, which had ties to SIU, and in April, Fuller was asked to address the planning committee for the extension of the university to a campus in Edwardsville, Illinois.

He spent much of his talk, which was published as *Education Automation*, predicting the end of the physical college itself. Describing himself as a "comprehensivist," he proposed that schools offer remote learning through television to serve "the children of the mobile people." He noted that if an expert recorded one lecture, it could be distributed as widely as needed—an approach that he was already putting into practice with film projects for Robert Snyder and the documentarian Francis Thompson.

Fuller believed that this was only the beginning. Given the technology of his time, he focused less on computers than on television networks, believing that the latter could allow documentaries to be retrieved on any topic. "There is no reason why everyone should be interested in the geography of Venezuela on the same day and hour unless there is some 'news' event there, such as a revolution," Fuller observed. "However, most of us are going to be interested in the geography of Venezuela at some time."

Along the way, Fuller anticipated many aspects of the Internet, including the economic impact of what would eventually be known as information technology: "You are faced with a future in which education is going to be number one amongst the great world industries, within which will flourish an educational machine technology that will provide tools such as the individually selected and articulated two-way TV and an intercontinentally networked documentaries call-up system, operative over any home two-way TV set."

He concluded that "the *infra-* and the *ultrasensorial* frequencies of the electromagnetic spectrum" had driven all the major scientific discoveries since World War I, which called for an advanced education that would enable mankind to "control the invisible events of nature." Remembering belatedly that he was advising on a new campus, he engaged in a lovely reverie:

> A circus is a transformable environment. You get an enclosure against "weather" that you can put up in a hurry, within which you can put up all kinds of apparatus—high trapezes, platforms, rings, nets, etc. You can knock it down in a few minutes. That is the way the modern laboratory goes. . . . You can put the right things together very fast, rig them up, get through the experiment, knock it down. It's one clean space again. You want clean spaces. The circus concept is very important for you.

As for specifics, he described a portable dome theater that Synergetics had designed for Ford, which would appear a year later at the 1962 Seattle World's Fair. Fuller recommended housing the Edwardsville campus under a clearspan subdivided by "delicate occulting membranes, possibly rose bushes or soap bubbles or smoke screens." He closed by urging the school to buy aircraft and a "generalized computer setup with network connections," based on the assumption that "any dreamable vision of technical advance will be a reality."

In June 1961 Fuller spoke at a conference on the Edwardsville extension, held in an inflatable dome in East St. Louis. He defined education as society's effort to reach "a comprehensive understanding of its obligations," which hinted at his private plans. Now that he had a base of operations, he could build his student following into a worldwide movement that could sustain itself without his direct involvement. If his existing network was comparable to a geodesic dome, this one would be a tensegrity structure, with its pieces widely separated, and he wanted to see how large he could make it.

It required a correspondingly vast idea, and Fuller found it in the challenge of using the earth's resources for the welfare of all mankind—a legitimately urgent and worthwhile goal that would have greater appeal to the young than the military and industrial applications he had pursued in the past. In July he spoke in London at the World Congress of the International Union of Architects (UIA), attacking the "you or me" ideology of the English economist Thomas Malthus. Industrialization trended toward universal wealth, but because capitalists and politicians acted only in war, Fuller contended that power ought to be given to students instead.

By arguing that architectural schools should take responsibility for encouraging radical design solutions in peacetime, Fuller was really just changing the name of the patron, which was a role that he intended to assume himself. It required powerful supporters, and at the congress ball, he danced with Monica Pidgeon, the influential editor of *Architectural Design*. Pidgeon had dreaded being drawn into one of his verbal labyrinths, but he charmed her so completely that she promised to "stuff the results" of his work down everyone's throats.

He followed up with a formal recommendation to the UIA. For the congress that was scheduled for 1965, he proposed that students design "a facility for displaying a comprehensive inventory of the world's raw and organized resources," which currently provided a high standard of living for only 40 percent of humanity. As one acceptable solution, he recommended that participants compile statistics and build models of the Geoscope. These dramatic presentations would lead to a consensus on the world's problems that would guide further studies over the next ten years.

It amounted to a global takeover of the architectural curriculum, but for now, it was only a proposal, and Fuller left for commitments in Greece, India, and Japan. In Calcutta, he oversaw the construction of three bamboo domes for the Ford Foundation, before flying out to conduct damage control on his ballpark in Tokyo. He had asked Hideo Koike, a fellow SIU faculty member, to refine the plan with a friend named Al Johnson, who noticed issues with the design. "Al had run the math through some mainframe computers in Tokyo and found the flaws," Koike recalled. "He told Bucky, 'Don't do it because it won't work.' Bucky didn't like that, so he fired Al."

Fuller insisted later that the aspension dome was sound, although not "efficient," but he urgently needed an alternative, and after returning from Tokyo, he shifted to a "deresonated" tensegrity dome that was stabilized by fastening together its struts. On his visit to Moscow in 1959, he had met Don Moore, an engineer for the Whirlpool Corporation, who built a prototype at Bear Island. Fuller named it the Sunclipse Dome, after the word that he proposed as a replacement for the misleading term *sunset*. His enthusiasm for the commission was waning, however, and he was distracted by a proposal by William Zeckendorf to install a gargantuan dome over Yonkers Raceway.

Despite his lack of realized projects, Fuller basked in the esteem of the establishment. At a January 1962 exhibit of his work at the US embassy in London, he received a surprise visit from Princess Margaret and her husband, Lord Snowdon, for whom he later arranged for the construction of a tensegrity chandelier. The following month, Fuller commenced his most meaningful lectureship of all. At the invitation of McGeorge Bundy, the former dean of the Faculty of Arts and Sciences at Harvard, Fuller would serve as the Charles Eliot Norton Professor of Poetry.

The Norton lectureship, which he shared with the architects Félix Candela and Pier Luigi Nervi, encompassed what press coverage described as "poetry in its broadest sense," but it would widely be taken to legitimize Fuller's status as a poet. In advance, however, he was uncharacteristically nervous, and he asked Arthur Penn for advice at a dinner party, where they reviewed the acting techniques that he had learned as Baron Medusa. Penn described it as "an analytical session, in that I was really sort of restoring some of the liberating experiences—a sort of retransference again."

Fuller was reinvigorated by the spirit of Black Mountain. In one talk, he provided a list of future trends, including the rise of computers, which were doubling in number every year. Computers, he predicted, would replace humans as specialists, but not in asking original questions, and mankind's success would be enabled by innovative designers and architects, with greater breakthroughs yet to come: "Medical science, through development of interchangeable human parts, both organic and inorganic, may be about to develop the continuous or deathless man."

He was becoming increasingly interested in biology, largely because of the virologist Donald Caspar, who arranged to see him at Harvard. Along with biophysicist Aaron Klug, Caspar was preparing a landmark paper, "Physical Principles in the Construction of Regular Viruses," for the Cold Spring Harbor Symposium. After his conversation with Fuller in London, Klug had read *The Dymaxion World of Buckminster Fuller*, which encouraged him to explore the parallels between energetic geometry and the shape of certain viruses, while his colleague Caspar used a children's construction toy called Geodestix to investigate a long-standing problem in virology.

Chemical analysis had shown that the shells of many viruses, including polio, were made of identical protein units. Since it was often impossible to arrange a specific number of molecules in a precisely uniform pattern, Caspar and Klug arrived at an analogy to geodesics. In a dome, the triangles all looked the same, but they actually varied imperceptibly in shape, as defined by their chord factors, which might be equally true of the structure of viruses. As the paper explained: "The basic assumption is that [the] shell is held together by the same type of bonds throughout, but that these bonds may be deformed in slightly different ways."

Robert Horne, who oversaw electron microscope research at Cambridge University, left no question regarding their inspiration: "Somebody told us about Fuller's book. We opened it, and there it was, all worked out." Fuller saw their research as a vindication. He was delighted when they confirmed his equation for the number of nodes in a viral shell, and he spoke effusively of the typical "comprehensive thinker" in their field: "Any good virologist sitting in with us today would find no trouble in participation in your subject. In fact, he would probably be able to lead you in formulating a grand strategy of attacking your special problem."

Elsewhere, though, he was facing challenges of his own. In Tokyo, he presented Matsutarō Shōriki with a model of a "monohex" dome that he had developed years earlier at SIU, featuring circular windows and uniform pieces that could nest together for storage. He also reached a deal for a domed golf clubhouse at the Yomiuri Country Club, with structural engineering provided by Geometrics. It came close to his ideal of the Garden of Eden, complete with waterfalls and miniature pines, and took six years to build at a cost of $400,000.

The stadium was a lost cause, however, and his hopes of a monument elsewhere were fading. In Houston, the geodesic design was found to be "relatively uneconomical," and Roof Structures of St. Louis emerged as the front-runner. Accusing the firm of trying to "invade" his field with a low bid, Fuller claimed that Judge Hofheinz had avoided paying him royalties by hiring contractors who "modified their designs so that their design would not read on my patents." He implied further that they were backed by U.S. Steel, which wanted to discourage the use of aluminum.

None of this was true. Synergetics had provided quotes for both aluminum and steel, and the Roof Structures dome used a lamella grid, not a geodesic one, which arranged its members in diamonds. The firm's principal designer, Gustel Kiewitt, whom Fuller dismissed as one of the "dome building aspirants" competing for his business, had executed a lamella clearspan for the indoor St. Louis Arena a full three decades earlier. Fuller said later that Hofheinz's son Fred, who, like his father, became the mayor of Houston, credited him privately with the design of the Astrodome, although Fuller continued to publicly criticize the stadium as "an engineering blunder."

In reality, it was a bitter failure, and the outlook on the whole was bleak. Two thousand domes had been built, but most were radomes. The government was moving on to more effective designs, civilian orders were nonexistent, and the Kaiser line was no longer yielding royalty payments. Fuller blamed "the number of licensees," which only indicted his own decisions, and was disheartened that the dome industry had peaked less than ten years after it began.

It was at this difficult point in September 1962 that Fuller learned that his proposal for a global student project had been approved by the International Union of Architects. The news came at just the right time, and it left him with little doubt of the way forward. His future would lie with the young, in a global movement that he could direct from a

distance, and he confidently prepared to inaugurate it as the World
Design Science Decade.

Early in the afternoon of November 9, 1962, a team of workers was
repairing the dome of the Ford Rotunda in Dearborn, Michigan. To
seal its persistent leaks in advance of an upcoming winter festival, two
men covered the seams in translucent tar, which they then diluted with
gasoline and heated with a propane torch. Looking up, they saw that
a number of small fires had sprouted on the north side. They tried
putting out the flames with handheld extinguishers, but to no avail,
and as they escaped down the inside stairway, the entire roof ignited.

It was an inferno. Within moments, the roof was consumed, and
as the fire spread downward into the draperies, the windows exploded
outward, turning the building into a huge chimney. The steel framework buckled from the heat, causing fissures to open in the concrete,
and the dome plummeted to the ground. Within forty minutes, the
whole structure collapsed. Apart from the fact that it had never been
successfully waterproofed, the dome itself was never implicated, but its
destruction was a reminder of how fragile it had been all along.

Fuller was building a more lasting legacy elsewhere. In Carbondale, he
established a global resources inventory, hoping that students at other
colleges would follow suit, and oversaw various books. SIU published
his *Education Automation* and *No More Secondhand God* over the
objections of the university press director, who protested that they weren't
written in English. A Prentice-Hall anthology, *Ideas and Integrities: A
Spontaneous Autobiographical Disclosure*, reflected his concern with
"problems of intellectual integrity," and he contracted with Macmillan
for five titles, including one on energetic geometry.

In April 1963 Fuller headed for London, where he spent an afternoon with the writer C. P. Snow. Four years earlier, Snow had given
a famous lecture, "The Two Cultures," on the divide between science
and the humanities, which Fuller supposedly persuaded him to reconsider.
Snow did publish another essay that noted the emergence of a third

culture based on the social sciences, including architecture, but he never mentioned Fuller by name.

Another notable meeting was arranged by the American poet Jonathan Williams, whose small press in North Carolina had released Fuller's previously unpublished *Untitled Epic Poem on the History of Industrialization*. In Hampstead, Williams invited Fuller to dinner with the postmodernist writer William S. Burroughs. Neither had ever heard of the other, but when Fuller mentioned the subject of virus structure, Burroughs brightened slightly and said, "I happen to have secret information that streams of alien viruses are filtering into the solar system at this very moment from the galaxy Boybutt."

In June Fuller attended a congress of architectural students in Barcelona, Spain, where he abruptly announced that he was revising the plan that he had proposed just two years ago. If the task of compiling statistics were left to the students, Fuller declared, "their lack of economic experience would either delay or altogether frustrate timely inauguration of the whole undertaking." According to Fuller, they all agreed with him that it would be impractical to proceed "by the students' own uncoordinated efforts at separate schools," and when he offered to conduct the research himself, he claimed that they became "enthusiastic."

The change effectively limited the students to the construction of Geoscope models. Instead of emerging from a collective effort, the resource data would be compiled in Carbondale, with Fuller and SIU covering the expense jointly. The shift in plans was also enabled by his access to a gifted collaborator. John McHale, who had written a book on his work, had become a visiting professor at the university, where he took over operations of the World Design Science Decade.

Carbondale had benefited from Fuller's ability to attract students and funding. He described the Mercury space capsule approvingly as "a little house" that was the first to be designed scientifically as an autonomous package, and he consulted on a contract that the school secured from NASA. Relying on a talented graduate student named Joseph Clinton, he explored structural concepts for space, including a "moon house" inspired by the Flying Seed Pod.

He was equally concerned with issues closer to home, thanks largely to Constantinos Doxiadis, a Greek architect best known for the planning of Pakistan's new capital city, Islamabad. After a frustrating conference at the United Nations, Doxiadis decided that a less formal setting would allow for a more productive dialogue on the science of human settlement, which he called "ekistics."

It resulted in the creation of the Delos Symposium, in which an assortment of public intellectuals cruised among the Greek islands aboard the ship *New Hellas*. Doxiadis invited Fuller to the first session in July 1963, where he met many influencers outside his usual circle, much as he had at Black Mountain. He had trouble following the discussions—he had begun to suffer from the partial deafness that would plague him for the rest of his life, despite a series of hearing aids—and was more comfortable lecturing into the night, but he was still treated as a celebrity.

The guests included the cultural anthropologist Margaret Mead, the British economist Barbara Ward, and an exceptional figure who connected with Fuller on the first day. Hearing someone call his name, Fuller turned to see Marshall McLuhan, who was brandishing *Nine Chains* and *No More Secondhand God*. "I am your disciple," McLuhan reportedly said. "I've joined your conspiracy."

They quickly hit it off. Both stood out on the dance floor, and McLuhan, who recalled that Fuller drank copiously, inscribed a copy of his 1962 book *The Gutenberg Galaxy* to "the glorious Bucky who provides us all with a foretaste of the extension of consciousness that is near in this Electric Age." Afterward, they exchanged letters on how information itself would emerge as the dominant industry of the future. McLuhan asked for recommendations on texts on "technology as creator of environment," and he wrote to Fuller that supersonic flight made domed cities "a necessity."

Fuller was sometimes inclined to put his new friend in his place. "McLuhan has never made any bones about his indebtedness to me as the original source of most of his ideas," Fuller said privately. "The 'global village' indeed was my concept. I don't think he has an original idea. Not one. McLuhan says so himself." He dismissed the philosopher's most famous aphorism, "the medium is the message," as "the message only of yesterday's middle-class elite," and he insisted that McLuhan had always conceded, "Bucky is my master. I am only his disciple."

In September John McHale and his team of students finished the first of six resource reports, concluding that a modest increase in mechanical efficiency—from 4 percent to 12 percent—could turn all of the world's "have-nots" into "haves," which Fuller defined as nations with one hundred or more energy slaves per family. The report also featured his essay "New Forms vs. Reforms," which he built around the image of the trim tab, a tiny piece on the trailing edge of a rudder that adjusted a ship's orientation, implying that the real work began after it seemed to be "all over."

He planned to present his findings at the upcoming UIA Congress in Cuba, which was complicated by the political situation under Fidel Castro. Americans were asked to boycott the event, and Fuller appealed unsuccessfully to McGeorge Bundy, who had been named the national security advisor to President John F. Kennedy. He also approached Dean Acheson, the former US secretary of state, and Henry Luce, arguing that he could promote "the positive potentials of United States moral and industrial leadership" to students, and he would say later that his absence at the congress merely proved to those in attendance that an American "was not a free citizen at all."

Fuller viewed the official stance purely as a personal inconvenience, and he claimed that it ended student participation in his resources project, when, in fact, he had really closed it off months before. In October he attended a sister conference in Mexico City for Americans who had been unable to travel to Havana, which the press implied misleadingly had been arranged for him alone, and attacked the great "pirates" of capitalism in a speech that lasted for three hours.

His travels in Mexico took him to the city of Puerto Vallarta, where director John Huston—an even more outrageous personality—was filming *The Night of the Iguana*. The shoot was the object of feverish tabloid speculation around its star, Richard Burton, and his frequent companion Elizabeth Taylor, who were conducting a celebrated affair that would soon culminate in his becoming the actress's fifth husband.

On his arrival, Fuller briefly took center stage as Huston presented him to the cast and crew, along with an extravagant plan to build a resort with a dome covered in vines. "We could all live under this immense dome," Huston enthused, "sleeping as the Japanese sleep, on slabs of polished wood."

Fuller became good friends with Huston, whom he visited again later on the set of *The Bible*. In Mexico, he showed renderings of the resort dome to Burton and Huston, who advised him to tell Taylor as well: "She loved the idea too," Fuller said, "but I don't know if anything will come of it." Cinerama, the company founded by Taylor's third husband, the late Mike Todd, had separately announced a chain of dome theaters, and Fuller was at an event at the Cinerama Dome in Hollywood when President Kennedy was assassinated on November 22, 1963.

After the release of *Ideas and Integrities*, Fuller was pleased to hear from John Cage for the first time in years. Although his commitment to surrendering control of the creative process contrasted sharply with Fuller's philosophy, Cage agreed with his belief that design was more effective than politics. "Your work and thought are now with me constantly," Cage said, and he went on to devote a series of public diaries to Fuller, praising him as "without doubt the greatest living man."

Fuller had reached a new level of cultural renown, which was underlined when he appeared on the cover of *Time* on January 10, 1964. The article was by Douglas Auchincloss, the magazine's associate editor for religion, whose wife, Lily, was a patron of the arts with connections in architectural circles. A correspondent, Miriam Rumwell, conducted most of the research, and after spending forty hours interviewing Fuller, she concluded that he was "the most fascinating man I ever met."

Even more notable than the profile itself was the cover portrait by the illustrator Boris Artzybasheff, who arranged for closeups of Fuller's eyes—which the Russian-born artist saw as the most distinctive feature in any person's face—to be photographed in Carbondale. Artzybasheff turned Fuller's bald head into an enormous radome, surrounded by symbols of his ideas, with two tiny spectators looking upward in awe. Fuller wrote to the magazine, "If my life provided nought else but legends to ultimately inspire Artzybasheff's cover, my life is fully justified."

The article was full of such legends, some newly invented for the occasion. Auchincloss wrote that geodesic domes had enclosed a greater area worldwide than any other form of architecture—a gross overstatement—and gave the last word to Fuller: "Today the world is my backyard. 'Where do you live?' and 'What are you?' are progressively less sensible questions. I live on earth at present, and I don't know what I am. I know that I am not a category. I am not a thing—a noun. I seem to be a verb, an evolutionary process—an integral function of the universe."

Fuller privately disliked "The Dymaxion American," dismissing the *Time* article as "an untrue and arrogant picture," but it marked a high point in his fame. The story ran while he was in Europe with Harold Cohen, whom he asked to read it aloud. "When dinner was finished, he lay his head on his dinner plate and slept," Cohen recalled. By then, he was constantly on the move. He had told Rumwell that most of his annual income of $200,000 came from dome royalties, but, in truth, he supported himself in large part by speaking alone.

As his travel schedule expanded, Fuller began wearing two watches, which later grew to three: one set for home, another for his current location, the last for wherever he was going. He spent the first third of 1964 on another international voyage, including a month in the African nation of Ghana to design a tensegrity dome "to protect clusters of native huts." In Bombay, India, he discussed a defense contract with the physicist Vikram Sarabhai, whose brother Gautam had built domes as showrooms for his textile empire, and then headed to Ahmedabad and New Delhi.

During this period, Fuller had emerged as a distinguished fixture in India, and he had finally managed to have a real conversation with Jawaharlal Nehru. On a previous trip, they had met in Kashmir, where Fuller apologized for talking while the prime minister was trying to rest. "I like your words," Nehru reportedly replied. Fuller offered to leave some articles for him to review at his leisure, and he never tired of recounting Nehru's response: "I read every word of yours I can."

He would have welcomed seeing the prime minister again, but Nehru had suffered a stroke, and Fuller reunited instead with Indira

Gandhi. Seeing that she was distraught, he asked if she intended to continue her father's work. "I would not think of doing so," Gandhi supposedly said. "I really have no aptitude." After Nehru died in May at the age of seventy-four, Fuller sent Gandhi a letter of condolence praising her father: "No one made clearer to me what our prime challenges are."

From Asia, Fuller proceeded to Australia and New Zealand, which sparked an interest in Maori culture, and even after returning home, he rarely remained in one place for long. When Shoji Sadao pursued a deal for domes with General Electric, Fuller flew out for a meeting in New York, where he visited the 1964 World's Fair. He had once hoped to cover the fairground in Flushing, Queens, with a gigantic dome, but instead, it featured a geodesic pavilion by T. C. Howard, which Fuller and Robert Moses later discussed turning into an exhibition hall. The GE venture went nowhere, but soon afterward, Fuller arranged for Sadao to join his staff in Carbondale.

In July Fuller flew to Leningrad at the invitation of Norman Cousins, the editor of the *Saturday Review*, which was preparing to print one of Fuller's first references to "a little automated spaceship called earth." Cousins had established the Dartmouth Conference in 1961 to encourage a dialogue with the Soviets, and a member from each delegation was asked to speak about the future. The esteemed geophysicist E. K. Fyodorov talked for fifteen minutes, while Fuller lectured for two hours. At the end, Fyodorov whispered to Cousins, "I give up. Fuller wins."

While there, Fuller identified another attendee, David Rockefeller, the president of Chase Manhattan Bank, as a potential patron. Instead of his wife, Margaret, who was weary of traveling, Rockefeller had brought his twenty-year-old daughter Neva, who was "absolutely intrigued" by Fuller. On their return, Fuller met with Rockefeller at Chase, and when he talked past their scheduled time, he was invited to the family estate in Tarrytown. With Fuller, one observer noted later, Rockefeller "would turn into a twelve-year-old boy."

Leningrad also marked a shift in Fuller's attitude toward politics. Despite his military and government connections, he presented himself as officially neutral, but in August he was named an organizer of Scientists and Engineers for Johnson and Humphrey. "Everything I disapprove of is in one basket this time," Fuller told the press. "Our integrity as a democracy is at stake. The Goldwater people believe we should look out for ourselves. They don't understand that they themselves cannot be successful unless the rest of the world is successful."

His involvement coincided with a letter from Donald M. Wilson of the USIA, who invited him to consult on the American Pavilion for the 1967 International and Universal Exposition in Montreal. On October 6, 1964, Fuller attended an intimate gathering at the White House to discuss science and technology programs with President Johnson. A few hours later, he saw Jack Masey, who happened to be the American chief of design for the Montreal Expo.

He owed the opportunity in large part to Serge Chermayeff's son Peter, who had met Fuller as a boy in the late forties. Peter and his brother Ivan were principals in the recently founded design firm Cambridge Seven Associates, which Masey had selected to oversee the exhibits in Montreal. Because Masey wanted the team to partner from the beginning with a more experienced architect, who would handle the exterior pavilion, he presented them with a list of three candidates—Philip Johnson, Paul Rudolph, and John M. Johansen. None of these options struck the group as particularly appealing, and after a tense pause, Peter asked if they could propose an alternative: "Would Buckminster Fuller be a possibility?"

Masey was receptive to approaching Fuller, who established the architectural office of Fuller & Sadao to formally pursue the project. As a collaborator, Sadao had been chosen for his perceived lack of interest in taking credit for inventions that emerged from his "proximity" to Fuller, who made his "dictatorial authority" over the partnership plain: "You understand that you are, in effect, becoming a part of the individual me." Sadao prepared drawings of the ambitious structure that Fuller had in mind, and they were granted the commission in November.

On December 2 Fuller was joined by Sadao and Bill Wainwright of Geometrics for a meeting with the planning committee in Montreal, where he flew with Peter Chermayeff by helicopter over the site on Saint Helen's Island. At that stage, Fuller favored a design for the pavilion based on a gigantic space frame supported above a Dymaxion Map that was the size of a football field, with the outlines of countries formed by landscape gardening. He had privately dreamed of a far more ambitious display in which a massive globe would slowly unfold into the map, but even his relatively straightforward official proposal presented daunting challenges.

Earlier that year in Carbondale, Fuller had installed a huge floor map to display votes by delegates at a Model United Nations event, and what he envisioned now was an immense expansion of the same idea. It centered on a high-stakes game called "Let's Make the World Work." Attendees, using a computer database, were to submit plans for the allocation of global resources. Wrong answers would lead to a simulated war, while valid strategies could shape the creation of actual policy. Fuller argued that American credibility, which he felt was at "its lowest point," would be enhanced by a peaceful reinvention of the military war game, and Masey was favorably impressed.

As they developed the design with Peter Floyd and Bill Ahern of Geometrics, however, Peter Chermayeff expressed strong reservations. In basing the exhibition on a resource allocation game, Fuller was essentially claiming control of the interior, as well as the pavilion, and a flat roof supported by four columns was unsuited for the climate in Montreal, which demanded an enclosed space. Instead of adding a separate set of walls, which would undermine its structural elegance, Chermayeff suggested that they replace it entirely with a single integrated dome. In response, Fuller said flatly, "Peter, I am so tired of being known as the dome man."

To address his objections, Chermayeff pitched a variation on Fuller's trademark design: "Let's do a big dome, a three-quarter sphere, a bubble that would appear to be rising from the ground and give us a great interior space, like the dome you proposed for a cotton gin." Opening a copy of The Dymaxion World, Chermayeff turned to an image of the geodesic cotton mill. It was enough to overcome Fuller's resistance to the concept, which the team effectively adopted by January 1965.

Despite this breakthrough, the group debated the details well into the following year. Fuller pushed for an untested tensegrity system for which he had a patent pending, and after he was overruled, he pivoted to a dome bolted out of aluminum tubes, allowing it to be disassembled and resold. After it became clear that this was structurally impossible, a permanent framework of welded steel was approved instead. In the end, every major decision was made against Fuller's wishes.

His real task was to provide the project with a public face, and he embraced the role enthusiastically. After the $4 million plan was revealed, the critic Peter Blake viewed it as a safe choice that pointed to a lack of "design creativity" in younger architects—confirming that Fuller was no longer seen as an outsider. When the writer Susan Sontag named Fuller as a member of the "new establishment," along with McLuhan and Cage, in a piece for *Mademoiselle* magazine, she was drawing a contrast with the conservatism of literary intellectuals, but it still represented a profound change in his cultural status.

For all his prominence, Fuller lacked the money to patent his latest invention, a machine to produce components for a tensegrity truss, and he was paying for much of his Carbondale operation himself. His office over a travel agency was divided into two rooms: one for the World Resources Inventory overseen by John McHale, the other for Fuller and his executive assistant, Naomi Wallace. Students were often there as well, conflating his academic and personal activities, and there was no institutional structure in place to formalize his relationship with collaborators.

Bill Perk, the head of the design department at SIU, encountered this issue firsthand. Perk was a longtime fan—he had promoted a project to use the Geoscope for the Strategic Air Command while working as a systems researcher at Lockheed Aircraft—and had been hired on Fuller's recommendation. When he arrived on campus, Fuller was traveling, and Perk learned that students were distributing a summary of energetic geometry, including the chord factors for the domes. They assumed that Fuller would approve, but when he returned, he furiously ordered McHale to confiscate every copy. Faced with the disclosure of his proprietary information, Perk recalled, Fuller went "apeshit."

Harold Cohen noted the same tensions in Fuller's interactions with his students. "He was abusive, which is a terrible thing to say," Cohen said. "He could be terrifying. He'd think that people who were listening to him at a lecture were stealing from him." On balance, Cohen thought that Fuller had made a positive contribution to the school, but he was unsentimental about the trade-offs: "He relied on other people to finish what he started. . . . He was a man that was twisted."

A student named Robert Williams, who went on to conduct important work in geometry, also witnessed this less attractive side. When Williams mentioned an idea for domes based on clusters of soap bubbles, Fuller claimed that he was developing a patent along the same lines. Afterward, Williams mentioned the exchange to his friend Dale Klaus, an undergraduate staff member in Fuller's office. Klaus looked at him and said, "He's not doing any patents at all." Until then, Williams had fantasized about working with Fuller, but now he decided to keep his distance: "I wasn't a person who would trust Fuller after my interview with him."

Fuller needed the energy of the young, which also brought him into contact with the civil rights movement. On a bitterly cold day early in 1965, he rode through Chicago with the broadcaster and author Studs Terkel and the Latino activist José Cha Cha Jiménez, the founder of the Young Lords Organization. Fuller supported the group's campaign for fair housing, and as he looked out at the neighborhood, he reflected on his own life: "This is where I lived during my worst days."

A protest over evictions was being held that afternoon at a Puerto Rican church. When Fuller asked to go there, Terkel felt uneasy, but they drove to the unheated chapel, which was packed with people in blankets. Fuller launched into a talk on urban renewal, linking it to a world where technology would create wealth for all. According to Terkel, the audience followed every word. "Bucky Fuller, Bertrand Russell, and Albert Einstein were and are on the same wavelength—yours and mine," Terkel wrote. "That's the big one. Are we ready for what the man of the future has requested of us?"

Another unusual project reflected Fuller's undiminished stature among the younger generation. After riots in Harlem that summer, the Black poet and feminist June Jordan recalled that she had been "filled with hatred for everything and everyone white," and she tried to overcome her inner rage by reaching out to Fuller. At a meeting in New York, they agreed to collaborate on an *Esquire* magazine article about a redesign of Harlem, which Jordan—who perceptively explored architecture as a means of addressing racial injustice—saw as "a form of federal reparations."

Jordan labored over every detail, but the inspiration for the published concept was credited solely to Fuller, who predicted that it would cause "integration in reverse," as white families rushed into minority neighborhoods. The piece that Jordan had called "Skyrise for Harlem" was retitled "Instant Slum Clearance," and her thoughtful prose was overwhelmed by Shoji Sadao's stark drawing of what resembled a grid of nuclear cooling towers. Although Jordan was disappointed by its presentation, she remained close to Fuller, whose support allowed her to pursue her studies of community planning "for participation by Harlem residents in the birth of their new reality."

In an era when the journalist Jane Jacobs was exposing the failures of urban projects in *The Death and Life of Great American Cities*, Fuller continued to think in terms of megastructures, which kept him in the public eye at the expense of his vision of decentralization. By contrast, many of his other views were strikingly progressive. Fuller said that America was already "the most socialized of all countries in the world," although it concealed this by underwriting corporations rather than individuals. He advocated using the benefits of technology to grant fellowships for people to do whatever they liked, even "fly-fishing," and argued that all children were born geniuses, only to be "swiftly degeniused" by their environment.

Fuller planned to publicize these ideas at the upcoming UIA Congress in Paris, which marked the culmination of the first stage of his student program. In July it opened with exhibits from seventeen schools, including Geoscope models by the University of Colorado and the University of Nottingham. Fuller claimed that the students had acted on their own initiative: "They call their project the Design Science Decade." By uncharacteristically removing himself from the narrative, he presented his plan as a spontaneous movement, one that he hoped would discourage young people from "using their heads as punching bags" in demonstrations.

Earlier that year, Fuller had participated in founding the World Society of Ekistics with Doxiadis, to whom he sent a long letter on his methods, "Design Strategy," in which he said that clients approached him only "when everything else they had tried had failed." On another Delos cruise, Fuller felt his age—he turned an ankle while dancing the twist—but also received a tribute to his influence on the young. At dinner, Doxiadis told the guests, including the historian Arnold J. Toynbee, that Fuller ought to be introduced by one of his contemporaries, and the architect handed the platform over to his own daughter, who had graduated recently from Swarthmore College.

Fuller took it as a compliment, but it romanticized the nature of his relationship with his younger followers, whom he had always needed. In the collection *Utopia or Oblivion: The Prospects for Humanity*, he presented a stark choice between violent upheaval, caused by rising inequality, and a bloodless design revolution: "Either war is obsolete, or men are." His solution depended on "the more or less freewheeling student world," which he assumed would take his side without question. "Because of the students' intuition and youth," Fuller concluded, "the chances are good!"

In August 1965 Calvin Tomkins visited Bear Island. The thirty-nine-year-old critic had already written a *Newsweek* piece on Fuller, whose ideas had "intoxicated" him in college, and they occasionally crossed paths, as Rosamond lived near Tomkins in Snedens Landing, New York. On his four-day trip to the island, Tomkins went swimming with Fuller. Despite having just turned seventy, Bucky taught him to jump into the frigid water, emerge immediately, warm up, and repeat until the process was marginally less painful. When Fuller saw that Tomkins's teeth were chattering, they returned to the main house, pausing only to look for tetrahedral pebbles.

Tomkins was now writing for *The New Yorker*, and the magazine published his article, "In the Outlaw Area," the following January. Fuller's deafness made it hard to have a conversation—he was using a bullhorn as a kind of ear trumpet—but he gladly played the design legend for Tomkins, stating that Gropius had seen engineering only as "an aesthetic" and praising Mies as "the most perceptive of all of them." Some of his pronouncements endured as aphorisms: "I never work with aesthetic considerations in mind. But I have a test: if something isn't beautiful when I get finished with it, it's no good."

Using a favorite image, Fuller referred to himself as a "guinea pig," and he spoke approvingly of seafarers who ventured into the unknown, tying them into his personal theory of history. After Tomkins asked him about the name of the *Nagala*, a beautiful sloop that he had been obliged to sell, Fuller said that it had been inspired by Naga, the South Asian serpent god, which symbolized a ship's path as it tacked against the wind. Tomkins listened carefully to his explanation:

> At the beginning, he said, when men first put to sea, on rafts, they just drifted away from the mainland of Asia on the Japan current, going where God seemed to will. . . . Then, after many centuries, successors of these earliest seagoing drifters who had found their way to the South Seas, and who may very well have been Maoris, learned how to steer their rafts. . . . They became the high priests and witch doctors and spiritual leaders of their people, and they continued to develop their secret mathematical knowledge.

Fuller said that he planned to write a book on the subject for Macmillan, and that John Huston was interested in adapting it as a movie.

The Naga story was only slightly more factual than the tales that he and his friends had spun decades earlier about the fictional explorer Spongee, who had supposedly discovered America before Columbus. Fuller came to believe that the human race had arisen in the atolls of the South Pacific, which seemed more favorable than Africa for a species born "naked and helpless," and had been beamed there from elsewhere in the universe. Back in the twenties, he had written that the notion of evolution from monkeys was "disgustingly theorized," and he continued to question Charles Darwin's assertion that simple forms could develop into complex organisms.

As he became publicly known as a Renaissance man, Fuller often spoke on subjects of which he had minimal knowledge, using scraps of information gleaned from newspapers and his own itinerary. He also told Tomkins that he had been marked by travel in another way: "I gained twenty pounds on this last trip. It's one of the really big problems on this kind of schedule—big, rich dinners everywhere, and all that airline food." His weight, which normally hovered around 165 pounds, was up to nearly 200. Fuller had always been conscious of how it looked on his short frame, and after experimenting with the chalky low-calorie supplement Metrecal, he decided to base his diet entirely on beef.

He justified this approach by arguing that solar energy was impounded by plants and concentrated by herbivores, as his colleague Edgar J. Applewhite explained in his memoir *Cosmic Fishing*: "The cows are eating much more vegetation and converting it to protein than he could possibly cope with at first hand." This was ecologically nonsensical—the resources needed for a pound of beef were much greater than the equivalent amount of plant protein—and Fuller may have been equally influenced by the late Arctic explorer Vilhjalmur Stefansson, his old friend, who had experimented with an Inuit diet of meat and fish.

Fuller arrived at an idiosyncratic menu of his own. "The cut of beef he prefers, because it has the least fat, is London broil," Alden Hatch wrote. "It is usually served in thin strips because it is so tough, but Bucky cheerfully chomps great hunks of it with his magnificent teeth." Fuller consumed it at every meal, cooking it himself when he was staying with others, along with a salad at lunch and dinner, supplemented with prunes and Jell-O. He lost weight, but dietary considerations led him to add cooked fruit, and he ate more conventionally while traveling.

His other staple was tea. Fuller consumed forty to fifty cups every day, although he clarified to Hatch, "Hot water would do as well, but it is so insipid. I just put in enough tea to give it a faint flavor." Elsewhere, Applewhite hinted that his consumption was less abstentious than he claimed. On one visit by Fuller to his house, Applewhite recalled, "After two sweltering hot nights, we found his pillowcases had pink stains that could be diagnosed only as tannic acid having seeped through his scalp."

Applewhite observed that Fuller's diet caused "dismay" among his younger admirers, though they still welcomed him on college campuses. After meeting with the leaders of the Free Speech Movement at the University of California at Berkeley, Fuller hoped that their comfort with technology, especially television, would make them receptive to his message. Young artists were taking his ideas in novel directions, including the experimental filmmaker Stan VanDerBeek, who built a psychedelic dome theater from a silo top, while the sculptors Forrest Myers and Sol LeWitt reimagined his geometry in what the land art pioneer Robert Smithson called "the most astounding manner."

The most significant development came in the spring of 1965. A circle of aspiring artists, the Droppers, had formed around the students Clark Richert and Gene Bernofsky, who met at the University of Kansas. After acquiring seven acres on a former goat pasture for a commune in Colorado, they were driving home one day when they saw a geodesic greenhouse. Jumping the fence, they took measurements—the chord factors for the domes were not yet widely available—and were told by the owner that he had obtained the plans from Fuller himself.

On returning to Boulder, they noticed a flyer on a telephone pole announcing an event with Fuller at the Conference on World Affairs, an annual forum of panel discussions hosted by the University of Colorado. Taking this as a sign, they cornered him afterward to ask for his advice on constructing their first dome, which they proceeded to build out of the "garbage of America," using scrap wood, homemade cement, bottle caps, and aluminum paint. They later switched to a design based on old car roofs, cutting them out with an axe in junkyards or, in one case, from a tempting Cadillac in a motel parking lot.

Many of the structures were the work of an Amherst College dropout named Steve Baer, the inventor of Zomes, which were based on geometric shapes that allowed domes to be stretched or joined in clusters. At its peak, the Drop City colony had ten dwellings described variously as "some Buck Rogers Indian village" or a landscape of "landed UFOs." Members gave talks at schools, and as word spread of what became known as the first hippie commune, it turned into a popular stopover on the way to San Francisco—further seeding the dome as a spiritual emblem of the counterculture.

When Bernofsky confessed that the group was running out of money, Fuller responded with a check for $500, which he wrote off on his taxes by referring to it as "the Dymaxion Award." McHale proposed that they operate a construction business "as your own local industry to help maintain the city," but the young people whose support Fuller needed were taking the dome into politically fraught territory, as a structure embraced by the government came to symbolize a community united by contempt for the Johnson administration.

Fuller was reminded of the stakes on September 15, 1965, when the Federal Bureau of Investigation interviewed him about his contacts with a Soviet official at the United Nations, whom he had approached two years earlier at the Conference on World Affairs. Hoping to gain support for the World Design Science Decade, he had paid for his contact to travel to Carbondale, and they had met up again at the Dartmouth Conference in Leningrad, where Fuller and Anne had lunch with the man he assumed "was apparently home to Russia on leave."

Although the contact's name was redacted from Fuller's FBI file, it was almost certainly Victor Lessiovski, who served as the personal assistant to U Thant, the secretary-general of the United Nations. Lessiovski, who met with Fuller in Carbondale on June 21, 1964, was secretly a KGB officer specializing in "influence operations," with a focus on "inducing influential foreigners to do, wittingly or unwittingly, what the Soviet Union wanted them to do." He was known for his charm, and he clearly took an interest in Fuller, who was sympathetic to the Soviets.

Fuller admitted to the FBI that he had been "playing politics," adding that the Soviet Union had never been responsive toward his work, possibly because he had criticized the country in print so often. In fact, while he sometimes framed his ideas as an alternative to Marxism, he had been inordinately complimentary of Khrushchev, but the matter was dropped, with the agents noting that he seemed more interested in talking about his "geosocial revolution."

If anything, Fuller's ambitions to shape policy had only increased. At an SIU design conference, he stated that if the world's industrial infrastructure disappeared, two billion people would die, while if all politicians were banished "by rocket ship," the planet might be better off. He announced impulsively that his computer simulation, now called "How to Make the World Work," would be operating within five years, and in the meantime, he experimented with a transparent set of concentric spheres to allow players to visualize the movement of resources. When John Cage asked if he played chess, Fuller replied that he played only "the biggest game."

In November Fuller traveled to Ghana with the artist Keith Critchlow and a London architectural student named Michael Ben-Eli. The country's move into bauxite mining inspired the team to build the Accra Dome, an aluminum structure with vents to produce the cooling effect, which led students to call it the "chilly machine." Fuller first saw it the following year, when it was lowered over him by helicopter on a soccer field, and he was moved to tears: "To me, it is the most beautiful of all the geodesic domes that have ever been built."

All the while, a far more prominent dome was slowly rising on the other side of the world. Early in 1966 work commenced on the US Pavilion at the Montreal Expo on Saint Helen's Island. It was only one component of a World's Fair with contributions from sixty countries, and the Canadians organized the immense undertaking using the "critical path" approach, in which a computer model of deadlines generated an optimal sequence of steps. The method directly guided the construction of the dome, and Fuller later adopted the term for himself.

Fuller described the Montreal Expo Dome as a completely rational structure conceived for "high performance per unit of invested weight, time, and energy" and derived from pure mathematics: "The aesthetics of such an undertaking take care of themselves." In reality, it was the other way around. To facilitate its predetermined shape, every aspect of the dome had to be calibrated separately by the team led by Peter Floyd and Sadao, who said explicitly that it had been "designed for visual impact and aesthetic considerations rather than for optimum use of materials." Aesthetics drove the math, not vice versa, and its symmetry was a painstaking illusion.

In collaboration with the structural engineer Frank Heger of the firm Simpson Gumpertz & Heger, Floyd and Sadao arrived at a hybrid design. Above the equator, the triangulated outer shell was an alternate grid with hollow steel tubes, while the lower, weight-bearing half consisted of horizontal lesser circles of solid struts. Inside was a second framework, made of hexagons, that was joined by spokes to the outer layer. To regulate temperature and brightness, it was covered with retractable sunshades, each of which had six triangular sections controlled by light sensors. Its skin was fashioned of hexagonal plastic bubbles, and its maximum diameter of 250 feet was based on the largest sheet of acrylic available.

Fuller stated incorrectly afterward that no construction drawings were needed, apart from tables of chord factors; that it used his "star tensegrity" truss, when that framework had actually been discarded; and that its parts, which really encompassed multiple strut and hub types, were "completely interchangeable." Leaving the execution to others, he contented himself with a version that existed only in his imagination. As Sadao noted, "He was always careful to make sure that his name was associated with it, even though he may not have done the physical work."

If nothing else, Fuller was too busy traveling to have much input on the details. In January 1966 he consulted on projects in Mexico, including an Olympic dome designed by a team led by Félix Candela. The following month, he and Anne headed to San Jose State, where he was to lecture throughout March. The college erected a dome on campus to serve as a lecture hall, and, according to the critic Theodore Roszak, it was there that Fuller became "one of the prophetical voices of the American counterculture."

Fuller often compared the precessional side effects of human actions to a bee's inadvertent pollination of flowers, and in this case, he had little sense of what might blossom in San Jose. In one talk, he said, "The drive in humanity to reproduce as prodigally as possible decreases considerably. This may be reflected in social behaviors—when all the girls begin to look like boys, and boys and girls wear the same clothes." He concluded that society had to revisit its assumptions: "Our viewpoints on homosexuality, for example, may have to be reconsidered and more wisely adjusted."

His most enthusiastic admirer was Nelson Van Judah, a graphic arts professor whose classroom housed a dome equipped with "an array of multisensory enhancements," prompting one participant to pronounce, "We're inside Bucky's brain!" Other visitors to Fuller's classes included Robert McKim of the Stanford University Product Design Program, as well as a curious young man named Stewart Brand. Fuller saw it as just another lectureship, and he left with no awareness that he had been launched on what Roszak called "the final and most spectacular phase of his career."

For now, he merely resumed his travels. He had tea with Salvador Dalí, who wanted a dome for his museum in Figueres, Spain, and in July he saw the arts patron Caresse Crosby in Cyprus. As a young woman, Caresse had received the first patent for the modern bra, and she had published important works of modernism with her husband, Harry Crosby, who died in a grisly suicide pact with his mistress in 1929. At Romany Marie's, she had allegedly engaged in a "brief affair" with Fuller, whom she had more recently recruited for an idealistic project of her own.

Crosby's mission was to build "an international peace center" in Cyprus to host congresses for independent groups of artists and scientists outside the jurisdiction of any nation. Given the ongoing conflict with the Turkish Cypriots, it was an unlikely location, and Fuller caused a stir with his offhand suggestion that locals might be allowed "to cross the line of sovereign nation control" into what he called the World Man Center, which he proposed covering with the inevitable dome.

Fuller implied later that he had been instrumental in persuading Cyprus to offer the land, although his role was only minimal. He hoped to recruit writer John Steinbeck as a trustee, but despite the support of Dalí, Noguchi, and Willem de Kooning, who donated art for auction, the endeavor was doomed from the start. "I feel our undertaking to be one of the most important in my whole life," Fuller told Crosby, but he was equally attracted by its combination of publicity value with a vanishingly small chance of any meaningful commitment.

Earlier that year, Matsutarō Shōriki had reached out to him with a similarly grandiose proposal. Out of concern over a planned housing development in Tokyo that threatened to overshadow his Yomiuriland amusement park, Shōriki wanted to respond with a tower that would be higher than Mount Fuji. At first, it was called the World Peace Prayer Tower, but Fuller, objecting that "no physical device can serve as a prayer," renamed it the Tower of World Man.

At more than twelve thousand feet in height—ten times higher than the roof of the Empire State Building—it would be unfathomably taller than any building before or since. Fuller passed along the commission to Sadao, who determined with Geometrics that it was theoretically possible. Its challenges included violent winds, snow loads, and low oxygen levels at the top, and their report included a discussion of helicopter rescues in case the elevators failed. The final plan secured the tower with guylines to three enormous tetrahedral supports, with an estimated cost of five times the initial $300 million budget.

Although the group produced specifications for an alternative version—a mere eight thousand feet high—that could be built for the original price, the project continued to expand. In August Fuller cabled Sadao with the news that it would incorporate a vertical city large enough for a million people, and, later that month, he wrote up a more detailed description in a letter from Hong Kong. Sadao recalled that it was "almost indecipherable, as if his hand were trying to keep up with the frenzied flow of ideas emanating from his mind."

Fuller's proposed city was a tetrahedron two miles high, providing the maximum surface area on the outside for apartments, which could be added or removed as necessary. It could be built on land or anchored in water, and Sadao worked dutifully on versions for both the mainland and Tokyo Bay. Fuller received an annual fee of $25,000 to develop Tetrahedron City, which replaced the tower altogether. Renderings were later published in *Playboy* of what the critic Lewis

Mumford called a monstrosity "big enough to entomb the population of a whole town."

While the most famous structure with which he would ever be associated was under way in Montreal, Fuller spent most of his time on ideas that were unlikely to be realized. Another was the Minnesota Experimental City, which the geophysicist Athelstan Spilhaus conceived as a futuristic community with zones for jet packs. With support from Hubert Humphrey, Fuller served on the steering committee, which he found politically frustrating, and he only recommended the usual dome.

He was more industrious in building up his entry in *Who's Who in America*, which he proudly noted was the longest ever—although he inflated it by including domes by others and minor items in his résumé. As an index of his power as a celebrity, Fuller systematically tracked his press clippings, and they skyrocketed as interest rose in the Montreal Expo. He appeared at another Delos Symposium, was interviewed by CBS-TV news anchor Walter Cronkite, and attended a reception with Anne at the White House, where he later presented a model to President Johnson.

Fuller was rarely at the Cambridge office, where the design work on the pavilion took place, and he was no longer informed of fundamental engineering decisions. He visited Montreal only for a ceremony to raise the flag in October, leaving after less than a day, and in his absence, construction continued through the winter. Installing the plastic took longer than expected, and Sadao worried about finishing on deadline. When snow and ice collected on the exterior, the uneven expansion caused leaks, which were barely cleaned up in time.

In early 1967 Fuller established a research arm, Tetrahelix Inc., to look into applications of spiral arrangements of tetrahedra, while the *New York Times* printed a glowing article by David Jacobs—who later created the primetime soap opera *Dallas*—hailing the Expo Dome as "perhaps a prototype for an enclosed community of the future." At a press preview in April, Fuller told reporters that the frame, which in reality had been engineered painstakingly by Frank Heger, had been welded together in a shortsighted attempt to save money, and media coverage continued to present him as the sole author of the project.

His failure to acknowledge his collaborators dismayed Peter Chermayeff, who felt that Fuller had monopolized the public acclaim "after almost no personal involvement in our two-year process of design, construction, and installation." Chermayeff thought that the other team members had been "treated badly" by Fuller, and he was left with mixed feelings about the experience: "It was a great privilege for all of us to do this magnificent project with him, but his monumental ego led him to over and over again beat his drum, distort the facts, take all the credit, and ignore the many others who played major roles in the work."

On April 28, 1967, the Montreal Expo officially opened on Saint Helen's Island, where Fuller prepared to seize the spotlight on the largest stage of his career. He stationed himself with Anne outside the tunnel of the metro station to watch as the visitors arrived. The dome, which Fuller called a "skybreak bubble," was the first structure that everyone saw, and he strolled around to eavesdrop discreetly on their exclamations over its beauty.

Exterior view of the US Pavilion for Expo 67 in Montreal.

The dome's lightness made for a pleasant contrast with the ponderous Soviet Pavilion across the channel, and it was a world apart from the blind colossus of Tetrahedron City. Its only real flaw was the sunshade system, which failed immediately: the motors were underpowered, and when the cables twisted, the panels were stuck in random positions. Fuller said that his preferred mechanism would have been fine, and he blamed it again on budget cuts, but the effect was oddly pleasing. The shades formed a complex grid against the light, like a Geoscope with the continents replaced by abstract patterns, yielding his private sky at last.

Inside the dome, where a monorail passed high above the floor, an escalator carried guests past the exhibits, including portraits of Andy Warhol and a painting of the Dymaxion Map by the artist Jasper Johns, who had asked for permission through John Cage. Johns had selected the subject to honor the pavilion, hoping to hinge the triangles to change the configuration "should it ever have to be shown in different environments," although he was unable to build one that folded into an icosahedron. Fuller complained afterward that Johns had sold another version for $150,000 but never thought "of offering any of it to me."

Warhol dismissed most of the exhibits as "an official acknowledgement that people would rather see media celebrities than anything else." Perhaps the most perceptive visitor was the scholar Umberto Eco, who observed, "The dome was aesthetically the strongest element of the pavilion, and it was so full of nuance, so open to different interpretations, that it affected the symbols inside and added depth to their easily identifiable, more superficial qualities." Warhol put it more emphatically: "The US building at Expo is so beautiful, they should have left it empty."

At night, the dome glowed softly, and its seemingly optimistic message was seen as a respite from the turmoil in the United States. Fuller rightly viewed it as the monument that he had sought, although he downplayed it in later accounts of his work. His office had lost money on the project, but Sadao, who drew no salary, hoped that it would generate future commissions. In fact, it was the climax of the dome era, as well as a subtle refutation of the solitary genius embodied by Fuller, who could have been removed from the process at any point without affecting the result.

Yet nothing else that he made in the real world would turn out to be as lasting. Its permanence, which he derided as a mistake, became a virtue, and many of those who saw it in person would go on to intersect with him in important ways. Afterward, Fuller began to describe himself more often as an architect, and he felt vindicated for the risks that he had taken.

Fuller sensed that Anne understood the message as well. As their fiftieth anniversary approached, he believed that she saw the Expo Dome as "validation to the extraordinary backing she had given me." He was reminded of what his mother had told him about the Taj Mahal, which she had called the world's most beautiful building because it was inspired by love. Turning to Anne, he spoke as much from the heart as he could: "I have inadvertently brought about the production and installation of our own Taj Mahal as pure fallout of my love for you."

On May 2, 1967, Fuller and Anne returned to New York, where they took a cab from Kennedy Airport to the St. Regis Hotel. It was a wet morning, and their taxi skidded in the rain, crashing against a bridge abutment and bouncing across the highway. Anne hit her head, but she insisted that she was fine, and they continued to their destination, where Fuller attended a lunch for the World Man Center. The next day, Anne was examined by doctors, who found a concussion and severe bruising.

Fuller was at the mercy of his speaking schedule, and he flew to Lebanon to deliver an address at the American University of Beirut. On his arrival, he received a telegram saying that Anne's condition had deteriorated. Suffering from headaches and disorientation, she was taken to the hospital, where she was diagnosed with a brain clot. Fuller was reassured that the blockage was clearing, but after proceeding to Baghdad, Damascus, and Lebanon again, he heard that Anne was in decline.

Flying back to New York, Fuller was driven to St. Joseph's Hospital in Far Rockaway to see his wife, who complained, "This gets continually worse." After staying the night in Lawrence, he was told that her recovery was "very slow but satisfactory." Fuller left for Arizona, only to be informed that Anne had undergone still another reversal. An emergency brain operation for two hemorrhages was successful, but she experienced paralysis on one side of her body, along with lung complications, and was left with a visible indentation in her skull.

Fuller's momentum carried him onward to Salt Lake City, and he finally returned on May 20. Anne remained in intensive care, and although he stuck to his relentless series of events, he saw her regularly. He had never forgotten the automotive accidents of his past, including the time that he had overturned the Dymaxion Car with his family inside, and he was terrified that he might lose her.

On the day that Anne was moved to a private hospital room, Fuller sent an update on her condition to the poet Gene Fowler, a former inmate at San Quentin with whom he had corresponded. Fuller treasured the letter that Fowler wrote in response, which he carried in his briefcase for the rest of his life:

> Anne is a woman of considerable strength: she will not leave you to continue alone. . . . It makes sense that this skybreak bubble should be, not a monument, but an embodiment of your love of her. . . . Human beings are fragile skybreak bubbles, as vulnerable and quickly gone as a child's soap bubbles in a bath. Yet, without vulnerability, there is no courage. Without mortality, there is no beauty or love, no reason to reach out and touch.

PART FOUR

WORLD GAME

1967–1983

What is the minimum number of stars needed to divide the Universe into outwardness and inwardness? I find it takes a minimum of four; you can't do it with three. . . . If I can at first discover only three stars in a thought challenge, there must be at least a fourth star lurking somewhere in the critical neighborhood. In fact, I discover that the total number of stars that could possibly be related is always subdivisible by four.

—BUCKMINSTER FULLER, *SYNERGETICS*

TEN

WHOLE EARTH

1967–1973

It is like the childhood game of Twenty Questions: You start by saying, "Is it physical or metaphysical?" Next: "Is it animate or inanimate?" (One *bit*.) "Is it big or little?" (Two *bits*.) "Is it hot or cold?" (Three *bits*.) It takes only a few bits to find out what you want. When we use bit subdivision to ferret out the components of our problems, we do exactly what the computer is designed to do.

—BUCKMINSTER FULLER, *SYNERGETICS*

By the early summer of 1967, Anne had recovered enough from her accident to leave the hospital. After she went to stay with one of her sisters on Long Island, Fuller resumed his travels, which seemed more charged with significance than ever. The United States Pavilion in Montreal was drawing huge crowds, his celebrity around the world was at its peak, and he knew this was the best chance that he would ever have to secure a lasting monument for his ideas.

Fuller was widely regarded as one of the few older Americans capable of bridging the generation gap, but the path ahead for the nation as a whole remained unclear. One portent of the contradictions to come was the movie *Bonnie and Clyde*, starring Warren Beatty and

Faye Dunaway, which had its world premiere that summer at the film festival at the Montreal Expo. It marked the rise of Arthur Penn as a major director, and by fusing the memory of the Great Depression with the images coming out of Vietnam, it exposed tensions that Fuller, who hated violent movies, tended to ignore.

Without Priscilla Morgan, Isamu Noguchi's lover and Penn's former agent at the William Morris Agency, the film might never have happened at all. Years earlier, Morgan had introduced Penn to Beatty, in perhaps the most famous instance of her abilities as a cultural matchmaker, which made her a valuable asset to Fuller as he sought to build on his growing fame. She was Romany Marie on a global scale, and he praised her as "the twentieth century's last half's greatest and most effective shepherdess of many of life's most significant creative regenerations."

At his fiftieth Harvard reunion in June, Fuller received an honorary Phi Beta Kappa key, which he proudly wore afterward to all his public events. Later that month, he flew to Rome to meet Morgan, who drove him north to the Festival of Two Worlds in Spoleto. Morgan worked closely with the composer Gian Carlo Menotti on the annual festival, which was a notable meeting ground for American and European artists, and they had arranged for it to feature Fuller's latest dome.

Sadao had overseen its construction, and over the objections of the local historical society, it served as a venue for plays, recitals, and an exhibit of Fuller's artifacts. Among the attendees was Ezra Pound, although the two men interacted only briefly that summer. In Venice, Robert Snyder had screened a documentary on Fuller for the octogenarian poet, who in the fifties had devoted three lines in *The Cantos* to a complimentary reference to "Buckie."

On July 6, 1967, Fuller left Rome for the UIA Congress in Prague, Czechoslovakia. In the aftermath of his failed coup of the World Design Science Decade, which the congress president described darkly as "a parallel manifestation of private character," the atmosphere was less than hospitable. Proceeding to Helsinki, Finland, he met the designer Victor Papanek and Armi Ratia, the founder of the Marimekko textiles company. At seventy-two, Fuller still managed to beat Ratia's son in a footrace from their house to the beach.

He was feeling the effects of the car accident, however, and he some-times experienced dizziness and a warm sensation at the back of his head. A few deep breaths were usually enough to recover, but he had briefly lost consciousness the month before, and it happened again on July 15 in London, where projects from the World Design Science Decade were exhibited in Bloomsbury Square. An examination revealed thickened arteries in his neck, but he made no changes to his schedule.

After the Delos cruise, Fuller returned to Bear Island, where Noguchi introduced the family to the pleasures of raw sea urchins. In Anne's absence, Fuller occupied himself with his "rowing needles," a refine-ment of his catamaran from the forties, with two parallel hulls that would prevent it from capsizing. Don Moore built a prototype, which Fuller later called his favorite invention.

Back in Carbondale, Fuller was urging the construction of a World Resources Center to coincide with Southern Illinois University's centen-nial. As a venue for "How to Make the World Work," it would include a computer that could be rented out for revenue, an electronic data link, and a colossal, animated globe. One proposal involved moving the Expo Dome to house it—which was possibly the real reason that Fuller opposed a permanent structure in Montreal—but $4 million of state bonds were approved instead for a new facility, on the condition that three times that amount could be raised in donations.

Fuller was confident that he could obtain it, and in the meantime, he recruited Calvin Tomkins for his Naga film. Tomkins had consid-ered writing a biography of Fuller, but he doubted that he could handle the geometry, and he was pulled into the movie instead. In September Fuller wrote to John Huston, his director of choice, that it might be the greatest motion picture ever made: "It would be pretty much a detective game in which we would have to count on our ability to discover and develop clues. Fortunately, I have been doing this for half a century."

Tomkins drafted a pitch that resembled an epic version of one of Fuller's educational programs. It opened with Fuller in Maine, intercut with "Maoris sailing between the islands, the navigators using their primitive methods of triangulation, the tribal ceremonies, the various kinds of sailing boats, the 'roaring forties,' the star-sighting towers in

India, the Indian Ocean dhows and their captains, the overland crossings, the excavated Minoan palaces on Crete—all with constant, visual reference to the old, scrotum-tightening sea." Tomkins hoped that the film would play in theaters and on television, but it never made it off the ground.

Fuller would find greater success with *Operating Manual for Spaceship Earth*, which was based on a lecture that he delivered to the American Institute of Planners in October. When published in book form, it became one of his most popular works, partly because he built it around his most evocative metaphor:

> I've often heard people say, "I wonder what it would be like to be on board a spaceship," and the answer is very simple. What *does* it *feel* like? That's all we have ever experienced. We are all astronauts.
>
> I know you are paying attention, but I'm sure you don't immediately agree and say, "Yes, that's right, I am an astronaut." I'm sure that you don't really sense yourself to be aboard a fantastically real spaceship—our spherical Spaceship Earth.

As the book's memorable title implied, Fuller treated the planet as a mechanical vehicle that came with no instruction manual and a limited amount of fuel. Because humanity had been granted barely enough resources to reach the next stage, Fuller proposed a redefinition of wealth as "the number of forward days for a specific number of people we are physically prepared to sustain." It was best, he concluded, to address any problem starting at the level of the universe itself, which ensured that nothing important was overlooked.

This approach required vast computing power, and the entire book amounted to a veiled argument for his World Resources Center. "You may very appropriately want to ask me how we are going to resolve the ever-acceleratingly dangerous impasse of world-opposed politicians and ideological dogmas," Fuller wrote. "I answer, it will be resolved by the computer." With their apparently objective results, computers would allow politicians to change their positions without losing face—

although Fuller secretly intended to populate the system with his own ideas "at the outset before lines of alternative and less alternative stratagems have been hardened."

The major obstacle was cost. An ephemeralized operation was capable of building as many small domes as he liked, but the processing power for the World Resources Center required an investment of millions of dollars. In fact, an alternative model did exist, and if he had seen the theoretical potential of an interlinked network of less expensive terminals, constructed along the lines of the virtual company that had developed the dome, his final years might have been very different.

Fuller never saw it coming, despite a valuable clue that fell into his lap at just the right time. In November 1967 Fuller attended several events in San Francisco, including an appearance at Golden Gate Park's Hippie Hill, a popular gathering place for flower children, among whom he cut an incongruous figure in his usual dark suit. Later that day, he went to speak at the Esalen Institute, the scenic retreat for the study of human potential in Big Sur, where he met a twenty-eight-year-old named Stewart Brand.

Brand had striking affinities to Fuller. He was born into a privileged family in Illinois, and his grandfather had generously endowed him with holdings in the Eastman Kodak Company. After attending Phillips Exeter Academy, the exclusive private boarding school in New Hampshire, followed by Stanford, Brand served as a parachutist in the army, an experience he credited with sharpening his organizational skills. From there, he wandered into art circles, including USCO, an experimental collective of poets, filmmakers, and other artists in Garnerville, New York, at which he encountered Fuller's ideas in depth for the first time.

After reading the novel *One Flew Over the Cuckoo's Nest*, Brand connected in California with the author Ken Kesey, who dreamed of holding psychedelic events with lights and music in a giant dome. Brand collaborated with Kesey's Merry Pranksters, a roving band of outsiders devoted to the liberating effects of hallucinogenic drugs, to produce the celebrated Trips Festival in San Francisco in 1966. Although Tom Wolfe dismissed Brand as "an Indian freak" in his chronicle of this period, *The Electric Kool-Aid Acid Test*, his interests ranged widely, and his first great insight came almost immediately after he saw Fuller speak at San Jose State:

> I was bored in the spring of 1966, and on my rooftop in North Beach, San Francisco, I took a half a dose—maybe 150 micrograms—of LSD, and just had a nice, long, thoughtful afternoon that was colored by having listened to a fair number of Buckminster Fuller lectures and having read his books over the previous months. He had said that people assume that the resources of the earth are infinite, and in their mind, the earth is flat. If they would just understand that the earth is really a sphere with a limited amount of surface, resources, and every- thing else, then they would behave better.

After realizing that no image of the entire earth had ever been released by NASA, Brand distributed buttons printed with a simple question: "Why haven't we seen a photograph of the whole Earth yet?" He sent them to public intellectuals, including Marshall McLuhan, and the only one to reply was Fuller, who pointed out that it was possible only to see half the earth at once.

Brand was encouraged. He was planning an education fair for the Portola Institute, a nonprofit founded by Dick Raymond that consulted on computer and music classes for high schools, and he invited Fuller to attend what he described as an event on "powerful tools technology." Before Fuller could write back, Brand spotted him at Esalen. Taking a seat across from him at lunch, he handed a button to Fuller, who looked at it. "Oh, yes, I wrote to that guy."

"I'm the guy," Brand promptly responded. "So what do you think? What kind of difference do you think it will make when we actually get photographs of the earth from space?"

Fuller paused to absorb this. Finally, he asked, "Dear boy, how can I help you?"

It was a chance meeting as fateful as the Colorado lecture that had inspired the domes of Drop City, and it would profoundly shape Brand, who for now was just one of many young people competing for Fuller's attention. Listening carefully to Brand's description of the fair, Fuller advised him to clarify its intentions, and he expressed concerns that it would become overly political. When Brand asked if a photo of the whole earth might change the world's perspective, Fuller said, "You're right. I'll take back what I said in my letter."

On January 3, 1968, Brand wrote to Fuller about research being conducted in Menlo Park by an engineer named Douglas Engelbart. "The program is operational now using constant on-line intense interaction with computer-driven cathode ray tube displays of text, graphs, and link-node arrays," Brand explained. "A number of persons at remote consoles can work in interactive concert on a mutually generated display, with mutual access to the computer memory." It was a massively useful hint for the future, although Fuller was too preoccupied with his own concept of computing to recognize it. In the end, the education fair was canceled, but the connection would pay off enormously for both men.

Fuller was more focused on his work with Sadao, who had been left in a peculiar position after the Montreal Expo. The dome attracted millions of visitors and was honored by the American Institute of Architects, but the anticipated rush of commissions had yet to materialize. An approach from the South Carolina Tricentennial Commission was passed along to Synergetics Inc., but the result—a gigantic, skewed cube known as the Tetron—was doomed by structural inadequacies. Most of the office's hopes rested on the World Resources Center in Carbondale, for which none of the necessary matching funds had been raised.

In Japan, Tetrahedron City had foundered. Fuller said later that the plans had faltered after his patron died in 1966—although Matsutarō Shōriki really lived for another three years. To keep their relationship alive, Fuller praised the mogul in the *Times* as a virtual copy of himself: "He wants to do it better than it has ever been done, incorporating all that's been learned about humanity and about the principles that are operative in the universe. . . . He's a very inspired old man."

Fuller forwarded the article to Shōriki, but the venture ended in what Sadao described as "several grave misunderstandings." An unexpected chance to revive the concept came from Charles Haar, assistant secretary of the US Department of Housing and Urban Development, who contacted Fuller about a program to convert water areas near cities into floating communities. To apply for a grant, Fuller established the Triton Foundation with Sadao and Peter Floyd, and they were awarded $30,000 to rework Tetrahedron City into a more practical form.

The result, Triton City, consisted of a triangular waterborne module with sloping terraces of apartments, at an estimated cost per resident that was competitive with existing renewal projects. Haar verified the figures with the navy, which was interested in the concept for base housing, while Baltimore studied it for the Chesapeake Bay. Fuller blamed its eventual abandonment on Richard Nixon's election in 1968, but questions over hurricane safety were never resolved, and it survived only as a model at the Lyndon Johnson presidential library.

As Fuller's name came to stand for futuristic urban planning, such opportunities became more frequent. The Canadian sculptor Gerald Gladstone convinced a local media baron and sports franchise owner named John Bassett to consult Fuller on the revitalization of Toronto.

Model of Triton City (1968) displayed at the Lyndon Baines Johnson Library and Museum in Austin, Texas.

A report by Sadao and Geometrics characterized the city on Lake Ontario as a "second order metropolitan area" that could distinguish itself with a waterfront university, an enclosed galleria, a crystal pyramid, and three floating residential islands.

None of these designs was ever more than a dream, but if Fuller had wanted to build domes for real, a viable model was close at hand. Don Richter, his most practical protégé, had founded the company Temcor under license from Fuller and Kaiser, and he landed commissions almost at once, including one for a Cinerama dome in Las Vegas. Fuller served on the board, but his ambitions were taking him increasingly far from problems of actual construction.

At Norman Cousins's request, he wrote a piece for the *Saturday Review*, "What I Am Trying to Do," that laid out his objectives—including his determination "to reform the environment instead of trying to reform humans"—in a single long sentence. In a talk for the American Association of Neurological Surgeons, he held forth on the alleged difference between the brain, which handled special cases, and the mind, which was able to generalize. He made regular television appearances—the talk-show host Dick Cavett joked that his guest didn't look a day under seventy—and received Gold Medal awards from the National Institute of Arts and Letters and the Royal Institute of British Architects.

All the while, his influence on the younger generation was moving along very different lines. One path had been initiated almost by accident two years earlier, when *Popular Science* magazine printed an article about the Sun Dome, conceived by Fuller as a structure that could be marketed to hobbyists. Readers who sent five dollars would receive a license and plans for building a dome that was positioned as a playroom or greenhouse. As a way to raise money, it was easier than selling kits, but Fuller didn't realize that by including the chord factors for the dome, he had effectively shared its proprietary formula with anyone who wanted it.

Fuller's secrets were taken definitively out of his hands in 1968 by the underground *Dome Cookbook* by Steve Baer, who published straightforward instructions for designing domes. In rescuing geodesics from their dependence on mass production, Baer, who had refined his designs at Drop City, reversed Fuller's historical progression from craft to industry—while also undermining the dome as a legitimate business forever. Baer respected Fuller, but not his assumptions, and he criticized Spaceship Earth, with its analogy between the planet and a machine, as "the work of small time ingrates."

An even more pivotal development came through Stewart Brand, whose deepest insight emerged, like Fuller's, from a private crisis. In March 1968, flying home from his father's "long dying and funeral" in Nebraska, Brand was reading the book *Spaceship Earth* by the economist Barbara Ward, who had met Fuller on the Delos cruise. As he looked out the plane's window, he fell into a reverie on how to give people access to information and tools to realize their dreams of independent living. One possibility, he thought, was a mobile truck store, while another was a catalog in the tradition of L. L. Bean.

"Amid the fever I was in by this time," Brand recalled, "I remembered Fuller's admonition that you have about ten minutes to act on an idea before it recedes back into dreamland." On the endpapers of Ward's book, he scrawled notes, including "Techniques and tools of access acceleration, for the self-motivated." The next day, he met with Dick Raymond about using space at the Portola Institute. When the educational consultant asked him later what the project was called, Brand's mind randomly turned to his button campaign. "I dunno, *Whole Earth Catalog*, or something."

Brand was driven by many of the impulses that had led Fuller to acquire *Shelter*, but he had greater financial resources, and although he was equally controlling, he was careful to incorporate a wider range of voices. After compiling promising titles at bookstores, he opened a truck store to conduct market research with his wife, Lois Jennings, at communes in Colorado and New Mexico. He told Fuller that he hoped to sell the Dymaxion Map and plans for domes, with the ultimate goal of connecting readers "with manufacturers, suppliers, authors, inventors."

In October 1968 Brand launched the catalog as a start-up in a garage at the Ortega Park Teachers Laboratory, which was housed in the former Rancho Diablo near Menlo Park. He based its layout on Steve Baer's *Dome Cookbook*, and he often reviewed books without reading them—he could generally gather from the illustrations and index whether or not they were worth recommending. The first printing had an initial run of a thousand copies, and although he started with just fifty subscribers, he sold tens of thousands by the end of the following year.

The book's tone was set by its famous opening sentence: "We *are* as gods and might as well get used to it." Brand began with two pages on Fuller, whose technological philosophy influenced the emphasis that the *Catalog* placed on "access to tools," and he made his admiration clear:

> People who beef about Fuller mainly complain about his repetition—the same ideas again and again, it's embarrassing. It is embarrassing, also illuminating, because the same notions take on different uses when re-approached from different angles or with different contexts. Fuller's lectures have a raga quality of rich, nonlinear, endless improvisation full of convergent surprises.
>
> Some are put off by his language, which makes demands on your head like suddenly discovering an extra engine in your car—if you don't let it drive you faster, it'll drag you. Fuller won't wait.

The *Whole Earth Catalog* made a compelling case to the counterculture for the man whom Brand called "one of the most original personalities and functional intellects of the age." Its collage of texts amounted to a composite portrait of the comprehensive designer, and browsing through it made readers feel like generalists. Unlike the student activists of the New Left, who emphasized politics and protest, its fans—described by one of Brand's colleagues as "baling wire hippies"—were drawn to technology, and they would advance far beyond Fuller's sense of what computers could be.

On December 9, 1968, Brand assisted with a talk at the Joint Computer Conference in San Francisco by Douglas Engelbart, who treated computers as tools for communication and information retrieval, rather than for data processing alone. Along with advising on logistics, Brand operated a camera that provided a live feed from Menlo Park as Engelbart demonstrated windows, hypertext, and the mouse. At first, its impact was limited to a handful of researchers, but the presentation would be known one day as the Mother of All Demos.

After the Montreal Expo committee rejected "How to Make the World Work," Fuller had persistently promoted it elsewhere, and his plan for the World Resources Center was the closest that he ever came to its realization. The name was cumbersome, however, and he wanted something more memorable as he turned seriously to fundraising. John McHale—who had left to form his own center for futures studies with his wife, Magda, after chafing against the situation in Carbondale—preferred "the Spaceship Earth Game," but Fuller had other ideas, and by the time the Expo Dome opened, he was publicly calling it the World Game.

Over the next two years, he had concluded that his concepts were needed more urgently than ever. In January 1969 he attended a second Dartmouth Conference at the Westchester Country Club in New York. Despite his prior brush with the FBI, Fuller became friendly with the Soviets there, who told him that their country wanted only to improve its standard of living. The United States, they said, had forced them into an expensive arms race, which kept them from "realizing a lifestyle for our socialist economy equal to or better than that already enjoyed by capitalism."

Fuller accepted this statement without question, and he felt that the World Game was his best chance to shape global policy. Access to elected officials was part of the equation, but he could also strengthen his case by maximizing his support among the younger generation, which was his greatest advantage as a public figure. "The young will lead the old in swiftly increasing degree," Fuller wrote, and while he was uniquely positioned to pass between these two groups, he needed representatives who were closer to the youth movement.

He crossed paths with thousands of young people each year, but apart from the occasional outlier such as Brand—who had sought Fuller's blessing before acting on his own—he rarely paid much attention until they became impossible to ignore. When a real opening arose, however, he was quick to respond. In February, for example, a secretary named Pat White sent Fuller a letter explaining that her employers were looking into a prefabricated house, asking, "They are wondering if you would be at all interested in designing one for them?"

The clients were John Lennon and Yoko Ono. Fuller was already a familiar name to the Beatles, who had been photographed recently in a dome that Paul McCartney built in his garden as a meditation platform. Fuller cabled back that he would be "delighted." A month and a half later, he followed up through Constance Abernathy, his assistant in New York, who knew Ono socially: "Dr. Fuller found [it] most significant that you two, who command the respect and leadership of today's young world because of the high quality of your work, should choose a prototype, prefabricated dwelling. We hope to be able to do something wonderful."

No further communication ensued, and Fuller focused on more conventional lines of attack. On March 4, at the invitation of Senator Edmund Muskie of Maine, he testified before Congress regarding the establishment of a committee on technology and the environment, although he devoted most of his remarks to promoting his pet project: "I have learned from a major American publisher that the news rights on this World Game are going to be extremely valuable."

In the absence of the game itself, Fuller wanted to produce a film as a proof of concept—essentially a shoestring version of his unrealized movie with John Huston. At Yale, he had teamed up that semester with the innovative Swiss-born photographer Herbert Matter to capture test footage of his lectures. For the next phase, they enlisted Matter's wife, Mercedes, who had founded the New York Studio School, to arrange for a workshop to be filmed there that summer, and Fuller already had someone in mind to run it.

If Brand was the most visible member of his new wave of admirers, the twenty-three-year-old Edwin Schlossberg was perhaps the most brilliant. As a student at Columbia, he had befriended John Cage and Merce Cunningham through the nonprofit Foundation for Contemporary Performing Arts, for which Fuller had given a talk on April 16, 1966. Over dinner, Fuller spoke for ninety minutes with Schlossberg, who found the experience "transformative." On the young man's subsequent visits to New York, they would meet for hours, and their philosophies were so closely aligned that Fuller briefly suspected Schlossberg of stealing his ideas.

As the end of college approached, Schlossberg planned to attend graduate school for a combined doctorate in physics and literature, but Cage advised him to consult Fuller first. After hearing his plans, Fuller responded, "That seems really interesting, but I would rather that you'd come to Carbondale. You can teach at the design school and we can work together." Schlossberg was flattered, and although he won prestigious fellowships elsewhere, he decided to head for SIU. His family, he recalled, thought that he was "insane."

In the fall of 1967 Schlossberg drove to Carbondale, only to learn from Naomi Wallace that Fuller would be gone for a month. In his absence, Schlossberg began teaching, despite a lack of formal design training, and he waited eagerly for Fuller's return. At last, he was back on campus, but when Schlossberg hurried to his office to greet him, Fuller just stared at him blankly. "I don't know who you are," he snapped. "I've never seen you before. Get out of my office."

Schlossberg was stunned. No one understood what had happened—Cage tried in vain to find out—and he felt foolish for having ignored his family's warnings. He stayed in Carbondale, but his feelings about Fuller had changed. "That I could survive it was very important," Schlossberg reflected. "I thought it was a gigantic mistake for a month. I was furious with him and never wanted to see him again."

At SIU, Schlossberg taught design courses with a teaching assistant named Jon Dieges, who moved with him to Berkeley to continue their work with students. Shortly afterward, he met again with Fuller, who finally explained his bewildering behavior. After making the offer to Schlossberg, Fuller had been in his car accident with Anne, which had supposedly affected his memory for months. As a result, he had forgotten

Schlossberg completely. This might not have been entirely true—Fuller's own granddaughter suspected that he was "exaggerating"—but the two men soon reconciled.

They saw each other again at a conference for the United States Student Press Association in Washington, DC, which was taken over by what Peter Rabbit, a resident of Drop City, called "an incredible list of freaks," including the radicals Abbie Hoffman and Jerry Rubin, the cofounders of the Yippies. After a wild weekend of debates over the Vietnam War, Fuller was given the last speaking slot. Peter Rabbit was amazed by his ability to hold the crowd for two hours: "He brought it all together—stood up there, reared back, closed his eyes, touched his body with his hands, and laid it out beautifully."

In their correspondence, Schlossberg expressed "my continued interest and willingness to help you in any way that I can," and Fuller took him at his word. At a meeting at Manhattan's Plaza Hotel, he told Schlossberg about the program at the New York Studio School, which Fuller's travel schedule would prevent him from overseeing himself. Because Schlossberg's teaching experience included leading classes for teenagers on Fuller's work, he seemed like the ideal candidate to supervise it on a daily basis.

It was a compelling vision, and Schlossberg signed up three months in advance. After securing funding from IBM and the Rockefeller Brothers Fund, he mailed flyers to design schools. Of fifty applicants, he ended up with twenty-four, ranging in age from nineteen to forty-six, including students in art, architecture, physics, biology, and anthropology. Medard Gabel, a former carpenter's apprentice who had moved to Carbondale in hopes of meeting Fuller, was enlisted as his right-hand man.

Clearing out the painting studio, they installed blackboards and two Dymaxion Maps in time for the workshop's opening on June 12, 1969. Herbert Matter filmed three days of lectures from Fuller, who challenged the participants to explore his concept for a world energy grid that could transmit surplus power between time zones to accommodate shifts in demand. A key segment would cross the Bering Strait, which made it a test of relations with the Soviets.

Fuller lecturing at the World Game workshop at the New York Studio School (1969).
Two large copies of the icosahedral version of the Dymaxion Map hang on the walls.

When he was finished, Fuller told Schlossberg, "You're in charge." Faced with filling time until Fuller returned from a lecture tour, Schlossberg decided to conduct research on resources. Drawing on his experience as a camp counselor, he walked to the school early each morning and worked for the next twelve hours. Groups laboriously gathered statistics from the United Nations and the New York Public Library, which they then displayed on maps and a triangulated data grid, and they learned to leave the garage doors open in the sweltering studio to create a hint of a breeze.

Schlossberg was exhilarated by what they were doing: "It was really fun." Fuller seemed pleased with their progress whenever he was in New York, but at the beginning of July, a contract for the television rights fell through with CBS. Taking responsibility for the "disaster," Mercedes Matter offered to sell a valuable Alberto Giacometti sculpture from her collection to raise cash, and she reassured Fuller that they would film as much of the seminar as possible.

The students were absorbed in their work, and they felt confirmed in its importance on July 20, when they viewed coverage of the moon landing on a giant screen in Central Park, which featured an audio message for the crowd from Fuller himself. Fuller was understandably thrilled by the Apollo space program—the ultimate example of how humans became more inventive "as the degree of gravitational hazard is visually evidenced." Later, at a reception with writers Kurt Vonnegut and Arthur C. Clarke, he watched Neil Armstrong and Buzz Aldrin become the first men to set foot on the moon's surface.

His example was an equally powerful inspiration to the students at the Studio School. They included the artist Mira Lehr; John Storyk, a recording studio designer who worked with Jimi Hendrix; and Stephen E. Selkowitz, a Harvard physics major who was encouraged by the experience to enter environmental design. At the last meeting, Fuller told the participants that they were intelligent enough to run the world themselves. John Cage, an attendee, was initially skeptical: "They looked like a bunch of hippies with some older oddballs thrown in. But while they spoke, [I] did as I do at the movies when it's clear everything'll turn out all right. I wept."

Their final report was built around the group's concept of the "bare maximum," or the resources needed for a person to reach "not his minimum potential but his maximum potential." They discussed the merits of a universal basic income, and Fuller was told that he had correctly anticipated everything that they had found, including the importance of a global energy network, which they called "our first move in the World Game." Growing emotional, Fuller said to the group, "I feel I have seen what has been going on in my head for years, but it is more beautiful than I imagined."

Like a scientist whose simulation had generated the results he wanted, Fuller felt that the seminar had fulfilled the promise of the World Design Science Decade, which he had set back five years by insisting on total control. His absence freed the students to be energized by Schlossberg, who succeeded by following a rule that Fuller had shared in one of his talks. Instead of waiting for a patron, it was best to simply get to work. As Fuller put it: "My whole strategy is—and I've been able to live now on the frontier without anybody guaranteeing me anything or telling me what to do—taking the economic initiative and trying to find out what nature needs to be done."

The World Game workshop itself was a surprise solution that had been developed by others, and although it had been conceived as a pretext for filming, it became a landmark in its own right. Apart from its name, it had no gamelike elements, but the concept itself excited people, and the idea of holding additional seminars, which resembled similar efforts by other futurists, might never have occurred to anyone if Schlossberg hadn't risen to the occasion so magnificently.

As always, Fuller kept his options open, as he noted in a lecture that summer: "When you can't go any further with one thing, then you look over to another one that needs attending to, and you keep shifting from one to the other. So, at any rate, I've learned how to survive there." A few commissions were handled by the indispensable Shoji Sadao, including a geodesic water tower in Kuwait; a hall for a kibbutz in Israel; and a domed religious center at SIU Edwardsville with colored outlines of the continents—the closest that Fuller ever came to the Geoscope.

None of these projects had an exceptionally high profile, but an opportunity emerged over the summer that was everything he might have wished. It was conceived by the playwright Francis Warner, a tutor in English literature at St. Peter's College at Oxford University. Envisioning a theater named for the novelist and playwright Samuel Beckett, Warner decided to build it underground due to the lack of space on campus. As a consultant, he brought on the theater director Herbert Marshall, a professor at SIU Carbondale, who recommended hiring Fuller as well.

It was the showpiece commission that had long eluded Fuller, and he agreed to work for free in partnership with a British architect. Fuller's representative in the United Kingdom, the graphic designer James Meller, suggested approaching his associate Norman Foster, whose studio was known for advanced industrial concepts. Foster, who was thirty-four, met them in London for lunch in the paneled dining room of the Institute for Contemporary Arts. Without even glancing at the younger man's portfolio, Fuller decided immediately that they would collaborate.

"It must have been like hearing a disciple," Foster recalled. As he worked with Synergetics on the proposed site, located just five hundred yards from the Thames River, he realized that any underground structure would have to take into account the high water table. At a discussion with the design team, Fuller rushed to the bathroom and returned holding a cake of soap: "This is what it would look like." He proudly signed it, and the bar appeared in promotional materials for the watertight theater that they described as "an underground submarine."

At a dinner conference on July 22, 1969, Fuller presented the plan to backers, as well as to Beckett himself. In anticipation of the meeting, Warner sent one of Fuller's essays to the author of *Waiting for Godot*. Beckett read it admiringly, although, he told Warner in a letter, "while slightly quailing before the optimism." Fuller was pleased by their encounter: "He's a skeptical playwright—he uses lines like 'It takes so long to die'—and I thought he would identify all the negative aspects of life with technology. But Beckett had read me. And I have the feeling that he did not think technology was so threatening."

Fuller told the guests that the Beckett Theatre would take ephemeralization "to the ultimate: invisibility." Its underground auditorium would be shaped like a ship's hull, with a pair of cylindrical stairways rising to the surface, and its two hundred seats could be rearranged to accommodate different productions. "The principle of mobility and lightness and adaptability is excellent," Beckett remarked. "If you had asked me before this discussion what I felt about the whole thing, I wouldn't have been sure. Now I am convinced."

It felt like a sign of greater triumphs to come. In the fall, Fuller was named Humanist of the Year by the American Humanist Association, and at the invitation of Indira Gandhi, who had been elected prime minister in 1966, he delivered the Jawaharlal Nehru Memorial Lecture in New Delhi. At a meeting earlier that year, Fuller had inscribed a copy of the Dymaxion Map, "To Indira, in whose integrity God is entrusting much of the evolutionary success of humanity and with utter safety." Gandhi confided in a letter that it left her "deeply touched and also somewhat embarrassed."

Fuller was confident that Indira Gandhi had assumed power out of dedication to her father's ideals rather than "the usual drive that politicians have." At his lecture in November, she introduced him as "a remarkable person," and he focused on his geometry during the talk, which was twice as long as any other speech in the series. "I have done so simply because I knew Nehru," Fuller explained, "and having experienced the integrity of his mind, I am confident that I am doing exactly that which he would have me do on this occasion, no matter how long it may take."

His connections in India resulted in a commission through the architect Charles Correa, who had attended his classes at the University of Michigan. While designing a resort in the state of Kerala, Correa proposed that Fuller provide a dome for an airport. After talks with Gandhi and Karan Singh, the minister of tourism, Fuller and Sadao were hired to consult on the future of aviation, which would guide work by the operating firm Aéroports de Paris on airports in New Delhi, Madras, and Bombay.

Although Fuller welcomed these clients, his real legacy lay with resourceful associates such as Schlossberg and Brand, who took aspects of his thinking in directions of their own. An older representative of this group was Edgar J. Applewhite, who had worked for Fuller Houses in Wichita. Afterward, he had joined the Central Intelligence Agency, serving as an intelligence assistant to Secretary of Defense Robert Mc-Namara. Upon retiring at fifty, he decided to reconnect more seriously with Fuller.

Applewhite went to Carbondale, where Fuller had moved his center of operations to the Ira E. Parrish Building. On the first floor was the bustling office in which Fuller oversaw the World Resources Center and numerous other projects, including a concept for computer memory based on the isotropic vector matrix. The basement was devoted to the next stage of the World Game, which Fuller predicted might "become the whole curriculum of the university itself." Applewhite was offered the chance to run either operation, but he declined politely.

What intrigued Applewhite was Fuller's endlessly gestating book on geometry. Drafts had been under way for years with such collaborators as Duncan Stuart and the designer Peter Pearce, but it had never seen the light of day, despite a grant from the Graham Foundation and the support of the editor John Entenza. Even Schlossberg had taken a crack at it, but most of his contributions had been discarded on a recent trip to Bear Island, and Applewhite agreed to assume responsibility for the project that ultimately became *Synergetics*.

Another new member of the team was John Huston's son Tony. When the invitation came to join Fuller's staff, the young man had been attending the University of London without much enthusiasm, and he accepted happily. Tony served as a travel companion, went to meetings concerning the Beckett Theatre, and taped lectures for Fuller, whom he admired unreservedly. "He is the only truly great man I have met in my life," Tony said decades later. "Everybody else pales in comparison. Particularly today, when we are surrounded by the abysmal. Bucky didn't have the usual sides that human beings have. He was not petty in any way."

As his staff grew to nearly thirty employees, Fuller continued to exert varying degrees of influence on autonomous movements. With so many factions, it resembled a more anarchic version of the Fuller Research Foundation of the fifties, and it still centered on the dome, which carried different meanings for various players. For the hippies, its circular outline stood for independence; for the British conceptual architectural group Archigram, it pointed toward a radical architecture of technology; and for Ant Farm, an experimental practice that "kidnapped" Fuller at the Houston airport—they brought him to an exhibition of the Dymaxion Car—it embodied design for its own sake.

One of the dome's most insightful advocates saw it as an entry point for reconsidering shelter itself. Lloyd Kahn, born in 1935, had quit his job in insurance to become a builder in Big Sur, and he was working on a redwood house when his carpenter friends told him about Fuller. Kahn, intrigued, went to see him at Esalen, where a storm knocked out the power, forcing Fuller to lecture by candlelight. Until then, Kahn had been hoisting beams into place with a crane, and now he began to fantasize about turning them into struts for a lightweight dome.

As he tinkered with geodesic models, Kahn was advised to contact Jay Baldwin, a design teacher in San Francisco. Years earlier, Baldwin had encountered Fuller in Ann Arbor, where he was impressed by the notion that a person who could identify urgent design problems didn't need to earn a conventional living. "I've found this to be true," Baldwin remembered. "I've never looked for a job since then." Though he was "very straight," he got along with hippies and students, and Fuller had asked him to teach classes in Carbondale in his absence.

Baldwin recommended the chord factors in the Sun Dome to Kahn, who used them to build a greenhouse out of recycled timber. As dome enthusiasts began to send him mail, Kahn thought that a mimeographed newsletter would save him from writing the same letter multiple times. "There's someone you should meet," Baldwin said, and they went to see Stewart Brand. The three men cemented their partnership at Alloy, a conference organized by Steve Baer at a disused tile plant in New Mexico, where spirited debates on building methods were held under a dome.

Brand said later that the attendees at Alloy represented everything he wanted to achieve: "Persons in their late twenties or early thirties, mostly. Havers of families, many of them. Outlaws, dope fiends, and fanatics, naturally. Doers, primarily, with a functional grimy grasp on the world. World thinkers, dropouts from specialization. Hope freaks." He recruited Kahn as the shelter editor for the *Whole Earth Catalog*, while Baldwin became increasingly disillusioned with the culture at Carbondale.

Abandoning the "second-rate" design department, Baldwin started a building company with Kahn, who hoped to construct a dome for the Wild West Festival in San Francisco. In Big Sur, they were hauling around their model for the event, which was eventually canceled, when they ran into some teachers at a graduation party for Pacific High School. It was an experimental program in the Santa Cruz Mountains that aimed to grant students complete independence, and the founders wanted to turn it into a boarding school to save money on busing.

After Kahn and Baldwin showed their dome model to the teachers, they were hired to build houses. At the end of the summer, the lumber arrived along with the students, who camped out in the woods to learn construction skills as they went along. Kahn ended up completing eight plywood-and-aluminum structures of various degrees of waterproofness, while Baldwin concentrated on a dome made of plastic pillows inflated with nitrogen—an idea that had come to him in a dream.

Kahn also returned to the notion of publishing a book. Borrowing equipment from the *Whole Earth Catalog*, he assembled it with Jonathan Kanter, a fifteen-year-old who wrote the overview of geodesic geometry on a paper bag. A section featuring the chord factors was credited to Joe Clinton, the graduate student who had worked with Fuller on the NASA contract. In Carbondale, Baldwin had asked for a copy of the numerical tables that described the structure of the domes, which Clinton was reluctant to provide before the underlying program had been debugged. Baldwin simply headed to the data lab to take one of the uncorrected computer runs, and it was included in the book without permission.

On January 27, 1970, Fuller gave a talk at Pacific High School and explored the grounds in a borrowed pair of galoshes. Kahn showed him a draft of the book, which had been cut and taped into a scroll twelve feet long, but Fuller was more taken by the Pillow Dome. Inside was a waterbed, the modern version of which had been invented as a master's project by Charles Hall, one of Baldwin's design students in San Francisco. After lying down to look up through the skylight, Fuller hired Baldwin to build a dome that summer at Bear Island. He made no comment on Kahn's manuscript, and he never noticed the chord factors that would be prominently reprinted in *Domebook One*.

For now, Fuller felt encouraged by what he saw as the rising generation's receptivity to his ideas. He told *New York* magazine, "The youth of today are absolutely right in recognizing this nonsense of earning a living. We keep inventing jobs because of this false idea that everybody has to be employed at some kind of drudgery." Criticizing the notion that human beings had to justify their right to exist, he concluded that technology would liberate people to follow their dreams: "The true business of people should be to go back to school and think about whatever it was they were thinking about before somebody came along and told them they had to earn a living."

His actual relationship with younger activists was more complicated. A few weeks after presiding at a memorial for Caresse Crosby, whose death in January 1970 ended any prospect of the World Man Center, Fuller visited Oxford University for a benefit performance of Beckett's short play *Breath*. He planned to hand out leaflets promoting his newest book, *The Buckminster Fuller Reader*, edited by James Meller, but discovered that his presence on campus was controversial. A leftist architecture magazine had distributed flyers showing his head as a geodesic structure, a helicopter carrying a dome superimposed over troops in Vietnam, and a word balloon emerging from his mouth: "Welcome to the Buckminster Führer Show."

Fuller had no desire to be associated with an unpopular war, but it had only been a matter of time. A year and a half earlier, in a speech at Auburn University in Alabama, he had noted that solutions to problems tended to become unworkably large until they were replaced by an entirely new art form. A jet carried more passengers in one year than an ocean liner; a satellite outperformed thousands of miles of underwater cable; and in war, the hydrogen bomb was too devastating to ever be used, making it more effective to undermine the enemy's ideology instead:

> Just as ephemeralization . . . took technology out of the limited ranges of the human senses . . . so too has major warfaring almost disappeared from the visible contacts of human soldiery and entered into the realm of invisible psychology. In the new invisible miniaturization phase of major world warring, both sides carry on an attention-focusing guerrilla warfare, as now in Vietnam, while making their most powerful attacks through subversion, vandalism, or skillful agitation of any and all possible areas of discontent within the formally assumed enemy's home economics.

This was a remarkable insight, but he was discouraged from sharing it with his most receptive audience by the conclusion that he drew. In war, Fuller said, an enemy would incite the opposing side's youth to engage in demonstrations, and protests were deadlier than a death ray because they did "everything with nothing."

Fuller believed that this was the true cause of the antiwar movement. Contending that protesters had been manipulated by foreign nations into "gratifying their own personal discontent," he issued a warning: "Because almost everyone has at least one discontent, a single, well-trained, conscious agent can evoke the effective but unwitting agency of hundreds of other discontents, promoters, and joiners. . . . The highly idealistic youth of college age, convinced that they are demonstrating against war, are in fact the frontline soldiers operating as unwitting shock troops."

These words exposed fissures that had been there all along. Fuller had campaigned for Johnson, worked on shelters for the Marine Corps, and presented his bamboo domes—or "bambassadors"—as a solution to Vietnam. For many of his admirers, this had been easy to overlook, and even Abbie Hoffman had thought little of inviting Fuller to appear with the Yippies. Fuller, for his part, doubted that most activists even wanted to benefit humanity, arguing that they were "agitating simply because they have come to enjoy a sense of power and importance by doing so."

His blind spots were especially glaring when it came to race. He had nothing useful to say about institutionalized racism, which he dismissed by noting that there was no such thing as race at all, apart from superficial differences based on geography: "Whites are bleached out colored people who are the normal people." Fuller felt that racism itself was being "swiftly eradicated," as evidenced by sports, and when an audience of Black activists listed the broken promises of white society, his response was utterly tone deaf: "I have never met you before, and I have never promised you anything."

These issues came to the surface during his visit to Oxford. Fuller wrote later to Francis Warner, "I am confident that if my detractors knew me better, they would not only terminate their campaign, but, contrariwise, would spontaneously support my work." On a flight back to Carbondale, Fuller said that the use of domes by soldiers didn't make him a militarist, any more than it did the farmers who produced their food. He concluded, "Marines eat rice too, you know."

One of the sympathetic listeners on the plane was Gene Young-blood, a critic for the underground *Los Angeles Free Press* who had written about the World Game. Tom Turner, director of the program at SIU, invited him to attend a workshop and fund-raiser, but Young-blood was more impressed by Fuller, and they talked until late at night. "A concrete scientific alternative to politics now exists," Youngblood marveled in print. "For the first time in history, it is now possible for society to shape its destiny completely outside the realm of political activity as we know it."

Fuller hoped to establish a simulation center under a dome in Ed-wardsville, complete with a vast computerized display, and federal legislation was introduced to obtain funding from NASA. The bill never passed, however, and his dreams of a new era of understanding between generations were about to be destroyed for good. After the Kent State shootings, demonstrations in Carbondale resulted in four hundred arrests, and the National Guard arrived at the school, which imposed a rule against gatherings of five or more people. On May 11, 1970, the same day that Fuller evaluated a workshop on campus, tear gas was used to disperse protesters.

After the troops and state police withdrew, the situation flared up again. A rally at the library in defiance of the order against group meet-ings swelled into a march of more than three thousand demonstrators, which ended at President Delyte Morris's house. Standing on his lawn, they shattered windows and broke into the office next door, prompting Morris to close the campus entirely. When SIU's chancellor announced the decision, the crowd erupted in cheers.

The next day, Fuller left town for an event upstate, then flew from there to Australia. From his dome house, he had seen the youth move-ment assume a frightening form, and he put the blame on a country that he had previously courted, claiming that the students had been "brilliantly exploited" by the Soviets. In an ironic twist on his own call for independent thought, Fuller wrote that "the determination of the students to do their own thinking was readily exploited in attacking the university presidents," who, in the end, had been nothing but "sitting ducks."

Like many members of the counterculture, Fuller's followers tended to be white and relatively privileged, which allowed them to undertake drastic life changes with minimal risk. This had been equally true of Fuller himself, who rarely acknowledged this fact when he discussed his fateful decision to forget about earning a living. As the new decade began, his audience became more diverse, but Fuller was still unable to see its concerns from any perspective but his own.

On March 27, 1968, Fuller had met in New York with the Real Great Society, a group of community organizers founded by Carlos "Chino" García and Angelo Gonzalez Jr., two former gang leaders from the Lower East Side, who were surprised when he accepted their invitation to speak. At his talk in a loft on East Seventh Street, Gonzalez noted that many of the attendees were skeptical: "He was supposedly coming in to tell us what to do, and people got really turned off by him."

After the lecture, a few sympathetic listeners spoke to Fuller over his usual steak dinner, and a core team—Chino García, Humberto Crespo, Angelo Gonzalez, Roy Battiste, Anthony Figueroa, and Sal Becker—combined their first initials into the acronym CHARAS. According to the photographer Syeus Mottel, who wrote a book on the group, they hoped to reshape their own lives using the ideas of Fuller and other thinkers, and they met later with Ed Schlossberg, "who basically spent time rapping about his systems of thought."

The following year, Fuller returned to encourage them to start a dome business, entrusting the program to the Israeli-born architect Michael Ben-Eli, who often met with students to discuss Fuller's ideas. The environment at CHARAS was very different, and Ben-Eli became frustrated by the "completely crazy situation." Until then, he had rarely thought about race, and his pupils were doubtful of his motivations:

"They could not see what the connection was of all the things I was talking about—all the triangles and cosines—and the one thing they wanted to do: namely, to build a house."

Eventually, they constructed domes using wood and canvas. There was talk of building one on a property that Fuller offered them in Woodstock, but they decided to begin at a vacant lot near the East River. Sensibly, they moved away from the notion of launching a dome company, which was unlikely to succeed in the city, toward a more abstract goal. "It's more about involvement of people who never thought they would be involved in anything or included into anything," a member named James Echevarria explained. "Not just in domes, but in anything."

Fuller was only peripherally involved with CHARAS, which he casually described as "a New York City Lower East Side gang," and the same was true with regard to a separate effort to broaden his appeal in a decidedly different direction. It originated with an independent book editor and packager named Jerome Agel, who worked with the graphic designer Quentin Fiore on a series of innovative paperbacks, including *The Medium Is the Massage* by Marshall McLuhan. Agel had met with Fuller to discuss a potential collaboration, and when the project was abruptly reactivated by Bantam Books, the team had only nine months to finish.

I Seem to Be a Verb, published in the summer of 1970, was put together with less input than usual from Fiore, who disliked Fuller's "turn of mind," leaving Agel to assemble a collage of hundreds of images and quotations that could be read in two directions. By juxtaposing Fuller with the Yippies and other countercultural icons, it made him seem hipper than he really was. In addition, only about a quarter of the book consisted of his own words, although it also featured one of his most memorable lines: "I always say to myself, What is the most important thing we can think about at this extraordinary moment?"

Despite his status as what the writer Harvey Wheeler called one of the youth culture's "reigning sex symbols," Fuller maintained the itinerary of a respectable public intellectual. In recognition of "his contribution to mankind's environment as probably the world's greatest living architect," Francis Warner nominated him for a Nobel Peace Prize, and he was honored with the Gold Medal of the American Institute of Architects, which described the dome as "the strongest, lightest, and most efficient means of enclosing space yet devised by man."

In private, Fuller was entering one of the most troubled periods of his career. Delyte Morris, suffering from the early stages of Alzheimer's disease, had resigned as SIU president following the demonstrations in Carbondale, which had left him "a broken man." He was replaced by an administrative council that was unlikely to accommodate Fuller, who had been named a distinguished university professor. The real cause of Morris's downfall had been his use of research funds to build a presidential house, and the new regime would look closely at Fuller's spending.

Fuller himself was rarely on campus. In September 1970 he and Anne attended an architectural colloquium in Iran, which he praised as a "strife-free center of international good-neighboring" in a letter and poem to Farah Pahlavi, the wife of the Shah, although he failed to add her to his roster of powerful women admirers. Fuller had better luck at a banquet in his honor in Canada, where he met with Pierre Trudeau. He claimed afterward that the prime minister passed along his idea of a world energy network to Soviet leader Leonid Brezhnev, whose advisors allegedly found it "feasible."

His most intriguing encounter that year occurred far from the centers of power. In October Fuller appeared at the International University of Art, in Venice, Italy, speaking for eight hours each day. In the front row for every lecture was Ezra Pound, whom Fuller had briefly encountered at Spoleto. During a tour of the lagoon with Anne, Noguchi, Priscilla Morgan, and Pound's longtime mistress, concert violinist Olga Rudge, they shared pomegranates for lunch, and in the afternoon, they were filmed on the island of San Giorgio Maggiore, where Fuller commented on the beauty of the pinecones.

In the evening, Fuller and Anne visited Pound and Rudge at their home on the cul-de-sac of the Calle Querini. After they left, they were preparing to cross over to the ferry when Rudge caught up to give them a volume of *The Cantos*, which Pound had inscribed with a few lines on "the gold from the pine cones—the pomegranate." Rudge read the Nehru lecture aloud to Pound late into the night, and Fuller saw himself in this "tremendously hurt man" who had fallen publicly silent after being ravaged by his own mistakes. The scholar Guy Davenport, a mutual friend, commented later, "That conversation should have been going on for thirty years. Exile has its disadvantages."

As always, Fuller's greatest priority in his own life was to retain control. World Game centers were established at colleges, with his associates Medard Gabel, Ed Hauben, Mark Victor Hansen, and Val R. Winsey spreading the gospel among students, while Fuller delivered talks at workshops in Boston and San Francisco, although he was displeased by their emphasis on "egocentric soul-searching." He advised John Cage against making an appearance, and he complained that participants were "exploited" by a program that conflated him with the *Whole Earth Catalog.*

Fuller placed much of the blame on Gene Youngblood, who had enthusiastically promoted the World Game. As Youngblood recalled, "I thought if I exaggerated just a little bit—oh, we'll have all these World Games here and there—that it would have a positive effect." Fuller warned Youngblood that his actions were "greatly weakening, if not completely emasculating, the work which I have been doing," and he took steps to rein in the seminars, which might otherwise have evolved to resemble the motivational movements to come.

Looking to reach a wider audience on his own terms, Fuller wrote opinion pieces for the *New York Times,* in one of which he observed, "Humanity is acquiring the right technology for all the wrong reasons." He also granted a candid interview to the journalist Barry Farrell for *Playboy* magazine. "I'm the only man I know who *can* sin," Fuller said. "I find everybody else too innocent. They don't know what they're doing. . . . I have no desire to sin, I assure you. The point is: I know how. There are many things I've done in my life that would be sinful if I did them today."

Fuller told Farrell that he saw history as a progression from ancient Egypt—which was concerned only with the afterlife of the pharaoh—to a society that could provide for everyone. The interview's most famous passage would be reproduced widely elsewhere, although the second half was often omitted:

> Something hit me very hard once, thinking about what one little man could do. Think of the *Queen Elizabeth* again: the whole ship goes by, and then comes the rudder. And there's a tiny thing on the edge of the rudder called a trim tab. It's a miniature rudder. Just moving that little trim tab builds a low pressure that pulls the rudder around. It takes almost no effort at all. So I said that the individual can be a trim tab. Society thinks it's going right by you, that it's left you altogether. But if you're doing dynamic things mentally, the fact is that you can just put your foot out like that, and the whole ship of state is going to turn around. So I said, "Call me Trim Tab."
>
> The truth is that you get the low pressure to do things, rather than getting on the other side and trying to push the bow of the ship around. And you build that low pressure by getting rid of a little nonsense, getting rid of things that don't work and aren't true until you start to get that trim-tab motion. It works every time. That's the grand strategy you're going for. So I'm positive that what you do with yourself, just the little things you do yourself, these are the things that count. To be a real trim tab, you've got to start with yourself, and soon you'll feel that low pressure, and suddenly things begin to work in a beautiful way. Of course, they happen only when you're dealing with really great integrity: you must be helping evolution.

He was still working on the Beckett Theatre, which led to his appointment as an honorary fellow of Oxford, and when funding stalled, he seized the chance to propose a radical change. Instead of locating it underground, he advocated replacing or supplementing it with a "demountable" auditorium, using an unproven asymmetrical tensegrity structure that could be transported by air and assembled as needed. Naturally, this only complicated the plans for the theater—its original estimated budget had tripled—and its prospects remained uncertain even after Beckett won the Nobel Prize for Literature.

Although Fuller had grown gentler with age, his need for control remained undiminished, and this ultimately undermined his relationship with Lloyd Kahn. Fuller was initially supportive of *Domebook One*, and his assistant even asked Kahn to send them 5 percent of the royalties. As soon as he realized that the book disclosed the chord factors, however, his feelings changed, and when Joe Clinton truthfully said that he had no idea how they had been obtained, Fuller took action.

On January 12, 1971, Fuller made a second visit to Pacific High School, and its tone was very different from the first. Leaving Anne in the car in the cold, Fuller laid down the law to Kahn for an hour and a half: "Your engineer who wrote this section on geodesic geometry didn't give me proper credit."

Kahn thought that Fuller was referring to Jonathan Kanter, the teenager who had written the section on geodesic math. He pointed at Kanter, who was lying on the floor. "This is the engineer."

They were talking past each other, and Kahn felt that Fuller was excessively "pissed," but they agreed that the second edition would include a corrected section of chord factors, which were scheduled to be released to the public by NASA. Joe Clinton recalled that "everyone concerned, even Bucky, was pleased with the final outcome." Even before this confrontation, however, Kahn had become increasingly doubtful about the dome, as he wrote later in a book with the significant title *Shelter*:

> What's good about 90° walls: they don't catch dust, rain doesn't sit on them, easy to add to; gravity, not tension, holds them in place. It's easy to build in counters, shelves, arrange furniture, bathtubs, beds. *We* are 90° to the earth.
>
> Not important how much a building weighs. It *is* important how much a bird weighs, but a building doesn't have to move or fly.

Engaging with Fuller on his own terms, Kahn argued that the weight of a shelter mattered less than the energy consumed by its materials, especially plastic, and rejected the premise of ephemeralization, stating that instead of doing more with less, the goal should be to do "*less* with less." He concluded famously: "Our work, though perhaps smart, was by no means *wise*."

Fuller was more preoccupied with an undertaking inspired by Katherine Dunham, the pioneering Black dancer and activist who had established the dance program at SIU Edwardsville. Dunham, now in her early sixties, knew Fuller through her husband's brother, a professor at Carbondale, and she was involved in plans for a "demonstrative urban conservation park" for East St. Louis, Illinois. According to Fuller, Dunham said to him, "Bucky, I think you've been in Africa so much you really understand a whole lot about the feelings of the African, about community and so forth. Could you not design something for us that really would be acceptable to the community?"

Whether or not this accurately reflected Dunham's views, Fuller agreed to contribute a design, despite his unfamiliarity with the social factors behind the city's bleak conditions, which he associated with the poverty of a developing nation: "There is nothing in Calcutta, Johannesburg, or Hong Kong that equals the squalor of the East St. Louis slums." Remembering his frustrations with the political side of the Minnesota Experimental City Project, Fuller covered his own expenses—giving himself a free hand—and he produced a few sketches of a terraced pyramidal structure under a geodesic dome a half mile in diameter.

For the details, Fuller recruited James Fitzgibbon, who reused elements of an existing proposal to cover Kuwait City with a geodesic umbrella. Fitzgibbon relocated the site to permit access to highways, and he acknowledged the influence of crusading journalist Jane Jacobs by dividing it into more livable quadrants. It still erased existing neighborhoods, but Fuller refused to compromise his vision, which he presented at a community forum in February. Cautioning that he would proceed only if they unanimously approved his plan for a "raceless and classless" development, Fuller unveiled a design that resembled a covered moon crater, which he called Old Man River's City.

Fuller said later that a skeptical faction feared the project "might be part of its social enemy's conspiracy to entrap them," but he won them over. "I said that what I would design must be so 'right' that the entire community would fall in love with it . . . or it would be dropped," Fuller recalled—a stance that also allowed him to avoid revising a proposal that he had conceived without any input from potential residents. Fuller discouraged them from taking money from the government or foundations, warning that to do so would undermine "our design solution," and the lack of funding effectively ensured that it would never be built at all.

He was less successful at fighting other forms of interference. Fuller's office was searched by the CIA in connection to the May 1970 Carbondale protests, and he had issues with students "using my phone, my credit cards." SIU had passed on covering a basketball stadium with a dome, and it asked him to assist in paying for maintaining his archives. In a final blow, the administrative council became concerned about his casual tendency to mix public funds with his own earnings.

Deciding that it would cause problems in the event of a state audit, the university abruptly withdrew its funding for the World Resources Inventory, which had annual costs of $100,000. Fuller was grappling with the situation on the day of the prizefight between Muhammad Ali and Joe Frazier, leading him to lament his lack of support when "these two fighters are getting millions of dollars just to spoil each other's brains." He retaliated by repurchasing his copyrights from the SIU Press, and, after a final staff meeting in June, he left Carbondale for good.

As Fuller looked for a new home, his outward activities remained largely unchanged. In Spoleto, he dined with Isamu Noguchi and Ezra Pound, who expressed regret over his past: "Every time humanity gives me a chance, I make such a mess of it." Pound inscribed another book for Fuller, extolling him as a "friend of the universe" and "liberator," but they never saw each other again. The following month, Fuller appeared at a symposium in Amherst with Maharishi Mahesh Yogi, the founder of Transcendental Meditation, whom he praised at a joint press conference: "You could not meet Maharishi without recognizing instantly his integrity."

Fuller withdrew that summer to Bear Island, where his beloved refuge was facing threats from the outside world. Earlier that year, he had sent Senator Edmund Muskie a lengthy telegram objecting to petroleum development in Maine, which was reprinted in the *New York Times* and the collection *Earth, Inc.* His concerns prompted him to research wind and tidal power, and he worked with Hans Meyer, a University of Michigan graduate, to prototype windmills on Little Spruce Head Island.

In the fall, Fuller left on a voyage that included five days in Bali. During a mountain hike, he was given a bamboo staff, and when his guide asked to keep it, Fuller was reminded of his daughter Alexandra, who had asked for a cane just before her death. "All those years and all those things that have happened since little Alexandra died," Fuller reflected. "Now somehow I feel intuitively that this is a kind of message. That she has her cane at last."

By 1971 a number of companies had entered the business of dome houses, and Fuller's original patent was expiring. In response, he prepared the Hexa-Pent Dome for *Popular Science*, with a structure by Sadao based on hexagons and pentagons, not triangles, which they hoped to sell to homeowners. Norman Foster worked with him on the Climatroffice, a concept for a multiple-level workspace enclosed by a space frame, but Foster later conceded that they lacked "the time or skills" to make it a reality.

None of these projects generated revenue, and the situation at SIU had deteriorated in his absence. Tom Turner, his academic administrator, wrote a memo on the "bizarre antics" of his staff, complaining that neither Fuller nor his employees knew how to work within the university system. Fuller lobbied successfully to reassign Turner, and he extracted an apology from the board, the members of which he dismissed as "complete strangers to me."

Apart from his speaker's fees, Fuller had no meaningful income outside of his salary, but he was still determined to cut ties. When President John Rendleman of SIU Edwardsville proposed that he move there, Fuller accepted. A series of layoffs had reduced his office to nine staffers, who heard about the relocation to the sister campus through a press release in September, and it was finalized five months later. When asked to comment, Fuller said only, "I live on planet Earth."

Another initiative was orchestrated by Herman Wolf, his former partner from Wichita, who advised Fuller on public relations after they unexpectedly reconciled. Wolf conceived of an organization that would posthumously perpetuate the concepts developed by Fuller, who granted that it made sense: "I could conk out any minute." The nonprofit Design Science Institute would be headed by Glenn Olds, the president of Kent State University, and its members included David Rockefeller's daughter Neva Kaiser, with Jonas Salk, Margaret Mead, and Arthur C. Clarke on its advisory board.

In 1972 Fuller met William F. Buckley Jr. at the University of Virginia's commencement ceremonies. "I would have assumed that I would not like you, but I find you charming," Fuller told the deeply conservative author, who joked that they should publish "a joint manifesto." Fuller had recently finished his latest declaration of intent, *Intuition*, named after a yacht that he had acquired several years earlier. Along with the title poem, it included two variations on the Lord's Prayer, which he mentally reworked whenever he went to sleep, reminding himself that the trespasses of others "are in fact the feedback of our own negatives."

On a practical level, he continued to sever connections to Carbondale. His office had been operating without him, but after "a financial intervention," Ed Applewhite fired most of the remaining staff. Accompanied by his new secretary, Shirley Swansen, Fuller paid his first visit there in more than a year to move his belongings. At Bear Island, he seemed tired and irritable, and he followed up by withdrawing his papers from SIU, estimating optimistically that the sale of a single manuscript would yield enough money to put his grandson through college.

Fuller considered moving his dome house to Edwardsville, but his salary of $27,000 was nowhere enough to cover his expenses, and the only available office space was located in a disused barracks off campus. With no prospect of raising funds for his World Resources Center, he considered a more definitive break, and Ed Schlossberg told Martin Meyerson of the University of Pennsylvania that Fuller was interested in moving to Philadelphia.

Meyerson came up with an unusual arrangement among Penn and three other colleges—Bryn Mawr, Haverford, and Swarthmore—to sponsor a professorship for Fuller based at the University City Science Center, which had been founded as an incubator for research. It was exactly the kind of setup that he needed, with shared funding, no teaching duties, and access to computers. Earlier that year, an environmental group known as the Club of Rome had used computer models to paint a pessimistic picture of mankind's ecological future, and he wanted badly to regain the initiative for the World Game.

On September 23, 1972, Fuller was publicly named a World Fellow in Residence in Philadelphia, although he continued to draw a salary from SIU. His contract ran for three years, complete with an office and archival space, as well as annual funds of $50,000 underwritten largely by a grant from the Rockefellers. Comparing the city's newest resident to Benjamin Franklin, the *Philadelphia Inquirer* hailed it as "a coup roughly of the same magnitude as the Philadelphia Eagles acquiring last year's entire Pro Bowl team for a couple of seventh-round draft choices."

Fuller's benefactors found him an apartment at Society Hill Towers on Locust Street, which had a pleasant view of the Delaware River. While Anne said that Philadelphia was the only city that was interesting enough to keep her husband at home, this was wishful thinking, even as Fuller began to suffer from the effects of his endless movement. On November 14 he collapsed in St. Louis due to a lung infection and had to be hospitalized for four days. He occupied himself by dictating to Swansen, who learned to take shorthand for phrases such as "the interaccommodative, comprehensibly rememberable Scheherazade number."

By the following week, Fuller recovered enough to move to Philadelphia. The transition coincided with the release of *Buckminster Fuller to Children of Earth*, a collection of his aphorisms with photo illustrations by Cam Smith, a former student of Fuller's at San Jose State. She and her husband, Tom Solari, the book's editor, were members of the Church of Scientology, which had been founded by L. Ron Hubbard. In her letters to Fuller, Smith praised the belief system as "a tremendous tool," and she described it in terms that would appeal to him: "Things could and should be made to work right for us on this planet."

Shortly afterward, Fuller received a letter from Charlene Sprang of the Celebrity Center Mission in San Francisco, who ventured the opinion that he and Hubbard were "in line with each other." Swansen responded noncommittally, "Dr. Fuller is familiar with the study of Scientology and regrets that other commitments prevent him from answering your specific questions." Despite the brush-off, Fuller agreed to work on Smith's book, with part of the proceeds donated to Narconon, a nonprofit "dedicated to reducing crime and drug abuse, for their youth program." In reality, it was a vehicle for Hubbard's theories and affiliated closely with Scientology.

As the futurist Alvin Toffler characterized Fuller as "the leader of something approaching a religious movement," other groups were turning his ideas to more positive ends. Domes went up at the Libre and Red Rockers communes in Colorado; teenagers built a dome as a youth center in the Bronx; and CHARAS had been working quietly all this time on the Lower East Side. The group constructed two domes out of cardboard and cement, and although the smaller one was smashed by firefighters responding to an emergency call—a homeless person had evidently built a fire inside—it planned to use the larger structure as a meeting place.

On January 14, 1973, Fuller promoted *Children of Earth* in a headlining appearance at Carnegie Hall, which featured a short montage by Cam Smith. After learning that CHARAS had finished its dome that afternoon, he went there the following day with Anne, Swansen, Michael Ben-Eli, and photographer Syeus Mottel. Emerging from the cab, Fuller went to study the dome through the fence. He said softly, "It's beautiful, isn't it?"

When the CHARAS members arrived, Fuller greeted them with enthusiasm and followed them into the lot. After examining the dome up close, he gave it an affectionate pat and proclaimed the structure "a triumph." A passerby approaching for a better look was astonished to see Fuller, whose hand he shook with delight. The cab driver, whom Fuller had asked to join them, leaned over to whisper to a bystander. "Hey, is this man important or something?"

ELEVEN

SYNERGY

1973–1977

Fictional history. Historical architecture. Crabs walk sideways; but only human society keeps its eyes on the past as it backs into its future. Madison Avenue aesthetics and ethics. Comic strips and cartoons . . . truth emergent, laughing at self-deception . . . momentary, fleeting glimpses of the glory, inadvertently revealed through faithful accuracy of observation—lucid conceptioning—spoken of as the music of the stars, inadequate to the mystery of integrity.

—BUCKMINSTER FULLER, *SYNERGETICS*

By the early seventies, Fuller was famous across the world, both as a celebrity and as a symbol. His effortlessly caricatured features—the glasses, the dark suit, the bald dome—made him a virtual cartoon even in person, and he was instantly recognizable elsewhere. In the opening number of Stephen Schwartz's hit rock musical *Godspell*, he "appeared" as one of eight singing philosophers, alongside Socrates and Sartre, with lyrics taken straight from the source: "I seem to be a verb, an evolutionary process, an integral function of the universe."

The irreverent humor magazine *National Lampoon* parodied him more savagely in "Buckminster Fuller's Repair Manual for the Entire

Universe" and "Utopia Four Comix." In the latter, Fuller was depicted as part of a team of countercultural superheroes—along with Marshall McLuhan, the feminist Kate Millett, and the academic Charles A. Reich—who turn out to be utterly incapable of dealing with the realities of the inner city. As "Super-Bucky" wanders in "a visionary trance," he delivers a demented monologue: "What this area needs is a Slum Game! A billion-dollar electronic map to tell us at a glance how many people are unemployed at any given moment!"

His most celebrated artifact, the geodesic dome, was simultaneously a vernacular icon of hippie culture and a signifier of futuristic technology. Radomes near Oslo, Norway, appeared in the establishing shots of the James Bond film *You Only Live Twice*, and the protagonist of the movie *Slaughterhouse-Five,* based on the novel by Kurt Vonnegut, was held in a dome by the aliens of Tralfamadore. The spaceship in *Silent Running* featured greenhouses inspired by the Climatron, and the model was reused by the television series *Battlestar Galactica*, which also shot an episode at the Expo Dome. Fuller himself advised on the CBS-TV movie *Earth II* and was asked to consult on John Huston's unrealized adaptation of Aldous Huxley's dystopian novel *Brave New World*.

Back on earth, one dome had achieved success in a setting worthy of science fiction. Fuller had always been fascinated by the poles, and a permanent dome finally arrived in Antarctica as an enclosure for the Amundsen-Scott South Pole Station. It was designed by Don Richter, who was making a concerted push at Temcor into the kinds of major commissions that Fuller rarely managed to land. To advance his case, Richter published a computer analysis showing that geodesic structures were superior to other domes at distributing stresses—a basic selling point that Fuller had never thought to use to attract clients.

Temcor came up with an aluminum dome to shield three buildings at the research station from the elements, and, for once, it was perfectly suited for its intended use. Its curved sides could withstand asymmetrical snow loads and high winds; it was light enough to ship by air; and it could be assembled by a crew encumbered by bulky clothing and with limited access to equipment. The dome was finished on January 8,

1973, and although it was expected to stand for only a decade, it lasted for fifteen years without major damage.

Fuller's own surroundings were considerably more ordinary, but still enviable. At the time of his move into the University City Science Center in Philadelphia, which provided him with nearly three thousand square feet of space, his staff had been reduced to Shirley Swansen and a second secretary. In theory, Fuller could have maintained his operation at this scale, but he immediately ramped it up to what the World Game workshop might have called his bare maximum: the minimum that he needed to live up to his full potential.

After being laid off in Carbondale, Medard Gabel had partnered with another World Game veteran, Howard Brown, to organize seminars through the nonprofit Earth Metabolic Design, and he was rehired to catalog the six tons of papers that Fuller regarded as his most valuable asset. Gabel recruited a University of Cincinnati graduate named Tim Wessels, while Fuller impulsively hired another staffer, Robert D. Kahn, after meeting him at an event that summer.

Fuller's overhead rose to $200,000 a year, just a quarter of which was covered by his fellowship, but he unhesitatingly increased it to the level that only a schedule of nonstop lecturing could accommodate. He wanted a full team of researchers at his constant disposal, even if they were inactive during his absences—which grew longer as he accepted more invitations. As a result, he was stuck at a permanently high rate of travel, and he persisted in conducting marathon lectures, despite back pain that started whenever he had been talking for over an hour.

It was a machine that existed largely to keep itself in motion. Fuller's standard quote for a talk was $5,000, an amount he received perhaps once a year, and he was usually content with one-fifth his supposed asking price. When Fuller was away, he assigned busywork to his staff members, who were expected to keep the bathrooms spotless at all times. Swansen struggled heroically to control his expenditures, which he took for granted. "I have absolutely no budget," Fuller said serenely. "I carry it all in my head. And by knowing that if I have paid the bills and then excess money comes in, then I know I can support new research."

He continued to explore ideas for alternative energy, including a patent for a floatable breakwater for areas without natural harbors, which took the form of a hydraulic tube that could generate tidal power. With his instinctive preference for wind and water, Fuller advised Windworks, Hans Meyer's firm, on utilizing windmills to convert water to hydrogen and oxygen for fuel, and he tried to convince the Rockefellers to fund related projects in Maine. He saw environmental pollution as nothing but a concentration of resources in the wrong place, and he strongly opposed nuclear plants, while downplaying solar as effective only "when the sun is at a favorable angle."

Fuller had actively fought the ecological threat to his beloved Bear Island, where his rustic lifestyle stood in stark contrast to his reputation as a technologist, but he was unsympathetic toward environmentalists. When the journalist Rasa Gustaitis challenged his concept of Spaceship Earth, he was unable to conceal his irritation. "A spaceship is built by man," Gustaitis said. "All the factors have been designed for a particular known purpose. The earth is an organic thing on which we live."

"The planet is designed—superbly designed—by a greater intellect than that of man," Fuller responded sharply. "And it's moving through space."

Gustaitis pursued her point. "But are we passengers or an organic part of it?"

"We've got three billion and a half passengers on board." When Gustaitis followed up by asking if he included other living beings in that category, Fuller was dismissive of animal rights. "What happened to those dragons? What happened to the flying creatures of a few million years ago? What happened about their rights? What about the rights of the volcanoes when the whole earth was volcanic?"

The publication of this exchange coincided with the release of the critic Hugh Kenner's book *Bucky: A Guided Tour of Buckminster Fuller*. Kenner, a longtime associate of Marshall McLuhan, had befriended Fuller on Ezra Pound's advice that "one has an obligation to visit the great men of one's own time," and he respected his subject enough to ask hard questions:

[Fuller] reiterates: there is no such thing as race. "We simply have humanity aboard this space ship," with more important things to give time to than how many children of *this* color to ship across town to the enclaves of *that* color. He is confident that the synergetic view has identified a false problem.

Anyone involved in human misery here and now is apt to retort that a good deal of big-picture talk has accomplished nothing except the reassurance of (a) white supremacists, (b) people who for other reasons either prefer things to stay as they are, or else doubt that interventions can improve anything, really. A prescription for doing nothing?

A few lines later, Kenner recalled an event at which Fuller addressed "some sharp kids from the streets," who wondered aloud what geometry had to do with the rats in their bedrooms. When Kenner asked afterward how he would have replied, Fuller said, "I wouldn't have. I never argue."

Fuller's more prominent projects had stalled. He and Sadao were still working on airports for Indira Gandhi, but his deafness—he had to constantly cup his hand behind one ear—made meetings awkward, and he was exhausted by the heat in India. He had requested payment only for his lost research and speaking time, but when he was told that his compensation would be taxed at the standard rate, he insisted that he was making no profit at all. Since his operation automatically expanded to consume all his available income, it was a shaky argument, and he was granted nothing but reimbursement for his travel expenses.

His ideas for the design itself had been poorly received. Fuller said to Ed Applewhite, "If anyone were to ask you, 'How did you like the airport?' the best answer would be, 'What airport?' An airport should work so well that you wouldn't know it's there." When Habib Rahman, an Indian architect, grumbled that Fuller had contributed nothing useful to the project, Indira Gandhi sent a cautionary note to Minister of Tourism Karan Singh: "I hope no irrevocable commitments are being made without giving careful consideration to the points Rahman has urged." Sadao thought that the final design "visibly had little to do with Bucky," and no terminals were ever constructed.

When not surrounded by his staff or family, Fuller was happiest with old friends. Norman Cousins recruited him as a board member of the magazine *World*, along with former secretary-general of the United Nations U Thant, who had met with Fuller to discuss the World Man Center and the Geoscope. Fuller contributed articles that Michael Ben-Eli had compiled from his earlier work, and he pressured another collaborator, Robert W. Marks, for the rights to *The Dymaxion World*. When the book was reprinted with Fuller listed first as author, Marks felt "terribly hurt."

Fuller needed more editorial assistance than ever. He worked on an overview of geodesic math with Joe Clinton, some of which would be incorporated into a second book by Kenner, while the rest went to Applewhite, who was laboring tirelessly on *Synergetics*. "He had proposed all his life to write a book attempting to describe all physical and metaphysical experiences in terms of the tetrahedron," Applewhite recalled. "What I proposed was to help him complete this task and to discover whether I would become a convert in the process."

Reviewing the material that already existed, Applewhite compiled an immense file of alphabetized excerpts on index cards, and at their first working session, he agreed to assemble the extracts into a rough draft, with the goal of producing the print equivalent of Fuller's "thinking out loud." Fuller had spent his life assigning impossible tasks to collaborators, leaving them to put together thousands of pieces, and Applewhite—like Sadao—neither took a salary nor sought credit as his coauthor.

Their partnership would test all of Applewhite's considerable ingenuity. At his own expense, he joined the roster of travel companions on whom Fuller depended, which often generated insights in itself. On one occasion, before his move to Philadelphia, Fuller's bags were misplaced at Washington National Airport. When the airline agent innocently asked for his address, Fuller grew red and banged his fists on the counter: "Address! That isn't the right question. Young man, I live on Planet Earth! Man was born with legs, not roots!"

Applewhite noted the exchange quietly for his card index. Much of the writing was done in transit, with one section sketched out on a menu—Fuller loved drawing domes on placemats—at a coffee shop at the Boston airport. Fuller dictated passages by phone from as far away as Tokyo, and they met several times a month. Usually, the only advance warning that Applewhite received was "an unexpected shipment of aluminum rods or wooden dowels or plastic struts," and they worked long hours on the sunporch of Applewhite's home office, with Fuller dressed in a kimono and drinking endless cups of tea.

Although Applewhite typed their drafts with generous spacing for edits, Fuller filled every sheet with additions and drawings, so that establishing a coherent sequence became a constant challenge. When he rose in the morning, Applewhite often found notes from Fuller, who once wrote, "I am feeling myself physically precessed into a strange ephemeral condition. It may be I just need a little sleep." Applewhite saw his role as "confronting the author with himself," and, by repeatedly asking Fuller to revisit his geometry, he created a feedback cycle that made it even more extravagant.

As always, Fuller's vocabulary presented special difficulties, and he wanted certain words and phrases set entirely in upper case, an idiosyncrasy that Applewhite discouraged. They capitalized many terms, including *Universe*, but not *nature*, which Applewhite explained was a deliberate distinction: "For Fuller, nature—like integrity and principle—is abstract and generalized and thus never capitalized. Universe is a place like Pleasantville or Baltimore and deserves a capital letter."

They had planned to incorporate sections by others, including Louis I. Kahn's associate Anne Tyng, but the only outside contributor to the final version was Arthur L. Loeb, whom Fuller saw as "the bridge builder from me to the rest of science." The Dutch-born scientist, a Harvard crystallographer, thought that Fuller had made genuine discoveries, including the equation for the number of spheres in a vector equilibrium, and he provided some of the formal proofs that their creator disdained.

While Loeb lamented the book's "digressions into metaphysics," he believed in *Synergetics* enough to propose hiring his friend M. C.

Escher, a fellow Dutchman known for his elaborate geometrical art, for the illustrations. Escher's participation would have increased the project's appeal, but his involvement was derailed by an old dispute. In the sixties, Escher's son had contracted with the US Air Force to build radomes in Canada, and Fuller had tried to extract royalties from the deal. "My own lay conclusion," Escher told Loeb, "is that there are indeed charlatan-like facets to this man."

Fuller's agent and attorney, Gerald Dickler, had renegotiated their contract with Macmillan, which agreed to pay $25,000 to review a third of the new manuscript, with the option of making equivalent payments for future installments. After receiving four chapters, which went to Hugh Kenner for comments, Macmillan committed in full. Another chunk followed a half year later, and the complete draft was submitted—to Applewhite's considerable relief—in March 1973.

When the galleys arrived in the fall, however, their impact on Fuller was "galvanic." Authors were discouraged from revising extensively at the proof stage, when a book should be all but ready to go to press, but Fuller indulged in his usual habit of rewriting every page at his own expense. "His imagination is triggered by what the eye frames in front of him," Applewhite noted in his 1977 book *Cosmic Fishing: An Account of Writing* Synergetics *with Buckminster Fuller.* "It was the same with manuscript pages: he never liked to turn them over or continue to another sheet. Page = unit of thought. So his mind was retriggered with every galley and its quite arbitrary increment of thought from the composing process."

The question remained of how Macmillan would respond to the result. If Applewhite had played the role of a partner like Sadao, the publisher had been cast as the patron Fuller expected to cover the cost of his uncompromising vision. He once told Applewhite that Thornton Wilder had advised him in the forties that a book on his geometry might be more important than Newton's *Principia Mathematica,* but only if it were proven correct. "Now we know that it is right," Fuller concluded, "and so tactically it is more important than the *Principia*—the Greatest Book."

Applewhite described *Synergetics* as "a book about models," and it certainly ranked among the strangest works ever produced by a commercial publisher. Fuller dedicated it to H. S. M. Coxeter, "*the* geometer of our bestirring twentieth century," who had dominated the field of geometry for decades. In the fifties, Coxeter had sent a letter asserting priority on aspects of the dome to Fuller, who, "after much deliberation," never responded. They finally connected after Coxeter visited the Montreal Expo Dome, which impressed him, although he was mystified by its hybrid design.

In 1968 the two men met in Carbondale, but Fuller resisted corresponding in depth, claiming that he wanted to keep his "foolhardy freedom to explore." Coxeter loved the domes, relating them to the structure of specific viruses, and he called the formula for the cubic packing of spheres "a remarkable discovery" by Fuller, whom he praised for his intuition. On the other hand, he also walked out of a Fuller lecture in disgust, characterizing him afterward as "a brilliant architect and engineer who knew very little mathematics but was very

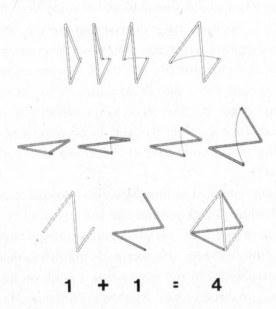

The first illustration in the text of *Synergetics*. "Two triangles may be combined in such a manner as to create the tetrahedron, a figure volumetrically embraced by four triangles. Therefore, one plus one seemingly equals four."

proud of himself," and he dismissed the work dedicated to him as "a lot of nonsense."

After advising its readers to "dare to be naïve," the book began with a definition of its central concept: "Synergy means behavior of whole systems unpredicted by the behavior of their parts taken separately." Like Margaret Fuller's *Woman in the Nineteenth Century*, it opened with an illustration of two interlocked triangles, one black and one white, which formed the four triangular faces of a tetrahedron when their ends were opened. "You may say that we had no right to break the triangles open in order to add them together," Fuller wrote, "but the triangles were in fact never closed because no line can ever come completely back into itself."

In a book with countless byways to explore, the triangle offered one possible way through. It was the simplest polygon that could divide an area into an inside and an outside, which implied in turn that it was impossible to draw fewer than four triangles: "When we draw a triangle on the surface of Earth . . . we divide Earth's surface into two areas on either side of the line. . . . There is a concave small triangle and a concave big triangle, as viewed from inside, and a convex small triangle and a convex big triangle, as viewed from outside. Concave and convex are not the same, so, at minimum, there always are inherently four triangles."

A triangle was also the minimum stable structure, which Fuller demonstrated using a necklace made of short tubes that were connected at the ends. Six tubes linked into a hexagonal necklace could be loosely draped around one's neck, as could the pentagonal and square versions formed by removing one tube at a time. When reduced to three tubes, the necklace became rigid. Fuller concluded, "Only the triangle produces structure and structure means only triangle."

By comparison, a cube drawn on a blackboard was a fiction that would collapse in real life. When it was stabilized using the smallest number of struts, the diagonals across its faces made a tetrahedron, which was the simplest stable system. Nature preferred to operate at 60 degrees, rather than right angles, and the tetrahedron served as a universal building block. It combined maximum surface area with minimal volume, optimizing its strength, and a series of tetrahedra could be joined in a spiral, which Fuller associated with the double helix.

On a more abstract level, Fuller saw the tetrahedron as "the minimum thinkable set." He related it to the number of possible relationships between a given number of objects, such as "the number of telephone lines necessary to interequip various numbers of individuals so that any two individuals will always have their unique private telephone line." There was only one possible phone line between two people, whereas three users could be connected in three different ways, four in six, and so on, as expressed by the formula $(N^2-N) / 2$.

As early as the forties, Fuller had noted that the result was always a triangular number, which could be visualized as dots arranged in an equilateral triangle. The same principle could apply to any set of relationships, including the number of connections between experiences in a person's mind, so that the sum of one's memories could be stacked progressively into an incomprehensibly huge tetrahedron. A human being

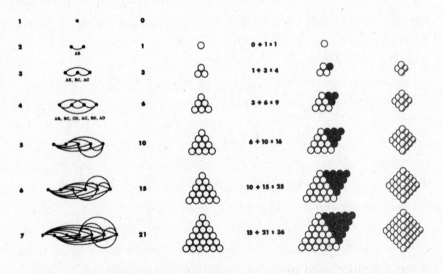

"Underlying Order in Randomness," illustrating how the number of relationships between events can be visualized as triangular and tetrahedronal numbers. The original version was copyrighted by Fuller in 1965.

was a kind of tetrahedron moving through time, and Fuller titled one section "Tetrahedron Discovers Itself and Universe," which might have served as the subtitle for the entire work.

Applewhite saw *Synergetics* as "a business of stark homage to the tetrahedron," but its central theme was obscured by a mass of other material. Loeb thought that its prose recalled Gertrude Stein, and it sometimes resembled the catechism of "Ithaca," the penultimate chapter in James Joyce's *Ulysses*. As Fuller himself observed, the modernists had arrived at their unconventional language for similar reasons: "James Joyce to a mild extent and Gertrude Stein to a considerable extent attempted to go along with the brush and chisel abstractionists in following the scientists into nonconceptual validity of reasoning." Ezra Pound had put it even more memorably: "Make it new."

In the most successful examples of his architecture, Fuller had achieved this goal, and *Synergetics* was the literary counterpart of the Expo Dome. It was the ultimate expression of his ideas, but less rational than it seemed, and its precisely numbered paragraphs were the equivalent of the visual sleight of hand that Sadao and Floyd had used in Montreal. In both cases, Fuller had been unable to do it alone, and the result could overwhelm even his collaborators. "There are to this day certain passages in this book that I would be hard put to explain to someone else," Applewhite confessed, and his final judgment anticipated the reaction of numerous readers to come: "I find certain passages sublime, others that verge on madness."

According to Applewhite, Fuller's compulsion to fill every sheet of paper with the maximum amount of information had a simple motivation: "Maybe the reader will never get to that next page." In practice, it prevented many readers from finishing even a single paragraph, and it was equally driven by Fuller's awareness of his age, which led him to prioritize sections by importance. "If we had only two days of work before Bucky was going to be hit by a bus," Applewhite remembered asking, "what would we most have wished to put his energies on?"

Fuller agreed, saying, "I always assume I've only got a few more minutes." As he drew close to eighty, he was confronted by what Applewhite called "intimations of mortality." Lincoln Pierce, his best friend from boyhood, had died a few years earlier, and a health scare sent Anne to the emergency room on March 2, 1974. Because Fuller was out of town, Shirley Swansen stayed at the hospital overnight. The next morning, she headed straight to the office, where a man in the elevator remarked, "Boy, this must be a nice job, where you can just roll in late."

Two weeks later, Louis I. Kahn died of a heart attack alone in a restroom at Penn Station. He had just returned from a trip to India, and it took two days for his family to learn the news. Fuller, who had advised Anne Tyng on her thesis on synergetics, had last seen Kahn a few months earlier at the University of Pennsylvania. After attending his funeral, Fuller became distressed at the prospect of a similar fate, which underlined his need for a traveling companion at all times.

He was disinclined to believe in the death of the soul with the body, comparing it to the relationship between a telephone and the speaker on the other end, and said that there was insufficient information to rule out the possibility that the sequence of experiences in life might be followed by yet another event that would reveal an even larger pattern. All the same, he was often moved to nostalgically revisit his past. In Washington, DC, he had dinner with the former Evelyn Schwartz, and he began to see the recently widowed Cynthia Lacey Floyd again in California.

When Fuller looked at the faces of the women he had loved, he was reminded of aspects of his life that were unimaginable to his younger admirers. A selective accounting of his formative years was provided by the publication of Alden Hatch's affectionate biography *Buckminster Fuller: At Home in the Universe*, which offered an authorized version of his career. Fuller thought that Hatch had been influenced by Anne's perspective as well, and he joked that it should have been called "Lady Anne and Her Little Old Man."

He continued to receive tributes, including the honorary title of world president of the high-IQ society Mensa, but he worried about Macmillan. The publisher had scheduled *Synergetics* as its lead title for the fall, even though almost no one there had read it. As the finished book grew to nearly nine hundred pages with color plates, its retail price had to be increased, and deeper misgivings came to the surface as the galleys began to circulate around the office. When a secretary in the publicity department declared that it was incomprehensible, the feeling mounted that the book was a potential disaster.

Bruce Carrick, their editor, told Fuller and Applewhite that the editorial board was considering pushing back publication in order to "give the reviewers more time"—or even repositioning it as a textbook. Over the summer, it was taken out of production. After Carrick left the company, their phone calls went unanswered, and as they considered legal action, Applewhite used his intelligence training to investigate what had gone wrong. Work on the title finally resumed in November 1974, thanks to the intervention of Macmillan's chairman, but its illustrations were limited to black and white.

As Fuller caught a second wind, he was awarded an honorary architectural license by the state of New York, and new friends emerged in surprising places. One was the author Annie Dillard, who would win the 1975 Pulitzer Prize for her nonfiction book *Pilgrim at Tinker Creek* shortly before her thirtieth birthday. During a reception in their honor at the Folger Shakespeare Library in Washington, DC, Fuller whispered to her, "You're very . . . *universe*." After he briefly left their table at dinner, Dillard borrowed his chair to speak to their host. When Fuller returned, he sat down in her lap—supposedly by mistake—and it evidently took him a while to notice.

Dillard recalled, "After he sat on me—which was legitimate, because I'd taken his seat in the dark—and as he felt around under himself as if to make sure he had indeed sat on a person, he was a gentleman. I guess one could 'feel around' horribly or well. It was not horrible; he was entirely decent, or as decent as possible. I didn't feel mauled. I was probably laughing." Later, Fuller wrote her "scores of incomprehensible letters," and she was intrigued by his view of mankind as a force against entropy, which implied that acts of the imagination "carry real weight in the universe."

But another valued supporter, Stewart Brand, had grown more skeptical of Fuller's most famous artifact. "A dome really might not leak," Brand said. "But they always do. A single 'perfect' technological skin doesn't make it—one pinhole and you're wet." Years later, he wrote:

> As a major propagandist for Fuller domes in my *Whole Earth Catalogs*, I can report with mixed chagrin and glee that they were a massive, total failure. . . . Domes leaked, always. The angles between the facets could never be sealed successfully. . . . The inside was basically one big room, impossible to subdivide, with too much space wasted up high. . . . Even the vaunted advantage of saving on materials with a dome didn't work out, because cutting triangles and pentagons from rectangular sheets of plywood left enormous waste. . . . Worst of all, domes couldn't grow or adapt. . . . When my generation outgrew the domes, we simply left them empty, like hatchlings leaving their eggshells.

Brand remained on cordial terms with Fuller, and he worked with Jann Wenner, the publisher of *Rolling Stone* magazine, to release a book by Medard Gabel based on the World Game. At a conference with Fuller and Brand, Wenner wondered aloud, "How do you convince people to change?" Brand thought that it was best to encourage them to adapt in advance, which he put to the test after a progressive Democrat, Jerry Brown, was elected governor of California in 1974. Brown invoked Spaceship Earth in his speeches, and he used Brand as a matchmaker to connect him with intellectuals, including Fuller, who met the governor for the first time on January 19, 1975.

After the meeting, Fuller flew to Philadelphia, where he prepared to cast off the constraints that his literary partner had imposed. As Applewhite conceded later, "*Synergetics* excludes his economics and industrial design programs, Spaceship Earth, and the World Game themes. This was a deliberate strategy and virtually a practical necessity, but Fuller regards this as an artistic deficiency of the book."

Fuller wanted to complete the picture, but he was unable to tackle another book of the same size by himself, and with Applewhite preoccupied with a sequel volume, he met the challenge in the only way that he could. At a studio at the Bell Telephone Building in Philadelphia, he embarked on an ambitious return to the idea of recorded lectures, on a scale now made possible by videotape.

On January 20 Fuller began a series of sessions that ultimately amounted to forty-two hours. Seated before a screen filled with the image of a galaxy, he opened with a basic fact: "All humanity has always been born naked, absolutely helpless, for months, and though with beautiful equipment, as we learn later on, with no experience, and therefore, absolutely ignorant."

The gargantuan lecture, which became known as *Everything I Know*, combined autobiography, engineering, and geometry in a nonstop verbal torrent. Along the way, Fuller credited Kenneth Snelson with the insights that had led to tensegrity, saying that he felt "tremendously tender" toward the student whom he rarely mentioned: "People would say, 'Bucky is stealing your things,' and so forth. He doesn't think so anymore. He really appreciates what we are doing."

As usual, his strangest statements frequently yielded genuine insights, which Fuller insisted was intentional: "When you can't find the logical way, take the most absurd way." On the future of computing, he was prescient: "When the kids can really get information from all around the world, with all the computers really being inter-hooked . . . they're going to go and get it." And on the differences between sexes, he said, "I don't see any pure males or pure females in human beings. . . . Often, males can get to be quite attractive as well."

One of the listeners in the studio audience was Michael Denneny, his new editor, who was in his first year in publishing. Other editors at Macmillan had given him their problem books, including *Synergetics*, of which he remembered, "When the bound galleys came in, it was something like nine hundred pages of mathematics and geometry. Everyone went into panic mode." He traveled to Philadelphia on weekends to be tutored by Fuller, and he advised his sales team to simply repeat the description of its author as "the Leonardo da Vinci of the twentieth century."

On its release later that year, *Synergetics* earned respectful reviews from critics, although they often weren't able to finish it. Fuller sent it to friends, including Annie Dillard, who thanked him affectionately in a letter: "You sent me a world, a universe—what an Olympian courtship!" Denneny was relieved when it went into a second printing and sold thirty thousand copies, but Fuller disliked its reception as his magnum opus, complaining that it had "many flaws."

His loved ones were concerned that the book's publication had left him adrift, and Ed Schlossberg wanted to do something special for Fuller's upcoming eightieth birthday. Schlossberg had collaborated on a volume of his own poetry with Tatyana Grosman, a Russian immigrant known as a brilliant printer and lithographer. Resolving to introduce these two remarkable people, he mailed a parcel of Fuller's art to Grosman, who was charmed by a set of illustrations of a child that he had made with Anne and Allegra.

On February 25 Schlossberg drove to West Islip, New York, as Fuller napped in the front seat. They arrived that evening at Universal Limited Art Editions, which was based out of the garage of Grosman's Victorian house. At dinner, Schlossberg sat beside the nearly deaf Fuller, repeating every word into his ear. Afterward, they went upstairs to a viewing room and examined artworks that the shop had produced for the abstract expressionists Robert Motherwell and Barnett Newman.

When Fuller asked to draw, he was handed a lithographic stone and crayon. As he sketched a dome, he fell in love with the medium, exclaiming, "The stone invites me to work!" The next day, he returned to announce that he would do a book, with the seventy-year-old Grosman serving as yet another collaborator who could handle the logistics as his imagination ranged freely. He decided to revisit the Goldilocks stories that he had once told to Allegra, which he described to Applewhite as "the empirical, the scientific way to present the argument of synergetics."

What felt like a sentimental choice soon became deeply strange. On the first stone, he sketched the constellations Ursa Major, Ursa Minor, and Cassiopeia, but because he was unfamiliar with lithography, he drew one of them backward. It was an unintentional homage to James Hewlett's reversal of the celestial vault of Grand Central Station, and Fuller declared that he wanted to leave it in. As he packed the book with geometry and his nautical theory of history, it expanded inexorably, and Grosman was so exhausted after their first session that she stayed in bed for two days.

"The Red Sea is still open," Fuller said of his sudden rush of inspiration. "We have to get this book out before it closes." He traveled from his hotel in New York to West Islip for one or two days at a time, with Schlossberg helping with the diagrams, while Grosman and her team were left to figure out how to make triangular leaves that could be arranged into a flat volume, a tetrahedron, or a sequence of solids and planes. The finished version, twenty-six large pages printed on gray fabric, was forty feet in length when fully extended, and Fuller proudly called it *Tetrascroll*.

Fuller left soon afterward for Asia, where he worked with the architect Sumet Jumsai on a proposal for a dome at an archaeological site in Thailand. In New Delhi, he met with Indira Gandhi, who was at a crossroads in her political career. After a high court nullified her election as prime minister on grounds of electoral malpractice, Gandhi declared a state of emergency that would ultimately lead to suppression of the press, arrests of opponents, and widespread abuses of civil rights.

After Fuller expressed his faith in Gandhi, she sent a letter to his hotel to thank him. Stating that she had seen "all kinds of undemocratic and unconstitutional acts" after the ruling, she claimed that she had made her decision for the sake of the nation: "Suddenly it seemed as if the country were disintegrating before our very eyes. That is why I had to act and declare an emergency." Fuller, who accused the *New York Times* of "treachery of information to humanity" by printing "propaganda" against Gandhi, accepted this without question.

In July a final Delos conference was held in memory of Constantinos Doxiadis, who had died the month before. On the cruise, Fuller celebrated his eightieth birthday, and he agreed later to serve as president of the World Society of Ekistics. The role was inconsistent with his belief that the city itself was a transitional stage on the way to "human unsettlement," and he gradually scaled back his participation.

Fuller's involvement with the proposed Beckett Theatre was also drawing to a close. The project had endured largely because of Richard Burton, a friend of Francis Warner, who contributed £100,000. That fall, Burton paid another £17,500 for a sliver of space by the Oxford Playhouse, which Fuller described as "very diminutive." In the end, the theater was never built, and its primary legacy was Fuller's lasting friendship with Norman Foster, who looked into opening a London office, Buckminster Fuller Associates, to handle deals in Europe.

Foster's efforts fell through, though, and Fuller found himself drifting apart from another collaborator. The invaluable Shoji Sadao had moved from Cambridge to Isamu Noguchi's studio in Long Island City, Queens—a seemingly practical decision that concealed a deeper divide. In recent years, Sadao had worked closely with Noguchi, who, he suspected, would bring in more money. "What I had surmised about Isamu's ability to attract commissions was correct," Sadao recalled, "and the lack of any projects from Bucky was more than made up by projects that Isamu secured."

Fuller was suffering from a shortage of paying work. As a professor emeritus at SIU, he no longer received a salary, and in the fall of 1975, his contract expired as a World Fellow in Residence in Philadelphia. It was renewed for two more years, but the college sponsors were unable to raise outside funds, making the situation even more fragile. If the arrangement ended, he would have to rely entirely on his speaking schedule, which he doubted he could maintain.

The financial pressures made it hard to be selective, and the consequences were sometimes embarrassing. Along with Marshall McLuhan, Jonas Salk, and Norman Cousins, Fuller agreed to participate in the fourth International Conference on the Unity of the Sciences, which was scheduled for New York in November. None of them was aware that it was sponsored by the Korean evangelist Reverend Sun Myung Moon, the head of the controversial Unification Church, and after a run of negative coverage, they belatedly withdrew before the event took place.

By then, Fuller's legacy had already been complicated by his association with other dubious causes. One troubling example was Synanon, a drug rehabilitation program founded by Charles E. Dederich Sr. in Santa Monica, California. Dederich began by offering residencies that lasted for two years, but after deciding that recovery from addiction was impossible, he embraced a model in which clients could never leave, and it acquired all the characteristics of a cult.

At his facility in Tomales Bay, California, Dederich oversaw the Game, a form of group therapy based on verbal abuse, and punished members for infractions by ordering them to shave their heads. Dederich also used tensegrity models to represent a perfect society built on "vectors and valences," as the residents spent hours "gluing their fingers into tortured geometrical configurations," and Fuller himself was praised as a great thinker on the level of Lao-tzu and Ralph Waldo Emerson.

In 1968 a writer named David Cole Gordon had urged Fuller to pay a visit to Synanon. "Chuck has installed a hobby-size workshop on one end of the large table he uses for a desk," Gordon wrote, "and is building tensegrity systems and coming to grips with your geometry." Fuller agreed to tour the center, where he was impressed by how its use of dance "allowed for individual expression while cultivating a group consciousness."

Two years later, Fuller's associates Mark Victor Hansen and Gene Youngblood gave a talk to Synanon residents about the World Game. Hansen told a young correspondent, "The process of World Game is just that—a process. Synanon is part of our infancy stage." Fuller praised a pamphlet about the program as "an extraordinary philosophic treatise," and he left no doubt of his feelings toward its founder: "I am an admirer of Chuck Dederich, and I think Synanon is magnificent."

After hearing more about the group's repressive practices, Fuller broke off ties with Synanon, which had greater success cultivating Cesar Chavez, the celebrated labor organizer and head of the United Farm Workers of America. Dederich's organization built domes in Tomales Bay, but its methods devolved into beatings and compulsory vasectomies, culminating in a 1980 incident in which members put a rattlesnake in an attorney's mailbox. Dederich pled no contest to charges of conspiracy to murder, which ended his control over the organization.

Fuller remained willing to spread his message through existing networks, but he tended to overlook the possible costs. Another test arrived in the unlikely form of Roger Stoller, an avid surfer in his early twenties, who served as his traveling companion. Roger and his twin brother, Jonathan, had been friends with Jaime Snyder since grade school, when Allegra's family moved to Pacific Palisades, California. Roger was entranced by Fuller, and they agreed to work together for a year.

While Roger was living in Santa Cruz, an acquaintance told him during an LSD trip about an organization that offered workshops to increase human potential. After attending one of the sessions, Roger advised Jaime to try it as well. Jaime, who was twenty, took part in the program, which made him think that Fuller might benefit by providing a similarly intensive encounter for fans. After mulling it over, he approached Fuller with what seemed like a straightforward proposal. "Grandpa," Jaime said, "I think you should meet Werner Erhard."

At the time of his first meeting with Fuller, Werner Erhard was at the height of his fame. His birth name was Jack Rosenberg, which he changed after reading an article in *Esquire* magazine that mentioned the economist Ludwig Erhard and the physicist Werner Heisenberg. After initially working in sales, he was inspired by the human potential movement to look into various disciplines and therapies, including dianetics, before founding Erhard Seminars Training in the early seventies.

Erhard's program, known as est, offered seminars that were held all day over the course of two weekends, combined with a handful of evening sessions. It amounted to a kind of interactive theater that challenged its participants to discard the roles that they had assumed in life. Attendees were required to surrender their watches, avoid talking to others, and refrain from using the bathroom except during designated breaks. The lectures, which lasted for hours, featured profane verbal abuse, with trainers often referring to trainees as "assholes."

When Jaime told his grandfather about Erhard, who was forty, more than sixty thousand people had taken the training, and its celebrity fans included the folk singer John Denver and the actress Valerie Harper. Fuller was unaware of most of this, and although he recognized the value of Erhard's network, he said only, "Jaime, I don't really have an interest in meeting him, but if you want to introduce me to him, I have great confidence in you."

On December 18, 1975, Jaime and Fuller went to Erhard's handsome house in San Francisco, which served as the headquarters of est. For two hours, Fuller spoke in the parlor with Erhard, who recalled it much later as "a great conversation with an authentic human being." At that stage, they made no definite commitments, but the benefits of an ongoing relationship were clear.

Despite what some of his supporters claimed, Fuller was far from naïve about Erhard. He had held his own against George Gurdjieff, L. Ron Hubbard, and Frank Lloyd Wright, and compared with such master manipulators, Erhard seemed relatively innocuous. Fuller himself had built his career on a form of performance art that could change lives, and Erhard offered him the chance to scale up what he did best. Like Stewart Brand, Erhard was his ambassador to a valuable audience, and its changing nature reflected the wider cultural shift of the second half of the seventies.

The partnership was facilitated by the expanding role of Fuller's grandson. Fuller had always been affectionate toward Alexandra and Jaime, whom Robert Snyder had filmed listening intently to their grandfather as children. At Bear Island, Fuller taught Jaime sailing, and he eventually came to see him as his successor in the family. Allegra had her own career as an ethnologist at UCLA, where she headed the dance department, and Fuller's conventional notions of gender at home led him to pass over Alexandra, who recalled simply, "My brother was the male descendant."

Jaime had assumed a more prominent position in Fuller's life several years earlier. As a freshman at the University of California in Santa Cruz, he had felt depressed after a breakup, and he confessed to his grandfather that he was homesick: "Grandpa, how do you deal with that?"

"You know, Jaime, what I've noticed is that my backyard just keeps getting bigger," Fuller said. It was a line drawn straight from his lectures, but the conversation moved him to propose that Jaime become one of his companions. Taking off four quarters from college, Jaime accompanied Fuller around the world, and after their meeting with Erhard, he partnered with the public radio show *New Dimensions* to plan a seminar called "Being with Bucky."

In the meantime, Fuller received an approach from a very different circle. On January 14, 1976, he was visited at his hotel in Santa Monica by four employees of the outdoor equipment company The North Face. As he watched with interest, they pitched an unusual object on his balcony. It was the first practical dome tent for backpackers—and it had been directly inspired by his work.

A similar design for a dome tent had been developed by Jeffrey Lindsay for Patrick Baird more than two decades earlier, but the seed of its most successful version was planted in 1970. Fuller had received a letter about tents from Don Butts, an associate of The North Face whose father, J. Arch Butts, had been the treasurer of Fuller Houses in Wichita. "I am writing to you now because I have always assumed that you were capable of anything," Butts told Fuller, who replied that he had thought for a long time about shelter for "the single occupant."

The project was revived a few years later by Bruce Hamilton, a North Face employee who had attended a World Game workshop and experimented with tensegrity structures. When he showed a model to Kenneth "Hap" Klopp, the firm's founder, he was authorized to turn it into a tent. Working with Jim Shirley, Hamilton built a prototype, while a patent licensed from an independent designer, Bob Gillis, contributed the idea of threading flexible poles through rings.

It resulted in the Oval Intention, the first commercial dome tent, which was built by Hamilton and Mark Erickson. The North Face catalog listed it with a quote from *Synergetics*: "It is no aesthetic accident that nature encased our brains and regenerative organs in compoundly curvilinear structures. There are no cubical heads, eggs, nuts, or planets." In Santa Monica, it delighted Fuller, who sketched out a tensegrity tent of his own. His design was never realized, but the dome tent became one of the most lasting incarnations of his ideas.

Fuller gave more of his attention to idiosyncratic side projects, including a recording of a live performance of his own songs at a Philadelphia folk club, and to cultivating new patrons. One was a Malaysian architect named Lim Chong Keat, the wealthy younger brother of the chief minister of the state of Penang. Lim enlisted Fuller, whom he met through the Indian airport project, in his plans for Komtar, a gargantuan urban center for the capital city of George Town that would include a geodesic auditorium. Fuller had always been most comfortable with benefactors who were close to power, but not at its exact center, and according to his secretary, he came to consider Lim "his best friend."

In March 1976 Lim organized the Campuan Conference on Global Issues as an informal meeting for Fuller's admirers at a village in Bali. Lim's opulent house provided a scenic backdrop for the conference, which Fuller said had been conceived "in awesome review and inventory of the magnificent and exquisite competence of universe to solve its own problems." The main event was the construction of a dome out of bamboo and palm rope. When a grass roof was added, the structure cracked, but the gathering was pleasant enough to inspire a second one the following year.

Another key colleague during this period was Carl Solway, an art dealer in Cincinnati who had worked with John Cage to sell a series of sculptural objects and prints. It was a welcome infusion of financial support for Cage, who told Solway, "Carl, I have a good friend who could use the kind of help that you have provided for me." He was referring to Fuller, and Solway contacted him about designing a sculpture for an office tower in Columbus, Ohio. Fuller agreed to build a hanging piece for the atrium, which Sadao put together as a motorized complex of jitterbugs.

After their submission was rejected because the Ohio Building Authority wanted to award the commission to a local artist, Solway proposed that they proceed with a three-foot version for collectors. Fuller had long insisted that he had no interest in producing such works, which he contrasted pointedly with artists such as Kenneth Snelson and Alexander Calder: "I've seen time and again I could exploit what I was doing as an art, and that might bring some admiration, and it might make some money, but that's not [where] I'm at."

In the end, his need for income was more decisive. Twenty-five sculptures were produced under Sadao's supervision at the Solway Gallery, and the earnings took some of the pressure off Fuller's travel schedule, which became less sustainable as he entered his eighties. His effort to move toward larger audiences was part of the same strategy, as Jaime took the lead in organizing elaborate events that didn't need an outside sponsor to sell tickets.

At the end of May, the first "Being with Bucky" seminar was held in San Francisco. The day before it began, the Montreal Expo Dome—rechristened the Biosphere—caught fire during renovations. While two welders were sealing a gap, a spark ignited a blaze, and their fire extinguisher turned out to be empty. As the flames raged out of control, the acrylic melted off in sheets, leaving shreds of molten plastic on the ground. Within fifteen minutes, the skin was gone. No one was hurt, and the frame was intact, but it was a disastrous blow to Fuller's monument.

Fuller insisted that the calamity vindicated the underlying framework: "It almost seemed as though the nonstructural skin of the great unharmed geodesic dome had been set afire by some mystical evolutionary wisdom to remind the world of geodesics' very high structural performance." All the same, it raised questions about safety, as a correspondent wrote to Lloyd Kahn: "What would have happened if the fire had occurred in a dome covering a town as Fuller suggests?"

The news had minimal impact on the San Francisco event. On account of concerns that Fuller would ramble, Jaime was persuaded to join him onstage, where he assisted with interactions with a crowd that included Erhard, Ruth Asawa, and the neuroscientist John Lilly. "I don't see this as some sort of side exercise," Fuller told his listeners, "but as part of the absolute frontier—the frontier of whether we are going to survive on our planet or not."

Three days later, he left for the United Nations Conference on Human Settlements in Vancouver. In addition to a meeting with Pierre and Margaret Trudeau, the trip focused on the Now House, an autonomous model home funded by Neva Kaiser and coordinated by the British-born architect Tony Gwilliam. Four domes were occupied by ten volunteers, who installed hydroponic tanks and endured heavy rains. The North Face donated two tents, but Hamilton and Gillis fought over the rights to the ring system, forcing the company to switch to a more cumbersome arrangement of threaded sleeves.

Fuller remained busy elsewhere. He was one of the founding editors for the Pushcart Prize, the annual anthology featuring authors, poets, and essayists published by small presses, and participated in celebrations of the US bicentennial. On July 6, 1976, he attended a reception held by Queen Elizabeth II and Prince Philip on the Royal Yacht *Britannia* in Philadelphia. When Anthony Crosland, the British foreign secretary, asked why younger Americans were interested in the royals, Fuller replied, "The young people recognize the integrity of the queen, in contrast to the corruptibility of all the big politicos."

Jaime had been working on Werner Erhard's first appearance with his grandfather, which took place in New York in September. Although it was advertised only in the est newsletter, it attracted a crowd of 1,700, including Peter Senge, the future author of the business best seller *The Fifth Discipline*. As the audience rose repeatedly to applaud, Jaime and Fuller held forth with Erhard, who often broke in to rephrase concepts in terms of est: "In other words, Bucky, what you're saying is that each person should take responsibility for his own life."

Erhard's reframing of the material only pointed to how effortlessly it could be adopted by others, and much of the evening session was devoted to mutual praise. "Computers can be run by ten-year-olds, and that's a good sign," Fuller concluded. "I'm quite optimistic, because of three things: youth, truth, and love. My grandson's a youth, Werner is truth and love." As they received another ovation, they all kissed, and Fuller went to embrace a man in the front row.

Even after a portion of the proceeds was shared with a foundation associated with est, it was Fuller's most lucrative appearance in years, as well as the exact opposite of the situation that he had resisted with the World Game: instead of an independent group of enthusiasts interpreting his ideas for a receptive audience, he had spoken directly to listeners with minimal knowledge of his work. Many of his staffers remained suspicious of the relationship, and Howard Brown made an even blunter accusation: "Bucky hated Werner Erhard—he thought he was a charlatan. Erhard really wanted to get the World Game and control it."

In reality, Fuller saw his new collaborator for precisely what he was. Unlike some of Fuller's admirers, Erhard had no desire to impose his own ideas, and Fuller described him only as "a good human being," categorizing the est founder as one of the countless patrons who had passed through his career: "New York was one of many, many meetings in my life and—like many others—it was very exciting."

If Erhard made great emotional demands on his adherents, Fuller did much the same with his own staff members. For the reopening of the Cooper Hewitt Museum at the historic uptown mansion built by industrialist Andrew Carnegie, the Austrian architect Hans Hollein wanted designers to present aspects of the field "in a surprising and brilliant way." He offered space in the library to Fuller, who seized the opportunity to build an exhibit exclusively around his geometry rather than the dome.

The exhibition consisted of wall panels and models assembled by volunteers at Noguchi's studio, including Alexandra Snyder, Roger and Jonathan Stoller, and twenty-one-year-old Guy Nordenson, who later became a prominent structural engineer. Under Sadao's supervision, Jonathan and a staffer named Rob Grip worked on fixed structures, while Roger had to decipher a sketch of a jitterbug made of bicycle wheels. This led to a strange confrontation when Fuller returned: "He did one of his freakouts," Roger recalled. "I don't know if he actually believed this, but he thought I was trying to steal his thunder by taking on all of these models."

Before the show opened, Fuller and Noguchi had discussed partnering on a design for a theater for Martha Graham. It tested their friendship, as Fuller wrote: "I would have gone on loving you but would have lost some confidence in our ability to communicate and understand." When Graham withdrew, the Cooper Hewitt exhibit turned out to be Fuller's last major project with Sadao, who recalled, "Despite our best efforts, no new commissions were to come our way."

The exhibition overlapped with the publication of *And It Came to Pass—Not to Stay*, a poetry collection that Michael Denneny had extracted from one of Fuller's enormous manuscripts. In its title poem, he alluded to his efforts to enable a "planetary revolution," which extended in the fall to his single most unusual brush with power. It fit neatly into Fuller's sequence of encounters with powerful women, but it was by far the most audacious, as he prepared to meet Ferdinand Marcos, the dictatorial president of the Philippines, at the invitation of his wife, Imelda.

Fuller liked to approach patrons through their spouses, and, as a prospect for commissions, the First Lady of the Philippines was both enticing and outrageous. She had built gigantic structures in the monolithic Brutalist style, which inspired critics to speak contemptuously of her "edifice complex," and her use of foreign loans contributed to a debt crisis that allowed her husband to declare martial law, giving them uncontested control over a country from which they stole billions.

"Within one of the great dictatorships," Fuller wrote later, "I might gain the physical means of realizing my technological artifacts by first persuading the dictator that his only militarily sustainable plans should be abandoned because they were diametrically opposed to my concepts." He stated that this approach was "not pursued," but, in practice, he was willing to work with any potential supporter capable of granting unconditional approval.

On October 21, 1976, Fuller arrived in Manila, where he stayed in the opulent Malacañang Palace. After meeting with Imelda about a housing program, he lectured at a convention center that she packed with attendees—including buses of schoolchildren, who left early from what a diplomatic observer described as "an only barely audible and completely incomprehensible discourse." The next day, Fuller had lunch with President Marcos, while Imelda hinted that work could begin within months on "geodesic communities" of thousands of domes.

Fuller promised to prepare plans for an entire dwelling industry, which Imelda celebrated by sending gifts for him to Philadelphia. According to his travel companion, Tom Vinetz, who sensed a "definite fear" of Imelda among her circle of associates, Fuller expressed no concerns about the political side: "This is not to say that he was unaware of the reputation of the Marcoses," Vinetz explained, "but he would go where he felt he could do the most good for humanity."

After departing Manila, Fuller passed through Southeast Asia on his way to New Delhi, where he saw Indira Gandhi again. On his return home, he worked with Jaime on a letter to follow up on his "memorably delightful" meetings with Imelda. Fuller wanted to present his proposal for a "dome farm" with twelve different prototypes that could be exported throughout Asia, as well as larger development along the lines of Old Man River's City, all for a massive fee of $320,000.

In April 1977 he returned to the Philippines for a second set of meetings with the Marcoses. Imelda approved of his plans in principle, but her advisors were concerned by his ownership of the technology generated by the program, as well as a lack of information regarding his obligations and the overall cost. After they advised further study, the initiative was shelved, and when Fuller attempted to arrange another visit, he received no response.

Despite his involvement with such grandiose prospects, Fuller often stood in the way of ideas that seemed on the verge of fruition. The Design Science Institute had met regularly, but no real projects had materialized, in part because Fuller hated the notion that he needed anyone to "promote" his work. He had asked for the offending word to be deleted from a press release, informing Ed Applewhite that he put himself forward "only when asked by others so to do."

Fuller kept the group from building up any momentum, and he was more interested in diverting its resources into questionable ideas, including a "fuselage cartridge" for travelers that resembled a personal shipping container with a bed. He proposed that it could be loaded onto a plane or train with its passenger onboard, detached at each transfer point, and recombined with others for the most efficient route, which reflected his ideal of travel, although it reminded Neva Kaiser of a coffin.

Kaiser wanted to expand the World Game, but a computerized version was no closer than before. She felt that Fuller put "a spoke in the wheels" whenever a project got traction, and she was chagrined to discover that two precursor organizations had labored for years with a similar lack of success. Applewhite told her that Fuller feared that his ideas would fail in practice, and he was clearly more comfortable continuing to appear with Werner Erhard.

"Making sense and making money are mutually exclusive," Fuller told his audiences, and his relationship with benefactors was inevitably strained. Kaiser felt that her financial contributions had been wasted by Fuller, who wrote to her criticizing "the world's politicos, business-men, and their master bankers"—which implicitly included her father. Kaiser reflected dryly on his certainty that the universe would provide for his work: "I thought that perhaps as a member of a rich family that I was a nominee for the universe's source of support."

In a final effort to keep going, they approached the real estate developer and political advisor Philip Klutznick about financing a proposal for a dwelling industry. Glenn Olds, who brokered the arrangement, recalled, "One never knows why the final negotiations of so delicate a deal—negotiations between a man whose second nature was privacy but whose ideas were genuinely universal and invited implementation and an exceedingly successful public servant, business entrepreneur, and social engineer—should fail at the last moment, but they did."

According to Jaime, the real issue was the board's insensitivity to Fuller's reluctance to ask for support, which led to an unfortunate scene: "Everybody sat down at the huge conference table, with Klutznick and Bucky across from each other, and Klutznick started by saying, 'Well, Bucky, what'cha got for me?' Bucky immediately hit the table hard with his right hand and said, 'I don't need anything from you.' And that basically ended the meeting."

Shortly afterward, Kaiser wrote to Olds that she intended to resign: "The necessary restrictions inherent in operating through a tax-exempt institution are inimical to Fuller's lifelong style of complete independence and total responsibility only to his own highly developed sense of what the Universe requires of him." There was talk of placing Jaime in charge, and Fuller's recent successes with Erhard made him confident that they could raise millions.

In the end, Norman Cousins took over from Olds, and the Design Science Institute revised its mandate to focus on Fuller's archives—effectively rendering it inactive. Kaiser eventually agreed to remain, but she came to feel that the real obstacle was Fuller himself, who had once informed John Cage that he intended to live to be 120 years old. His granddaughter, Alexandra Snyder, saw the situation clearly: "He talked about dying, but I don't know how much he was really thinking in a strategic way about what happened after he died, in any way at all."

TWELVE

EQUILIBRIUM

1977–1983

We can put the touchable things in the ground, but we can't put the thinking and thinkable you in the ground. The fact that I see you only as the touchable you keeps shocking me.

—BUCKMINSTER FULLER, *SYNERGETICS*

Although Fuller's staff was concerned by his relationship with est, it had an even greater effect on Werner Erhard. On October 13, 1977, Fuller attended an event in San Francisco where Erhard unveiled a plan to eliminate starvation within two decades. Supporters were asked to fast for one day and donate the amount that they otherwise would have spent on food and alcohol, but instead of going directly to those in need, the money would be used to create a "context" for ending hunger.

The Hunger Project amounted to a stealth integration of Fuller's methods into est. In a booklet, Erhard stated that he presented the concept to his board "after considerable research and discussions with authorities such as Buckminster Fuller." Its ambitious deadline recalled similar targets set by Fuller himself, who had said that "anything dreamable" could be achieved by the early seventies, and it was allegedly based on one of his favorite questions: "What can the little individual do?"

Fuller's involvement was limited to lending his reputation to the Hunger Project's advisory board, but his work provided a crucial precedent. According to Edwin Schlossberg, the World Game workshop had been "a morally religious idea at its core," based less on specifics than on an awareness of what was possible: "If you could see it, everyone would do it." Erhard stripped away the statistics, but his inspirational language came directly from the source. "You and I want to contribute to the quality of life," Erhard wrote. "We want to make the world work."

With its motivational rhetoric abstracted from the underlying data, the Hunger Project went beyond even Fuller himself—it ephemeralized the World Game. Erhard organized it as an independent nonprofit, but it was deeply entwined with est, and within a year and a half, it raised more than $800,000, of which only $2,500 reached the global poverty charity Oxfam International. The rest went mostly to overhead and promotion, leading *Mother Jones* magazine to call it "a thinly veiled recruitment arm for est," but Erhard still profited from its air of legitimacy, drawing respectful attention from President Jimmy Carter.

It was a lesson that Erhard had learned from Fuller, who had masterfully advanced his personal ambitions by convincing his followers that they could change the world. If his ideas were uniquely vulnerable to this kind of assimilation, it was a precessional effect of ephemeralization itself, which produced strategies that could be repurposed at minimal cost. In this case, Fuller was content to receive public credit, along with a grant to cover his expenses from Erhard, who personally introduced him to Douglas Engelbart, a member of the est board.

Fuller himself would speak to any organization willing to pay. He claimed to never use an agent or promoter, but he was represented by Speakers Unlimited—an agency that advertised his availability alongside its other clients, such as astronomer Carl Sagan and an Austrian body builder named Arnold Schwarzenegger—and he hired a second firm to book lectures on campuses. Despite being in his early eighties, Fuller was almost constantly on the move, and he learned to keep warm on planes by lining the inside of his suit with the newspaper that he had just finished reading.

Apart from the design of a temporary dome for a nuclear plant in North Perry, Ohio, Fuller landed few commissions. He made overtures to the actress and activist Jane Fonda, whose "revolutionary courage" he admired, and hoped that Prince Shahram Pahlavi-Nia, nephew of the Shah of Iran, would underwrite a television show on synergetics. In March 1978 he attended a conference that the prince convened in the Seychelles islands for fifteen environmentalists, including Jacques Cousteau and James Park Morton, the dean of the Cathedral of St. John the Divine in New York.

From there he flew to Sri Lanka to visit the science-fiction author Arthur C. Clarke, best known for collaborating with Stanley Kubrick on *2001: A Space Odyssey*. At their first meeting at a symposium years earlier, Clarke had been amused by the bullhorn that Fuller used as a hearing aid, and they occasionally appeared together as experts on the future. Fuller saw Clarke as "one of the important exceptions" to specialization, which was fatal to both species and societies, while Clarke hailed him as "one of the world's most valuable natural resources" for promoting science to young people, adding, "I think he's the world's first engineer-saint."

Clarke, who was gay, had left his native England to settle in Sri Lanka in 1956. For two days, he took Fuller around the island by car and helicopter, focusing on locations from his upcoming novel, *The Fountains of Paradise*, which revolved around the construction of a space elevator. Fuller felt characteristically compelled to claim priority, telling Clarke that he had devised an analogous idea years ago for a tensegrity ring that would float above the equator, with "traffic vertically ascending to the bridge."

In June Fuller spoke at the Cathedral of St. John. Before the sermon, he picked up a maple seed from the cathedral grounds, and he tossed it toward the congregation to symbolize how nature had "many alternate plantings" on other planets in case mankind failed its final examination. Later that month, he met with Morton to discuss a plan for the cathedral roof, which would feature a tower topped with a geodesic sphere. Morton's advisors rejected a submission by Sadao as "wildly impractical," and the project faltered after Shahram Pahlavi-Nia, who had provided funding, fled to London to escape the Iranian Revolution that removed his family from power.

Fuller had independently drawn attention from a much more controversial church. In the seventies, Scientology was under FBI investigation for obstruction of justice, burglary, and other crimes, but his contacts with its members continued. He was a tempting target for recruitment by its celebrity program, and he kept up his correspondence with Cam Smith. In a letter from 1978, she wrote, "I'm glad you liked hearing from me and getting the statement of Ron Hubbard's philosophy."

He was also in touch with a Scientologist named Mary, who had worked on one of his seminars in California. The following year, she suggested that he deliver a talk for "people in my environment" for his standard fee of $5,000. After he accepted, she sent another letter with a much darker tone. She had discovered that Fuller was involved with Erhard, whom she accused of stealing ideas from Hubbard, and she proposed that he educate himself further:

> I have learned, however, that you have in fact met Ron Hubbard, and that perhaps your opinion of Scientology is not an extremely high one. I don't know what has happened, but it distresses me to know this. . . . I feel that there must be some misunderstanding between the two of you for such a separation to occur. I know that human conflicts can be resolved through communication, and I greatly look forward to seeing this take place between yourself and Ron Hubbard.

Mary said later that she and Cam Smith had acted out of "intellectual curiosity" about the affinities between Hubbard and Fuller, rather than as part of a coordinated effort to recruit him. On August 7, 1978, however, she met with Fuller to discuss Hubbard and est, and he seemed on the verge of growing closer to Scientology, which he evidently saw as just another network.

Fuller and Anne, now both in their eighties, were slowing down. An operation for intestinal cancer left Anne with swallowing difficulties, while Fuller was ordered to abandon his steak diet, prompting him to experiment with "various seaweed things." Because their trips to Bear Island had become harder to manage, they spent the summer in Sunset, Maine, where their relaxation was disrupted when Fuller suffered chest pains. Taken to a hospital in Bangor, he was diagnosed with pericarditis, an inflammation of the membrane encasing the heart.

After his release, a lack of energy led him to cancel all events for two weeks, which turned out to be a blessing in disguise. Fuller never delivered his talk to the Church of Scientology, sparing him from any further involvement. Afterward, their contacts ceased, although Fuller reportedly inscribed a copy of *Synergetics* to Hubbard, who praised him in a bulletin for members as "an engineer and architect of some renown."

Fuller found funding elsewhere, including a paid testimonial for the Honda Civic hatchback: "Its handling feels better to me than any other car I've ever owned—except my Dymaxion." His monthly income from speaking often reached $30,000, but it was consumed immediately by his expenses, and so his schedule remained as relentless as ever, despite the support of his famous admirers.

The year before, he had appeared with Erhard at an event in Pasadena, where the television star Valerie Harper thought that Fuller looked "adorable." Erhard treated Fuller to dinner at Trader Vic's with the singer-songwriter John Denver, one of his most prominent fans, who sang along with the music playing at the tiki bar. Denver was fascinated by Fuller, whom he enlisted in his existing campaign to popularize sustainable energy sources, and they formed an unlikely friendship.

Fuller became remarkably close to Denver, who was photographed kissing him on the cheek at an alternative energy fair in Colorado. Denver visited the office in Philadelphia, where he sat quietly in the corner, and persuaded Fuller to reconsider his skepticism toward solar power. Fuller wrote gratefully that Denver and Erhard had entered his life at precisely the right time: "You brought me through my second crisis. . . . I told you about the first crisis occurring in the Bangor hospital. I have since weathered the third such crisis on my own."

This third test was the financial catastrophe that had been looming for months, as the consortium of four colleges that sponsored Fuller in

Philadelphia failed to secure outside funding. Control of his office had shifted from the University of Pennsylvania to the Science Center—a nonprofit that could hardly be expected to donate the space. An annual rental of $30,000 was more than Fuller could afford, but his backers had extended their agreement for another year.

On October 4, 1978, Martin Meyerson informed Fuller that their arrangement was ending. In response, Fuller lamented that he was faced with starting "a whole new life program" in old age: "I feel a deep responsibility to keep my operation healthy and capable of swift expansion as emergencies may dictate. I cannot just fold up Philadelphia and crawl into a hole in Maine. Our mail and telephone calls from around the world make clear that our headquarters are going to be turned to ever more frequently for guidance in the emergency transitions."

Denver and Erhard hired a financial advisor to talk with his secretary, the newly married Shirley Sharkey, but the writing was on the wall. In November, as he flew to England to attend the wedding of Tony Huston and Lady Margot Cholmondeley, the papers reported that Fuller's funding had been terminated. In a later article, the president of Swarthmore called the relationship "of marginal benefit," while a Penn instructor observed that Fuller was "not the blue star he once was."

Fuller said that he had no alternative but to consider "two very attractive offers" from Amherst College in Massachusetts and the State University of New York at Buffalo. He eventually signed a lease in Philadelphia for another year, which he financed with a large loan, and the Science Center agreed to cover the cost of one assistant. Sharkey spoke more bluntly to the press than her employer might have liked: "He has not made a penny out of licensing for the domes or other inventions. He has no money in the bank."

All the while, Fuller was uncomfortably aware that the most recognizable work ever based on his ideas was under way in Florida. It had originated with Walt Disney, who had bought over forty square miles of land near Orlando in 1965. The project included a park, but Disney was more interested in a futuristic planned community, as the general manager of Disneyland recalled: "He expected a house that would be completely self-sufficient, its own power plant, its own electricity."

In the original plan, the town center was covered by a dome—a concept that was based partially on the Houston Astrodome, which had made a strong impression on Disney. Beneath this megastructure, conceived to enclose fifty acres, internal buildings such as a hotel tower, stores, and theaters would benefit from total climate control, allowing for a variety of architectural styles, while emissions would be minimized by an electric transit system.

Disney called it the Experimental Prototype Community of Tomorrow, or Epcot. After his death, the idea was shelved, but aspects were revived for the expansion of Disney World in the seventies. The park would include a new area called Future World, where a domed structure survived in a more manageable form. In a detailed scale model, the key feature at the entrance was a "giant golden geodesic dome," which would house a theatrical ride called Spaceship Earth.

It was obviously inspired by Fuller, and its preliminary design strongly resembled the Montreal Expo Dome, which had once been proposed as a location for a Disney park in Canada. Although later accounts claimed that Fuller served as an advisor, he was never consulted in any meaningful way. Disney designer and director John Hench, who oversaw the ride, said that a sphere was an element at an early stage, and while the Disney "Imagineers" in research and development were "familiar" with Fuller, they treated his ideas as though they were in the public domain.

Fuller heard about Spaceship Earth purely by chance. In the fall of 1978, at a meeting with the studio to discuss an unrelated project, John Denver was shown conceptual art for Future World. He recognized immediately its debt to Fuller, which led to a belated invitation for a meeting from Martin A. Sklar, the vice president of concepts and planning, although they had clearly never intended to contact Fuller at all.

On December 4 Fuller met the Imagineers in Glendale, California. He gave a lecture and examined the models that had already been made, but there was no talk of any further collaboration. It was the most stunning instance imaginable of the assimilation of his ideas into popular culture, as well as proof that he had definitively lost the control that he had fought so hard to keep.

Fuller also mourned a more personal loss. Of all the celebrities he had encountered on the Delos cruise, he had been closest to Margaret Mead. As early as World War II, he had wanted to meet the noted "woman anthropologist," who later spent time with him in Carbondale. After Mead died, Fuller used his connections to arrange an event in her memory at the United Nations. At a second memorial at the American Museum of Natural History in New York on January 20, 1979, he delivered perhaps his shortest speech ever: "I know she can hear me. We love you, Margaret."

His financial problems remained unresolved. Fuller had high hopes for a proposal to cover fourteen acres of the 1982 World's Fair in Knoxville, Tennessee, with an elongated tensegrity structure, but Norman Foster, who developed it, acknowledged afterward that it was "obviously too radical for the organizers to accept." As his options stalled, Fuller turned to Werner Erhard, the only patron on whom he could consistently depend. Erhard suggested three joint speaking appearances, with Fuller receiving all the profits, beginning with one in Berkeley in February 1979.

Before the event, they held a press conference. According to Erhard's briefing materials, which were leaked to the *San Francisco Examiner*, he was discouraged from placing a hand on Fuller's shoulder, as that might imply that he had him in his "hip pocket." For his own part, Fuller dismissed the notion that he needed "some kind of charity," and he candidly shared his views on Erhard: "I have quite a few million people who listen to me, and I say Werner Erhard is honest. He may prove untrustworthy, and if he does, then I'll say so."

Their appearance in Berkeley was followed by a more extravagant event on February 24 at Radio City Music Hall, where Fuller lectured for hours beneath a projection screen, alongside Erhard and Jaime Snyder. The evening closed with a song by ten est graduates in honor of their "Great Grandfather," which moved Fuller, and its practical impact was even more welcome. Jaime called it "a huge influx of support at a time when they really needed it," and Fuller reported that the events—which would include a third in Washington—had earned him $63,000. It was the last significant infusion of cash he ever received.

The next day, Fuller and his granddaughter, Alexandra, were picked up for an unusual appointment at the apartment of author Truman Capote, whose young boyfriend, Robert MacBride, was writing a profile on Fuller for Andy Warhol's *Interview* magazine. After MacBride took Fuller into a separate room, Alexandra ended up alone with Capote and Warhol. Capote had recently undergone a facelift, leaving him with visible stitches running along his hairline, and he seemed eager to show off some plastic placemats that he had made by ironing butterfly decals. In his diary, Warhol noted briefly of Fuller, "He's eighty-three, he can't hear so well, he was cute."

Afterward, they all had lunch at the UN Plaza, where Alexandra was annoyed that the others directed their questions to her—"Would Mr. Fuller like some more tea?"—instead of asking her grandfather directly. She was more amused by the resemblance between Fuller and Capote, who were nearly the same height, with round heads, "like little mirror images of each other." When it was over, Alexandra and Fuller headed by themselves for the elevator. As the doors shut, Fuller turned to his granddaughter. "Who was that little man, darling?"

"All of humanity now has the option to 'make it' successfully and sustainably, by virtue of our having minds, discovering principles, and being able to employ the principles to do more with less," Fuller wrote in 1979. "We have that option, but humanity has been set against itself by thinking that it's against technology." It was the same gospel that he had preached for decades, but as his energy began to fade, he was left with the troubling sense that he was almost out of time.

Although Fuller appreciated the cash from his appearances with Erhard, it was gone within months. Mindful of his failure to leave any savings for his family should he die suddenly, he returned to the idea that he had always seen as the ultimate solution to his problems—a global dwelling service. The latest version centered on the Fly's Eye Dome, a refinement of his earlier monohex shelter with circular windows and nesting parts, which he still dreamed of transporting by helicopter.

He handed the project to John Warren, a surfboard designer who had developed his own concept for a fiberglass dome. Warren erected a prototype of his planned Turtle Dome in Robert Snyder's backyard in Pacific Palisades. When he first saw it, Fuller responded to the dome warmly, saying, "I've been thinking about you for years." Ordering several for himself, he hired Warren to refine the Fly's Eye Dome at Temcor, where he personally funded his last serious effort to achieve his vision of shelter.

Fuller's ideals of mobility and universal housing were unchanged, but when his building in Philadelphia converted to condominiums, he couldn't afford to buy his own apartment. For a man who had never ceased advocating for a dwelling rental industry, it was the final straw. In one of his last books, Fuller wrote darkly, "Reorganizing all its strategies, capitalism has now unloaded its real estate property onto the people by refusing to rent and forcing people to buy their condo or coop homes."

At first, the practical impact was minimal, since he was rarely in town. In May Fuller toured China, which he had long wanted to visit. Decades earlier, he claimed, China had extended an offer through Canada for him to advise on aluminum construction, but the United States had refused. More recently, he had blamed student drug use on "Chinese psycho-guerrilla warfare," yet he still saw the country—like India and Japan—as a crucial frontier for his work.

Although China lacked an obvious government figure such as Indira Gandhi on whom he could lavish his charms, he met officials who were reportedly "anxious" to use his ideas. According to Fuller, Gao Yuan, the director of the Council on Science and Technology, asked him to disclose his "grand strategy of problem solving." After two days of lectures, Gao allegedly said, "You are China. Please return."

Fuller had been suffering from prostate problems, which were worsened by his consumption of tea. He said that he concentrated best when he needed to use the toilet, but he acknowledged its inconveniences: "I get so preoccupied with my thoughts that I cannot find where my fly is, and I make a mess in the bathroom." There were rumors that he used an astronaut diaper system or would discreetly urinate down his trousers while lecturing, but the prosaic truth was that he sometimes spoke with a "busman's leg," a urine collection bag intended for long drives.

In July Fuller finally underwent his prostate operation, which he had hoped to postpone through acupuncture and "standing on head after meals." It was his most grueling procedure to date, and a long recovery in Bangor forced him to contemplate his financial situation. Fuller was eager to sell his papers, and he had told Martin Meyerson that one university had inquired whether he "would be willing to put a price" on his archives of $1.5 million.

He was misrepresenting the facts. After the University of Rochester in New York expressed interest in an acquisition, he had arrived at the quoted figure himself by indexing the value of his Chronofile based on the price of gold. By the time he met with Charles Sachs, a dealer in rare documents in Beverly Hills, California, Fuller had increased his price to $2 million, and even after it failed to find a buyer, he refused to lower his estimate of the worth of his life's work. "I do have the know-how for humanity to make it here on board Spaceship Earth," Fuller told Sachs. "Gold can't buy that!"

In September 1979 Fuller flew to Copenhagen, Denmark, to be invested in the Order of the Knights of St. John of Jerusalem, an honor that he owed to Werner Erhard, who had received it the year before. His companion, Jonathan Stoller, followed the example of Fuller's other associates by paying his own way. Jonathan had considered establishing a stipend for a travel companion—and John Denver agreed to provide it—but Fuller responded predictably: "That's fine, but just give me the money, and I'll decide what to do with it."

The passage of time had done nothing to reduce his financial pressures. A year earlier, four of his staff members had founded a small company to sell his books, his map, and tensegrity models made of dowels and surgical tubing. They worked on occasional commissions that Fuller sent along, including a bank sculpture in Dayton, Ohio, that they had to design themselves, but most of their activities were intended only to pay the bills, and they discovered that more ambitious projects could threaten their relationship with their employer.

Tim Wessels learned this lesson firsthand. He and Rob Grip had worked with Ernest C. Okress, a physicist at the interdisciplinary Franklin Research Institute in Philadelphia, on a floating tensegrity sphere that could serve as "a large manned stratospheric station" to monitor or modify the weather. It was a rare instance of one of Fuller's ideas being put to its intended use, but he bristled at not being consulted, which amounted to the unforgivable sin of "promoting" him without his knowledge. He fired Wessels, who had been on his team for six years, and the program ended quietly.

Fuller got along better with Grip and a staffer named Chris Kitrick, who knew how to walk that delicate line. They developed a mathematical transformation for the Dymaxion Map—a step that Fuller had never taken—and spent a year plotting features by hand on huge triangles of paper. As a side project, Kitrick came up with a tensegrity truss of octahedra, which Fuller paid to be patented. When Grip realized that Fuller had miscalculated the relative volumes of a tetrahedron and sphere in his geometry, he filled two models with water to demonstrate that he had been wrong.

Grip and Kitrick also drafted illustrations for the book *Synergetics 2*, which was less a sequel than a corrective to its predecessor's shortcomings, including its lack of an index. Fuller indulged in long autobiographical asides, as well as a tribute to an old friend: "The music of John Cage is preoccupied with the silent intervals; his growing audience constitutes the dawning of the transition of all humanity into synchronization with the metaphysical rather than the physical."

His most valued colleague, Shoji Sadao, was now working exclusively with Noguchi. Looking for a licensed partner to take his place, Fuller turned to Thomas Zung, who was born in Shanghai in 1933. Zung had met Fuller in the fifties as a student at Virginia Polytechnic, and he had served as the architect for a college gymnasium with a space frame vault by Synergetics. At the announcement of the Cleveland office of Buckminster Fuller, Sadao & Zung, they were asked how they planned to address urban blight. "We've been doing some things about it," Fuller replied, "but we want to come to some bigger solutions."

This proved to be overly optimistic. "I shall never let you down," Zung told Fuller, but in the absence of larger commissions, he spent more time on pieces for Carl Solway. "Even though Thomas was very good about contacting people and trying to pursue projects," Sadao remembered, "nothing's ever come out of the association with Bucky. It's strange that, you know, there would be no other projects with Bucky, but that's the way things turned out."

Fuller's dearth of clients stood in contrast to the success of his nemesis Philip Johnson, who was landing major contracts in old age. As Fuller contented himself with his acceptance by the art world, a lingering issue resurfaced almost by accident. While Fuller was out of the office, his staffers decided to fulfill a sculpture order by building a tensegrity sphere with connectors that had been patented by Kenneth Snelson, who promptly complained that it infringed on his work. Although Fuller claimed that he ordered them to stop, Snelson suspected that they were proceeding with a pirated version, rekindling their long-standing feud over tensegrity itself.

Snelson forwarded a copy of a letter that Fuller had written decades earlier, crediting him with the concept's "original demonstration," and asked why he had failed to attribute it properly ever since: "You even resorted to renaming your earlier work 'tensegrity' to obfuscate the fact that I showed you in 1949 an entirely new principle." Observing that Fuller had previously denied any interest in sculpture, Snelson ended on a challenge: "I would ask you please to explain to me at last—directly, not through an aide—why you have been so purposefully dishonest in this entire matter."

He closed by offering his best wishes to Anne, but he had thrown down the gauntlet, and Fuller struck back on March 1, 1980. After writing that he had credited Snelson repeatedly for his contributions, he accused him of "hundreds of deceitfully negative private and public statements about me," especially to MOMA curator Arthur Drexler: "He has concluded that I am a charlatan, and so to be treated. I am not a charlatan. Nothing could be more abhorrent. If I thought there was an iota of charlatanry in me, I would immediately commit suicide."

The letter went on for nearly thirty pages. Along with giving his own account of the development of tensegrity, Fuller alternately praised and criticized Snelson on startlingly personal terms:

> You used to tell me with tears in your eyes about the unhappy times you had experienced with your family. As years went on, your mother heard me speaking on the stage in Oregon. . . . Since then, we have exchanged letters. I find her to be your tenderly loving mother. How you could have said what you did about her, I don't know. . . . Whenever you met anyone who knew me—I was told by those who later told me—you would take them aside and, often bursting into tears, would tell them how treacherous I had been to you—Harold Cohen, for instance.

An earlier draft included a passage that Fuller deleted: "Until then, please don't send your regards to my wife or family through me. Anne saw your hateful letter. She had never, ever heard me say a negative thing about you. She had, through the third of a century since you and I met, heard me say only positive things about you. She was shocked." In the final version, he stated that if Snelson wanted to reconcile, "all you will have to do is say so." The offer went unacknowledged.

Fuller had attacked Snelson's claims so savagely because they undermined an aspect of his career that he believed to be safe, even as much of his life remained unstable, including his living situation. When Jaime proposed that he move out west to be closer to Allegra, it sounded fine to Anne, who missed her daughter and disliked the Philadelphia winters. After finding a house with a pool that was a block and a half from the Snyders, they moved in March to 1147 Via de la Paz in Pacific Palisades, just north of Santa Monica, where Jaime would live with his grandparents.

Since none of his staff members could afford to relocate to California, Fuller kept his Philadelphia office, which continued to drain his finances.

He was usually traveling, so his new address left his routine mostly unchanged. Soon after the move, he headed for Brazil to present the report on its industrial development that he had written during World War II, while another trip took him to Winooski, Vermont, a declining mill town that had proposed covering itself in a dome.

Fuller was constantly looking for inexpensive ways to advance his research. Roger Stoller had enrolled in the ArtCenter College of Design in Pasadena, where Fuller offered to speak for free if its students worked on his fog gun and packaging toilet. However, he failed to understand that the program focused on style, not engineering, and when the class was unable to deliver functional prototypes intended for the Chinese market, Roger felt that Fuller became "completely irrational."

Feeling cheated, Fuller demanded his full fee, which he had raised to $10,000 to "discourage the volume of requests." In practice, he rarely received that amount, and Roger had barely resolved the dispute before he was pulled into another project. John Denver had purchased a thousand acres near Snowmass, Colorado, for his Windstar Foundation. The environmental organization sponsored conferences and a demonstration lab for alternative energy, and Fuller wanted to exhibit the Fly's Eye Dome there.

Roger had been installing the dome's bubble openings in Pasadena, and he paid for it to be driven to Colorado. A few of its windows cracked on the way—it was too fragile to be transported far by truck, much less by helicopter—and Fuller threatened to force the ArtCenter to cover the damage. More welcome was the sight of Dymaxion Car #2, which the casino operator William F. Harrah, of the famous Harrah's Hotel, had purchased for his automobile museum in Reno. Denver arranged for it to be brought to Snowmass, where Fuller and Anne posed with the car that he had once overturned with her and their daughter inside.

Fuller's mood improved at a party at Windstar for his upcoming eighty-fifth birthday. Inside the Fly's Eye Dome, Denver performed a wistful ballad that he had composed for the occasion, "What One Man Can Do," which portrayed Fuller as a benevolent grandfather figure. When he finished, Denver kissed Fuller, stroked his neck, and murmured, "I love you, Bucky."

Fuller and the fifty-foot version of the Fly's Eye Dome installed
at the Los Angeles Bicentennial in December 1980.

Another birthday tribute came from John Cage, who wrote a
mesostic poem—with the letters down the center spelling Fuller's
name—in his honor. A different version appeared in Cage's book
Themes and Variations, which he built around the names of his heroes,
including Fuller, who had reassured him that the use of randomness in
art was perfectly compatible with design science. "We'll be remembered,"
Cage said elsewhere, "as those who lived in the age of Buckminster
Fuller."

Many of his other loved ones were gone. Anne's sister Anx, whom
Fuller had courted first, died in March, shortly after Rosamond suffered a stroke. His younger sister passed away in October, and Fuller
attended her burial in the graveyard on Bear Island. He found consolation in old friends, including Cynthia Floyd and Ruth Asawa, and
more recent acquaintances such as astronaut Buzz Aldrin, who fondly
recalled their talks "on tensegrity and cosmic energy."

His artifacts appeared occasionally in public—a Fly's Eye Dome
served as an information center at the Los Angeles Bicentennial—but it

was easier to maintain his literary output. He found an indispensable collaborator in Kiyoshi Kuromiya, who was born in a Japanese internment camp in Wyoming in 1943. In the sixties, he had marched for civil rights in Montgomery, Alabama, where he was beaten by law enforcement, and had grown close to Martin Luther King Jr.

Kuromiya continued his activism in Philadelphia. In 1969 he cofounded a chapter of the Gay Liberation Front in response to the brutal antigay policies of Frank Rizzo, the police commissioner and future mayor, who separately honored Fuller as a "positive personality" of the city. When *Everything I Know* was screened for local audiences, Kuromiya had wanted to attend a showing, but he couldn't afford a ticket. Instead, he volunteered as a typist at the office, and he was eventually recruited to replace Tom Vinetz as Bucky's "adjuvant," or writing assistant, on a book that ballooned to thousands of pages.

It evolved into one of Fuller's last major statements, along with the book *Buckminster Fuller: An Autobiographical Monologue/Scenario*, in which Robert Snyder assembled film stills and interviews into an impressionistic version of his life. Fuller and Kuromiya's collaboration became *Critical Path*, which Michael Denneny, now at St. Martin's Press, recognized would be "fucking impossible" to print in its entirety. The editor carved out a chunk that stood on its own, and although he confirmed that Fuller was "totally" involved, much of it consisted of lecture transcripts that Kuromiya had lightly revised.

Fuller informed Ed Applewhite that it was his "best and most important book because it has everything you wouldn't let me put in *Synergetics*." He opened it with his memories of the poet E. E. Cummings, who he hoped "would feel happy that I had written it." Its title was inspired by the critical path method used by the Montreal Expo, which Fuller associated with the Apollo program. "Two million things" had to occur in advance of the moon landing, and he presented the book as a similar sequence of actions that were necessary to save mankind.

According to Fuller, the obstacles were a lack of scientific understanding and the conflicts between superpowers, although his treatment of both subjects was very uneven: he repeated his assertion that the human species had been beamed to earth from space, while describing the Soviet Union as "a very reluctant player" in the Cold War. Fuller predicted that the United States of America would cease to exist as the globe was desovereignized, culminating in a traumatic bankruptcy that would motivate false patriots to "get out their guns and start a Nazi movement."

The book also featured moments of vast lucidity, including an unambiguous warning about climate change. Global warming had been under serious discussion for more than a decade, but Fuller and Kuromiya's treatment of rising carbon dioxide levels, which supported their overall argument, was admirably direct: "The 'greenhouse' effect from the sun's heat and increasing amounts of this otherwise harmless gas could send average global temperatures soaring by as much as 6 degrees Fahrenheit within fifty years, according to a US government study. This unprecedented global environmental catastrophe would be virtually irreversible for centuries."

Fuller framed his proposals as an "icebreaker" to inspire further discussion, but most were inextricably linked to his artifacts, which would succeed by rendering existing systems obsolete. The first move was to be a global energy network, accompanied by a reduction in fossil fuels. He provided a classification of resources used in the World Game, ranging from "friendliness" to "moon rocks," and said that people would renounce physical ownership once they saw "that the greatest luxury in the world is to be able to live unencumbered while able to get any information you want in split seconds and any desirable environmental condition you want in a day."

As usual, the most compelling sections emerged from his life, which Fuller presented as a case study: "The larger the number for whom I worked, the more positively effective I became. Thus, it became obvious that if I worked always and only for all humanity, I would be optimally effective." At his best, he offered the kind of advice that had inspired countless others to follow his example. "What you want for yourself may never be gratified," Fuller wrote. "What you want for everybody, because you can see the total benefits that can accrue, is usually reasonable and technologically gratifiable, and to be realized possibly within your own generation."

Critical Path appeared in the spring of 1981, and despite its factual errors and a lack of organization, it turned out to be one of his most readable books. Denneny accompanied Fuller to a sales conference at which he mentioned his list of predictions in *Nine Chains*, which had been removed from recent printings. In the ensuing decades, Fuller said, all of them had come true, except for one. He had stated that there would be "no change in the 'way of a man with a maid' and vice versa." Looking around the room, Fuller said mildly, "That's not the case."

On July 7, 1981, Fuller's staff threw a surprise party for his eighty-sixth birthday at the Franklin Institute Science Museum, where he had been told that he would be giving a private tour that night to the actress Ellen Burstyn. Years earlier, Burstyn had attended a lecture by Fuller at Carnegie Hall while she was shooting *The Exorcist*. She had been interested in starring in a movie about Margaret Fuller, and when she called to arrange a meeting, she was informed that she could see him for five hours during a layover at O'Hare International Airport in Chicago.

After their first conversation over breakfast, Burstyn had asked him if she could smoke a cigarette. "Oh, I don't mind for myself, dear," Fuller replied. "I mind for you." As for the habit itself, he made his feelings clear: "I—being the most sensitive sending and receiving mechanism ever designed—don't want to do anything to interfere with my sensitivity." Burstyn later gave up smoking, and Fuller hoped to advise on her film, which he viewed as evidence that she was "mystically inspired" by his great-aunt.

When they arrived at the museum that night for the surprise party, Fuller was startled to see dozens of familiar faces, including John Cage, Merce Cunningham, and Bil Baird, with congratulatory telegrams on hand from Jerry Brown, Jonas Salk, and Woody Allen. At the after-party, Burstyn lamented to a reporter from the *New York Times*, "It's criminal that he doesn't have a position now. At Princeton, for example. Or Harvard, for God's sake. It's unthinkable that a mind of that magnitude should not have a home."

For many of his other associates, Fuller's reputation and expectations could be a burden. His tendency toward embellishment extended to everyone around him, and he rarely resisted the temptation to introduce Roger Stoller, for instance, as "a famous surfer," which subtly undermined his actual achievements. His own granddaughter felt this acutely as she entered her late twenties. Decades later, Alexandra recalled, "I did not like the way my grandfather perceived me."

At home, Fuller was always the center of attention, which affected Alexandra's parents more than they were willing to admit, and she knew all too well that her grandfather had passed her over for Jaime. One day in Maine, Alexandra mentioned that she had been sailing. Brightening at once, Fuller began to rave about his new engineer, Amy Edmondson, who had sailed with Walter Cronkite: "She could really teach you how to sail. She's one of the greatest sailors."

Alexandra hinted that her feelings were wounded. "That doesn't make me feel very good,"

Fuller suddenly stamped his foot. "I will *never* talk to you about Amy Edmondson again. One word of hate ruins years of love."

For the next few days, Fuller was cold to her. They finally reconciled after Alexandra made tea at the house, sensing as she passed it around the table that this was the kind of "sweet, supportive granddaughter" that Fuller wanted to have. She told herself that it was better than the alternative, but she was saddened that he was unable to see her in any other light.

The incident also reflected the growing prominence of Edmondson, a Harvard student who first encountered Fuller's ideas in a class taught by Arthur L. Loeb. After seeing Fuller speak at MIT, Edmondson wrote to ask how she could best put his geometry to use. To her surprise, Fuller responded, "I would like to take advantage of your offer to come and work with me."

Edmondson signed up to serve as his assistant engineer, and she was often amazed by his energy. At an event at a ski resort in Kirkwood,

California, in the summer of 1981, he decided impulsively to build a dome based on a design that Edmondson had finished only recently. She feared that it was impossible, but she soon learned otherwise: "I hadn't yet seen Bucky Fuller in action; I didn't know that the sight of all these people would evoke irresistible images of past projects, of campuses and students and domes built, it would seem, out of thin air." Within a day, the materials for the dome had materialized as if by magic.

The Kirkwood conference was significant in other ways. It was sponsored by Burklyn Business School, an organization founded by Marshall Thurber, a real estate investor and est graduate who had met with Fuller the year before. Thurber had purchased a historic estate as a conference center in Vermont, where he spoke of hiring Fuller to consult on solar panels and windmills. As a motivational speaker, Thurber's greatest success was the seminar "Money and You," in which he reassured the children of the sixties that it was perfectly fine to be rich.

Thurber mentored a generation of gurus that included Robert Kiyosaki, the future author of *Rich Dad, Poor Dad*, who operated a camera at the Kirkwood event, and the controversial life coach Tony Robbins. Another member of this cohort was Mark Victor Hansen, a former member of Fuller's staff in Carbondale. Hansen went on to run a dome company, with a sideline as a speaker on Fuller's ideas, and he later partnered with Jack Canfield on the enormously successful *Chicken Soup for the Soul* motivational book series. The next wave of the human potential movement would draw extensively on Fuller's methods, but instead of proposing radical life changes, it often advised its listeners to invest in real estate.

For now, Fuller took comfort in his prominent friends. He met for the last time with Indira Gandhi, who had been swept back into power, and saw John Huston in California. On November 14 Edwin Schlossberg attended an exhibit on Fuller at the University City Science Center in Philadelphia with his girlfriend, Caroline Kennedy, whom he had met while she was working at the Metropolitan Museum of Art. Within five years, they were married, which unexpectedly ushered Schlossberg into the most famous family in America.

Later that fall, Fuller and Anne had dinner in Honolulu with Clare Boothe Luce, who was nearing the end of her unprecedented career as a politician and diplomat. Another guest, a retired admiral in the navy, found himself debating wind energy with Fuller. Afterward, he complained to Luce, "You invited me to talk with the genius of the world, and I spent the evening arguing with a stubborn old man hipped on windmills." Some time later, however, the admiral was asked to chair a task force on wind power, and he confessed to Luce that Fuller might have had a point.

Fuller busied himself with a hexagonal hanging bookshelf that gained stability as its weight increased, which became his last patent, and he asked his Cleveland office—to which he owed tens of thousands of dollars—to work on a huge model of a megastructure called the Gigundo Dome. With Norman Foster, he designed the Autonomous House, a rigid "double deresonated dome" with concentric shells that could rotate to regulate the capture of solar energy. Fuller hoped that a prototype could become his home in Los Angeles, as well as a house in Wiltshire County, England, for Foster's family.

He sometimes struck others as lonely. Applewhite noted that Fuller maintained both an unlisted number and a listing in the New York phone book—a reflection of his dual impulses toward "self-isolation and the need to leave out the latchstring for some wayward stranger." On trips to Philadelphia, he stayed at the modestly priced Holiday Inn, where a waitress became fond of the peculiar old man who always ate alone. At his table, he did drawings on notepaper, which he signed before handing them over: "You never know, they might be worth something someday."

Fuller had been suffering from pain in his right hip, and when Jaime brought him to a doctor, he was diagnosed with osteoarthritis. After hearing the news, Fuller said, "I think I'd like to have some hip surgery done now." The physician warned him that it would be a serious operation, with a 20 percent chance of survival, but Fuller saw it as a way to obtain a few more active years.

On December 9 Fuller received a steel hip, which required a month of hospitalization. After the operation, he remained under sedation with "a mild but definite encephalopathy" that could bring on delirium, and he dreamed about an award for individuals who demonstrated the highest degree of integrity. It was how he wanted to be remembered, and after he described it to his grandson in a call from the hospital, Jaime resolved to make it happen. They would hold one last series of events, which they decided to call Integrity Day.

After several days in intensive care at Good Samaritan Hospital in Los Angeles, Fuller was transferred to a private room to recover, which gave him a chance to reflect on his finances. His vulnerability was reflected in his relationship with John Denver, his most famous fan, who had commissioned a script with original songs for a television special about the life of the man he admired so much. Fuller was flattered, but he wrote to Denver that he was concerned by its inaccuracies: "It does not seem to me that John Denver would want to produce a play in which he 'invented' a character called Bucky Fuller."

He was also troubled by the perception that Denver was helping him financially. Noting that he had personally put millions of dollars into his projects over the years, Fuller complained that Denver's staff had implied that he was "operating only as a beneficiary of your generosity." He closed with a gentle warning: "If it seems easier to clear the space between us by terminating all financial support, feel free to do so." The ultimatum was disregarded.

After Fuller went home to recover, he left many of his business matters to Jaime, who recalled this period as his grandfather's "longest time in one place." All of his boats had been sold except one, the *Frangipani*, and they began to face the problem of continuing his work. He partnered with an investor, James Patterson, on a research center bearing his name in DeLand, Florida, while Jaime put together a proposal for the Friends of Buckminster Fuller, a volunteer foundation in Pacific Palisades that would raise money for his projects.

Left unresolved was the World Game, which was his most valuable idea, as well as his best chance to leave something for Anne. Fuller had repeatedly postponed the issue, but he finally told Medard Gabel and Howard Brown, who ran the World Game Institute, that he was giving everything to his family, with one notable exception: "There's one thing that's extremely important to me that I'm leaving to you two. And that is this trademark."

Not long after the meeting, he abruptly changed his mind. Brown suspected that Fuller felt guilty about providing no meaningful inheritance to his family, especially because he had poured so much of Anne's money into his work. "When we went back to sign the agreement his lawyer had written up," Brown said, "it left the ownership of the trademark to his estate—basically to his grandson, Jaime—and the control over it to us. And, like everything else in his life, all that did was create conflict, and it kept anything from ever happening with it."

Jaime disagreed with this characterization of the agreement, feeling that it was structured to license Gabel and Brown to use the trademark "in an intelligent way." Fuller preferred to focus on symbolic artifacts, including a Dymaxion Map the size of a basketball court, which he displayed for members of the United States House of Representatives. It was little more than a shadow of the electronic version that Fuller had once envisioned, however, and he was painfully aware that his former collaborators were undertaking ambitious projects on their own.

On one side of the country was a dome to house the *Spruce Goose*, the wooden flying boat constructed in the forties by the eccentric millionaire Howard Hughes, based on a proposal by Henry J. Kaiser. Along with the ship the *Queen Mary*, now permanently moored in Long Beach, California, it had been leased by entertainment entrepreneur Jack Wrather, who commissioned a hangar from Don Richter at Temcor.

With a diameter of 415 feet, which was the practical limit for its aluminum frame, it was the most massive dome ever built along the lines that Fuller had established. Fuller declined an invitation to witness the plane's installation in February 1982, although he toured it the following month. In another world, it might have been his final monument, but instead, he was stuck watching it from the *Queen Mary*. He was upset that its panels hid the underlying structure, and when he demanded that it be changed, Richter said only, "It's not ours. It belongs to them."

Even more infuriating was Spaceship Earth, which was under construction at Epcot. The geodesic sphere was designed by Peter Floyd and David Wallace, formerly of Geometrics, with engineering by Simpson Gumpertz & Heger, a firm that had consulted on the Expo Dome and the unrealized tower in Japan for Matsutarō Shōriki. With its upper and lower sections supported separately on an elevated platform, and an inner steel frame joined to a cosmetic outer shell, it owed as much to clever sleight of hand as its inspiration had in Montreal. Fuller hoped that it would feature a World Game, and he was disappointed when it turned out to be nothing but a ride.

He also fumed that his name was nowhere to be found. Inquiries about his involvement were answered by a form letter that credited him with the idea of geodesics and the phrase "Spaceship Earth," only to end tersely, "We regret to inform you that no visits by Dr. Fuller are planned at this time." Fuller publicly attributed the structure to "my Cambridge, Massachusetts, engineers," and he noted pointedly that he had built a geodesic sphere at Cornell in the fifties. He concluded of Disney, "It must have been their business people who thought their company might be advantaged by avoiding identifying me with my work."

Fuller was right to be furious, as well as concerned by the ease with which he had been written out of a narrative from which he had excluded so many others. A few domes were associated with him more prominently, including the Home of the Future at the Knoxville World's Fair and a greenhouse with a closed ecosystem at the New Alchemy Institute on Cape Cod, Massachusetts, which he toured with Jay Baldwin. During the visit, he seemed worn out, but he perked up on the drive back. Turning to Baldwin, Fuller joked, "Well, old man, when are you planning to retire?"

As Fuller aged, his mind often turned to his romantic history. In May he saw Cynthia Lacey and her new husband, Keith Asbury, and he met over the summer with the former Evelyn Schwartz. Neva Kaiser recalled that Fuller once apologized to her "in case I was expecting him to make a pass at me," and she was amused by his explanation: "God had told him to expect that women would be attracted to him, and that he shouldn't respond if it wasn't his wife." When Kaiser asked why he had been talking to God, Fuller replied, "Who else would grade my papers?"

He remained capable of drawing important new admirers. In France, he met the saxophonist Ornette Coleman, who had pioneered the art of free jazz. Coleman had first seen him speak in the fifties, and he was reminded of his approach to music as Fuller demonstrated the properties of the tetrahedron. "He manipulated the model, turned it inside out, made it dance—but the corners never touched," Coleman remembered, and he composed a piece for string quartet and drums inspired by Fuller, whom he described as his "number one hero."

In October Fuller delivered a eulogy for Tatyana Grosman, who had died in July. His reflective mood was evident at an event in San Diego, where Anwar Dil, a professor of linguistics, asked if he would consent to an interview. The following day, Fuller and Dil engaged in a long conversation that would be published posthumously as *Humans in Universe*. Ranging freely over his career, Fuller was candid about his affairs, which he compared to the wanderings of Odysseus:

> I have always had the experience of some, to me, enchanting female coming into my life concurrently with the scientific discovery. . . . Despite flirtations and sex with others, I have never stopped loving my wife above all others. What I am getting at here is the *Iliad*- and *Odyssey*- or *Aeneid*-like sirens or other females' interception of the explorer's route. . . . Every time I am about to make a discovery and am developing a high sensitivity of thinking, along comes, to me, an exceptionally charming female with whom I find myself tending to fall in love.

Like Penelope, who faithfully waited twenty years for Odysseus to return, Anne was left behind while Fuller traveled, and as her strength diminished, she stayed at home, serving as her husband's interpreter on phone calls. One evening, she sat with Jaime to watch Fuller on *Two on the Town*, a local news program, as she sipped a glass of sherry before dinner. When a commercial came on, Anne glanced at her grandson and said, "You know, that Bucky really does know his stuff."

During a trip to Penang in 1983, Fuller learned that he was to be awarded the Presidential Medal of Freedom. The poet Archibald Mac-Leish, hailing him as among "the very few by whom the greatness of our society would be judged," had furnished a statement of support: "If there is a living contemporary who truly knows what it is to be an American, Buckminster Fuller is that man."

On February 23 Fuller received the medal from President Ronald Reagan. With Anne unable to travel, he was accompanied by Allegra, Jaime, Gabel, Applewhite, Calvin Tomkins, and his secretary Ann Mintz. The other recipients included the Reverend Billy Graham and, remarkably, Clare Boothe Luce, who had resisted his advances nearly five decades earlier. At the ceremony, Reagan praised Fuller as "a true Renaissance man, and one of the greatest minds of our times," as well as proof "that America is a land of pioneers, haven for innovative thinking and the free expression of ideas."

Fuller's feelings about Reagan were very different. Two years earlier, he had briefly watched the inauguration before turning off the television. "A farce," he grumbled. "America doesn't exist." In print, he dismissed Reagan as "a dumb actor who can memorize his lines, and no more," and his views were widely publicized by an accident of timing. The day after the event, reviewers obtained copies of *Grunch of Giants*, which Michael Denneny had extracted from Fuller and Kuromiya's mass of manuscript material. *Grunch* stood for "Gross Universe Cash Heist," which Fuller used to identify a global conspiracy of banks, corporations, and politicians that included Reagan himself.

According to Fuller, millions of dollars had been spent on "the most negatively momentous presidential election in nearly a third of a century," and Reagan's only objective was to postpone America's bankruptcy for as long as possible, allowing the conspirators to enrich themselves with military contracts. Fuller criticized "the Reagan government's reduction of taxes for the rich money-makers which might otherwise have been applied to saving the nation's human security system," and he arrived at a bleak conclusion: "What was once the Republican Party of Lincoln is now the party of unmitigated selfishness of big money."

Any worthwhile points were undermined by his tone, which was a return to the bitterly satirical voice of his earliest writings. Fuller had always been critical of capitalists, despite his willingness to take their money whenever he could, and he argued that the only solution lay in adopting his own artifacts, along with a worldwide computer network to make the information behind policy decisions universally available. It was fortunate, Fuller said, that he had commenced his design-science revolution a half century ago: "It would take another fifty years to do the critical path work, and we now have only fifty months."

As the pessimistic residue of *Critical Path*, it was one of his darkest books. It surprised fans who saw Fuller as a utopian optimist, which was the only version of him that they had ever seen, but he preferred to be known as a realist, and his long-standing sense of disillusionment had been further exposed by illness and his failures with politicians. He had said that he would resign if appointed to a position of power, but he had always sought access to policy makers, and *Grunch* was a warning of the cost of their indifference.

Fuller was slightly heartened by the first Integrity Day event at the Los Angeles Convention Center, which drew more than a thousand attendees for eight hours of lectures, including a demonstration of the Big Map. He was reminded that his hopeful mode was more persuasive than a negative tone, which he noted "reduces the credibility of my wisdom," and that his public persona was still the one asset completely under his control. When Werner Erhard asked him around this time what business structure he used, Fuller simply said, "An individual."

He prepared to publish a more uplifting book built around his twenty-eight "patented works," including both incarnations of the Dymaxion House, for which he had never filed a successful claim; his forty-seven honorary degree citations; and the Basic Biography maintained by his office, an endless listing of honors, press, and appearances that Denneny compared to "a curriculum vitae gone mad." The editor convinced him to omit it, but they included the doctorates, which were described as "experimental data" on the value of his life.

Inventions: The Patented Works of R. Buckminster Fuller opened with "Guinea Pig B," his last major essay, which characterized his career as an experiment to see what a single man could accomplish more effectively than conventional institutions. He said that he had always resisted "publicly declaring yourself to be a special 'son of God' or a divinely ordained mystic leader," and that he had relied on "miracles that occur just when I need something, but not until the absolutely last second." Observing that no one had ever profited from his ideas, he added, "Fortunately, this gambling on the financial success of my work despite my warnings seems altogether to have ceased a third of a century ago."

Fuller never saw any text as a final statement, and he was already working on another manuscript with Kuromiya, but time was running out. Anne's cancer had returned, and she underwent another operation on May 13, 1983. Two days later, while seated at her bedside, Fuller wrote himself a note on the isotropic vector matrix, signing it with a line of praise "to God, the eternal sum of all truths." It was the last significant piece of writing that he ever produced.

After Anne returned home that month, her problems with swallowing persisted. Her mouth would fill with saliva, but she refused to acknowledge it, prompting Jaime to tell her gently, "Grandma, you have to spit that out." Allegra, who had been teaching at New York University, returned to be with her mother, spending the night with her while Fuller was away. Although Anne was given a liquid protein diet, her digestive problems led her to lose weight and hallucinate in her sleep, and Allegra concluded that she "was basically starving to death."

Fuller was told that he could maintain his usual schedule, and he went to Boston with Amy Edmonson for Harvard commencement in June. At the subway stop at Park Street, where he had disembarked long ago to see the chorus girls at the Shubert Theatre, he was caught on a stalled escalator, "ran out of wind," and collapsed. After Edmonson got him to a bench, he was checked by paramedics and discharged at the hospital. The incident shook him, and Sharkey wrote to Tomkins,

"Bucky did *not* ask me to write to you about this, but I know he's feeling rather anxious about whether you are really going to do his biography or not."

After Anne returned to Good Samaritan Hospital in Los Angeles, her status was seen as serious but not critical, and Fuller flew to London to sponsor the presentation of the Royal Gold Medal for Architecture to Norman Foster. During a conversation at the architect's office, Fuller said that he still intended to live in the Autonomous House, and when Sharkey pointed out that Anne's health made this inadvisable, he broke off, flustered, and agreed.

The morning after the event, Fuller went to another meeting with Foster, where he made an unusual request—he insisted that it be recorded. In what Foster recalled as "instructions from the master," Fuller advised him to dream "very much bigger" for the future: "I think it is your world that you are going to lead in." Fuller confessed that he felt tired, and before they parted, he said casually that he could "pull the plug" on himself at any time.

On June 23 Fuller took the Concorde supersonic jet back for a presentation at the White House. From there, he headed to Chicago for a dinner for subjects of *Time* magazine covers, including the Reverend Jesse Jackson and Donald Rumsfeld. Sharkey remembered it as "the longest, most exhausting day" that she ever spent as his secretary, but Fuller reminisced fondly as they traveled about their years together, and he struck her as "the sweetest, most sentimental I had ever seen him."

Fuller returned the next day to California, where Jaime picked him up at the airport and drove him directly to the hospital. The doctors had tried to put a feeding tube down Anne's throat, in which they encountered the blockage that had caused her problems with swallowing. In the process, they accidentally punctured her esophagus, leaving them without any good options. At an Integrity Day event the next morning, Fuller wondered aloud if he and his wife would leave the world together.

He visited her again the following day. Anne was still experiencing hallucinations, and she had once pointed at an empty spot in the air and asked, "Do you see that? Those little angels floating there?" Otherwise, she seemed lucid, although vaguely embarrassed by her shakiness, and when her visitors left, she was still coherent. When they returned a day later, she was nonresponsive, and everyone understood that it was only a matter of time.

Fuller continued to make public appearances, but he was often at Anne's side. He had never lost his guilt over leaving her without any money, despite placing so many ventures in her name, as if to compensate for his long absences. As long as she was still alive, he refused to rest, but he didn't want her to cling to life for his sake. In his spare moments, Jaime noticed, he was reading a book about a girl who died at the age of seven. It was called *Mister God, This Is Anna*.

On July 1, 1983, Allegra learned that her mother's vital signs had worsened, and she called Jaime to tell him to bring his grandfather to the hospital. When they arrived, Fuller sat down to take Anne's hand, as he had done a lifetime ago with their daughter Alexandra.

As Jaime left to deposit a check, Fuller and Allegra kept their vigil. Frustrated that he was unable to talk to his wife, Fuller said softly, "I think she's worried to go without me." Allegra said nothing in response, but Fuller suddenly turned to her: "She's squeezing my hand—I know she's squeezing my hand!"

It was as though she had sent the signal that he was awaiting. Fuller rose to his feet and, at that instant, suffered a massive heart attack.

He tried to reach the bathroom. A nurse rushed inside, ordering Allegra to leave, and an emergency team arrived. It was a quarter to three in the afternoon. A minute later, Jaime emerged from the elevator to see a man being wheeled off on a gurney. He thought that it might be his grandfather, but he wasn't sure, and it was only when he saw Allegra that he learned what had happened.

The efforts to revive him failed, and Fuller was pronounced dead at 4:45 p.m. "After he had died, he had an exquisitely happy smile on his face," Allegra remembered. "I think he felt that perhaps Mother was sort of hanging on the edge because she was afraid to leave without his going too, and that he should usher the way for them both to depart together."

Thirty-six hours later, early in the morning on July 3, Anne died. She had never regained consciousness.

As many of their loved ones would observe, neither Fuller nor Anne had to live with the knowledge of the other's death, and it was equally miraculous that he had been with her at the end, rather than halfway across the world. He had died just short of his eighty-eighth birthday, which would have been their sixty-sixth anniversary. John Cage felt that they had passed away "so beautifully," and his granddaughter, Alexandra, was startled by it: "It said to me things about their relationship that I don't think I would have recognized before they died together."

On Fuller's desk, his staff found his final manuscript on geometry, along with a note: "If something happens to me and I die suddenly, I want you to know of the extraordinary importance of my now being written book *Cosmography*." It went unpublished for years, but Kiyoshi Kuromiya never abandoned it. In its earliest form, it was written in the second person, as a letter to Annie Dillard, and its last sentence came as close as Fuller ever would to a final statement: "Dear reader, traditional human power structures and their reign of darkness are about to be rendered obsolete."

Fuller's front-page obituary in the *New York Times* described him as "a thoroughgoing original who for many years was dismissed by many as something of a crackpot," and it mentioned his "cultlike following," although it left no doubt of his importance. In the next day's *Times*, critic Paul Goldberger published an evaluation in which he called Fuller "a somewhat curious cult figure for this moment in history, when our culture has lost much of its faith in the ability to solve problems by any means, rational or otherwise."

On July 8 a funeral was held at Mount Auburn Cemetery in Cambridge, near Caroline Fuller's gravesite and the memorial to Margaret Fuller. Two hundred mourners paid their respects to Fuller and Anne, including many of his staff members, Noguchi, Erhard, and John Denver, who performed "What One Man Can Do." John Cage felt that it was "sentimental and small" compared with Fuller's achievements, and he was relieved when it was over.

Fuller's most spectacular memorial was held at the Cathedral of St. John the Divine in New York on September 27, which testified to his enormous following. Alexandra and Arthur Penn gave readings, while Harris L. Wofford Jr., the former president of Bryn Mawr College, delivered the eulogy. Wofford quoted from a letter that Fuller had sent to a friend whose wife had died, whom he reassured that human beings were immortal: "I go right on seeing them. There they are."

Ellen Burstyn, who had once hoped to play Margaret Fuller, read the closing prayer: "Almighty and everlasting God, who didst make the universe with all its marvelous order, its atoms, worlds, and galaxies, and the infinite complexity of living creatures; grant that, as we probe the mysteries of thy creation and the laws behind all structure, we may, even as thy servant Buckminster, come to know thee more truly, and as certainly show forth the mind of our Maker; through Jesus Christ our Lord, Amen." The service ended with Fuller's favorite naval hymn:

Eternal Father, strong to save,
Whose arm hath bound the restless wave,
Who bidd'st the mighty ocean deep
Its own appointed limits keep:
O hear us when we cry to thee
For those in peril on the sea.

Fuller and Anne were buried in a shared plot, with a headstone that was decked at the funeral with roses, frangipani, and greens from Bear Island. A geodesic sphere was engraved next to Fuller's name, and his epitaph was inscribed underneath: *CALL ME TRIMTAB*. For Anne, there was the image of a single rose, like the ones that he had sent to her when their life together was still just a dream of the future. It had taken them oceans apart, but now they lay beside each other forever, almost close enough to touch. "The behavior of 'Universe' can only be shown with a minimum of two pictures," Fuller had said. "Unity is plural and at minimum two."

EPILOGUE

Tetrahedron Discovers Itself and Universe

It could be that by traveling mentally backward in history as far as we have any information, humans could—like drawing a bowstring—impel our thoughts effectively into the future.

—**BUCKMINSTER FULLER,** *SYNERGETICS*

In the fall of 1976 a Yale senior named Donald E. Ingber noticed a group of art students carrying cardboard sculptures across campus. Looking more closely at the models in their hands, he realized that they were in the shapes of various polyhedra, which reminded him of viral structures. When he asked for more information, he was told that they were an assignment from the Austrian sculptor Erwin Hauer, who taught a course on three-dimensional design.

Ingber was a molecular biophysics and biochemistry major, but he impulsively decided to enroll. At their first meeting, Hauer offered him a crushing handshake and asked bluntly why he wanted to take the class. Ingber launched into a prepared explanation: "Because life *is* three-dimensional design. Every time you contract your muscle, it is because individual molecules change their three-dimensional shape. When protein called myosin bends, it pulls on a long actin fiber . . ."

In the face of so much enthusiasm, Hauer was persuaded, and Ingber took instantly to the subject matter. One day the professor told his students to buy dowels and fishing line to make a sculpture that could stand without any of the sticks touching. He showed them tensegrity models by members of Fuller's circle, which were powerful teaching tools, and Ingber—who had been "wowed" by the Montreal Expo Dome at age eleven—became entranced by tensegrity.

When Hauer pressed down on a rounded tensegrity structure, which sprang back up when released, Ingber connected it to something that he had seen recently in the laboratory. To transfer cells between culture dishes, he had added an enzyme to clip their anchoring points to the substrate. At once, the cells, which were normally flat, had rounded and "bounced up off the dish."

Ingber assumed immediately that they were tensegrity structures. A few days later, he was in the lab, looking at another group of cells, when he noticed that their shape was affected by a drug in his tests. Without thinking, he exclaimed, "The tensegrity must have changed!"

The postdoctoral student standing next to him glanced over in surprise: "What did you say?" After Ingber told him about Fuller, the other man turned aside. "Well, never say that again."

Accepting this as a challenge, Ingber headed to the library over winter break and found a copy of *Structure in Art and in Science*, a 1965 anthology edited by the design theorist György Kepes, who was one of Fuller's greatest supporters. In one chapter, Fuller reproduced photos that illustrated the similarities between geodesics and cellular tissue, and he closed with a call to action: "Those who are versed in fundamental structuring principles will be the next half century's *most needed* men."

Ingber remembered this visit to the library long afterward as a personal turning point. In his postgraduate studies, he carried out his dissertation in the cell biology department, where he conducted experiments using tissue from the rat pancreas. Usually the cells were arranged in a neat epithelium, in which they were packed into a thin layer anchored to the extracellular matrix, but they became disorganized and irregular whenever cancer appeared.

Examining the data, Ingber speculated that the cell's shape was regulated by tensegrity, with the cytoskeleton—a web of tensed fibers that resisted deformation—providing a supporting framework. Its microfilaments were obviously tension structures, while Ingber thought that its hollow elements, the microtubules, might function as compression units. An instability in this network could possibly lead to a loss of shape control, followed by the uncontrolled growth of cancer.

On April 5, 1983, Ingber attended a party at Randolph Hall, the Yale art and architecture building, where he ran into a woman he had met years earlier in Hauer's class. She asked if he was going to see Buckminster Fuller, who was scheduled to appear the next day at Atticus Bookstore in New Haven. Ingber hadn't even known that Fuller was coming to town, and he rushed home to write a letter in case they didn't have a chance to speak there in person.

Enclosing an article about his work on tumor formation, Ingber wrote that he believed that tensegrity could explain structures in cells and tissues: "Cancer may then be viewed as the opposite of life resulting from a breakdown of this geometric hierarchy of synergetic arrangements." He closed by hinting that he regarded Fuller as a hero: "I would just like to thank you personally for holding true to your way of looking at life in a world full of followers and technocrats."

Ingber slid his letter and research paper into a manila envelope. The following morning, he went to the bookstore, ordered a coffee, and sat down to wait. Shortly before eleven o'clock, a car pulled up and Fuller emerged. He was eighty-seven and visibly frail, but he was mobbed by admirers at once, and Ingber assumed that there was no chance of a real conversation.

A second later, he recognized the woman at Fuller's side, standing back to avoid the crush of fans. It was Amy Edmondson, whom Ingber had met through a summer sailing class. Reintroducing himself, he gave her the envelope, which she promised to share with Fuller after the signing. When she led him over to Fuller, they shook hands, but there was no time for the talk that Ingber wanted: "The whole encounter passed by in an instant, and I felt almost no connection."

Later that month, however, Ingber was at the campus medical clinic when his beeper rang. Returning the call, he found that a woman in California was looking for the book in which his research appeared— Fuller had mentioned it in a lecture. Ingber was amazed to learn that Fuller had been carrying his letter everywhere, and he often read it aloud to audiences.

Fuller always denied that his designs were based on nature, but he welcomed confirmation from the sciences, which Ingber had taken further than anybody since Caspar and Klug. On April 11 Fuller wrote to Ingber to say that he was "elated" by his ideas, which were the first serious attempt to connect tensegrity to cell biology. They kept in touch, and Ingber raised the possibility of "setting up a network of individuals in different fields who have used synergetics."

On May 18 Fuller responded with a remarkable letter: "You are the first one I have met who is undertaking to do what I undertook to do—that is to educate myself to be a comprehensivist." Fuller hoped that Ingber would work for him one day, but he advised him in the meantime to find a job in aviation, which would allow him to learn modern production methods. He often gave this advice to students, and he observed that the only one who had ever followed it was Don Richter.

Fuller wanted to turn Ingber into his ideal protégé, which he came closer to embodying than anyone since Edwin Schlossberg, and the offer assumed that they had plenty of time. Before Ingber could reply, Fuller died, leaving the younger man to brood over his final message. "I believe he was telling me that to gain fundamental insight into the world in which we live," Ingber said, "and to change the world for the better, we must span the divide that separates different disciplines and work at the interface, as he had done before me."

After Fuller's death, Ingber began to look into tensegrity more carefully, seeing it as a valuable tool that had been neglected by others. Like many of his predecessors, he started with models. Ingber was working without the benefit of Fuller's guidance, so he found what he needed in *Synergetics*, and he received input from Edmondson as he systematically explored these constructions.

One was a tensegrity icosahedron made from elastic thread and wooden applicator sticks, which Ingber fastened to a sheet of paper. When the pins that held it down were removed, it pulled the paper up into wrinkles, which he suspected could explain a familiar phenomenon in cell biology. Cells attached to a rigid dish flattened themselves, but when they were anchored on rubber, the surface puckered, which was exactly what would happen if they were tensegrity structures.

Ingber followed this model with a larger icosahedron made of shock cords from a sailboat and aluminum struts that he had stolen from a chemistry lab. To represent the nucleus of the cell, he used elastic to attach a smaller tensegrity sphere at the center. When Ingber pressed down on the main structure, the "nucleus" spread out and moved downward in its enclosure, which was the precise behavior observed in cells when they flattened on a substrate.

Photographing living cells under similar conditions, Ingber became certain that he was right. Cells were prestressed tensegrity structures, which he followed Fuller in seeing as the most efficient forms in nature; and, by acting as a system, they could even transmit information, as external forces were conveyed throughout the cytoskeleton. The conversion of a mechanical stimulus to a biochemical reaction, or mechanotransduction, might affect growth, differentiation, and gene expression, while a breakdown might lead to cancer.

After graduation, Ingber presented his work at a conference, but he misread the room, and two older scientists criticized him harshly over dinner. "My presentation had been totally inappropriate for a scientific audience," Ingber remembered them saying. "The idea that mechanical forces are informative had gone out with the notion of 'vitalism' at the turn of the last century. Chemicals and genes were the only important biological controllers."

Ingber refused to be dissuaded, and he found ingenious ways to subject cells to targeted stresses, which yielded results that pointed to the presence of tensegrity. He concluded that mechanics played as great a role as genetics, with tensegrity structures bridging the gap, but although he published in reputable journals, his hypothesis remained controversial. Ingber blamed this partly on skepticism toward Fuller's vocabulary, including the term *tensegrity* itself. "If you stop using the word," Ingber recalled being told, "you can get grants."

Like Fuller, he extended the concept of tensegrity to domes, triangulated structures, and prestressed systems of all kinds, and he was fond of sweeping statements, exposing himself to charges that he was overstating his case. Ingber consistently defended his views in print, however, and he used tools that Fuller would have admired to make significant discoveries, which came closer to validating the scientific claims of tensegrity than anyone else ever would.

In September 1992 Ingber was staying on Martha's Vineyard with his wife and son, whom he sent away so that he could concentrate on writing a journal article. As he read papers on cellular structure, his mind turned to a persistent problem. He had argued that tensegrity forces in cells arose from the tension produced by microfilaments and resisted by microtubules, but the latter were sometimes absent, which meant that there had to be another way to generate these effects.

In his reading, he found a paper from the seventies that discussed the triangulated microfilament networks inside spreading cells, part of which resembled a geodesic dome. Studying an electron micrograph and drawing, he was reminded of a bent version of what Fuller had called the isotropic vector matrix. The filaments were made of actin and myosin—the proteins that Ingber had mentioned in his very first conversation with Hauer—and if their resistance to compression varied locally, they could yield a tensegrity structure without any microtubules.

Ingber rushed to buy drinking straws and spools of elastic thread from a sewing supplies store. Back at the house, he made a model of the vector equilibrium. Folding it down in Fuller's jitterbug transformation, he ended up with a tetrahedron with edges consisting of multiple straws, which he thought might account for the triangles that were visible in the microfilament network. When he pulled on the unfolded model along an axis, as if it were contracting, it formed a linear arrangement of bundled struts, which exactly matched the description in the paper.

Using more straws and thread, he made a second model of the isotropic vector matrix, which he bent in his hands to replicate what would happen during cell spreading. It created a three-dimensional pattern that corresponded precisely with the electron micrograph. Because the bundling of structural members naturally increased their stiffness, the result also explained how compression elements could occur in cells with only microfilaments. The microtubules, he realized, might not be necessary at all. He had been right all along.

"It was like touching the heartstrings of the universe," Ingber recalled. Feeling a rush of discovery, he left the house and walked toward the shore in the rain. Even then, he knew that his conclusions undermined the conventional understanding of cell structure, and after they were published, he would be compelled to defend them against established models. He would never abandon his belief that tensegrity was the theory that best fit the facts, however, and he was never proven wrong.

By any standard, he was a hard man to dismiss. Ingber became the Judah Folkman Professor of Vascular Biology at Harvard Medical School, as well as the founding director of Harvard's Wyss Institute for Biologically Inspired Engineering. He emerged as Fuller's closest successor in the scientific establishment, and it was especially meaningful that he made his professional home at the university from which his hero had withdrawn twice.

Ingber's rise coincided with that of Amy Edmondson, who became the Novartis Professor of Leadership and Management at Harvard Business School, where she studied human interactions within institutions—a quality that had always been Fuller's weakness. Edmondson and Ingber learned to navigate within the academic world more capably than Fuller ever had, but both were guided by principles that he would have recognized as his own.

To honor his legacy, Edmondson and her husband, George Daley, the dean of Harvard Medical School, endowed the R. Buckminster Fuller Professorship of Design Science at the Harvard Graduate School of Design, while Ingber used his ideas as a form of lateral thinking to produce solutions that others overlooked. His one meeting with Fuller was enough for him to be changed, but not ensnared, and he emerged with just the touch of strangeness that he needed.

Ingber found himself occupying what he called the fruitful "middle zone" between science and art, and like many scientists, he found that Fuller's approach had real value in biology. Given his limited resources, Fuller had concentrated on structures that were simple and efficient enough to be reproduced at minimal cost, which meant that the geodesic dome, the shell of a virus, and the interior of a cell all seemed like expressions of the same natural laws. "In the end," Ingber said, "this is the most fundamental contribution he made that will live on: uncovering how nature builds."

On that rainy day in Martha's Vineyard, all of this lay in the future. Moving through a dense fog, Ingber walked alone to the beach, as Fuller had once gone to the edge of Lake Michigan, but instead of looking out at the waves, he plunged into the ocean. He was exhilarated from his discovery, and it took him a second to realize that he was almost out of sight of land. Coming to his senses, Ingber asked himself, "What the hell am I doing?" And then he swam back to shore.

"Imagine human beings have this tiny little band where you and I can tune in, and we find that that is less than a millionth of reality," Fuller once said. "Just think of it. This is reality—these are the realities—and you and I can see less than a millionth of reality." Since the early forties, he had marveled at the power of the spectroscope, which allowed researchers to analyze matter based on its interactions with radiation, and it was only fitting that it would provide the backdrop for the most lasting tribute he would ever receive.

The central player in Fuller's greatest moment of posthumous glory was Harold W. Kroto, who was born in England in 1939. Kroto studied chemistry at the University of Sheffield, but he seriously contemplated a career in architecture or graphic design. While working at Bell Labs, he made what his wife, Margaret, described as "a kind of pilgrimage" to the Montreal Expo Dome, where they pushed their infant son in a pram along the ramps, and he thought about writing to Fuller for a job conducting research "on the organized growth of massive urban structures."

If the offer had been made and accepted, Kroto might have occupied the same role in Fuller's life that Schlossberg did in the late sixties. Instead, in what was arguably a lucky break, a position at the University of Sussex enabled him to specialize in spectroscopic studies of long carbon chains in outer space, which he theorized were generated in the atmospheres of the aging stars known as red giants.

In 1984 Kroto saw a chance to test his ideas. At a conference, a microwave spectroscopist named Robert Curl told him about a huge machine at Rice University in Houston. The laser supersonic cluster beam apparatus, which had been designed by the physical chemist Richard Smalley, could produce aggregates of any element on demand. As a laser vaporized the material, helium blew it into a vacuum chamber, where it cooled into clusters that a mass spectrometer could measure to determine the amounts of each size of molecule.

Curl explained that it would work perfectly well with carbon, and by introducing different gases, it could re-create the conditions inside a red giant to see if the long chains appeared. He was especially interested in whether they might be responsible for the diffuse interstellar bands, which were dark lines in astronomical spectra caused by unidentified matter in the space between the stars.

An enthusiastic Kroto went to meet the team at Rice. Smalley was a few years younger than Kroto, but he was equally ambitious, and from the beginning, friction arose between the two men. According to Smalley, he was skeptical of Kroto's "cockamamie theory," and he suspected that carbon chains would simply break apart when exposed to light.

Smalley preferred to focus on applied research on elements such as silicon, but he decided to give Kroto a chance. The following summer, they scheduled a second visit, which Smalley still saw as an unwelcome interruption: "I thought Harry was sort of a loose nut," he admitted, "and I just wanted to get rid of him." To break the news to his graduate students Jim Heath and Sean O'Brien, he asked jokingly, "What's the worst possible thing that could happen?" They responded, "Harry's coming."

Their lack of excitement was due partly to their awareness that similar research had already been conducted at the Exxon corporation, which the Houston team attempted to replicate the week before Kroto's arrival. When they trained the laser on a graphite disk, it worked as expected, but the settings on the computer display led them to overlook a peak for C_{60}, or molecules of sixty carbon atoms. The spike was enormous, but although one member of the group—probably a graduate student named Yuan Liu—made a note of it, no one paid much attention.

Kroto officially started on September 1, 1985. The first stage was devoted to calibrating the equipment, with helium used as a carrier gas for the carbon, and when they ran a spectrometric analysis, they saw the C_{60} peak. Over the following days, they introduced hydrogen and nitrogen into the helium stream, which combined with carbon to produce a mostly incomprehensible mess of reactions.

The team observed that pronounced peaks continued to be seen for large clusters with even numbers of atoms, which were considerably less reactive. When they increased the backing pressure of the helium, a gigantic spike at sixty atoms was accompanied by a smaller but still prominent peak for C_{70}. This result was truly unexpected, and they paid close attention to it for the first time.

Under some conditions, C_{60} was dozens of times more abundant than most of the other clusters, implying that there was something special about the number 60 itself. The absence of reactivity pointed to a closed molecule that lacked dangling bonds, which Smalley called "the mother wodge" and Kroto dubbed "the god wodge," and the discussion eventually turned to the possibility of a sphere. Smalley glanced at the others. "Who was that guy who built those domes?"

Like Kroto, Smalley had seen the Montreal Expo Dome, but he had never given much thought to Fuller. The team didn't have a crystallographer, and no one could remember the details of the dome—they thought that it was made entirely of hexagons, which was consistent with the geometry of carbon bonds. Kroto had once constructed a paper star dome from a kit for his children, and he suspected that some of the faces might have been pentagons. It was in a box at his house in Sussex, and he considered calling his wife to check.

In the meantime, they decided to look into Fuller. "Here, after all, we had a hexagonal sheet," Smalley recalled, referring to the structure of graphite. "Maybe if we figured out how Buckminster Fuller did this, we could figure out how to curl these things around on each other." At the library, Smalley found *The Dymaxion World* by Robert Marks, which had inspired Donald Caspar and Aaron Klug years earlier. Leafing through it, Smalley somehow missed the numerous illustrations that showed pentagons in geodesic structures. He was most struck by the Union Tank Car Dome, which seemed at first glance to consist solely of hexagons.

The team continued the conversation over dinner, arguing over possible structures and drawing sketches on napkins. Afterward, Jim Heath bought a package of what would be variously described as gummy bears, jelly beans, or "Juicy Fruit gum balls." At home, he tried to build a structure out of toothpicks, using the candy pieces as connectors, in an unknowing echo of Fuller's experiments as a kindergartner. Heath found that it was impossible to make a sphere using only hexagons, which fell apart before they could form an enclosed shape.

Smalley was tackling the problem at his own house. After trying unsuccessfully for hours to write a program to generate a solution on his home computer, he started to cut hexagons out of paper. When he stuck them together with tape, they wouldn't produce a sphere without overlapping, and even if he cheated slightly, they refused to make a closed surface. Feeling discouraged, he went into the kitchen around midnight and cracked open a beer.

It was the kind of quiet moment that was ideal for creative insights, and he suddenly remembered what Kroto had said about the pentagons. He cut out a piece with five sides, and when he taped hexagons around it, it formed a shallow bowl—a fact that would have been common knowledge in Drop City. Adding one layer at a time, he ended up with a sphere with twenty hexagons and twelve pentagons. Its sixty vertices were all the same, indicating that it could form a hollow structure constructed out of identical carbon atoms, and it seemed remarkably stable. When Smalley dropped the paper ball on the floor, it bounced.

Since it was three in the morning, instead of telling the others, he fell asleep. The following day, he called the office from the car phone in his black Audi, relating his discovery to Curl's answering machine. When the group reconvened, Smalley tossed the sphere onto the coffee table. Kroto was "ecstatic and overtaken by its beauty," but Curl wanted to make sure that it alternated correctly between single and double bonds. Borrowing a stack of adhesive labels from Smalley's secretary, they stuck them on the model to confirm the pattern, and Curl was convinced.

Hoping to learn more about the polyhedron itself, Smalley placed a call to Bill Veech, the chairman of the mathematics department at Rice, who said that he would check with one of his students. When Veech called back, he found himself talking to Curl instead. "I could explain this to you in a number of ways," Veech said, "but what you've got there, boys, is a soccer ball."

None of them had seen the obvious. The arrangement of twelve pentagons and twenty hexagons—known more formally as a truncated icosahedron—was identical to the soccer ball's familiar pattern of black and white panels. Heath went to buy one from a sporting goods store, while another student rushed to purchase the campus bookshop's complete supply of molecular modeling kits, which they used to build the first of many models to come.

The next issue was what to call it. They pitched many possible names, mostly using the suffix -ene, which designated a structure based on a ring of alternating single and double bonds. The proposals included "ballene," "spherene," "soccerene," and "carbosoccer," while "footballene" was discarded because it would confuse Americans. While its origins were later disputed, Kroto was almost certainly the one who suggested "buckminsterfullerene," which the others accepted with misgivings. At one point, according to Kroto, Smalley said bluntly, "Your name sucks."

Although it evoked some of Fuller's more outrageous coinages, Kroto pointed out afterward that it was packed with information. It concisely conveyed lightness, strength, and hollowness, and when it was combined with the chemical formula, anyone with a modicum of ingenuity and knowledge of Fuller could deduce the structure. Smalley later noted that it had twenty letters, which corresponded with the faces of an icosahedron, and Curl came to see it as "an unforgettable name for the molecule—a name that stirred the imagination."

In mere days, the team had done what Fuller himself had never managed—it had used geometry to make a fundamental discovery about nature. It was a previously unknown third allotrope of carbon, along with graphite and diamond, in which the arrangement of the chemical bonds produced radically different characteristics. Fuller would have been pleased by the sequence of graphite, which was based on the hexagon; diamond, on the tetrahedron; and C_{60}, on a geodesic sphere. It was the most symmetrical molecule imaginable.

They wrote a short piece for *Nature*, the most prestigious scientific journal, in which they emphasized their approach to the problem of a closed molecule: "Only a spheroidal structure appears likely to satisfy this criterion, and thus Buckminster Fuller's studies were consulted." Buckminsterfullerene, they speculated, might yield a "super-lubricant," based on an analogy to the polymer of carbon and fluorine trademarked as Teflon. As a nucleus for novel compounds, it pointed to unexplored frontiers in chemistry, and since its stability implied that it was distributed throughout the universe, it might have played a role in the origin of life.

Before sending the paper off, they knelt on the lawn outside the Space Sciences building for a group photo, along with a model of buckminsterfullerene and a soccer ball. Kroto wanted to include a picture of Heath's candy framework, but another student had already eaten it. After flying back to Sussex that evening, Kroto immediately looked for his paper star dome, which he confirmed had twelve pentagons, twenty hexagons, and sixty vertices.

It was the most exciting breakthrough in chemistry in a generation. In the months and years that followed, the team learned that a similar structure had been theorized by others, notably the chemists David E. H. Jones and Eiji Osawa. The discovery of the allotrope in a laboratory setting had been a happy accident, like the "inadvertencies" that Fuller invoked, and it shook up the entire field. It also aroused widespread popular interest, and references to "buckyballs"—a name that the team members used among themselves—soon appeared in print.

In the end, it took five years to convince the skeptics, who suspected that it was an elegant hypothesis unsupported by the evidence, like many of Fuller's own assertions. The team performed additional tests, but it was hampered by the difficulty of making macroscopic samples of buckminsterfullerene, which would be a key factor in its case for the Nobel Prize. All the while, Smalley continued to consult the source: "We refined our knowledge in looking at Buckminster Fuller's printed work and the work that his followers have carried on."

Smalley eventually fell out with Kroto, who sometimes joked that he would have to "commit suicide" if they were proven wrong, but a flood of supporting evidence arrived almost all at once in 1990, when a second team demonstrated that buckminsterfullerene could be cheaply produced by heating graphite with electrodes and dissolving the resulting soot in benzene. As the price dropped, the number of publications skyrocketed, and the US Patent Office began to receive more inquiries on the subject than on all others combined.

Throughout the discussion, Fuller's name was justifiably prominent. Kroto acknowledged that his work had guided the team from the beginning: "The geodesic ideas associated with the constructs of Buckminster Fuller had been instrumental in arriving at a plausible structure." A passing mention of Fuller had put them on the right track, and the analogy to the dome, which was sometimes seen as fanciful, was really beautiful and exact. In all domes based on an icosahedral framework, most of the triangles formed hexagonal groups, but exactly twelve pentagons appeared at the vertices of the original icosahedron.

If the faces of a model of buckminsterfullerene were divided into triangles, the result was identical to many small domes, and Fuller and Sadao had made the underlying structure explicit in their Hexa-Pent Dome, which consisted of slightly more than half of a sphere with twelve pentagons and twenty hexagons. On a theoretical level, the analogies to geodesics were just as revealing. With twelve pentagons, a closed structure could be made of any number of hexagons except for one, which Fuller knew well. This property produced an entire family

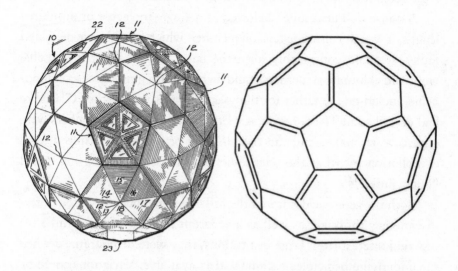

An overhead view of the Hexa-Pent Dome patented by Fuller and Sadao in 1974 (left)
corresponds exactly with the chemical structure of buckminsterfullerene.

of molecules, known collectively as fullerenes, that could take a vast variety of shapes.

Geometry led to other crucial insights. Not every closed carbon molecule was equally stable, and the formation process favored arrangements that eliminated dangling bonds, which Fuller might have characterized as the path of least resistance. When the pentagons touched, the molecules were under greater strain, and C_{60} was the simplest one in which all the pentagons were isolated. The next was C_{70}, which explained the second spike seen in the mass spectra. Kroto nicknamed the pair the Lone Ranger and Tonto, while others called them Bucky and Mrs. Bucky.

In 1996 Kroto, Smalley, and Curl were awarded the Nobel Prize in Chemistry, but the story of buckminsterfullerene had barely begun. For enthusiasts, it heralded the dawn of "round" chemistry, which ushered the science into three dimensions. Fantastic claims were made for its potential applications, and it encouraged researchers to investigate the chemistry of elemental carbon, with many proposals centering on the nanotubes that could be formed by adding rings to the sphere, which Kroto compared to a structure in Fuller's patents.

Because a "buckytube" behaved as a single molecule of unlimited length, it was the ultimate tension member, which would have delighted Fuller. It could conceivably form the basis for the finest imaginable wire, and elaborate structures could be built with nanotubes as components, including the tether for the space elevator that Arthur C. Clarke had described in *The Fountains of Paradise*. In practice, the fibers were expensive to make, and the risk of their proliferation inspired fears of pollution by what the science-fiction author Bruce Sterling called "buckyjunk."

Both the geodesic dome and the fullerenes had benefited from interdisciplinary thinking, as well as a generous amount of luck, and after drawing interest from large institutions, they were democratized when the underlying principles became widely available. Variations could be found by following a few rules, and their symmetry made it easy to theorize. Both were undeniably photogenic, and "buckminsterfullerene" itself was a masterstroke of branding, as Kroto noted: "The name has given it far more visibility across fields outside chemistry, from the arts to architecture, more than any other could have done."

Like the dome, the fullerenes were vulnerable to accusations of hype. The dome's applications turned out to be rather limited, and the fullerenes never lived up to their promise while Kroto and Smalley were alive. They were unsuitable as lubricants—breaking down at high temperatures—and other avenues were obstructed by the challenges of handling materials on a molecular level. If their time ever comes, Fuller will have a secure place in the history of nanotechnology, the ultimate in ephemeralization, which the fullerenes transformed into a serious field.

Buckminsterfullerene went further than Fuller himself ever did in justifying the fundamental nature of geodesics, and it finally made an impact where it all began—in outer space. In 2015, the year before Kroto's death, a spectrum of C_{60} ions was matched with two previously unidentified diffuse interstellar bands, which was subsequently confirmed by data from the Hubble Space Telescope. They were the largest molecules ever detected under such conditions, and the implications were profound, as Smalley had noted much earlier:

> If fullerenes have been present in interstellar space in large enough quantities, they could have provided the first real surfaces in the universe. . . . Molecules could accrete around C_{60} to form the first dust balls, dust balls could form large rocks, rocks could form asteroids, comets, and planets as well as new stars. How fitting if the geodesic shape that provides such flexible stability to modern manmade structures also turns out to have helped build the very first solid things.

Fuller would have seen it as the strongest possible confirmation of his intuitions, and it was a legacy that he truly deserved. As Smalley rightly concluded, "Buckminster Fuller would have loved it."

On January 6, 2019, after receiving the Cecil B. DeMille Award at the Golden Globes, the actor Jeff Bridges told the audience about his personal hero, whose ideas he had encountered through the Hunger Project: "One guy—he had nothing to do with the movies, but I've taken a lot of direction from him—that's Bucky Fuller." After relating the metaphor of the trim tab, Bridges concluded, "All of us are trim tabs. We might seem like we're not up to the task, but we are, man. We're alive! We can make a difference! We can turn this ship in the way we wanna go, man!"

Most viewers were merely bewildered. Fuller's legacy was a complicated one, but his public afterlife was certainly nothing like what he might have hoped. After his death, he left a network without a center, and it was compromised by the strategies—the expansive claims, the messianic language—that he had used to sustain it. Without the propulsive force of his presence, he faded from view, even as his ideas were invisibly absorbed by the culture. As a disillusioned former admirer told Hugh Kenner, "When he dies, it'll all come apart."

The institutions that had dedicated themselves to his memory played only a secondary role in what followed. After Fuller died, his Philadelphia staff lacked the funding to continue, and the Buckminster Fuller Institute relocated to Los Angeles, to be overseen by Jaime Snyder. A conflict erupted over the World Game, with Howard Brown and Medard Gabel facing off against the estate in what Alexandra Snyder called two "warring factions," undermining the program's expansion. Allegra eventually sold Fuller's archives to Stanford, which acquired 1,400 feet of material that took six years to catalog.

There were more prominent tributes elsewhere. Noguchi and Sadao collaborated on a tetrahelix memorial to the crew that perished in the 1986 explosion of the space shuttle *Challenger*, and a team led by Jay Baldwin moved the Wichita House, which had been occupied by raccoons, to the Ford Museum in Michigan. Fuller lent his name to a crater on Mercury; his life was explored onstage and in fiction; and Buzz Aldrin unveiled a postage stamp in his honor in 2004. When Don Ingber's wife asked a post office clerk if he had any for sale, he replied, "Piles of them. Nobody wants them. Who the hell is this guy?"

Although Fuller's influence remained visible by way of such unlikely figures as the New Age writer Carlos Castaneda and the director Oliver Stone, it faltered in the mainstream. After a 2008 exhibit at the Whitney Museum of American Art inspired a brief surge of interest, Nicolai Ouroussoff of the *New York Times* lamented that contemporary architects were preoccupied with "colossuses in China," and he concluded, "Fuller's brand of idealism seems more distant than ever." In reality, Fuller would have loved to build a Chinese colossus, and his impact on architecture endured in subtler ways.

His legacy was most obvious in the style of neo-futurism, where his leading protégé was Lord Norman Foster, who called it "a privilege to spend serious working time with Buckminster Fuller" during the formative years of his practice. He credited Fuller with influencing his Swiss Re Building, better known as the Gherkin, which became a fixture of the London skyline; his dome for the Reichstag building in Berlin; and the "diagrid" framework of the Hearst Tower in New York. Foster quietly referenced his mentor throughout his career, including in his work for Apple, and he even commissioned a loving replica of the Dymaxion Car.

Among architects, Fuller cast an unexpectedly long shadow. "The idea that somebody could actually talk about molecules and talk about buildings and structures and talk about space just amazed me," the environmental architect William McDonough recalled, while Renzo Piano admired him for questioning assumptions "in a very free and non-conformist way." Thom Mayne offered a more backhanded compliment: "Fuller did everything wrong. But he made that work for him."

In the sixties, Fuller predicted that the designer of the future would rely extensively on computing: "He will put the data into the information machines, and it will be processed by automation into physical realization of his effective thinking." This revolution was central to the art of Frank Gehry, the most famous architect of his time. Gehry paid an indirect homage to Fuller in his Dancing House in Prague, which he designed with Vlado Milunić: its roof was crowned with a twisted metal hemisphere, nicknamed Medusa, that resembled a dome caught halfway through an explosion, and it was lifted into place by a helicopter.

Fuller's fans in the design community included James Dyson, Yves Béhar, Roman Mars, and especially Olafur Eliasson, whose work unfolded at the intersection between art and science. For one project, Eliasson worked with Einar Thorsteinn, an Icelandic architect who had known Fuller well, to install two domes outside Copenhagen. At his studio in Berlin, he explored themes derived from Fuller, whom he saw as a model for a civilization that was confronting a series of crises: "He was organizing his present based on what he believed was the future."

Although Fuller influenced a generation of environmentalists, including Bill McKibben and Al Gore, his views on sustainability could have unintended consequences. One admirer was George P. Mitchell, whose energy company drilled thousands of oil wells in Texas. After speaking with Fuller at the Aspen Design Conference, Mitchell felt compelled to investigate the problem of resource scarcity. "A tremendous mind," Mitchell recalled of Fuller. "I was intrigued with him and his concept that the earth could handle only so many people."

As they parted ways in Colorado, Fuller asked, "What are you going to do about it?" Taking the question to heart, Mitchell concluded that alternative energy lacked the capacity to replace fossil fuels, and he pioneered the practice of hydraulic fracturing, in which bedrock is injected with fluid to release natural gas. He became known as "the father of fracking," which he argued was cleaner than coal, and it was entirely consistent with the principle of Spaceship Earth.

Fuller had an even greater impact on John P. Allen, the founder of the Institute of Ecotechnics. In 1982 Fuller and the biologist Richard Dawkins attended a conference at which the architect Phil Hawes presented a concept for a space colony filled with plants. It led to $150 million in funding from the businessman Ed Bass, with the goal of building a closed ecological system to house a human crew for two years. Allen credited its basic "epiphany" to Fuller: "The idea that Bucky had, that we developed, was to make a model of the Earth's biosphere."

They called it Biosphere 2. Peter Pearce, a former colleague of Fuller's, was hired to design and build a space frame that enclosed three acres in Arizona, with a pressure control system housed in two domes. In 1991 its crew of eight entered to considerable fanfare, but a rise in carbon dioxide forced them to artificially introduce oxygen. As Ghillean Prance, a botanist for the project, observed of the Biospherians, "They are visionaries. And maybe to fulfill their vision they have become somewhat cultlike. But they are not a cult per se."

In the space program, the octet truss appeared in NASA proposals, while tensegrity showed genuine potential in robotics. Geodesic structures were used in the Hawaii Space Exploration Analog and Simulation, the Mars Science City under construction outside Dubai, and even films like *The Martian*, but despite their visual appeal, domes would be poorly suited for Mars. Aside from issues with leaks, they would be difficult to connect or expand, and tensional forces would cause most large domes to rip themselves out of their foundations in low gravity.

On earth, the dome endured in popular culture, but it had relatively few practical applications. Apart from playgrounds, it could be seen at the annual Burning Man festival and in scientific projects that benefited from its advantages in cold climates, as well as in more doubtful settings. A glass sphere loomed over a factory that the Taiwanese technology company Foxconn promised but failed to establish in Wisconsin, while the notorious Fyre Festival offered lodging in "geodesic domes"—although they were actually repurposed tents from the Federal Emergency Management Agency.

Because of their usefulness as temporary housing, domes had an uncanny way of appearing in news coverage during events that Fuller would have seen as opportunities for emergence through emergency. They served as shelters after Hurricane Katrina and overflow morgues for victims of the coronavirus, which exposed social disparities that design alone was unable to address. When a restaurant in San Francisco installed domes to allow patrons to dine outdoors during the pandemic without being troubled by the homeless, the food critic Soleil Ho observed, "We're talking about domes because this country doesn't take care of us."

Not surprisingly, Fuller's example found its most lasting home in Silicon Valley. Adam Neumann, the founder of WeWork, pitched his coworking start-up using the language of ephemeralization: "We want to be known for being a company that does more with less." It held staff meditation sessions in a dome and announced a "future cities initiative" by the Israeli designer and artist Dror Benshetrit, who consulted on redeveloping the Montreal Biosphere, but a fumbled public offering resulted in the ouster of Neumann, who had warned his employees, "The universe does not allow waste."

As Jaron Lanier noted, during his lifetime, Fuller had been famous "maybe in the way of a figure like Elon Musk." Fuller and Musk both promoted themselves as outsiders who disrupted established fields—including the automotive industry—and presented a dream of innovation to counter the inability of existing systems to effect change. Both were also contemptuous of critics, obsessed with their press coverage, willing to pull projections out of thin air, and eager to portray themselves as martyrs, although Fuller concealed his flaws more capably than Musk, who has occasionally spoken of using geodesic domes on Mars.

The most fascinating parallel of all was with Amazon. On January 29, 2018, Jeff Bezos dedicated the centerpiece of the company's Seattle headquarters, which consisted of three domes "where employees can think and work differently surrounded by plants." He also partnered with Stewart Brand on a clock that would keep accurate time for ten thousand years, in the kind of giant pedagogical object that Fuller could never afford—a Geoscope for enormous timescales. "Ideally," Brand said of the Clock of the Long Now, "it would do for thinking about time what the photographs of Earth from space have done for thinking about the environment."

Amazon itself had even deeper affinities to Fuller. As a vast process company, it exemplified both ephemeralization—with a massive infrastructure that went unseen by its customers—and Fuller's conviction that economic forces favored gargantuan service industries. "Industrialization must be recognized and operated on a total world and total humanity basis, or it is nothing," Fuller said. "Industrialization involves all the resources of the earth, all the knowledge and all the experience of all men everywhere and involves everybody on earth as the logical clients."

Fuller's claim that he needed an entire distribution system to sell the Wichita House had been little more than a convenient excuse, but for Jeff Bezos—who represented the concentration of wealth that his predecessor saw as a danger to society—it was a practical strategy. Bezos wanted to control the supply chain from one end to the other, and his obsession with carrying packages over the last mile demanded a move from track to trackless. Amazon's smart devices, drones, fleet of vans, and wireless delivery of content all aimed at the same goal, while its most profitable business was selling ephemeralization in the form of cloud computing.

Although the success of Amazon highlighted the power of these ideas, it also exposed their negative consequences. Ephemeralization never produced the revolution in housing that Fuller wanted, but it came true for virtually everything else, and its damaging effects were often pushed out of sight. As Lloyd Kahn had foreseen, the question of doing more with less had to account for the systems and raw materials on which technological innovations depended, and the invisible costs of a cell phone, lithium battery, or unit of cryptocurrency were even greater than those of the dome.

Recent events have only confirmed that Fuller's idea of emergence through emergency was fundamentally incomplete. He predicted correctly that inequality would cause social disruption, but he focused on disparities between countries, not within them, and the results in America have been disheartening. After CHARAS outgrew its dome, it turned an abandoned school in New York into an arts and community center. Two decades later, the administration of Mayor Rudolph Giuliani auctioned it off, and the activists were evicted by police. The site remains vacant to this day.

These externalities become more obvious at times of crisis. The rise of the sharing, or gig, economy allowed companies to operate using minimal infrastructure and a workforce of contractors, which was close to Fuller's ideal, but it also diminished protections for employees, and a lack of meaningful regulation ensured that its price was extracted from the most vulnerable. Millennials ephemeralized against their will, with rent replacing ownership and housing costs contributing to a fall in mobility. As Airbnb caused prices in its neighborhoods to rise, its single most popular property was a geodesic cabin in California.

The greatest test of Fuller's philosophy was the coronavirus pandemic, which initially seemed like a definitive moment of emergence through emergency. Fuller blamed his daughter's death, which changed his life, on the conditions that contributed to the Spanish flu, and his views on decentralization, efficient manufacturing, online education, and remote working are more relevant now than ever. He advocated replacing doorknobs with sensors to reduce contact with germs; hoped to eliminate "seventy percent of local commuting" by the eighties; and produced ideas of immense value for the challenge of building a more dispersed and flexible society.

At the same time, the ephemeralization of lean manufacturing damaged supply chains and caused shortfalls of essential materials, and in the United States, the pandemic was politicized at once. Fuller believed that computers could persuade politicians to reconsider their positions, although he ignored the fact that elected officials prefer models that tell them what they want to hear, while programs and algorithms tend to reflect the biases of their creators. Even if an emergency leads to change, disaster capitalism favors existing lines of power, and ambitious individuals can use catastrophes to further their own goals—as Fuller did during two world wars.

As other crises, especially climate change and economic inequality, continue to resist design solutions, the answer evidently lies in political transformation. Fuller avoided this issue, which allowed his corporate sponsors to see him as provocative but unthreatening, and by portraying politicians as irrelevant, he implicitly absolved them of responsibility. He praised automation as a liberating force, but in the absence of social policy reform, it has led to layoffs, dehumanizing work, and a decline in real wages. The true design-science revolution depends on a culture that commits first to caring for its entire population, and not the other way around.

This is the exact opposite of what Fuller claimed, largely because his career demanded it, but the evidence is overwhelming. Between the Green Revolution, Moore's Law, and advances in energy efficiency, the accelerating breakthroughs that he predicted have come to pass, but instead of enabling mankind to make logical choices, they have deepened existing divides. As Fuller recognized, the machine of capitalism moves wealth in one direction, rewarding the very rich while discouraging the distribution of resources to those most in need.

More recently, technologists such as Musk and Andrew Yang have paired the issue of automation with a universal basic income, which looks increasingly like the best candidate for the first move in the World Game. "We should not be haunted by the specter of being automated out of work," the New York congresswoman Alexandria Ocasio-Cortez has said. "We should be excited by that. But the reason we're not excited by it is because we live in a society where if you don't have a job, you are left to die." This requires confronting long-standing obstacles of race and class, which calls for activists who can use the tools of ephemeralization in ways that Fuller never imagined.

Fuller's ideas are still powerful, but only when integrated into a view of human relations that expands on his moral universe. In the end, his most frightening insight may have been that a political emergency could catalyze the dissolution of nations, which he compared to "blood clots" that prevented the rational allocation of wealth. He argued for years that humanity's survival depended on desovereignization, which would occur in America after a traumatic social breakdown. Before Richard Nixon's reelection in 1972, Fuller said, "I feel that his lack of real vision means that the United States will come apart more rapidly. That's all. It has to go."

Yet there was an even more difficult renunciation that Fuller was inherently unable to endorse. His ideal of the design visionary, taken to another level by Steve Jobs and his successors, is no longer sufficient in itself to address the challenges of an increasingly complex and connected world. As nostalgic as we might be for an era of expansive solutions, our real goal should be a society that can endure without the cult of the comprehensive designer, even if that myth inspires some of us to do our best work. The ultimate ephemeralization lies in transcending men like Fuller altogether, and this final leap could take place only after he was gone.

Change often happens despite such figures, not because of them, and although Fuller offers an urgent case study, many other players in his career are equally instructive. Perhaps the most remarkable of all was Kiyoshi Kuromiya, who became a charter member of ACT UP Philadelphia after being diagnosed with HIV in 1989. Operating out of his apartment, he founded the Critical Path Project, which provided information and resources to people with AIDS through a newsletter and online bulletin board. He single-handedly realized the core premise of the World Game, and he pursued it tirelessly—using both politics and technology—until he died from complications of AIDS at fifty-seven.

Fuller himself remains indispensable, both as a role model and as a cautionary tale, and his story is far from over. As the critic William Kuhns once wrote, "Marx's ideas changed history only in the century after he lived and died. And who can say yet of Fuller?" His deepest insights deserve to be part of every thinking person's life, and he embodied all the contradictions of our future. "I do want you to understand that I do know what I'm talking about," Fuller said in the seventies. "This is a very critical period. Nature is trying very hard to make us succeed, but nature does not depend on us. We are not the only experiment."

ACKNOWLEDGMENTS

> Because the data constitutes a faithfully comprehensive record,
> I am now able to comment objectively upon my subjectively
> disclosed self, approximately as critically as though the sub-
> ject were another man. As with any book when my subject is
> prospering, I am glad, and when he is unprosperous, I am
> sad. That is the extent of my prejudice.
>
> —BUCKMINSTER FULLER

My greatest debt is to the late Allegra Fuller Snyder, as well as
to Jaime Snyder, Alexandra May, and Lucilla Fuller Marvel, who
were generous with their time, resources, and feedback, even when
they disagreed with my conclusions, which do not reflect the views
or opinions of the Buckminster Fuller estate. I am also grateful to John P. Allen, Taylor Barcroft, Chuck Beatty, Betty Beleny,
Lesley Beleny, Michael Ben-Eli, Trevor Blake, Howard J. Brown,
Peter Chermayeff, Hsiao-Yun Chu, Joseph D. Clinton, the late Harold Cohen, Robert Curl, Michael Denneny, Annie Dillard, Amy
Edmondson, David A. Edwards, Olafur Eliasson, Katrina Howard Fairley, C. J. Fearnley, Lee Felsenstein, Brian Ferree, Dick
Fischbeck, Lord Norman Foster, Medard Gabel, Karen Good-

man, Neva Goodwin, Che Gossett, Robert W. Gray, Sam Green, Rob Grip, Tony Gwilliam, Rich Heinemeyer, Paul Hendrickson, Andy Hertzfeld, Bill Higgins, Dan Hooper, Tony Huston, Donald E. Ingber, Lloyd Kahn, Barry M. Katz, Alan Kay, Dodie Kazanjian, Jonathon Keats, Lisa Kessler, Christopher Kitrick, Dale Klaus, Daniel Kottke, Lady Margaret Kroto, Jaron Lanier, Mira Lehr, Lore Devra Levin, David D. Levine, Lim Chong Keat, Casey Mack, Jerry Manock, John Markoff, Gillian Meller, Ann Mintz, Federico Neder, Bill Perk, Prop Anon, Michael Ravnitzky, Michael Reid, Hugh Ryan, the late Shoji Sadao, Edwin Schlossberg, Stephen E. Selkowitz, Shirley Sharkey, Deborah Snyder, Tom Solari, the late Carl Solway, Yrjö Sotamaa, Barbara Stevenson, Jonathan Stoller, Roger Stoller, Calvin Tomkins, Thomas Turman, Tom Vinetz, Tim Wessels, Robert Williams, Paul Young, the late Gene Youngblood, Thomas T. K. Zung, and many others.

On the scholarly side, I owe the most to Bonnie DeVarco, my unfailing guide to the Chronofile, and to Lionel Wolberger, whose meticulous engagement with the details of Fuller's life fills me with admiration. This book would have been unthinkable without the Stanford Libraries, particularly Tim Noakes and the staff at Special Collections, and my indefatigable research assistants, especially Delaney Holton, Kyla Walker, Michael Rendon, and Kimi Bito in Palo Alto, as well as Jennifer Gersten in New York and Gary Stein at the University of Southern California. John Ferry, Josh Pang, and Faith Flanigan of the Buckminster Fuller Institute provided valuable assistance, as did Jessica Tubis at the Beinecke Rare Book and Manuscript Library at Yale; the Benson Ford Research Center; the Cook County Clerk's Office; Aryn Glazier at the Dolph Briscoe Center for American History; Katy Harris and the rest of the staff at Foster + Partners; Indira Berndtson at the Frank Lloyd Wright Foundation; Virginia Mokslaveskas at the Getty Research Institute; the Harvard University Archives; Geo Rule at the Heinlein Archive; Laura Pearle at Milton Academy; Debbie Rafine at the Oak Park Public Library; Jennifer A. Bryan at the US Naval Academy; the Special Collections Research Center at the University of Chicago; and Anna Schmidt at the University of Illinois at Chicago. I have also benefited enormously from the efforts of previous scholars, notably Yunn Chii Wong and the late Loretta Lorance.

David Halpern, my agent, was as passionate an advocate as always, and I'm thankful for the support of Kathy Robbins, Janet Oshiro, everyone at the Robbins Office, and Jon Cassir at CAA. My editors, Carrie Thornton and Kate Napolitano, believed in this project and guided it superbly to completion, along with Rosy Tahan, Peter Kispert, Phil Bashe, Angela Boutin, Paul Miele-Herndon, David Palmer, Andrew Clark, Eliza Rosenberry, Erin Reback Cipiti, and the entire team at Dey Street Books and HarperCollins. This book is dedicated to my parents, who provided me with an upbringing that made this subject seem like a natural choice, and I feel more gratitude than ever for my friends and loved ones, especially my brother, his family, and all the Wongs; my wife, Wailin; and my daughter, Beatrix, who has heard more about Fuller than any fourth grader probably should.

As a teenager, I discovered Fuller in the pages of the *Whole Earth Catalog*, and ever since my freshman year in high school, when I read *Critical Path* and as much as I could manage of *Nine Chains to the Moon*, he has rarely been far from my mind. Writing this book has only made me more aware of how deeply his example shaped my own life. "It probably is synergetically advantageous to review swiftly the most comprehensive inventory of the most powerful human environment transforming events of our totally known and reasonably extended history," Fuller wrote in *Synergetics*, and I have approached this biography in the same spirit. As the members of the first World Game workshop might have said, it seemed like the bare maximum.

For further exploration, I recommend tracking down a used copy of *The Buckminster Fuller Reader* and beginning with "Designing a New Industry," which features Fuller at his most lucid and persuasive. Next would be "Design Strategy" in *Utopia or Oblivion*, followed by the riches of *Synergetics*, starting with the sections "Synergy," "Considerable Set," and "Tetrahedron." *Cosmic Fishing* by E. J. Applewhite is a delightful memoir by Fuller's most observant colleague, while the images in *The Dymaxion World of Buckminster Fuller* or the more recent *Your Private Sky* offer plenty of material for dreams. Finally, the *Whole Earth Catalog* remains unsurpassed as a guide for aspiring generalists, and I suspect that it still has the potential to change a curious person's life as profoundly as it once did for me.

NOTES

For the period before 1951, the dates and details of events in Fuller's life can be verified most directly by consulting the Dymaxion Chronofile at Stanford University, especially when paired with the collection's exhaustive finding aid. After the early fifties, when the expansion of the archives makes this approach less practical, the most useful starting point is Fuller's increasingly detailed itinerary in the Dymaxion Index ("Section 11. Schedule of Lectures, Projects and Days," R. Buckminster Fuller Papers, Department of Special Collections and University Archives, Green Library, Stanford, CA, M1090, Series 4, Subseries 13, Mixed Materials 6–7). To keep these notes to a manageable length, citations for the archives are provided only for direct quotations and when the location of the original source might be unclear to researchers with access to the finding aid and itinerary.

Unless indicated otherwise, all references to boxes, volumes, and folders in the Stanford archive refer to the Dymaxion Chronofile in the R. Buckminster Fuller Papers, M1090, Series 2. References to Applewhite's *Synergetics Dictionary* follow the numbering system of the version made available by R. W. Gray at "Synergetics Dictionary Online" (http://www.rwgrayprojects.com/SynergeticsDictionary/SD.html). Full transcripts of the *Everything I Know* lectures were accessible at the time of writing at the website of the Buckminster Fuller Institute (https://www.bfi.org). Except where necessary to convey aspects of the writer's thinking or personality, spelling and punctuation have generally been standardized in quotations from letters and other primary sources.

ABBREVIATIONS

AHF	Anne Hewlett Fuller
AMS	R. Buckminster Fuller, *An Autobiographical Monologue/Scenario*
Anthology	R. Buckminster Fuller, *Anthology for the Millennium*
BFR	R. Buckminster Fuller, *The Buckminster Fuller Reader*
"BV"	Loretta Lorance, "Building Values: Buckminster Fuller's Dymaxion House in Context" (dissertation)
CP	R. Buckminster Fuller, *Critical Path*
Discourse	R. Buckminster Fuller, *Your Private Sky: Discourse*
EIK	R. Buckminster Fuller, *Everything I Know*
4DT	R. Buckminster Fuller, *4D Time Lock*
Harvard file	Student folder of Richard Buckminster Fuller, Harvard University Archives, Undergraduate Student Records, UAIII 15.88.10 1890–1968, box 1690
Hatch	Alden Hatch, *Buckminster Fuller: At Home in the Universe*
"Influences"	RBF, "Influences on My Work," in *Ideas and Integrities*, 9–34
Itinerary	"Dymaxion Index—Section 11. Schedule of Lectures, Projects and Days" (R. Buckminster Fuller Papers, M1090, Series 4, Subseries 13, Mixed Materials 6–7)
Lorance	Loretta Lorance, *Becoming Bucky Fuller*
Marks	R. Buckminster Fuller and Robert Marks, *The Dymaxion World of Buckminster Fuller*
RBF	Richard Buckminster Fuller
Sadao	Shoji Sadao, *Buckminster Fuller and Isamu Noguchi: Best of Friends*
SD	R. Buckminster Fuller, *Synergetics Dictionary: The Mind of Buckminster Fuller*
UOO	R. Buckminster Fuller, *Utopia or Oblivion*

Wong   Yunn Chii Wong, "The Geodesic Works of Richard Buckminster Fuller" (dissertation)

PROLOGUE: GEODESIC MAN

1 **"I've got Bucky Fuller here":** Taylor Barcroft, interviewed by author, July 1, 2020.

1 **Apple Computer:** RBF's itinerary and correspondence with Taylor Barcroft (B426-F4) indicate that this tour was originally scheduled to take place at Atari. Barcroft decided shortly beforehand to visit Apple instead (Barcroft, author interview).

1 **"I knew Steve":** Barcroft, author interview.

2 **"Buckminster Fuller's here":** Daniel Kottke, interviewed by author, May 21, 2019.

2 **driver's license:** RBF driver's license, issued June 29, 1936, B35-V54.

3 **"He didn't believe it":** Barcroft, author interview.

3 **"In my early friendship":** Daniel Kottke, interviewed by author, March 20, 2019.

3 **"make the world work":** RBF, "Keynote Address at Vision 65," in *Utopia or Oblivion: The Prospects for Humanity* (New York: Bantam Books, 1969), 125. Referred to hereafter as *UOO*.

3 **"an emerging synthesis":** RBF, "The Comprehensive Designer," in *Your Private Sky: Discourse,* ed. Joachim Krausse and Claude Lichtenstein (Zurich: Lars Müller, 2001), 243. Referred to hereafter as *Discourse.*

3 **"one of the bibles":** Steve Jobs, Stanford commencement address, June 12, 2005, https://news.stanford.edu/2005/06/14/jobs-061505 (accessed January 2021).

4 **Apple lobby:** Jerry Manock, interviewed by author, March 13, 2019.

4 **"The insights":** Stewart Brand, *Whole Earth Catalog,* Fall 1968, 3.

4 **"Google in paperback form":** Jobs, Stanford commencement address, June 12, 2005.

4 **"That definitely comes":** Stewart Brand interview, *The Tim Ferriss Show,* February 3, 2018, https://tim.blog/2018/02/03/the-tim-ferriss-show-transcripts-stewart-brand (accessed January 2021).

4 **"The stuff came out":** Evgeny Morozov, "Making It," *New Yorker,* January 13, 2014, 70.

4 **Marc LeBrun:** John Markoff, *What the Dormouse Said* (New York: Penguin, 2006), 183–84.

4 **Lee Felsenstein:** Lee Felsenstein, email to author, December 18, 2020.

4 **"an architect and engineer":** Charles Raisch, "Pueblo in the City," *Mother Jones,* May 1976, 30.

4   **ideas for the Exploratorium:** Hilde Hein, *The Exploratorium: The Museum as Laboratory* (Washington, DC: Smithsonian Institution, 1990), 239n10.

5   **Efrem Lipkin:** Steven Levy, *Hackers: Heroes of the Computer Revolution* (Sebastopol, CA: O'Reilly Media, 2010), 164–65.

5   **Ken Colstad and Mark Szpakowski:** "St. Jude Memorial and Virtual Wake," https://people.well.com/conf/inkwell.vue/topics/190/St-Jude-Memorial -and-Virtual-Wak-page01.html (accessed April 2021).

5   **"a textbook example":** Levy, *Hackers*, 219.

5   **grounded in Fuller's philosophy:** Barry Katz, *Make It New: The History of Silicon Valley Design* (Cambridge, MA: MIT Press, 2015), 121. In the fifties, RBF lectured three times at Arnold's "Creative Engineering" course at MIT. R. Buckminster Fuller Papers M1090, Department of Special Collections and University Archives, Green Library, Stanford, CA, "Section 11. Schedule of Lectures, Projects and Days," Series 4, Subseries 13, Mixed Materials 6–7, June 25, 1954, June 25, 1955, and June 22, 1956. Referred to hereafter as Itinerary.

5   **"He wasn't interested":** Manock, author interview. When Manock opened his consulting business after college, he printed business cards that said "Manock, Comprehensive Design" (Manock, email to author, January 11, 2021).

6   **"He was kind of an intuitive":** Alan Kay, interviewed by author, December 3, 2019.

6   **"My dad was more":** Jennifer Kahn, "The Visionary," *New Yorker*, July 11, 2011, 53.

6   **"very much in reference to Fuller":** Jaron Lanier, interviewed by author, March 12, 2019.

6   **"The way everything is triangulated":** Ibid.

6   **"enchanted by Buckminster Fuller":** Jaron Lanier, *Dawn of the New Everything: Encounters with Reality and Virtual Reality* (New York: Henry Holt, 2017), 86.

6   **"For those who have context":** Lanier, author interview. Andy Hertzfeld of Apple recalled RBF as a more marginal presence: "I was vaguely familiar with Fuller's work and have a hazy recollection of folks talking about him at Apple, but I don't remember anything specific" (email to author, March 23, 2019). Another figure to cite RBF as an influence was Ted Nelson, the author of the 1974 book *Computer Lib*, who was inspired by the word *Dymaxion* to coin the term *hypertext* (Michael Fraase, *Macintosh Hypermedia*, vol. 1, *Reference Guide* [Glenview, IL: Scott, Foresman, 1990], 31).

6   **"prophetical voices":** Theodore Roszak, *From Satori to Silicon Valley* (San Francisco: Don't Call It Frisco, 1986), 18.

6   **"Fuller and his Bay Area disciples":** Ibid., 20–21.

6   "cut from the mold": Ibid., 35.

6   "the twentieth century's Leonardo da Vinci": Steve Wozniak, blurb for Lloyd Steven Sieden, ed., *A Fuller View: Buckminster Fuller's Vision of Hope and Abundance for All* (Studio City, CA: Divine Arts, 2011). Wozniak was echoing a frequent comparison, with the most common version— "the Leonardo da Vinci of the twentieth century"—often attributed to Marshall McLuhan, who actually praised RBF as "the twentieth century Pythagoras" (McLuhan to Bruce Carrick, September 10, 1974, B375-V6).

7   Janek Kaliczak: Itinerary, February 1, 1982.

7   "the Han Solo of the Apple world": David Hunter, "Graphics in the Sky with Diamond Vision," *Softalk*, July 1983, 88.

7   "a graphical display": George Madarasz to RBF, March 31, 1983, B482-F1.

7   "but maybe better": Steve Wozniak with Gina Smith, *iWoz: Computer Geek to Cult Icon—How I Invented the Personal Computer, Co-founded Apple, and Had Fun Doing It* (New York: W. W. Norton, 2006), 246.

7   Fuller and the World Game: Hunter, "Graphics in the Sky," 86.

8   legacy at Apple: Jobs later hired RBF's former associate Norman Foster to design the company's corporate headquarters. Foster thought that RBF was only "lurking in the background" of Apple Park (Norman Foster, interviewed by author, October 30, 2020), but he credited him afterward for inspiring the glass dome of the Apple Marina Bay Sands store in Singapore (Norman Foster, "Buckminster Fuller—File for Future," provided to the author by Foster).

8   "Here's to the crazy ones": "Apple—Think Different—Full Version," YouTube, 1:09, Harry Piotr, September 30, 2013, https://www.youtube .com/watch?v=5sMBhDv4sik.

8   request of Jobs himself: Luke Dormehl, *The Apple Revolution: The Real Story of How Steve Jobs and the Crazy Ones Took Over the World* (London: Virgin, 2012), 395, and Leander Kahney, *Inside Steve's Brain* (New York: Portfolio, 2008), 130.

8   "You can quote them": "Apple—Think Different—Full Version," YouTube, https://www.youtube.com/watch?v=5sMBhDv4sik.

8   "Apart from the temporary notoriety": Alden Hatch, *Buckminster Fuller: At Home in the Universe* (New York: Crown, 1974), 1. Referred to hereafter as Hatch.

9   Spaceship Earth: RBF compared the earth to a spaceship as early as 1951 ("Fuller Lectures on Motion Trend," *Kingston [NY] Daily Freeman*, July 25, 1951, 9). The first notable use of the exact phrase was evidently by James E. Webb, the head of NASA, in a speech in 1962: "We are all space travelers on the good spaceship Earth" (Jean Pearson, "You're a Spaceman, Too," *Detroit Free Press*, April 7, 1962, 24). Its popularization was overwhelmingly due to RBF, who began to use it frequently around 1967.

9   "My objective": RBF, with Jerome Agel and Quentin Fiore, *I Seem to Be a Verb* (New York: Bantam Books, 1970), 9.

10 All of these "artifacts": RBF, *Education Automation: Freeing the Scholar to Return to His Studies; A Discourse Before the Southern Illinois University, Edwardsville Campus Planning Committee, April 22, 1961* (Carbondale: Southern Illinois University Press, 1962), 12.

10 "They appear exciting": Lloyd Kahn and Bob Easton, *Shelter II* (Bolinas, CA: Shelter, 2010), 200.

11 created a convoluted language: One saying often credited to RBF is "Think globally, act locally" (for example, on the dust jacket of *Discourse*). It appears nowhere in his published work, and the attribution may have originated in a quote from his biographer Lloyd Steven Sieden: "There's a phrase today: 'Think globally, act locally,' and Fuller was talking about that 40 years ago" (Dennis McLellan, "Futurist Provides Fuller Explanation for Passengers on Spaceship Earth," *Los Angeles Times*, July 7, 1989, pt. 9, 6).

11 knew the word *synergy*: RBF, *Operating Manual for Spaceship Earth* (1969; New York: E. P. Dutton, 1978), 73.

11 "Nothing—an electromagnetic field": RBF, "Domes," in *Synergetics Dictionary: The Mind of Buckminster Fuller,* comp. and ed. E. J. Applewhite, 4 vols. (New York: Garland, 1986), 4186. Referred to hereafter as *SD*.

12 "Ephemeralization trends toward": RBF, *Critical Path*, Kiyoshi Kuromiya, adjuvant (New York: St. Martin's Press, 1981), 234. Referred to hereafter as *CP*.

12 "the increasing tendency": Paul Graham, "Tablets," December 2010, http://www.paulgraham.com/tablets.html (accessed January 2021).

12 page in his wallet: Rich Heinemeyer, interviewed by author, February 17, 2020.

12 "We owe the notion": Yves Béhar interview, "Buckminster Fuller: Utopia Rising," https://www.nowness.com/story/buckminster-fuller-utopia-rising (accessed February 2021).

13 "emergence through emergency": RBF, *Nine Chains to the Moon* (Philadelphia: J. B. Lippincott, 1938), 240–46.

13 "The individual can simply": RBF, "Bureaucracy," in *SD*, 01674.

13 largest of its kind: Robert Kahn and Medard Gabel, "The Fuller Archives and the Dymaxion Index," Series 4, B5-F1, 2.

13 "the ethereal world of software": Derek Thompson, "Where's My Flying Car?," *Atlantic*, January/February 2020.

13 most influential futurist: George Thomas Kurian and Graham T. T. Molitor, eds., *Encyclopedia of the Future,* vol. 2 (New York: Macmillan, 1996), 1077.

13 **"There hasn't been anybody":** Philip Johnson, interviewed in *Buckminster Fuller: Thinking Out Loud*.

14 **"to revert to saying":** RBF and Anwar Dil, *Humans in Universe* (New York: Mouton, 1983), 95.

14 **"above all a mystic":** Hatch, 2.

14 **"Blacks skillfully persuaded":** RBF, "Race," in *SD*, 14611.

14 **"I tell students to stop":** RBF, "The Designers and the Politicians," in *The Buckminster Fuller Reader*, ed. James Meller (Harmondsworth, UK: Penguin, 1972), 373. Referred to hereafter as *BFR*.

14 **"more square feet":** Douglas Auchincloss, "The Dymaxion American," *Time*, January 10, 1964, 46.

14 few thousand permanent structures: RBF stated that around two thousand geodesic domes were installed around the world by 1961 (*CP*, 391, and Yunn Chii Wong, "The Geodesic Works of Richard Buckminster Fuller, 1948–1968 (The Universe as a Home of Man)" [doctoral thesis, Massachusetts Institute of Technology, 1999], 395; referred to hereafter as Wong). By 1965 his estimate rose to six thousand, which included playdomes ("Emergent Humanity," in *R. Buckminster Fuller on Education*, ed. Peter H. Wagschal and Robert D. Kahn [Amherst: University of Massachusetts Press, 1979], 109). According to Calvin Tomkins, the number around this time was "more than three thousand" ("In the Outlaw Area," *New Yorker*, January 8, 1966, 73), while a 1969 profile raised the number to more than five thousand (Fred Warshofsky, "Meet Bucky Fuller, Ambassador from Tomorrow," *Reader's Digest*, November 1969, 199). RBF's original dome patent expired in 1971. These numbers contrast starkly with the unverified claims of two hundred thousand (RBF, "The Way We Live: Reflections and Projections," *Architectural Digest*, June 1978, 24) or even three hundred thousand domes, including a hundred thousand playdomes (RBF and Dil, *Humans in Universe*, 86).

15 his masterpiece: Kurt Soller and Michael Snyder, "The 25 Most Significant Works of Postwar Architecture," *New York Times Style Magazine*, August 2, 2021.

15 **"The way the world":** Edwin Schlossberg, interviewed by author, March 27, 2019.

15 **"Everyone who was close to him":** Harold Cohen, interviewed by author, March 6, 2021.

15 **"the planet's friendly genius":** William Overend, "Fuller, 'the Planet's Friendly Genius,'" *Los Angeles Times*, March 2, 1977, pt. 4, 1.

15 **"a public relations tool":** Loretta Lorance, *Becoming Bucky Fuller* (Cambridge, MA: MIT Press, 2009), 228. Referred to hereafter as Lorance.

15 overflowed with misinformation: In addition to the errors documented in the text, RBF incorrectly stated that the mass of the earth was increasing

due to its accumulation of "stardust," which would end in the planet becoming a star itself (RBF, *Intuition* [Garden City, NY: 1972], 105 and 115); that "the relative abundance of chemical elements" in the body was "congruent" with their percentages in the universe (RBF, *Everything I Know*, session 3, Amazon Kindle edition, 2019; referred to hereafter as *EIK*); and that ships were drawn to each other at close proximity because of gravitational attraction (RBF, *EIK*, session 6). For his errors in the geometry of chemical bonds, see Hugh Kenner, *Bucky: A Guided Tour of Buckminster Fuller* (New York: William Morrow, 1973), 118–20 and 277. On his misleading description of the pattern of buckling in spherical shells observed at General Dynamics, see Tibor Tarnai, "Geodesic Domes: Natural and Man-Made," *International Journal of Space Structures* 26, no. 3 (September 2011): 8.

16   **"mythologize his childhood":** E. J. Applewhite, *Cosmic Fishing: An Account of Writing* Synergetics *with Buckminster Fuller* (New York: Macmillan, 1977), 81.

16   **"In the future":** Lloyd Steven Sieden, *Buckminster Fuller's Universe: An Appreciation* (1989; Cambridge, MA: Perseus, 2000), xiii.

16   **world's unrealized ability:** The architectural theorist Sanford Kwinter described this challenge as "the marriage of fixed and intellectual capital," and he concluded, "To get there, we must pass through Fuller" (Kwinter, "Fuller Themselves," *ANY: Architecture New York* 62, no. 17 [1997]: 62).

16   **"It has always seemed to me":** Calvin Tomkins to RBF, January 5, 1977, Calvin Tomkins Papers, Museum of Modern Art, folder I.13.

17   **"When you've made":** Tomkins, "In the Outlaw Area," *New Yorker*, January 8, 1966, 78.

17   **"Because I live":** RBF, "Design Strategy," in *UOO*, 335–36.

PART ONE: ORIGINS

19   **"It is one of the strange facts":** RBF with E. J. Applewhite, *Synergetics: Explorations in the Geometry of Thinking* (New York: Macmillan, 1975), 278.

ONE: NEW ENGLAND

21   **"The tactile is very unreliable":** Ibid., 441.

21   **Richard Buckminster Fuller Jr.:** Sources consulted on RBF's childhood include Hatch, 11–30; Athena V. Lord, *Pilot for Spaceship Earth* (New York: Macmillan, 1978), 5–17; RBF, "What It Was Like to Grow Up in the Boston Area," May 1, 1979, Series 1, B2-F2; RBF, in Jeffrey L. Lant, ed.,

Our Harvard: Reflections on College Life by Twenty-two Distinguished Graduates (New York: Taplinger, 1982), 3–11; Miriam Rumwell, *Time* profile notes, Takes 1 and 2, December 23, 1963, B126-F4; and the chronology in *CP*, 379–80.

21 **"And down will come baby":** RBF, "Vertical Is to Live, Horizontal Is to Die," in *Discourse*, 280.

22 **his older sister, Leslie:** RBF's older sister was baptized Caroline Lesley Fuller. She used the spelling "Lesley" until at least 1916, but "Leslie" became more common in primary sources by the time of her second marriage in 1927. For the sake of consistency, "Leslie" will be used throughout this book.

22 **by their smells:** RBF, *EIK*, session 6.

22 **"So every time":** "R. B. Fuller Lecture at San Quentin Prison," January 31, 1959, Series 8, B15-V11, 12.

22 **"Slow and Solid":** RBF to J. Bryan III, July 1, 1939, B29-V46.

22 **Friedrich Froebel:** "Froebel's Gifts," *99% Invisible*, April 9, 2019, https://99percentinvisible.org/episode/froebels-gifts (accessed September 2021).

23 **"When I got to the triangle":** RBF, interviewed by Hans Meyer, in Lloyd Kahn, *Domebook 2* (Bolinas, CA: Pacific Domes, 1971), 90. As noted in chapter 7, RBF said elsewhere that he built a full "octet truss" (RBF with E. J. Applewhite, *Synergetics 2: Explorations in the Geometry of Thinking* [New York: Macmillan, 1979], 232). A handwritten note about his teacher, Grace Parker Hatch, refers more plausibly to his kindergarten construction as his "first tetra" (B1-V1).

24 **"I love and trust you":** RBF and Dil, *Humans in Universe*, 75.

24 **picture book:** Fred M. Taylor, "The 'Bubble House' and Other Summer Projects," *Student Publications of the School of Design*, North Carolina State College, Fall 1952, 24.

24 **"to be out in that green grass":** RBF, "Intuition of the Child," in *SD*, 8747.

24 **"the coming years":** RBF, *CP*, 57.

24 **raised as a Unitarian:** RBF himself would occasionally be described as a Unitarian, but he was raised in the Episcopal Church, and his beliefs in adulthood were too idiosyncratic to be easily categorized.

24 **smell of incense:** RBF, *EIK*, session 11.

24 **"a little liar":** RBF, *R. Buckminster Fuller: An Autobiographical Monologue/Scenario* (New York: St. Martin's Press, 1980), 9. Documented and edited by Robert Snyder. Referred to hereafter as *AMS*.

24 **"So to avoid":** Ibid.

24 **"the love of its conceiver":** RBF, "The Bear Island Story," Series 7, B6-F14, 33.

24 **Spartan boy:** Applewhite, *Cosmic Fishing*, 150.

24 **new moon:** Ibid., 103.

24 **burned up her nightgown:** RBF note to Ann Mintz (undated), Series 1, B2-F2.

25 **Great Chicago Fire:** Ibid.

25 **elementary school:** Educational timeline based on RBF résumé in B8-V16.

25 **"Bucky, when I was in first grade":** Hatch, 14.

25 **"an act which I have regretted":** RBF to David Travers, January 18, 1982, B489-F2.

25 **"he would pick up":** RBF, *Grunch of Giants* (1983; Santa Barbara, CA: Design Science, 2008), 56.

25 **family home:** Architectural background on the Milton house and the Big House on Bear Island is drawn from Margaret Henderson Floyd, *Architecture After Richardson: Regionalism Before Modernism— Longfellow, Alden, and Harlow in Boston and Pittsburgh* (University of Chicago, 1994), 123–25 and 430–35. The current address of the Milton house is 85 Columbine Road (RBF, "The Bear Island Story," Series 7, B6-F14, 12).

25 **"a rather large Gay Nineties Victorian":** *AMS*, 9.

26 **"to get rid of the evil fairies":** Hugh Kenner, "Bucky Fuller and the Final Exam," *New York Times Magazine*, July 6, 1975, 10.

26 **"gypsy camps":** RBF, "What It Was Like to Grow Up in the Boston Area," May 1, 1979, Series 1, B2-F2, 7.

26 **"The road curved":** Ibid., 1.

26 **carried knives:** RBF, "Guinea Pig B," in *Inventions: The Patented Works of R. Buckminster Fuller* (New York: St. Martin's Press, 1983), xviii.

26 **suits of armor:** RBF, "Child Sequence," in *SD*, 1968.

27 **Bear Island:** Descriptions of Bear Island and its history are drawn from *Bear Island Centennial Book, 1904–2004* (privately pub.), 31–34 and 41–71, and RBF, "The Bear Island Story," Series 7, B6-F14.

28 **"a pair of hinged oar blades":** RBF to J. Bryan III, July 1, 1939, B29-V46.

28 **"teepee-like":** RBF, "Influences on My Work," in *Ideas and Integrities: A Spontaneous Autobiographical Disclosure*, ed. Robert W. Marks (New York: Collier Books, 1969), 10. Referred to hereafter as "Influences."

29 **"mechanical jellyfish":** Hatch, 21.

29 **"If you are going to survive":** RBF, "Keynote Address at Vision 65," in *UOO*, 122. RBF gave various dates for this conversation, but the chronology and details point to here.

29 **recovered from appendicitis:** RBF résumé, B8-V16. RBF claimed to have been the recipient of "one of [the] first appendectomies by Dr. Dan Jones of Massachusetts General Hospital" (RBF, *CP*, 380).

29 **skipping stones:** RBF, *Cosmography: A Posthumous Scenario for the Future of Humanity*, Kiyoshi Kuromiya, adjuvant (New York: Macmillan, 1992), 121.

29 "He who calls his brother a fool": RBF, "Guinea Pig B," in *Inventions*, xii.

29 wooden playpen: *Your Private Sky: R. Buckminster Fuller—The Art of Design Science*, ed. Joachim Krausse and Claude Lichtenstein (Zurich: Lars Müller, 2017), 70.

29 a storm came: RBF and Dil, *Humans in Universe*, 78–79. RBF claimed incorrectly to have been nine years old at the time, which is contradicted by the chronology.

29 "These two ideas": RBF, handwritten notes, B24-V39.

30 brain tumor: Harvard Class of 1883, Thirtieth Anniversary Report, 70.

30 "when they were played": RBF, "The Bear Island Story," Series 7, B6-F14, 18.

30 "desissifying" treatment: RBF, "Lightful Houses," in Loretta Lorance, "Building Values: Buckminster Fuller's Dymaxion House in Context" (dissertation, City University of New York, 2004), 390. Referred to hereafter as "BV."

31 "She wanted me to get angry": RBF, "Anger," in *SD*, 408.

31 "would undoubtedly": RBF, "Lightful Houses," 391.

31 "horrid smells": RBF, "The Music of the New Life," in *UOO*, 32.

31 "straitjacket" of his desk: Ibid.

31 "a town-wide riot of joy": RBF, "Race," in *SD*, 14611.

31 "I whooped": Ibid.

31 "Dare to tell the truth": RBF, "Mistake Mystique," in *Discourse*, 295.

31 "One mark for Fuller": RBF with Applewhite, *Synergetics*, 68.

32 As the school "antifavorite": Ibid.

32 "I assumed that I was just a freak": RBF, *EIK*, session 2.

32 "I am now going to teach you": Tomkins, "In the Outlaw Area," *New Yorker*, January 8, 1966, 52.

32 "out the window": Ibid., 53.

32 "How old is it?": Ibid.

32 "I thought she was very pretty": Ibid.

32 first airplane: RBF, "Buckminster Fuller Chronofile," in *BFR*, 13.

32 Halley's comet: Jerred Metz, *Halley's Comet, 1910: Fire in the Sky* (St. Louis: Singing Bone, 1985), 108.

32 "great confidence": Wong, 505.

32 Grandmother Andrews had died: Allegra Fuller Snyder, "Bear Island Guest Book," in *Bear Island Centennial Book, 1904–2004*, 42.

32 "on his deathbed screaming": Margaret F. Eliot to "Mr. Appel," February 10, 1966, Series 1, B2-F3.

32 Dr. Pierce next door: Allegra Fuller Snyder, "Bear Island Guest Book," 43.

32 died of "apoplexy": *The Harvard Graduates Magazine*, September 1910, 144.

33 Jim Hardie: Lucilla Fuller Marvel, "The Hardie Family," in *Bear Island Centennial Book, 1904–2004*, 73–87.

33 "a most extraordinary": RBF, "The Bear Island Story," Series 7, B6-F14, 23.

33 Episcopal church: RBF confirmation certificate for Grace Church, January 22, 1911, B1-V1.

33 "such tension systems": RBF, "Influences," 10.

33 "darling" or "dear": RBF, "The Bear Island Story," Series 7, B6-F14, 17.

33 an estate of around $150,000: Rumwell, *Time* profile notes, Take 2, December 23, 1963, B126-F4.

33 "I'd forge": Ibid.

33 Rosamond remembered "shivering": Tomkins, "In the Outlaw Area," *New Yorker*, January 8, 1966, 85.

34 William Lusk Webster Field: RBF with Applewhite, *Synergetics 2*, 235.

34 Homer LeSourd: Ibid., 69, and Lord, *Pilot for Spaceship Earth*, 101.

34 Starling Burgess: Lord, *Pilot for Spaceship Earth*, 16.

34 "a song and dance man": Allegra Fuller Snyder, quoted in Jaime Snyder, "Experiments in Food and Festivity," afterword to *Synergetic Stew*, i.

34 broke a knee: RBF to J. Bryan III, July 1, 1939, B29-V46.

34 junior exams: Student folder of Richard Buckminster Fuller, Harvard University Archives, Undergraduate Student Records, UA III 15.88.10, 1890–1968, box 1690. Referred to hereafter as Harvard file.

34 "too pretentious": RBF to J. Bryan III, July 1, 1939, B29-V46.

34 "Richard B. Fuller": Allegra Fuller Snyder, "Bear Island Guest Book," in *Bear Island Centennial Book, 1904–2004*, 43.

35 "I think Einstein": RBF and Dil, *Humans in Universe*, 57.

35 sick with typhoid: Harvard Class of 1883, Secretary's Report No. 1, 1883, 92.

35 "no offices of profit": Harvard Class of 1883, Secretary's Report No. 2, June 1886, 85.

35 he called her Sandy: RBF Sr. to "Sandy" (Caroline Fuller), September 3, 1890, B43-F10.

35 maternal Wolcott side: Genealogical information drawn from "Record of Direct Parentage of Allegra Fuller," Series 1, B3-F5, and "Genealogy of Wolcott and Andrews Families," Series 1, B2-F2.

35 moved to Chicago: Hugh Kenner writes that "the wedding of [Caroline's] grandfather was the first to be solemnized in Chicago" (*Bucky*, 66). The man in question was really John Kinzie, whose daughter married the brother of Caroline's great-grandfather (RBF, "Record of Direct Parentage of Allegra Fuller," Series 1, B3-F5).

35 "freedom loving": RBF, *Untitled Epic Poem on the History of Industrialization* (New York: Simon and Schuster, 1962), 9.

36 "the black eyes": Richard F. Fuller, *Chaplain Fuller: Being a Life Sketch of a New England Clergyman and Army Chaplain* (Boston: Walker, Wise, 1863), 6.

36 Salem witch trials: Charles W. Upham, *Salem Witchcraft*, vol. 2 (Boston: Wiggin and Lunt, 1867), 25 and 172–73. Thomas Fuller Sr. and Jr. are listed as residents of Salem in vol 1., xix.

36 Reverend Timothy Fuller: Arthur Buckminster Fuller, *Historical Notices of Thomas Fuller and His Descendants, with a Genealogy of the Fuller Family, 1638–1902* (Cambridge, MA: 1902), 6–9.

36 "a rebellion of the students": Ibid., 9.

36 "an upright and decent posture": Edith Davenport Fuller, "Excerpts from the Diary of Timothy Fuller, Jr., An Undergraduate in Harvard College, 1798–1801," *Proceedings of the Cambridge Historical Society* 11 (1920): 51.

36 "too indolent or unenterprising": Megan Marshall, *Margaret Fuller: A New American Life* (New York: Houghton Mifflin Harcourt, 2013), 17.

36 two of his children: Another son, Richard Frederick Fuller, was the great-great grandfather of Timothy Geithner, the US secretary of the Treasury under Barack Obama. Geithner notes the connection through his mother in *Stress Test: Reflections on Financial Crises* (New York: Crown, 2014), 24. For a family tree, see https://gw.geneanet.org/tdowling?lang=en&n=geithner&oc=0&p=timothy+franz (accessed December 2020).

37 "I must do something": Richard F. Fuller, *Chaplain Fuller*, 303.

37 "the first important American woman": Susan Sontag, *Alice in Bed: A Play in Eight Scenes* (New York: Farrar, Straus & Giroux, 1993), 115.

37 "wonderful child": Marshall, *Margaret Fuller*, 24.

37 "nearsightedness, awkward manners": Ibid., 25.

37 "no natural childhood": Ibid., 21.

37 "And, even then": *Memoirs of Margaret Fuller Ossoli*, ed. Ralph Waldo Emerson, W. H. Channing, and James Freeman Clarke, vol. 1 (Boston: Emerson, Sampson, 1852), 141.

37 "We shall go to him": RBF, notes on family Bible, November 10, 1975, Series 1, B2-V4.

38 "I felt that the whole wealth": Ednah Dow Cheney, *Reminiscences of Ednah Dow Cheney* (Boston: Lee & Shepard, 1902), 205.

38 "I now know": *Memoirs of Margaret Fuller Ossoli*, ed. Emerson, Channing, and Clarke, vol. 1, 234.

38 "astonished and repelled": Ibid.

38 "It was such an awful joke": Donna Dickenson, *Margaret Fuller: Writing a Woman's Life* (London: Palgrave Macmillan, 1993), 32.

39 "the tendency to respect": Ralph Waldo Emerson, "The Transcendentalist."

39 "But if you ask me": Margaret Fuller, *Woman in the Nineteenth Century*, in *The Portable Margaret Fuller*, ed. Mary Kelley (New York: Penguin Books, 1994), 328–29.

39 "At present I look": Ibid., 317.

40 "what Margaret Fuller did": RBF, *4D Time Lock* (Albuquerque, NM: Lama Foundation, 1972), 147. Referred to hereafter as *4DT*.

40 "The destiny of each": Margaret Fuller, *The Letters of Margaret Fuller*, vol. 4, *1845–1847*, ed. Robert N. Hudspeth (Ithaca, NY: Cornell University Press, 1987), 95.

41 "to fill the ear": Marshall, *Margaret Fuller*, 80.

41 "in which we took": Frederick Augustus Braun, *Margaret Fuller and Goethe* (New York: Henry Holt, 1910), 7–8.

41 "the physical resources": RBF, "Margaret Fuller's Prophecy," in *BFR*, 41.

41 "I accept the universe": William James, *The Varieties of Religious Experience* (1902; New York: Modern Library, 1994), 47.

41 "I must start with the universe": *AMS*, 12.

41 "Chiefly the seashore": Ralph Waldo Emerson, "Civilization."

41 "The line of beauty": Ibid., "Beauty."

41 "Swedenborg approximated": Margaret Fuller, *Woman in the Nineteenth Century*, 297.

41 "The genius which": Ralph Waldo Emerson, "Swedenborg," *Representative Men*.

42 "This man, who appeared": Ibid.

42 "a freak": Samuel Rosenberg, *The Confessions of a Trivialist* (Baltimore: Penguin, 1972), 76. RBF recalled discussing Sidis with the cyberneticist Norbert Wiener, another former prodigy of the same era at Harvard (Amy Wallace, *The Prodigy* [New York: E. P. Dutton, 1986], 157).

42 "mythical Harvard": Hatch, 30.

42 Waldo Fuller: RBF, *CP*, 380.

42 "busted" his knee: RBF, "Later Development of My Work," in *BFR*, 75, and RBF, *EIK*, session 7. RBF referred to breaking his knee for the second time "shortly after freshman practice started at Harvard" (RBF to J. Bryan III, July 1, 1939, B29-V46).

42 leaping from one dormitory roof: RBF to J. Bryan III, July 1, 1939, B29-V46.

42 "Take it easy": RBF, in *Our Harvard*, ed. Lant, 13.

43 "helped me through": Ibid.

43 "I thought I'd take": Tomkins, "In the Outlaw Area," *New Yorker*, January 8, 1966, 53.

43 his schedule: Harvard file.

43 "very neglectful": Memo by C. E. Whitmore, c. January 12, 1915, Harvard file.

43 first girl he ever loved "jilted": *AMS*, 23. RBF dated this incident elsewhere to "just before entering college" (RBF, *Education Automation*, 4).

43   John P. Marquand: Stephen Birmingham, *The Late John Marquand: A Biography* (Philadelphia: J. B. Lippincott, 1972), 29–31.

43   Brooks Atkinson: *AMS*, 23.

43   "too much of an oddball": Hatch, 32.

44   "attracted some excellent members": RBF, in *Our Harvard*, ed. Lant, 11.

44   "for his rum": RBF, "'Life' of R. Buckminster Fuller," Series 8, B5-V2, 9.

44   "a harbinger": RBF, *CP*, 130.

44   "magically converted": RBF, in *Our Harvard*, ed. Lant, 14.

44   "hired mother": RBF, *EIK*, session 7.

44   debut in 1914: "'The Passing Show of 1914' Prospers at the Shubert," *Boston Sunday Globe*, October 18, 1914, 54.

45   most celebrated version: *AMS*, 24, and RBF, in *Our Harvard*, ed. Lant, 15.

45   William J. Burns: RBF to J. Bryan III, July 1, 1939, B29-V46.

45   "He secured a considerable sum": Dean Henry A. Yeomans statement on R.B. Fuller disciplinary case, February 13, 1914, Harvard University Archives, UAIII 5.33, Disciplinary Records Series, box 372, folder Fa-Fu.

45   he lived "riotously": Ibid.

46   "Your failure to take": Henry A. Yeomans to RBF, February 4, 1914, Harvard file.

46   "They can't fire you": RBF, "Later Development of My Work," in *BFR*, 75.

47   "practical men": E. W. Atkinson to Byron Hurlbut, September 25, 1914, Harvard file.

47   "to show him no favors": Ibid.

47   Château Frontenac: Cammie McAtee, "Geodesic Dreams: Jeffrey Lindsay and the Fuller Research Foundation Canadian Division," in *Montreal's Geodesic Dreams: Jeffrey Lindsay and the Fuller Research Foundation Canadian Division,* ed. McAtee (Halifax, NS: Dalhousie Architectural Press, 2017), 18.

47   In the finished mill: RBF, *CP*, 324.

47   "a burlap bag": RBF to J. Bryan III, July 1, 1939, B29-V46.

47   "self-tutored course": RBF, "Influences," 12.

48   costume ball: RBF, "The Bear Island Story," Series 7, B6-F14, 20.

48   "Fuller now appreciates": E. W. Atkinson to Byron Hurlbut, September 25, 1914, Harvard file.

48   check the dictionary: RBF, in *Our Harvard*, ed. Lant, 13–14.

48   "pretty bad on the whole": Note on report by E. Bernbaum, Harvard file.

49   "I do not believe": E. A. Cottrell, report, January 19, 1915, Harvard file.

49   "He . . . seems to make little effort": W. C. Heilman, report, January 20, 1915, Harvard file.

49   "It hardly seems": Henry A. Yeomans to RBF, January 22, 1915, Harvard file.

49 "lack of sustained interest": RBF, "Influences," 13.
49 "high scholastic grades": RBF, "Later Development of My Work," in
    BFR, 75.
49 "anguish and shame": RBF, "Influences," 13.
50 "Dear Dean Yeoman": RBF to Henry A. Yeomans, undated (1915),
    Harvard file.

TWO: IN LOVE AND WAR

51 "Life and Universe": RBF with Applewhite, *Synergetics*, 40.
51 "beef lugger": RBF, "Memorandum of Activities," Series 4, B6-F6.
51 "They were sure": RBF, "What It Was Like to Grow Up in the Boston
    Area," May 1, 1979, Series 1, B2-F2, 4.
52 "It is where almost everyone": Arthur Meeker to William Hazard,
    February 2, 1915, B1-V1.
52 Germans in the market "exulted": RBF to J. Bryan III, July 1, 1939,
    B29-V46.
52 "an unusually ambitious": J. C. Leddy to Auditors Headquarters, May 22,
    1915, B1-V1.
52 "big men": Wolcott Fuller to RBF, undated (1915), B1-V1.
52 "ankle-length": Hatch, 41.
52 "best girl": Ibid.
53 "by being the dumbest, gloomiest": RBF to Kenneth Phillips, October 20,
    1915, B43-F13.
53 "a semi-invalid": Hatch, 42.
53 Packer Collegiate Institute: Educational information from "Record of
    Direct Parentage of Allegra Fuller," Series 1, B3-F5.
53 illustrator of children's books: Rumwell, *Time* profile notes, Take 5,
    December 26, 1963, B126-F4. AHF later worked with Allegra on the
    illustrations for an unpublished children's book, *The Proud Shirtfront*,
    with text by Christopher Morley (correspondence in B39-V63).
54 "casting out nines": RBF with Applewhite, *Synergetics*, 765.
54 trial balance: RBF, "Trial Balance Inventory," in *SD*, 19011.
54 "She made it perfectly clear": Hatch, 43.
54 "You can come": AHF to RBF, undated (1915), B1-V2.
54 "You *are* the wife": RBF to AHF, December 1, 1915, B45-F1.
55 Mark Twain: Hatch, 46.
55 "a bribe or a lever": RBF, *EIK*, session 12.
55 "to make the girl feel good": Ibid.
55 "a little story": RBF to AHF, January 9, 1916, B45-F1.
55 "the most wonderful girl": Ibid.
55 "Gracious God": Ibid., January 23, 1916, B45-F2.

55 "good news for us": AHF to RBF, "Thursday afternoon" (January 13, 1916), B1-V2.

56 "I wouldn't even like you": Ibid., "Thursday a.m." (1916), B1-V2.

56 "sort of dismissed": Hatch, 46.

56 "Oh, I think he's *very* attractive": AHF to RBF, "Easter 1916" (April 23, 1916), B1-V2.

56 "the Butcher Boy": Hatch, 44.

56 arranged to meet "unexpectedly": AHF to RBF, May 23, 1916, B1-V2.

56 "irresponsible, thoughtless, and selfish": Hatch, 47.

57 "small stock": RBF to William Hazard, June 16, 1919, B8-V16.

57 Plattsburg military camp: In contemporary sources, the name of both the town and the camp is generally spelled without the *h* at the end.

57 "I'm so sick of listening": RBF to William Hazard, May 9, 1916, B1-V2.

57 "We lay there": RBF to AHF, July 13, 1916, B43-F4.

57 "the war bride": AHF to RBF, July 6, 1916, B2-V3.

57 "a sour old fool": RBF to AHF, July 14, 1916, B43-F4.

58 "so she would not get": Ibid., July 23, 1916, B43-F8.

58 "I have taken them to supper": Ibid., July 25, 1916, B43-F11.

58 "Did you really think": AHF to RBF, "Wednesday a.m." (July 26, 1916), B2-V3.

58 "Lady Anne": Ibid., "Thursday before dinner" (July 20, 1916), B2-V3.

58 Beaux Arts Ball: Hatch, 49. Photographs of RBF and AHF in costume for the ball appear in B3-V6 and in the "Early Photos" section of Sidney Rosen, *Wizard of the Dome: R. Buckminster Fuller, Designer for the Future* (Boston: Little, Brown, 1969).

58 "with a suitcase of grape juice": RBF to J. Bryan III, July 1, 1939, B29-V46.

58 "an old Owl Club man": RBF to AHF, April 4, 1917, B43-F15.

59 "a good sport": Ibid., April 15, 1917, B47-F9.

59 "The doctor looked at my eyes": RBF to Caroline Fuller, April 21, 1917, B3-V6.

59 "the State of Maine Navy": Hatch, 51.

59 Machias Bay: RBF, log for *Wego*, June 21, 1917, B6-V11.

59 "German submarines": RBF to J. Bryan III, July 1, 1939, B29-V46.

59 At two in the morning: RBF, "The Bear Island Story," Series 7, B6-F14, 21.

60 in the afternoon: RBF to J. Bryan III, July 1, 1939, B29-V46.

60 "secret service work": RBF to AHF, July 13, 1920, B45-F6.

60 "knew all about": Ibid.

60 "I have never been": Wolcott Fuller to RBF, August 8, 1918, B8-V15.

60 refueled at Machiasport: RBF to J. Bryan III, July 1, 1939, B29-V46.

60 "It was not until after the war": RBF, "The Bear Island Story," Series 7, B6-F14, 21.

60 **fishing shack:** "Hidden Explosive Shakes Wide Area on Maine Coast," *New York Times*, October 16, 1916, 1.

61 **"Von Papen will join us":** "The Eagle's Eye," *Photoplay*, August 1918, 100.

61 **"a story of thrilling":** Advertisement for *The Eagle's Eye* in *Dramatic Mirror of Motion Pictures and the Stage*, January 26, 1918, 23.

61 **it was "ridiculous":** AHF to RBF, "Sunday" (May 6, 1917), B4-V7.

61 **"on the verge":** Ibid., "Thursday night" (June 7, 1917), B4-V8.

61 **"I wish I was dead":** Ibid., "Friday night" (June 22, 1917), B4-V8.

61 **"It must be quite something":** Hatch, 55.

62 **"the lieutenants of George Washington's day":** "Hewlett Wedding," *Brooklyn (NY) Daily Eagle*, July 15, 1917, 15.

62 **covered in crumbs:** *AMS*, 31.

62 **stories about Spongee:** RBF to AHF, October 12, 1917, B43-F13, and Lord, *Pilot for Spaceship Earth*, 51.

62 **a Fox Film Corporation silent epic:** *4DT*, 91, and "Queen of the Seas," *Baltimore Sun*, November 11, 1917, pt. 2, 6.

62 **"less inspiring":** RBF, "The Bear Island Story," Series 7, B6-F14, 33.

63 **"the best in the fleet":** RBF to AHF, October 17, 1917, B43-F13.

63 **"by beautiful girls":** Nels Nelson, "Who Was That Masked Man?," People Paper People, *Philadelphia Daily News*, October 25, 1974, 2.

63 **"struggling designers":** RBF, *Nine Chains*, 294.

64 **"awfully naughty":** AHF to RBF, "Very early Sunday morning" (February 10, 1918), B6-V12.

64 **range tests:** Lee de Forest, *Father of Radio: The Autobiography of Lee de Forest* (Chicago: Wilcox & Follett, 1950), 342–43.

64 **"established the first":** RBF, "Influences,"15.

64 **transmissions from ships:** *Introduction to Aviation* (Montgomery, AL: Air University Headquarters Air Force ROTC, 1953), 118.

64 **"fantastically absurd number":** RBF, "Conceptuality of Fundamental Structures," in *Structure in Art and in Science*, ed. György Kepes (New York: George Braziller, 1965), 71.

64 **"I said to myself":** *AMS*, 28.

65 **he played with Spaghetti:** RBF to Evelyn Schwartz, May 8, 1931, B24-V39.

65 **refilling his water tanks:** Hatch, 60–61.

65 **"all cut up":** RBF to AHF, February 13, 1918, B47-F12.

65 **"air pocket":** "Naval Flyer Killed," *Baltimore Sun*, March 14, 1918, 2.

65 **"I invented and had installed":** *AMS*, 27.

65 **"a seaplane rescue mast":** RBF, "Influences," 16.

66 **"hundreds of pilots' lives":** Sieden, *Buckminster Fuller's Universe*, 51.

66 **"an aviator a day":** RBF, *EIK*, session 9.

66 **only two fatalities:** One was Leslie M. MacNaughton; the other was Calvin Crawley, who died after his plane collided with a pile while landing

("Calvin Crawley of St. Louis is Killed," *St. Louis Globe-Democrat*, May 18, 1918, 1). No other air fatalities at Hampton Roads for this period appear in *Officers and Enlisted Men of the United States Navy Who Lost Their Lives During the World War, from April 6, 1917, to November 11, 1918* (US Government Printing Office, 1920).

66 **"won me an appointment":** RBF, "Influences," 16.

66 **carved a wooden map:** RBF to Patrick Bellinger, April 3, 1943, B55-F2.

66 **flew an airplane:** RBF, "Buckminster Fuller Chronofile," in *BFR*, 13.

66 **"jet stilts":** RBF, "Influences," 18.

66 **"navigation, ballistics":** RBF, in *Our Harvard*, ed. Lant, 16.

66 **"Our plan has been to equip":** Edward W. Eberle, "General Review of Naval Academy Activities During Year 1917–1918," United States Naval Academy. Provided to the author by Jennifer A. Bryan.

66 **"very *comprehensive* men":** RBF, *Education Automation*, 66.

67 **"horrible baby":** AHF to RBF, "Tuesday" (June 25, 1918), B7-V14.

67 **"to have to sacrifice everything":** Ibid.

67 **"Please try to get used":** RBF to AHF, June 27, 1918, B45-F4.

67 **"It doesn't matter how sick":** AHF to RBF, "Tuesday" (July 2, 1918), B7-V14.

67 **navy ensign:** Although some sources state that RBF reached the rank of lieutenant, junior grade, he was an ensign at the time of his resignation from the navy, and his request to enroll as a junior lieutenant in the Naval Reserve was granted only in 1921. See correspondence in B10-V20.

67 **"the finest cream":** RBF, *EIK*, session 4.

68 **"to always remember":** D. K. Romig, *The United States Ship Great Northern* (Brooklyn: Eagle, 1919), 70.

68 **named after the queen:** Rumwell, *Time* profile notes, Take 5, December 26, 1963, B126-F4.

68 **"infantile paralysis":** RBF to Arthur Meeker, April 29, 1921, B9-V18.

68 **medical records:** Edwin T. Wyman to RBF, July 15, 1922, B11-V22.

69 **"large teamwork maintenance":** RBF, "Influences," 15.

69 **"The human voice":** Ibid., 16.

69 **"I was there":** *AMS*, 30.

69 **Transatlantic radio messages:** Joseph Nathan Kane, *Famous First Facts* (New York: H. W. Wilson, 1964), 505.

69 **"the New Brunswick radiophone":** Harold H. Beverage, "Duplex Radiophone Receiver on USS *George Washington*," *General Electric Review*, October 1920, 810–11.

69 **"an enormous metallurgical war":** RBF, *EIK*, session 3.

70 **"that little Alexandra":** Caroline Fuller to RBF, August 15, 1919, B8-V16.

70 **Romany Marie:** Hatch, 69. RBF stated elsewhere that his first meeting with Romany Marie took place in the company of Vincent Astor in 1919 (Robert Schulman, *Romany Marie: The Queen of Greenwich Village*

[Louisville, KY: Butler Books, 2006], 104), which conflated two different visits.

70 *Beyond the Horizon*: Eugene O'Neill, *Selected Letters of Eugene O'Neill*, ed. Travis Bogard and Jackson R. Bryer (New York: Limelight Editions, 1994), 102.

70 "bigger work": William Hazard to Arthur Meeker, telegram, September 15, 1919, B8-V16.

71 "I have not told Anne": RBF to William Hazard, January 16, 1920, B9-V17.

71 Anne's brother Willets: "Hewlett" (death notice), *Brooklyn Life*, January 17, 1920.

71 acute kidney disease: "James Monroe Hewlett," http://jmh.wikifoundry .com/page/Timeline (accessed April 2021).

71 "a comfort and joy": James Hewlett to AHF, July 12, 1920, B9-V17.

71 "I seem just to go around": AHF to RBF, "Saturday night" (1920), B9-V17.

71 "Dearest little girl": RBF to AHF, July 20, 1920, B45-F7.

71 "We discovered": Richard J. Brenneman, *Fuller's Earth: A Day with Buckminster Fuller and the Kids* (New York: New Press, 2009), 98–99.

72 "one of the most material": *4DT*, 2.

72 "I thought of how": AHF to RBF, August 1, 1921, B10-V19.

72 "My heart certainly jumped": RBF to AHF, January 6, 1922, B11-V21.

73 "tearing up and down the deck": AHF to RBF, "Friday" (January 13, 1922), B11-V21.

73 with an "unspeakable": Ibid., "Monday" (January 16, 1922), B11-V21.

73 "a mixture of colored": Ibid., "Friday" (January 1922), B11-V21.

73 "Isn't that lovely": Ibid.

73 "As you know": RBF to AHF, January 24, 1922, B11-V21.

73 "big brother": Kelly-Springfield Motor Trucks catalog, B12-V23.

74 "When I'm bigger": AHF to RBF, "Wednesday night" (March 1, 1922), B11-V21.

74 "I am so happy": RBF to AHF, March 10, 1922, B11-V21.

74 "I don't mind your drinking": AHF to RBF, "Wednesday night" (March 1, 1922), B11-V21.

74 Police Reserve Aviation Division: "The New York City police commissioner [was made] an honorary officer in the Naval Reserve, in return for which he made all the captains honorary police officers" (Hatch, 73). In fact, the Police Reserve Aviation Division had been officially mustered into the Naval Reserve in its entirety ("Police Air Reserve Is Now Part of Navy," *New York Times*, May 6, 1922, 14). RBF applied for membership in the Police Reserve the following week (John F. Dwyer to RBF, May 10, 1922, B11-V22).

74 a display of "dishonesty": RBF to Charles Young and Edward McDonnell, April 5, 1922, B12-V23.

74 "bright young man": RBF, *EIK*, session 10.

74 "National users": Charles Young to James M. McCarthy, June 6, 1922, B12-V23.

75 "Eddie gave me a job": RBF, *Inventions*, 1.

75 references to the "termination": RBF, "Memorandum of Activities," Series 4, B6-F6.

75 "a ne'er-do-well": Hatch, 72.

75 started to drink: Ibid.

76 "Daddy will be crazy": AHF to RBF, "Tuesday" (August 1, 1922), B11-V22.

76 "our mutual interest": RBF to Vincent Astor, August 28, 1928, in *4DT*, 88.

76 "the efficiency": "Uses Astor Seaplane to Attend Weddings," *New York Times*, September 29, 1922, 7.

76 "after they found themselves": "Flies to Weddings," unidentified clipping, B11-V22.

76 "the first man ever": Schulman, *Romany Marie*, 104.

76 failed rescue: RBF to J. Bryan III, July 1, 1939, B29-V46.

76 brought along an oar: Lord, *Pilot for Spaceship Earth*, 38.

76 wedding in October: The chronology in *Your Private Sky*, ed. Krausse and Lichtenstein, 27, indicates that RBF took classes during this period at the Harvard Graduate School of Business Administration. Course materials appear in B12-V23, but no other evidence has been found for his enrollment.

76 "big and cunning": Susie E. Vooris to AHF, November 14, 1922, B45-F11.

76 World Series: Hatch, 73.

77 reunited with Astor: Hatch places this visit before the seaplane episode (74), but RBF dated it to Astor's return from his trip overseas (Schulman, *Romany Marie*, 105).

77 waved at games: RBF, narration in Mark Stouffer, teleplay for *What One Man Can Do*, B496-F7, 31.

77 "Daddy, will you bring me a cane?": Hatch, 76.

77 "Alexandra is very sick": Ibid. A similar account appears in RBF to J. Bryan III, July 1, 1939, B29-V46.

77 "contiguous displacements": RBF, "Influences," 22.

77 "the messy world": RBF, "Designing a New Industry," in *BFR*, 159.

77 "Daddy, did you bring me my cane?": Hatch, 76.

THREE: STOCKADE

78 "It is a basic principle": RBF with Applewhite, *Synergetics*, 114.

78 backdated his sketches: *Your Private Sky*, ed. Krausse and Lichtenstein, 85.

78 "Mr. Hewlett was the first man": Hatch, 81.

79 "I believe truly": AHF to "Ned," undated (1922), B45-F12.

79 "The greatest poem ever known": *Life*, April 7, 1921, 494.

79 "I was never": Hatch, 82–83.

80 Grand Central Station: RBF to Geoffrey T. Hellman, September 21, 1944, B59-F2.

80 unconventional background: *4DT*, 14.

81 Vincent Astor: RBF to Vincent Astor, August 28, 1928, in *4DT*, 88.

81 "He didn't seem to know": RBF, "Later Development of My Work," in *BFR*, 77.

82 "for concrete stanchions": "State Charters Granted Long Island Corporations," *Brooklyn (NY) Standard Union*, January 28, 1923, sec. 2, 9.

82 "seaweed from the Sargasso Sea": RBF, *EIK*, session 10.

82 "the way you'd pull spaghetti": Ibid.

83 felt manufacturer: Ibid.

83 Scott H. Perky: Hatch, 80.

83 "If he was enthusiastic": RBF, "Later Development of My Work," 80.

83 caught fire: "Fire at Summit Destroys Building Materials Plant," *Courier-News* (Bridgewater, NJ), September 1, 1923.

84 wooden staves: RBF to Kenneth Snelson, March 1, 1980, Series 8, B184-F4.

84 carbon black powder: RBF, *EIK*, session 7.

85 "I would have to": Ibid.

85 showing a profit: "New Fibre Brick Company Issues First Statement," *Orlando Morning Sentinel*, January 6, 1926, 15. For a failed attempt to establish an operation in Florida, see "Building Blocks Made from Palmetto Scrubs New Florida Industry," *Tampa Tribune*, October 14, 1925, 1-C.

85 "perhaps in speakeasies": Hatch, 83.

85 *The Great Gatsby*: Sarah Churchwell, "In Search of Bunny Burgess, or Down the Rabbit Hole: Anecdotes, Secret Histories, and *The Great Gatsby*," *Cambridge Quarterly* 46, no. 2 (June 2017): 140–61.

86 "There was absolutely no momentum": RBF, *EIK*, session 7.

86 "pretty much the center": RBF to Alfred and Pauline Fuller, January 9, 1928, quoted in Lorance, 28.

87 his brother-in-law had been "soft": RBF to AHF, August 17, 1926, B46-F1.

87 "The Chicago crowd": Ibid.

87 "a funny little Dutchman": Ibid., September 28, 1926, B46-F3.

87 Farley Hopkins: Biographical information from "Von Maur-Hopkins Engagement Announced," *Daily Times* (Davenport, IA), May 20, 1920. Additional background provided by Lionel Wolberger, email to author, January 15, 2021.

88 "that darned mad rush": RBF to AHF, July 9, 1926, B45-F16.

88 "bulging tummy": Ibid., September 28, 1926, B46-F3.

88   "I still ride high": Ibid., September 21, 1926, B46-F2.

88   "over a thousand": Hatch, 84.

88   known as the Stockade: "Den Burned by Vigilantes," *Chicago Tribune*, May 31, 1926.

88   "They seemed to be the only place": Hatch, 84.

88   one speakeasy: It may have been the Roadside Home, which was owned by Capone (Mezz Mezzrow and Bernard Wolfe, *Really the Blues* [New York: New York Review Books, 2016], 71–73).

88   "not a very attractive man": Hatch, 84.

88   "the latest machine gun gang murders": RBF to AHF, October 17, 1926, B46-F3.

88   from the trolley: *AMS*, 34.

88   "contacts with all sorts": RBF to William Delano, undated (1932), B26-V42.

89   "the warmest house": Lorance, 34.

89   construction seemed "cheap": RBF to AHF, September 22, 1926, B46-F2.

89   "a bitter fight": Hatch, 86.

89   "I am going to have": RBF to AHF, November 14, 1926, B46-F4.

89   "the elevation of materials": RBF to Kirkwood H. Donavin, December 30, 1926, B15-V27.

90   "a horrid, cheapish way": RBF to AHF, January 5, 1927, B46-F4.

90   "I don't think it would be safe": Ibid.

90   dirty song: Hatch, 86.

90   "tricky methods": James Hewlett to RBF, undated (1927), B15-V28.

90   "Being informed": RBF to Stockade board, February 10, 1927, B17-V32.

90   Anne happened to be in town: RBF to AHF, April 14, 1927 (misdated 1926), B45-F16.

91   "the only times": AHF to RBF, "Monday" (May 2, 1927), B15-V28.

91   "wantonly approached": *4DT*, 5.

91   fourteen thousand attendees: RBF, "Designing a New Industry," in *BFR*, 156.

91   "due to any derogatory remarks": RBF to William J. McCartney, February 16, 1927, B15-V28.

91   "keep your temper": Andrew King to RBF, telegram, February 17, 1927, B15-V28.

91   "Things are black": RBF to AHF, March 4, 1927, B46-F4.

92   "I pretended to myself": Ibid., March 8, 1927, B46-F5.

92   "misstatements or overstatements": Frances Freeman to RBF, March 14, 1927, quoted in Lorance, 42.

92   Hopkins was a "skunk": AHF to RBF, July 13, 1927, B16-V29.

92   "I could just stay": Ibid., "Tuesday" (May 3, 1927), B15-V28.

92   "Isabel [Burton]": Ibid., "Monday night" (May 16, 1927), B16-V29.

92 **"Darling, I am going":** RBF to AHF, March 13, 1927, B46-F5.

93 **"if he brought along a board of directors":** *4DT*, 126.

93 **"It would be absolutely ruinous":** RBF to AHF, July 25, 1927, B46-F6.

93 **"I dressed":** AHF, diary entry, August 7, 1927, in Lorance, "BV," 296. The original is archived in B16-V30.

93 **Virginia Hotel:** Lorance, 48–51.

93 **Raymond Hood:** Hatch, 49.

93 **"I don't believe business":** RBF to AHF, July 18, 1927, B46-F6.

94 **"the big grab":** RBF to Robert McAllister Lloyd, quoted in Lorance, 44.

94 **"a bad man":** Hatch, 86.

94 **"The thing that hurt":** RBF to Robert McAllister Lloyd, quoted in Lorance, 44–45.

94 **Lord Byron:** Hatch, 86.

94 **"happy girl":** RBF, *EIK*, session 12.

94 **Nettie Bonenkamp:** *4DT*, 130.

95 **would be "impersonal":** RBF to Robert McAllister Lloyd, quoted in Lorance, 44.

95 **loaned $700 to a sales manager:** RBF to J. Bryan III, July 1, 1939, B29-V46.

95 **"a colored taxi driver":** Ibid.

95 **150 homes:** RBF, *Nine Chains*, 37. In later accounts, RBF increased the number to 240 (RBF, "Influences," 24).

95 **November 20, 1927:** While the evidence strongly points to this date, there is another possible candidate. In an autobiographical letter to the journalist Joseph "J." Bryan III, RBF said that his mugging on Wabash Avenue took place after a friend invited him to dinner at the Blackstone Hotel "within a few hours of this realization" at the lake (RBF to J. Bryan III, July 1, 1939, B29-V46). In his growth chart for Allegra, RBF dated this incident to October 13, 1927 (B16-V30). Given the lack of any change in his behavior in October, as opposed to the steps that he took in November, it seems likely that he conflated the two events. Lionel Wolberger suggests that RBF frequently walked by the lake to think, which opens up the possibility that his account incorporates details from multiple dates (Wolberger, interviewed by author, December 23, 2020).

95 **"Cloudy drizzly day":** AHF, diary entry, November 20, 1927, 296–97. AHF wrote "Nellie," but the name "Nettie" appears in *4DT*, 130, and a household budget in B16-V30.

95 **the Harvard football squad:** RBF, growth chart for Allegra, November 20, 1927, B16-V30.

95 **Fuller attended a Thanksgiving service:** "Local Programs," *Chicago Tribune*, November 20, 1927, pt. 7, 12. The following week, RBF went to a service at St. Chrysostom's Episcopal Church, which he continued to attend throughout the period that followed.

95   **Thanksgiving 1831:** An account of this experience appears in Braun, *Margaret Fuller and Goethe*, 34–35, which RBF claimed to have read in 1927 (Rumwell, *Time* profile notes, Take 4, December 24, 1963, B126-F4).

96   **"I had adequate courage":** RBF and Dil, *Humans in Universe*, 95.

96   **"greater intellect":** RBF, "Later Development of My Work," in *BFR*, 78.

96   **"Do I know best?":** Ibid.

96   **"You do not have the right":** Ibid., 79.

96   **"work out his fate":** RBF to J. Bryan III, July 1, 1939, B29-V46.

96   arrived at an **"egocide":** RBF, "Guinea Pig B," in *Inventions*, xxviii.

97   **"a miniature picture":** "Visit the Ford Industrial Exposition," advertisement, *Chicago Tribune*, January 29, 1928, 21.

97   **"the Ford method":** Ibid.

97   **"marvelous exhibition":** RBF, diary entry, January 30, 1928, in Lorance, "BV," 304.

97   **"to kill him":** RBF to J. Bryan III, July 1, 1939, B29-V46.

97   **"I found myself":** *AMS*, 35.

98   **"From now on, you need never":** Ibid.

98   **"From now on, write down everything":** RBF to J. Bryan III, July 1, 1939, B29-V46.

98   **"much in need of mental diversion":** RBF, diary entry, February 1, 1928, 305.

98   **"After much philosophical thought":** Ibid.

98   **"Bucky, you are to be":** Barbara Marx Hubbard, "The Path of Social Evolution," in *Anthology for the Millennium,* ed. Thomas T. K. Zung (Carbondale: Southern Illinois University Press, 2014), 141.

98   **"all that my mother would like":** Applewhite, *Cosmic Fishing*, 153.

PART TWO: THE DYMAXION AGE

99   **"You cannot get out":** RBF with Applewhite, *Synergetics*, 85.

FOUR: THE FOURTH DIMENSION

101   **"We start playing":** RBF with Applewhite, *Synergetics*, 296.

101   **"I scarcely spoke":** *AMS*, 39.

101   **"the voluntary experiments":** Susan Sontag, *Styles of Radical Will* (New York: Farrar, Straus & Giroux, 1969), 20.

101   **"He did not seal up his mouth":** Hatch, 92.

101   **"Anne made all":** Ibid.

102   *How to Work Wonders*: North American Institute form letter, B17-V32.

102   **"blind date with principle":** RBF, "Influences," 24.

102 On November 21: Many dates in this section are based on the RBF and AHF diary in B16-V30, which is reprinted in Lorance, "BV," 296–312.

102 "I felt awfully depressed": AHF, diary entry, December 5, 1927, in "BV," 298.

102 new building: Lorance, 51.

102 "a nice red apple": Hatch, 91.

103 "slender single supports": RBF to J. Bryan III, July 1, 1939, B29-V46.

103 "like a man with ague": Ibid.

103 "in which each house": RBF and Robert W. Marks, *The Dymaxion World of Buckminster Fuller* (Garden City, NY: Anchor Books, 1973), 13. The 1960 edition (New York: Reinhold) was credited to Marks alone. Referred to hereafter as Marks.

104 "the house of his childhood": RBF to J. Bryan III, July 1, 1939, B29-V46.

104 "startled at coincidence": RBF, diary entry, January 30, 1928, in "BV," 303.

104 "The lesson of the airplane": Le Corbusier, *Towards a New Architecture*, trans. Frederick Etchells (1927; New York: Dover, 1986), 4.

105 "the majesty of solutions": Ibid., 97.

105 "A problem adequately stated": RBF, "Universal Architecture," *T-Square*, February 1932, 41.

105 "like a soap bubble": Le Corbusier, *Towards a New Architecture*, 181.

105 "parallel table": Ibid., 271.

105 "misses main philosophy": RBF, diary entry, January 30, 1928, in "BV," 303.

105 "the spiritual and abstract": *4DT*, 23.

105 "the philosophy": RBF, diary entry, February 23, 1928, in "BV," 308.

106 "garbled facts": RBF to James Hewlett, telegram, March 28, 1928, B18-V33.

106 cease operations: Print references to Stockade cease almost entirely after 1928. According to the *Index of Trade-marks Issued from the United States Patent Office*, a trademark was approved in 1930 for the Stockade Corporation, which issued a booklet that year for an acoustical tile called Soundex (*Building Age*, vol. 52, 104). By 1938 the Soundex brand had been acquired by Williams & Co. of Chicago ("Notice of a Ruling by the Building Inspector," *Hartford Courant*, April 5, 1939, 21), although RBF later identified Celotex as the buyer (RBF, *EIK*, session 7). He often stated elsewhere that Celotex acquired Stockade itself at the time of his departure, which is contradicted by the available sources (Lorance, 62).

106 "as a matter of integrity": RBF to J. Bryan III, July 1, 1939, B29-V46.

106 attorney's advice: RBF, *Inventions*, 11.

106 Moscow Palace of Labor: Undated clipping from *L'Architecture Vivante*, B18-V33.

106 Bodo and Heinz Rasch: Martin Pawley, *Buckminster Fuller* (London: Grafton / HarperCollins, 1992), 37–39.

107 "The living abodes": RBF, "Cosmopolitan Homes," in "BV," 322.

107 Fuller spoke of "debunking": Ibid., 330.

107 "one best way": Ibid., 331.

107 "It is the subjects": RBF, "Lightful Houses," in "BV," 334.

108 "As time is saved": Ibid., 385.

108 "theory of spheres": RBF, diary entry, February 1, 1928, in "BV," 305.

108 "All matter": RBF, "Lightful Houses," in "BV," 373.

108 "Figure out what they are talking": Ibid., 396.

108 "the demand be upon us": Ibid., 364.

108 "the application of time": Ibid., 341.

108 "the greatest time saver": Ibid., 360.

108 the coded pun of "4D": Wong, 34.

109 "with unexpected enthusiasm": RBF to Homer L. Dodge, April 24, 1944, B58-F2.

109 lack of faith: RBF to AHF, July 31, 1932, B47-F1.

110 "cycles of research": 4DT, 24.

110 "circular progression": Ibid., 15.

110 "only in hotels": Ibid., 21.

110 "ironing drudgery": Ibid.

110 "The trouble is the shirts": William Marlin, "The Evolution and Impact of a Teacher," Architectural Forum, January 1972, 71.

110 "The operator of all this": 4DT, 33.

110 "Philosophy to be effective": Ibid., 6.

110 "This chunky little fat boy": Marlin, "Evolution and Impact of a Teacher," 68.

110 "was greeted": RBF to Henry W. Tomlinson, May 21, 1928, in 4DT, 43.

111 "complacently befuddled": RBF, notes on AIA convention, B18-V33.

111 "peas-in-a-pod-like": Marlin, "Evolution and Impact of a Teacher," 68.

111 "Local characteristics": "Cities Becoming 'Peas of One Pod,' Architects Warn," St. Louis Star, May 17, 1928, 11.

111 Claude Bragdon: For the possible influence of Bragdon's ornamental designs on RBF's geometry, see Jonathan Massey, "Necessary Beauty," in New Views on R. Buckminster Fuller, ed. Hsiao-Yun Chu and Robert G. Trujillo (Stanford, CA: Stanford University Press, 2009), 114–23. Bragdon's work also introduced him to the writings of P. D. Ouspensky (4DT, 46), an associate of Gurdjieff, whose description of the spiral paths of the earth and moon may have informed RBF's characterization of the system as "spiro-orbital" (Ouspensky, A New Model of the Universe [London: Routledge & Kegan Paul, 1960], 449–50, and RBF with Applewhite, Synergetics 2, 452–53).

111 controlling interest: RBF to James Hewlett, June 8, 1928, in 4DT, 51–54.

111 "gold mine": RBF to Caroline Fuller, May 11, 1928, in 4DT, 43.

111 **"over my head"**: Caroline Fuller to RBF, May 23, 1928, in *4DT*, 48.

111 **"I honestly cannot"**: Wolcott Fuller to RBF, August 6, 1928, in *4DT*, 77.

111 **"That was the only time"**: Hatch, 93.

112 **experiments with sleep**: Ibid., 94, and RBF to J. Bryan III, July 1, 1939, B29-V46. See also "Science: Dymaxion Sleep," *Time*, October 11, 1943, 27.

112 **Leland Atwood**: RBF to Paul Nelson, August 10, 1928, in *4DT*, 86. For the allegation that RBF copied elements of the Dymaxion House from a 1926 design by Atwood, see Robert Boyce, *Keck and Keck* (New York: Princeton Architectural Press, 1993), 47.

112 **"for his exclusive ownership"**: *4DT*, 94.

112 **delivered by an airship**: Marks, 74–75.

112 **a "chronofile"**: *4DT*, 39.

112 **required for copyright**: Lorance, 153.

112 **vertical zeppelins**: In 1934 RBF met the zeppelin designer Karl Arnstein, who reportedly encouraged his work on the Dymaxion Car (RBF to J. Bryan III, July 1, 1939, B29-V46). In his list of references for *4DT*, he included an article on an airship designed by a team led by the engineer Barnes Wallis (Wong, 185n136). Wallis developed a "geodetic" construction method, which RBF described as "a geodesic spiral-counter-spiral framing" (RBF to Kenneth Snelson, March 1, 1980, Series 8, B184-F4), but there is no evidence that it influenced the development of the geodesic dome.

113 **"auto-airplane"**: *4DT*, 107.

113 **"the forgotten graves"**: Ibid., 133.

114 **"the greatest letter"**: RBF to George N. Buffington, August 31, 1928, in *4DT*, 120.

114 **"4D is my own creation"**: Ibid., 135.

114 **two structural rings**: Michael John Gorman, *Buckminster Fuller: Designing for Mobility* (Milan, It.: Skira Editore, 2005), 41.

114 **"We might arrange"**: RBF to Jean Toomer, June 15, 1928, in *4DT*, 70.

114 **"I have an increasing interest"**: Toomer to RBF, June 17, 1928, in *4DT*, 70.

115 **"I did not know"**: RBF, *Inventions*, 11.

115 **"to prevent their being exploited"**: RBF, *Nine Chains*, 348.

115 **six tons and cost $3,000**: Lorance, 188.

115 **Rudolph Weisenborn**: RBF, "Dymaxion House Exhibitions and Lectures," in Lorance, "BV," 417.

115 **Ray Schaeffer**: Leigh White, "Buck Fuller and the Dymaxion World," *Saturday Evening Post*, October 14, 1944, reprinted in *Your Private Sky*, ed. Krausse and Lichtenstein, 132.

115 **long-standing program**: Suzanne Strum, *The Ideal of Total Environmental Control: Knud Lönberg-Holm, Buckminster Fuller, and the SSA* (London: Routledge, 2017), 87.

115   coined the term *radio*: RBF, *EIK*, session 12.

115   *dynamism, maximum*: Marks, 21.

116   "the dynamically balanced home": RBF, "Lightful Houses," in "BV," 364.

116   "It seemed at first mouthing": RBF to Peggy LeBoutillier, December 21, 1936, B37-V57.

116   defined retroactively: The *Oxford English Dictionary* consulted RBF years later for its entry on *dymaxion*: "Yielding the greatest possible efficiency in terms of the available technology, 'doing the most with the least'" (2nd Ed., Vol. V, 6).

116   Howard Fisher and George Fred Keck: RBF lecture list, B49-F6.

116   "What's this?" RBF interview in Harold Drake, "Alfred Korzybski and Buckminster Fuller: A Study in Environmental Theories" (doctoral thesis, Southern Illinois University Press, 1972), 168.

116   "many famous Harvard professors": RBF, in *Our Harvard*, ed. Lant, 20.

117   "I learned vast amounts": Nicholas Fox Weber, *Patron Saints: Five Rebels Who Opened America to a New Art, 1928–1943* (New Haven, CT: Yale University Press, 1995), 68.

117   "absolutely comfortable clothes": *AMS*, 59.

117   "I showed up and rudely announced": Ibid.

117   "bank clerk's clothing": Ibid.

117   Lewis Mumford for Scribner: *Your Private Sky*, ed. Krausse and Lichtenstein, 82, and Gorman, *Designing for Mobility*, 49–52.

117   "a liar": RBF, *CP*, 269.

117   sympathetic architects: One was Ely Jacques Kahn, who later commissioned RBF to design a "tension room" to display sculptures by Isamu Noguchi, although it was never realized (*Your Private Sky*, ed. Krausse and Lichtenstein, 29, and Dana Miller, "Thought Patterns: Buckminster Fuller the Scientist-Artist," in *Buckminster Fuller: Starting with the Universe*, ed. K. Michael Hays and Dana A. Miller [New York: Whitney Museum of American Art, 2008], 26).

117   "thinking out loud": RBF, "The Bear Island Story," Series 7, B6-F14, 1.

118   "different labor-saving things": AHF to RBF, September 12, 1929, quoted in Lorance, 177.

118   "All our troubles": RBF to AHF, July 31, 1932, B47-F1.

118   "It was flight": Ibid., August 22, 1935, B36-V55.

118   Antonio Salemme: Antonio Salemme, interview in *Buckminster Fuller: Thinking Out Loud*.

118   "should lend itself to change": Elizabeth A. T. Smith, "The Comprehensivist: Buckminster Fuller and Contemporary Artists," in *Starting with the Universe*, 64.

119   "If you will give a hand": Schulman, *Romany Marie*, 106.

119 **"official talker"**: Shoji Sadao, *Buckminster Fuller and Isamu Noguchi: Best of Friends* (Long Island City, NY: Isamu Noguchi Foundation and Garden Museum, 2011), 77. Referred to hereafter as Sadao.

119 **"a core of people"**: Schulman, *Romany Marie*, 109.

119 **"every intellectual in the world"**: Sadao, 77. RBF's list of visitors to Romany Marie's appears in Series 8, B89-F7.

119 **Theodore Dreiser**: Dreiser expressed polite interest in RBF's ideas: "It would either require a change in society to accept them, or else, by slow degrees they would change society itself" (Dreiser to RBF, April 28, 1931, B23-V38). RBF said later that he also "may have met" the paranormal researcher Charles Fort through Dreiser (RBF, introduction to Damon Knight, *Charles Fort: Prophet of the Unexplained* [London: Victor Gollancz, 1971], xv), but he stated definitively elsewhere that he had not known Fort (Gene Keyes, in Stewart Brand, *The Last Whole Earth Catalog: Access to Tools* [Menlo Park, CA: Portola Institute, 1971], 312).

119 **"always exciting"**: Martha Graham to RBF, January 30, 1964, B125-F8.

119 **Salemme's studio**: "Factory-Made Houses to Hang on Masts," *Oakland Tribune*, December 8, 1929.

119 **Isamu Noguchi**: Much of the material on RBF's and Noguchi's early friendship is drawn from David Michaelis, *The Best of Friends: Profiles of Extraordinary Friendships* (New York: William Morrow, 1983), 41–73.

120 **study art in Paris**: One of Noguchi's works from this period, retroactively titled *Abstraction in Almost Discontinuous Tension* (1928), anticipates aspects of tensegrity, but there is no evidence that RBF ever saw it (Isamu Noguchi, *A Sculptor's World* [London: Thames and Hudson, 1967], 18).

120 **show of their work**: Adnan Morshed, *Impossible Heights: Skyscrapers, Flight, and the Master Builder* (Minneapolis: University of Minnesota Press, 2015), 126.

120 **"fondly as a period"**: Stewart Brand, afterword to Medard Gabel, *Energy, Earth, and Everyone: A Global Energy Strategy for Spaceship Earth*, with the World Game Laboratory (San Francisco: Straight Arrow Books, 1975), 155.

120 **"a fundamental invisibility of the surface"**: RBF, foreword to Noguchi, *Sculptor's World*, 8. Noguchi slightly qualified this claim: "It may have been Bucky's influence to make it chrome. But it must have been my volition because it was modeled for that" (Amy Wolf, *On Becoming an Artist: Isamu Noguchi and His Contemporaries, 1922–1960* [Long Island City, NY: Isamu Noguchi Foundation and Garden Museum, 2010], 47).

121 **"belong to their respective lands"**: RBF, foreword to Noguchi, *Sculptor's World*, 7.

121 **Kay Halle**: Benjamin Forgey, "Noguchi, Honing His Craft," *Washington Post*, April 15, 1989.

121   "Mr. Fuller": Michaelis, *Best of Friends*, 56.

121   "You can't be submerged": Ibid., 57.

121   "It was amazing": Ibid., 52.

121   "in love": Ibid., 69.

121   "Neither of us was homosexual": Ibid., 51.

121   Fox Movietone: Lorance, 184–86.

121   Phil Nowlan: RBF, *CP*, 262.

121   during a snowstorm: Michaelis, *Best of Friends*, 61. Michaelis misdates this trip to 1929.

121   "quite sad today": Hayden Herrera, *Listening to Stone: The Art and Life of Isamu Noguchi* (London: Thames & Hudson, 2015), 92.

121   "a spiritual portrait": Ibid., 83.

122   "who that Jap": Rumwell, *Time* profile notes, Take 5, December 26, 1963, B126-F4.

122   made strides: In June RBF proposed a factory tower for the Gillette Corporation (Wong, 77). The company's founder, King C. Gillette, had published a utopian manifesto in 1894 that anticipated ideas— megastructures, domed gardens, hexagonal layouts—that RBF later explored (King C. Gillette, *The Human Drift* [Boston: New Era, 1894], 84–109).

122   declined to attend: Itinerary, May 3, 1930.

122   "halfway between a gas tank": Wong, 55n130.

122   "Can you tell me?": RBF, "Designing a New Industry," in *BFR*, 169.

122   "a dripping rag": Frank Lloyd Wright, "A Week in New York," in *Collected Writings*, vol. 2, ed. Bruce Brooks Pfeiffer (New York: Rizzoli, 1992), 356.

122   "My young friend": RBF in Olgivanna Lloyd Wright, *The Shining Brow* (New York: Horizon, 1960), 133.

122   "Am to visit Frank Lloyd Wright": RBF to AHF, April 6, 1930, B46-F12.

122   "I know that he will no sooner": RBF to Paul Nelson, August 17, 1928, in *4DT*, 84.

123   Muriel Draper: Schulman, *Romany Marie*, 138.

123   dismissed Gurdjieff: Sadao, 82.

123   "beautiful expressions": RBF, "Gurdjieff," in *SD*, 7115.

123   "go to work writing": Ibid.

123   "let Gurdjieff take over": RBF, "Artist: Histrionics," in *SD*, 753.

123   "the cosmic religious sense": Albert Einstein, "Religion and Science," *New York Times Magazine*, November 9, 1930, 1.

123   new era in history: RBF, "The Leonardo Type," in *Earth, Inc.* (Garden City, NY: Anchor Press, 1973), 69.

123   On November 12: In a letter to Schwartz dated May 24, 1931, RBF stated that he had known her for 194 days (B24-V39).

123   "to which my very nature": RBF to Schwartz, undated (1932), B24-V40.

124 **"porch with wheels"**: RBF, "Epilogue," in *UOO*, 359. More than fifty years later, RBF sent Schwartz a birthday sketch of him driving her to Brooklyn, which few would have recognized as a reference to their romantic history (RBF to Evelyn Stefansson Nef, May 31, 1983, B487-F2).

124 **"Anne's siblings"**: Evelyn Stefansson Nef, *Finding My Way: The Autobiography of an Optimist* (Washington, DC: Francis Press, 2002), 53. The twins were Hope and Hester Hewlett.

124 **"Being an expert con man"**: Ibid.

124 **"The top, made of sheet aluminum"**: Ibid., 53–54.

124 **"freezing reception"**: Ibid., 54.

125 **"valuable notes of lectures"**: RBF to Herbert Biberman, May 11, 1931, B24-V39.

125 **"Our chaste good-bye kisses"**: Nef, *Finding My Way*, 55.

125 **"developing the potentialities"**: "Fuller Cited for Architect's Work," *Berkshire Eagle* (Pittsfield, MA), January 1, 1931.

125 **"which he predicts"**: "Home in 21st Century," New York *Daily News*, November 9, 1929.

125 **Metropolitan Life**: RBF, *EIK*, session 10.

125 **Soviet trade office**: RBF to David J. Saposs, August 2, 1932, reprinted in *Shelter*, November 1932, 134.

125 **"they would all want it"**: RBF, "Dymaxion House," in *SD*, 4344.

126 **Henry Wright**: Archibald MacLeish, "Five Questions . . . and a Striking Answer," *Fortune*, July 1932, 62.

126 **"a distinctly advanced"**: RBF to John Fistere, May 5, 1943, B5-F12.

126 **Vanderlip hired Neutra**: Ibid.

126 **"went dead"**: RBF to J. Bryan III, July 1, 1939, B29-V46.

126 **Pierce Foundation**: Much of the following account is based on Mark Wigley, *Buckminster Fuller Inc.: Architecture in the Age of Radio* (Zurich: Lars Müller, 2015), 97–107.

126 **"I have never been to a place"**: RBF to AHF, May 7, 1931, B46-F12.

127 **"if he weren't such a nut"**: RBF to Schwartz, May 18, 1931, quoted in Wong, 94.

127 **"an annoying bore"**: RBF to AHF, undated (1931), B47-F4.

127 **"blustering and blushing"**: RBF to Howard Myers, December 3, 1943, quoted in Wigley, *Buckminster Fuller Inc.*, 106n73.

127 **"ripen and harvest"**: Richard Neutra to RBF, c. July 1931, quoted in Wong, 94.

127 **"a nervous breakdown"**: RBF to Schwartz, undated (1932), B24-V40.

128 **"unfeeling cruelty"**: Nef, *Finding My Way*, 57.

128 **"two long months"**: Ibid.

128 **"intense and frustrating"**: Ibid., 56.

128 **Starrett-Lehigh Building:** Morshed, *Impossible Heights*, 132–38.

128 **"I finally gave in":** Nef, *Finding My Way*, 56. Nef recalled that these encounters took place before RBF's visit to Bear Island, which is contradicted by the timing of his move to the Starrett-Lehigh Building.

128 **"Having gone this far":** RBF to Schwartz, undated (1932), B24-V40.

128 **stalk her:** "I do not wait for you in the car at E.S. 34 or drive in front of your window late at night with the idea of tempting you to come to me or to arouse your sympathy. I do it . . . so that you may have no feeling of panicky abandonment" (RBF to Schwartz, undated [1931], B24-V40).

128 **"To the dearest nymph":** Nef, *Finding My Way*, 56.

128 **"in almost hysterics":** RBF, diary entry, December 27, 1931, B47-F2.

129 **"a direct contradiction":** RBF to Schwartz, undated (c. January 1, 1932), B24-V40.

129 **"waiting to torture me":** RBF, diary entry, January 2, 1932, B47-F2.

129 **"partial estrangement":** RBF to AHF, July 31, 1932, B47-F1.

129 **"tension theories":** Henry Churchill, "Structural Study Associates," *Shelter*, May 1932, 5.

129 **"peripatetic guide":** Ibid.

129 **4D Dymaxion:** SSA minutes, November 18, 1931, B26-V42.

129 **MacDougal Street:** Strum, *Ideal of Total Environmental Control*, 66.

129 **"colored shirt movements":** RBF to T. H. Gibbins, December 10, 1976, quoted in Wong, 84.

129 **"penthouse":** SSA blank letterhead, B1-V1.

130 **"He was quite large":** Hatch, 115.

130 **"capitalists, industrial leaders":** RBF, "Universal Architecture," *T-Square*, February 1932, 22.

130 **"nothing at all to do with architecture":** Wong, 115.

130 **ranking his possible endeavors:** RBF, notes, undated (1932), B26-V42, reprinted in Wong, 107.

130 **scraped together $800:** RBF to J. Bryan III, July 1, 1939, B29-V46.

130 **insurance policy:** RBF to Karl Jansen, November 8, 1935, B36-V55.

130 **"Bucky Fuller and I":** Philip Johnson and Robert A. M. Stern, *The Philip Johnson Tapes: Interviews by Robert A. M. Stern* (New York: Monacelli, 2008), 43.

131 **he had stolen it:** Hatch, 118.

131 **experimental house:** Strum, *Ideal of Total Environmental Control*, 165.

131 **slept on air mattresses:** Hatch, 118. Other sources refer to the location of this exhibit as the Roger Smith Hotel (Michaelis, *Best of Friends*, 41–43), which was the name of the Hotel Winthrop after 1938.

131 **"A Correlating Medium":** Masthead, *Shelter*, May 1932.

131 **designed by Lönberg-Holm:** Gorman, *Designing for Mobility*, 52.

131 "Progressing towards ephemerality": Photo caption, *Shelter*, May 1932, 10.

131 "Industrially reproducible compositions": RBF, "Universal Architecture, Essay 3," *Shelter*, May 1932, 35.

131 ideas were unintelligible: Philip Johnson to Max Levinson, May 27, 1932, quoted in Wong, 114.

131 "Bucky Fuller was no architect": Strum, *Ideal of Total Environmental Control*, 84.

131 "At the same time": RBF to George Howe, August 1, 1932, B47-F1.

132 SSA exhibition: RBF to the Pathé News, August 4, 1932, B46-F12.

132 Starling Burgess: RBF, "Our Intimate Journal of Summer Events," *Shelter*, November 1932, 21–23; RBF, "Dymaxion—Confidential," October 24, 1933, B30-V47; and Pawley, *Buckminster Fuller*, 52–53.

132 "some kind of abstract abortion": RBF to AHF, August 16, 1932, B47-F1.

132 "for black men and women": "House in Utopia," *Time*, August 22, 1932, 12.

132 Soviet farms: J. Baldwin states incorrectly that RBF traveled to "the Ural Mountains" in connection with this project (*BuckyWorks: Buckminster Fuller's Ideas for Today* [New York: John Wiley & Sons, 1996], 26). This trip was actually described by the architect Matthew Ehrlich in *Shelter*, November 1932, 93.

132 "in some shipyard": "House in Utopia," 12.

132 including his "cathedral": Wong, 505.

132 "War days": RBF to AHF, August 22, 1932, B47-F1.

132 homeless shelter: RBF, "Journal of the Space Hotel," *Shelter*, November 1932, 55–61.

132 Wiscasset, Maine: RBF, "Designer Cooperative," *Shelter*, November 1932, 23.

133 "the prototype": Archibald MacLeish, "Five Questions . . . and a Striking Answer," *Fortune*, July 1932, 62.

133 MacLeish was dropped: Wong, 114.

133 *Miss Expanding Universe:* Sadao, 103.

133 "A winter residence": RBF, "An SSA Credit Consultation," *Shelter*, November 1932, 33.

133 "The study": RBF, "Putting the House in Order," *Shelter*, November 1932, 2.

133 "Don't fight forces": Photo caption, *Shelter*, November 1932, 108. Simon Breines identified this saying as the "basic precept" of the SSA ("The Emergencies," *Shelter*, May 1932, 22).

133 "the greater mass in front": Le Corbusier, *Towards a New Architecture*, 146.

133 "A designer may": RBF, "Streamlining," *Shelter*, November 1932, 78.

134 "exploited for high gain": RBF to Karl Jansen, November 8, 1935, B36-V55.

134 Franklin D. Roosevelt's election: Hatch, 121.

134 "entailed too much scientific research": RBF, "Our Intimate Journal of Summer Events," *Shelter*, November 1932, 15.

134 "a simple matter": Leon Levinson to J. C. Folsom, December 12, 1932, quoted in Lorance, 196.

134 "another piece of design novelty": RBF to Howard Myers, October 28, 1943, B57-F2.

134 "The basic cost": Marks, 23.

134 "Certain that he was dealing": Ibid.

134 Pierce Foundation: Lorance, 198.

134 *Vanity Fair* magazine: "Hall of Fame," *Vanity Fair*, December 1932, quoted in Wong, 121.

135 "ten architects": Frank Lloyd Wright to RBF, August 30, 1932, Frank Lloyd Wright correspondence, 1900–1959, Getty Research Institute, F006C06.

135 Fuller and Noguchi teaching: Wright to RBF, November 29, 1932, B26-V42.

135 "I would certainly": RBF to Wright, December 10, 1932, B26-V42.

135 "my friend": "Fact and Comment," *Los Angeles Times*, January 22, 1933.

135 "a messiah of ideas": Auchincloss, "Dymaxion American," *Time*, January 10, 1964, 49.

135 "I found myself being followed": Hatch, 120.

135 "I premeditatedly began to drink": Ibid.

135 "to streamline man's competitive volition": RBF, "New Year's Resolution of Buckminster Fuller," January 1, 1933, B26-V44.

135 "stumbling out of bars": RBF, "Fuller, R. B.: On Drinking Liquor," in *SD*, 6320.

135 spiked eggnog: Jaime Snyder, "Experiments in Food and Festivity," afterword to *Synergetic Stew*, iii.

135 "my stupid time": Rumwell, *Time* profile notes, Take 5, December 26, 1963, B126-F4.

135 Alfred Reeves: RBF, "Dymaxion—Confidential," October 24, 1933, B30-V47.

136 "the customs and living conditions": "Lone Alaskan Traveler," *Pittsburgh Post-Gazette*, December 4, 1931, 26.

136 Vilhjalmur Stefansson: RBF to Schwartz, September 27, 1931, B24-V40.

136 "all sorts of things": RBF, "Dymaxion—Confidential," October 24, 1933, B30-V47.

136 had hosted a reception: "Social and Other Items of Interest," *Philadelphia Inquirer*, December 7, 1932, 13.

FIVE: STREAMLINES

137 "We think of ourselves": RBF with Applewhite, *Synergetics*, 619–20.

137 "The first motorcars": Le Corbusier, *Towards a New Architecture*, 138.

137   earliest sketches: *4DT*, 107.
137   "jet stilts": RBF, "Influences," 18.
137   duck beating its wings: Ibid.
138   pole vaulter: Marks, 25–26.
138   "ground taxiing": RBF, "Influences," 19.
138   the Zoomobile: Hatch, 122. Hatch spells it *zoommobile*, but the alternate spelling is found more frequently elsewhere.
138   "like hydra cells": RBF, "Epilogue," in *UOO*, 359.
138   "If I want to use": Marks, 26. Hatch's version refers to "ice cream sodas" (123).
139   4D Dymaxion Corporation: Much of the technical background in this section is drawn from Norman Foster and Jonathan Glancey, *Dymaxion Car*, ed. David Jenkins and Hsiao-Yun Chu (London: Ivorypress, 2010), 31–44.
139   a bank holiday: Hatch, 126.
139   "I had been starving": RBF, *EIK*, session 9.
139   "We waste so much": AHF to RBF, "Friday night" (March 24, 1933), B26-V44.
139   "death wobble": Foster and Glancey, *Dymaxion Car*, 31.
140   "Polish sheet metal experts": Marks, 27.
140   Emma Lu Davis: James M. Goode, *The Outdoor Sculpture of Washington, DC* (Washington, DC: Smithsonian Institution, 1974), 554.
140   "most of which have been sold": Foster and Glancey, *Dymaxion Car*, 38.
141   "It seems to me": RBF to Isamu Noguchi, June 28, 1933, quoted in Sadao, 88.
141   "vested in us": RBF to Starling Burgess, June 20, 1933, B47-F3.
141   instead of a "nobody": RBF, "Dymaxion—Confidential," October 24, 1933, B30-V47.
141   "aroused her maternal instinct": Ibid.
141   nineteen feet in length: By way of comparison, the 1932 Ford V-8 Model 18 was fourteen feet long, weighed 2,200 pounds, and retailed for $465–$655 ("1932 Ford V-8 Model 18 Technical Specifications and Dimensions," https://www.conceptcarz.com/s10206/ford-v-8-model-18.aspx [accessed January 2022]).
142   the first streamlined automobile: Jonathon Keats, *You Belong to the Universe: Buckminster Fuller and the Future* (New York: Oxford University Press, 2016), 32–35.
142   "air-cooled aviation engine": Foster and Glancey, *Dymaxion Car*, 35.
143   multiplied its specifications: Dan Neil, "A Test Drive of the Death-Trap Car Designed by Buckminster Fuller," *Wall Street Journal*, April 24, 2015.
143   "uneasy, oscillating swivels": Ibid. Neil's observations were based on a copy of Car #1 commissioned by Jeff Lane, the founder of the Lane Motor Museum in Nashville.

143 "I had to practically": Foster and Glancey, *Dymaxion Car*, 44.

143 Diego Rivera: Federico Neder, *Fuller Houses: R. Buckminster Fuller's Dymaxion Dwellings and Other Domestic Adventures* (Baden, Switzerland: Lars Müller, 2008), 122–23.

143 Seaside Park: "Three-Wheeled Auto Goes 70 M. Per Hour," *Boston Globe*, July 22, 1933, 9.

143 "was tired out": RBF, "Dymaxion—Confidential" memo, October 24, 1933, B30-V47.

144 "going mad": Ibid.

144 cited for speeding: Traffic summons (August 1, 1933) shown in *Buckminster Fuller: Thinking Out Loud*.

144 "midget racing car track": RBF, *EIK*, session 9.

144 impressed a policeman: Marks, 29.

144 At Roosevelt Field airport: "Test 3-Wheeled Car with Engine in Rear," *Brooklyn (NY) Daily Eagle*, August 11, 1933, 6.

144 "startling effect": RBF to Lincoln Pierce, September 9, 1933, B29-V46.

144 Walter Chrysler: RBF to J. Bryan III, July 1, 1939, B29-V46.

144 fixed a "bug": Ibid.

145 wanted to see the car: Information on the car's movements before the accident of October 27, 1933, is based primarily on materials in B30-V47 and Samuel Halsted's notes on the inquest, November 16, 1933, B30-V48.

145 Francis T. Turner: Biographical sources include death certificate for Francis T. Turner, November 22, 1933, State of Illinois Department of Public Health.

145 a lift to the airfield: If all had gone according to plan, Sempill was slated to fly from Chicago to Akron, Ohio, to catch up with the zeppelin, while Turner would have proceeded with the car to Detroit, where an air conditioner was scheduled to be installed by Chrysler ("Dymaxion—Report of Fuller's Chicago Trip," confidential report, October 30, 1933, B30-V47).

145 October 27: The account of the accident is based on multiple sources, including Samuel Halsted to RBF (October 28, 1933), RBF to Vilhjalmur Stefansson (October 28), notes on the scene by RBF and Halsted (October 29), and the confidential report "Dymaxion—Report of Fuller's Chicago Trip" (October 30), B30-V47; Samuel Halsted's notes to RBF on the inquest (November 16), B30-V48; and contemporary news reports, especially "2 European Air Experts Hurt in Crash; One Dead," *Chicago Tribune*, October 28, 1933, 3; Associated Press, "Driver Killed; Foreign Visitors Hurt When Three-Wheeled Car Overturns," *Salt Lake Tribune*, October 28, 1933; and "Swerve Caused Crash," *Birmingham News*, November 18, 1933.

145 Meyer Roth: "2 European Air Experts Hurt in Crash; One Dead," 3.

146 **"not disfigured"**: RBF to Vilhjalmur Stefansson, telegram, October 28, 1933, B30-V47.

146 **"Steel bows"**: Ibid.

146 **"I believed in the car"**: "Dymaxion—Report of Fuller's Chicago Trip," October 30, 1933, B30-V47.

147 **not to attend Turner's funeral**: Ibid.

147 **king of England**: RBF, *EIK*, session 9.

147 **sold it to an engineer**: For the car's fate, see Foster and Glancey, *Dymaxion Car*, 87.

147 **"not rich but not broke"**: Memo by RBF, November 28, 1933, B30-V48.

147 **"driving odd type car"**: Death certificate for Francis T. Turner, November 22, 1933, State of Illinois Department of Public Health.

147 **"apparently having a race"**: Vilhjalmur Stefansson to Lester D. Gardner, December 11, 1933, B30-V8.

148 **shoe salesman**: Roth's address was given as 4905 North Whipple Street in "2 European Air Experts Hurt in Crash," 3. Census data indicates that a shoe salesman of that name lived on Whipple Street in 1940.

148 **"no such collision"**: Vilhjalmur Stefansson to RBF, December 11, 1933, B30-V48.

148 **"the pressure of finances"**: Nannie Dale Burgess (Biddle) to RBF, February 24, 1938, B38-V59.

148 **"the thousands of dollars"**: Max Levinson to RBF, April 20, 1937, B35-V54.

148 **"In future, I'll stick to flying"**: UP, "Says He'll Stick to Air in Future," Baltimore *Evening Sun*, December 16, 1933, 3.

149 **excluded from the event**: RBF to J. Bryan III, July 1, 1939, B29-V46. Notes by Richard Hamilton on press coverage indicate that RBF knew before his arrival that the car would not be exhibited, and that he intended to park outside all along (B533-F7).

149 **"You produced the exact car"**: RBF, *EIK*, session 9.

149 **Leopold Stokowski**: RBF to J. Bryan III, July 1, 1939, B29-V46.

149 **sizable estate**: "Darien Son Left Share of Mrs. Fuller Estate," *Hartford Courant*, October 24, 1934.

149 **clapboard cottage**: *The 1936 Book of Small Houses* (New York: Simon & Schuster, 1936), 100–101.

149 **took credit**: RBF, "Universal Architecture Currently Appraised," 1935, B36-V55.

149 *Four Saints in Three Acts*: Foster and Glancey, *Dymaxion Car*, 100.

149 **Alexander Calder**: Jed Perl, *Calder: The Conquest of Time—The Early Years, 1898–1940* (New York: Alfred A. Knopf, 2017), 446. A letter from Calder to RBF from May 15, 1934, can be found in B31-V50.

150 **Thornton Wilder**: RBF to J. Bryan III, July 1, 1939, B29-V46.

150 "a smart young society matron": Clare Boothe Brokaw to RBF, February 14, 1934, B31-V49.

150 "a sudden longing": RBF to Brokaw, February 13, 1934, B31-V49.

150 "I am a tangle": Brokaw to RBF, "Monday night" (1934), B31-V49.

150 "Anyway, Bucky I am going away": Ibid.

150 misspelled her name "Claire": RBF to Brokaw, February 13, 1934, B31-V49.

150 "the most thrillingly beautiful": Ibid.

150 "the lovely love": Ibid., undated (1934), B31-V49.

150 "very silly and hysterical": Brokaw to RBF, February 14, 1934, B31-V49.

150 "I do not like the mutability": Ibid.

151 "hungry telephoning beaux": RBF to Brokaw, undated (1934), B31-V49.

151 "many times greater": Ibid.

151 Frank Morley: Louise Cochrane, *The Sense of Significance* (Great Britain: Sidaway and Barry, 2015), 14.

151 *Saturday Review*: Christopher Morley, "Streamlines (Thoughts in a Dymaxion Car)," *Saturday Review of Literature*, March 31, 1934.

151 heard of him through Stefansson: Vilhjalmur Stefansson to H. G. Wells, August 12, 1932, reprinted in *Shelter*, November 11, 1932, 128–29.

151 "pleasantly high": RBF to J. Bryan III, July 1, 1939, B29-V46.

151 *Things to Come*: Foster and Glancey, *Dymaxion Car*, 49. The film ultimately featured a streamlined flying vehicle, but the design was credited to others.

151 a newsreel: "1934 Dymaxion Car Invention Newsreel Archival Footage," YouTube, 0:19, PublicDomainFootage, October 6, 2013, https://www.youtube.com/watch?v=MxMg-391LQw.

151 "astonishing automobile": "Tadpole on Wheels," *New Yorker*, May 5, 1934, 94.

152 Dymaxion Tudor Sportster: RBF, *The Artifacts of R. Buckminster Fuller: A Comprehensive Collection of His Designs and Drawings*, vol. 1, ed. James Ward (New York: Garland, 1985), 171–73.

152 five hours: Foster and Glancey, *Dymaxion Car*, 50.

152 "Everybody in the stands": RBF, *EIK*, session 9.

152 "immensely interested": RBF to H. G. Kimball, January 4, 1936, B37-V56.

152 purchase an engine: RBF to J. Bryan III, July 1, 1939, B29-V46.

152 "the greatest artist": RBF, *Nine Chains*, 219.

152 Soviet trade office: RBF to J. Bryan III, July 1, 1939, B29-V46.

152 a Soviet deal for two cars: "Soviet Buys Two Dymaxions," *Bridgeport (CT) Times-Star*, July 23, 1934, and Baldwin, *BuckyWorks*, 99.

152 Frank Henry: RBF to J. Bryan III, July 1, 1939, B29-V46.

152 *Little Dipper*: Hatch, 131.

153 sheriff of Fairfield: Ibid., 133.

153 **two property seizures:** Foster and Glancey, *Dymaxion Car*, 51.

153 **a liquidation sale:** A. E. Magnell, "Hartford Stocks," *Hartford Courant*, April 13, 1935.

153 **Amelia Earhart:** RBF noted a visit from "Amelia Erhardt [*sic*]" in his diary for June 10, 1934 (B31-V50). A photograph of RBF, Earhart, and Sir Charles Kingsford Smith is reproduced in Foster and Glancey, *Dymaxion Car*, 65. RBF recounted his 1935 trip with Earhart in his letter to J. Bryan III, July 1, 1939, B29-V46. Another account appears in Doris L. Rich, *Amelia Earhart: A Biography* (Washington, DC: Smithsonian Institution, 1989), 200.

153 **let go of the steering wheel:** RBF to Bob Stuart, February 11, 1980, B411-F2.

153 **Mrs. Roosevelt "was not yet convinced":** Isabelle Florence Story, "Chatting with the First Lady," *Clarksville (TN) Leaf-Chronicle*, March 7, 1935, 2. RBF's friend Kay Halle offered a different account of Roosevelt's encounter with the car in "The Time of His Life," *Washington Post*, February 5, 1978: "[RBF] had asked me to invite Eleanor Roosevelt to drive with him on one of its first trial runs, which she graciously accepted. After they had gone a short distance, it nosed into the curb, and to Bucky's despair, stalled!"

153 **President Roosevelt:** RBF to J. Bryan III, July 1, 1939, B29-V46.

153 **letter to the First Lady:** RBF to Eleanor Roosevelt, February 8, 1936, B37-V56.

153 **minor accident in February:** A May 29, 1935, letter from the Hartford Accident and Indemnity Company to RBF referred to accidents on February 27 and May 27 (B36-V55).

153 **"I think he thought":** Foster and Glancey, *Dymaxion Car*, 51.

153 **returning to the damaged vehicle:** RBF to J. Bryan III, July 1, 1939, B29-V46.

153 **J. Mezes and Sons:** The proprietor was John Mezes Sr., whose son Theodore E. Mezes purchased Dymaxion #2 in July 1950. Mezes brought the car to Mesa, Arizona, where three students at Arizona State University acquired it in 1968, and it was later bought by William F. Harrah (Foster and Glancey, *Dymaxion Car*, 51–52 and 101).

154 **"for advertising hire":** RBF to J. Bryan III, July 1, 1939, B29-V46.

154 **Waterhouse Company:** *Bear Island Centennial Book*, 35 and 49, and Tomkins, "In the Outlaw Area," *New Yorker*, January 8, 1966, 86. Correspondence between RBF and his creditors can be found in B31-V50.

154 **"with the little money":** *AMS*, 72

154 **"She was the most stable car":** Lord, *Pilot for Spaceship Earth*, 82.

154 **Isamu Noguchi drove to California:** Herrera, *Listening to Stone*, 122, and Michaelis, *Best of Friends*, 65.

154 **"America becomes":** Noguchi to RBF, July 1, 1935, B36-V55.

155 **"Our wedding":** RBF, notes on Francise Clow, undated (1935), B36-V55.

155 "You should never have met Bucky": RBF to Clow, June 23, 1935, B36-V55.

155 "with animal fear": Clow to RBF, undated (1935), B36-V55.

155 "This route is perhaps": Ibid., June 29, 1935, B36-V55.

155 "cramming my mouth": Ibid., September 6, 1935, B36-V55.

155 Gilbert Stearns: RBF rebuked AHF for her trip upstate with Stearns in a letter dated July 31, 1932 (B47-F1). Stearns is identified as "Gee" in Hatch, 131.

155 his sins had been "myriad": RBF to AHF, July 28, 1935, B36-V55.

156 "a dog avoiding medicine": Ibid., August 22, 1935, B36-V55.

156 "at its most perilously feeble state": RBF, "The Bear Island Story," Series 7, B6-F14, 32.

156 "two vital affairs": RBF to Francise Clow, September 7, 1935, B36-V55.

156 James Joyce: *From the Library of R. Buckminster Fuller* (New York: Glenn Horowitz Bookseller, 2004), 7. "After the publication of Fuller's book *Nine Chains to the Moon* in 1938, Joyce himself wrote Fuller a testimonial, praising him for the clarity and comprehensiveness of his style" (Robert W. Marks, "Bucky Fuller's Dymaxion World," *Science Illustrated*, November 1948). No such letter has been found in the Chronofile.

156 "air map": RBF was publicly using this term at least by 1937 ("Buckminster Fuller Speaks at College," *Bennington [VT] Evening Banner*, April 30, 1937, 1).

156 *Eight Chains to the Moon*: Paul B. Hoeber to RBF, December 3, 1935, B36-V55.

157 his wife, Peggy: Correspondence with Peggy Osborn appears in B36-V55, along with letters addressed to RBF care of the Osborns.

157 "each verbal dose": RBF, *No More Secondhand God, and Other Writings* (Carbondale: Southern Illinois University Press, 1963), x.

157 "This is lucid": Ibid.

157 "Who were those poets?": Kenner, *Bucky*, 219.

157 Bell Telephone: RBF, "Earth, Inc.," in *BFR*, 248–49.

158 "quite dyslexic": Allegra Fuller Snyder, interviewed by author, March 26, 2019.

158 "One of you was the new world": Francise Clow to RBF, November 4, 1935, B36-V55.

158 a "veil": Ibid.

158 "We never had any conflict": Michaelis, *Best of Friends*, 59.

158 "Please wire me rush": RBF, *Nine Chains*, 73.

158 "Einstein's formula": RBF to Noguchi, telegram, undated (1935), B36-V55. Reprinted in RBF, *Nine Chains*, 73–74.

159 "You cannot imagine": Francise Clow to RBF, December 10, 1935, B36-V55.

159 "established the metallurgical principle": Marks, 32.

159 "They were really very scared": RBF, *EIK*, session 6.

159 the concept of "lag": RBF, *CP*, 282–84.

159 spike in the scrap supply: Ibid., 284. In fact, secondary production from old copper scrap increased only slightly from 267,300 short tons in 1938 to 286,900 short tons in 1939. By 1940 the figure had risen to 333,890, which was still considerably lower than the 408,900 short tons recorded in 1937. Figures based on US Bureau of the Census, *The Statistical History of the United States, from Colonial Times to the Present* (New York: Basic Books, 1976), 602.

159 "the highly mechanized dwelling": RBF to William Osborn, February 27, 1936, B37-V56.

160 his bathroom: Background and technical details drawn from Marks, 96–97, and Wigley, *Buckminster Fuller Inc.*, 108–15.

160 *Architectural Record*: Wigley, *Buckminster Fuller Inc.*, 108–9.

161 Richard Neutra: RBF to John Fistere, May 5, 1943, B55-F2.

161 Jasper Morgan: RBF to William Osborn, December 20, 1937, B38-V61.

161 "You are always trying": RBF, "Fuller, R. B.: A Propos Ben Franklin," in *SD*, 6341.

161 Ellen Tilton: Letters hinting at an intimate relationship can be found in B37-V57, B38-V59, and B38-V61. Tilton had previously designed a house consisting of "a flat, circular platform, covered in a glass dome" ("Mrs. Holmsen—Crusader," *American Weekly* [insert], *San Francisco Examiner*, January 27, 1935, 9).

161 Evelyn Schwartz: Nef, *Finding My Way*, 52.

161 "The Silver Prince himself": Tom Wolfe, *From Bauhaus to Our House* (New York: Farrar, Straus & Giroux, 1981), 45.

161 Fuller obtained his notes: RBF to Walter Gropius, December 31, 1937, B35-V54.

161 "So I rashly wrote back": *AMS*, 69.

161 Cooper's invitation: Viola Cooper to RBF, February 22, 1938, B35-V54.

162 Morris Fishbein: RBF referred to the host as "Dr. Fishbein" (RBF and Dil, *Humans in Universe*, 41), whom Hatch identifies as "Doctor Morris Fishbein" (141). Morris Fishbein was the editor of the *Journal of the American Medical Association*, and his detailed social journal for the period makes no mention of Einstein (Morris Fishbein Papers, University of Chicago, B95-F8).

162 "mystical aura": *AMS*, 69.

162 "horsepower" of everyday life: RBF, *Nine Chains*, 70–82.

162 "Young man, you amaze me": RBF marginalia in *Physical Science* by Sir James Jeans, quoted in Wong, 194n160.

162 **"Mr. Einstein's statement"**: Ibid.

162 **"Whether or not I shall be convinced"**: RBF to J. Bryan III, July 1, 1939, B29-V46.

162 **Fuller's tangled prose**: A chapter on the secret meaning of the alphabet, "Bible to Babble to Bible," was removed as "much too esoteric" (RBF, "Local vs. Comprehensive," in *SD*, 9701). A revised version, "The Game of Life," was made available to readers for $1 (RBF, *Nine Chains*, 368). Drafts can be found in Series 8, B3-F4.

162 **"big book"**: "Book Notes," *Morning Call* (Allentown, PA), June 12, 1938.

163 **"It may serve as a snare"**: Ruth Green Harris, "Art in Our Daily Lives," *New York Times*, May 1, 1938, sec. 10, 9.

163 **plumbers union**: RBF, *EIK*, session 10.

163 **had reacted positively**: A complimentary piece in the *Ladle* (April 1937), a publication of the New York State Association of Master Plumbers, is reprinted in Marks, 101.

163 **trip with Christopher Morley**: Cochrane, *Sense of Significance*, 45–46.

163 **László Moholy-Nagy**: Ibid., 129.

163 **called him "Buckling"**: RBF to Christopher Morley, December 3, 1940, B50-F2.

163 **Fuller and Morley traveled**: Cochrane, *Sense of Significance*, 47–54.

163 **"in stupor"**: Christopher Morley, "Ammonoosuc," reprinted in ibid., 52.

163 **"with Bucky at the wheel"**: Hatch, 147. RBF later denied this account: "The story of the potty on the truck (from Alden Hatch) is not true" (notes on the teleplay *What One Man Can Do* by Mark Stouffer, 46A, B496-V7).

164 **all his royalties**: Agreement between RBF and William Osborn, July 22, 1938, B38-V61.

164 **Bil Baird**: RBF to Joseph W. Lippincott, February 27, 1939, B39-V63, and Marks, 150.

164 **"The reward"**: RBF, *Nine Chains*, x.

164 **"A self-balancing"**: Ibid., 18–19.

164 **"phantom captain"**: Ibid., 19.

164 **"the universal extension"**: Ibid., 43.

165 **from track to "trackless"**: Ibid., 300–301.

165 **"This is a practical key"**: Ibid., 281.

165 **"energy slaves"**: RBF, "World Energy" map in "U.S. Industrialization," *Fortune*, February 1940.

165 **"rented" raw materials**: RBF, *Nine Chains*, 321.

165 **"utterly different"**: Ibid., 341.

165 **"plenitude" goods**: Ibid., 275.

165 **"through the effort of one man"**: Ibid., 40.

165 **amateurs for "publicity"**: Ibid., 313.

166 "perfect whore house": Ibid., 389.

166 "scientific dwelling machine": Ibid., 405. RBF dated this invention to 1927.

166 "way of a man with a maid": Ibid., xvii.

166 "emergence through emergency": Ibid., 240–46.

166 "receiving and sending apparatus": Ibid., 380.

166 "geometrical progression": Ibid., 370.

166 "The author has sound knowledge": William Marias Malisoff, "The Dymaxion Way," *New York Times Book Review*, October 9, 1938, 14.

166 "I know it will be": RBF to Frank Lloyd Wright, July 5, 1938, Frank Lloyd Wright correspondence, 1900–1959, Getty Research Institute, F031A02.

166 "the most sensible man": Frank Lloyd Wright, "Ideas for the Future," *Saturday Review of Literature*, September 17, 1938, 14.

166 "I still get a little chokey": RBF to Frank Lloyd Wright, September 26, 1938, B39-V62.

167 "something like heaven": RBF in Wright, *Shining Brow*, 134.

167 Eleanor Roosevelt's syndicated newspaper column: Eleanor Roosevelt, "My Day," *Knoxville (TN) News-Sentinel*, February 28, 1939, 17.

167 "R. Buckrogers Fuller": Clifton Fadiman, "R. Buckrogers Fuller," *New Yorker*, September 10, 1938, 93. In his review, Fadiman referred to RBF's "Leonardesque mind."

167 "by the thousands": "Building for Defense," *Architectural Forum*, June 1941.

167 binding his Chronofile: Allegra Fuller Snyder, "The R. Buckminster Fuller Archives," in *Anthology*, 321–22.

167 "A man has only one escape": Clare Boothe Luce, *The Women* (New York: Dramatists Play Service, 1966), 25.

167 *Chassis Fountain:* Sadao, 113.

168 "I remember very well": Herrera, *Listening to Stone*, 105.

168 Hale threw herself: "Dorothy Hale Dies in 16-Story Plunge," *New York Times*, October 22, 1938, 34. Other sources include Andrea Kettenmann, *Frida Kahlo, 1907–1954: Pain and Passion* (Cologne, Ger.: Taschen, 2003), 46–51, and Herrera, *Listening to Stone*, 105.

168 "Dorothy has given us": RBF and Noguchi to Mrs. Grant Mason (Jane), undated (1938), B39-V62.

168 "I have been terribly saddened": Francise Clow to RBF, October 25, 1938, B39-V62.

168 "a beribboned bomb": Geoffrey T. Hellman and Harold Ross, "Ribbon Around Bomb," The Talk of the Town, *New Yorker*, November 5, 1938, 19.

168 memorial portrait: Sylvia Jukes Morris, *Rage for Fame: The Ascent of Clare Boothe Luce* (New York: Random House, 1997), 330–32, 338–39.

168 "the idea market of the world": RBF to A. L Furth, memo, November 14, 1938, B39-V62.

169 "The New US Frontier": "The U.S. Frontier," *Fortune*, September 1939, 76.

169 "this inadvertent declaration": RBF to Max Schuster, February 7, 1939, B39-V63.

169 hockey stick: James Lonsdale-Bryans to RBF, November 8, 1938, B39-V62.

169 "trajectory through the ages": RBF, "Ballistics of Civilization," March 14, 1939, B39-V63.

169 move to Canada: RBF claimed that Henry Luce said that he was proven right when Britain transferred its secret archives to Ottawa in World War II (RBF, *EIK*, session 6).

169 "on the rebound from him": Wilfrid Sheed, *Clare Boothe Luce* (New York: Berkley Books, 1984), 67.

169 "The sewing or calculating machines": RBF, "A Speech to be Made by Henry Luce," 1939, B39-V63.

169 Gilbert Seldes: Marian Seldes, "My Best Friend's Father," in *Anthology*, 171–73.

169 "the learning of geography": "Television II," *Fortune*, May 1939, 70.

170 Al Hirschfeld: Al Hirschfeld to RBF, note, January 24, 1939, B39-V63.

170 Gertrude Vanderbilt Whitney: At the end of 1939, RBF was approached by Whitney to consult on color and lighting for a sculptural fountain for the World's Fair. RBF advocated "such new materials as neon tubing, Lucite, and Plexiglas," but the design went unused (B. H. Friedman, *Gertrude Vanderbilt Whitney: A Biography* [Garden City, NY: Doubleday, 1978], 627).

170 Arshile Gorky: Hayden Herrera, *Arshile Gorky: His Life and Work* (New York: Farrar, Straus & Giroux, 2003), 335 and 399.

170 Alexander King: RBF, *EIK*, session 3. The play was *Profound Bow to Memory* ("Solo Dramatist," *New Yorker*, September 23, 1939, 15).

170 "I believe that you could communicate": Hatch, 184, and RBF, *EIK*, session 12.

170 designs for the Dymaxion Car: Foster and Glancey, *Dymaxion Car*, 205. The designs dated from 1934 (RBF, *Artifacts of R. Buckminster Fuller*, vol. 1, 171–73).

170 L. Ron Hubbard: Sadao, 82. Hubbard mentions seeing Stefansson at the Explorers Club in a letter dated December 13, 1939 (L. Ron Hubbard, *Adventurer/Explorer: Daring Deeds & Unknown Realms* [Commerce, CA: Bridge Publications, 1996], 99).

170 Fuller's series of articles: A complete list of the articles to which RBF contributed can be found in RBF, "Memorandum of Activities," Series 4, B6-F6, 4.

170 "I read *Patent Gazette*": RBF, "Reading," in *SD*, 14848.

171 article on drug development: "Cure by Chemicals," *Fortune*, September 1939, 45.

171 **"money enough to wage his own war"**: "Revolution in Radio," *Fortune*, October 1939, 121.

171 **transmission methods**: "The Coming Boom in Kilowatts," *Fortune*, December 1939, 58–60, 106–12.

171 **Glenn L. Martin**: "Glenn L. Martin Co.," *Fortune*, December 1939, 73–77, 126–30.

171 **"that wants to revolutionize"**: "Plywood," *Fortune*, January 1940, 42.

171 **"expert on America's future"**: "The Future in the News," *Life*, June 5, 1939, 86.

171 **"the scope and daring of Leonardo"**: Russell Davenport, "Bucky Fuller's Notebook," in RBF, *Untitled Epic Poem*, vii. An original draft is preserved in Series 8, B5-F1.

171 **"I send for Bucky"**: Hatch, 151.

171 **"the most important nation"**: *Fortune*, contents page, February 1940.

171 **"You have set back"**: Peter Drucker, *Adventures of a Bystander* (New York: John Wiley & Sons, 1997), 246.

171 **"US Industrialization"**: *Fortune*, February 1940. Reprinted in RBF, *The World Game: Integrative Resource Utilization Planning Tool* (Carbondale: Southern Illinois University Press, 1971), 48–61.

172 **"energy slaves"**: "World Energy" map in "U.S. Industrialization," *Fortune*, February 1940.

172 **"No More Secondhand God"**: RBF dates the poem in *No More Secondhand God*, 3.

172 **"God is a verb"**: Ibid., 28.

172 **"in a hell of a hurry"**: Ibid., 30.

172 **Frank Lloyd Wright and E. E. Cummings**: List of recipients for "No More Secondhand God," 1940, B49-F2.

172 **"Untitled Epic Poem"**: RBF's manuscript list dates it to November 1939 (Series 4, B6-F5). He credited "about 10 percent of the wording" of the poem to Russell Davenport (*SD*, xxv).

172 **Sperry Corporation**: "Sperry: The Corporation" and "Sperry's Spinning Wheel," *Fortune*, May 1940, 52–64, 94–99, 112–21.

172 **"extensions of the sensorial spectrum range"**: *New Worlds in Engineering* (Detroit: Chrysler Corporation, 1940), 15.

172 **"butchered" by publicists**: RBF, note on front endpapers of *New Worlds in Engineering*, B536-F5.

173 **"It'll just make those fellows"**: RBF, "Corporation," in *SD*, 3033.

173 **"qualified for such mastery"**: RBF, "Machine Tools," in *No More Secondhand God*, 46. According to RBF, an excerpt was published in the December 1940 issue of *Fortune* (ibid., xi).

173 **"with some part"**: RBF, note, undated (1940), B50-F4.

173 **Mechanical Wing:** "The Mechanical Wing," *Architectural Forum*, October 1940, 273.

173 **leisurely ramble:** Cochrane, *Sense of Significance*, 72–74. Hatch dates the trip incorrectly to "the early summer" (157–58).

173 **poker game:** Geoffrey T. Hellman and Harold Ross, "Bomb Shelter," The Talk of the Town, *New Yorker*, October 4, 1941, 14.

174 **"any portable buildings at all":** Letter to RBF, March 11, 1941, B50-F8.

174 **scorching day:** RBF, *EIK*, session 10. Temperature information from Charles Chamberlain, "Tin House Newest Wrinkle in National Defense Boom," *Macon Chronicle-Herald* (MO), July 5, 1941.

175 **"More than any other man":** RBF to Geoffrey T. Hellman, September 21, 1944, B59-F2.

175 **museum's importance in wartime:** Beatriz Colomina, "DDU at MoMA," *ANY: Architecture New York*, no. 17 (1997): 52.

176 **photographed for *Vogue*:** Ibid., 53.

176 **"It might jump off the ground":** Hellman and Ross, "Bomb Shelter," 14.

176 **informal discussion group:** David Cort, *The Sin of Henry R. Luce: An Anatomy of Journalism* (Secaucus, NJ: Lyle Stuart, 1974), 290–91.

176 **"as light and as terrible":** RBF and David Cort, "Foot Pound Hitting Power of an Energy-Borne Economy," May 1942, B536-F3, 7.

176 **"liquid oxygen or uranium":** Ibid.

176 **"John, there's a war on":** John Fistere in *Synergetic Stew*, 85.

176 **"several technical devices":** RBF to Charles Edison, January 30, 1942, B52-F6.

176 **tracing geodesic lines:** Sadao, 133.

177 **"beautiful little clear crystal balls":** RBF, *EIK*, session 7.

177 **"six in its own plane":** Thompson, *On Growth and Form*, 328. RBF mentioned the book in his diary for 1943 (B59-F4), but he may have encountered it earlier. He also listed the 1948 edition in his "Live Book Squad" of books that he carried in a mobile footlocker (Series 4, B6-F6).

177 **It had appeared in his sketches:** RBF, undated sketch (1935), B36-V55.

177 **regular icosahedron:** RBF to Christopher Morley, October 9, 1943, B57-F2.

178 **a "presumptuous" gesture:** RBF, "Vector Equilibrium," in *SD*, 20064.

178 **less distortion:** RBF, "Fluid Geography," in *BFR*, 151.

179 **only four segments:** Sadao, 134.

179 **"three-way great circle grid":** RBF, "Noah's Ark #2," in *Discourse*, 181.

179 **son of the architectural professor Edwin Park:** The late David Park stated that RBF presented him with this version of the Dymaxion Map "around 1935" (Sadao, 133), which contradicts the available evidence. In all likelihood, RBF initially gave Park a copy of his earlier "air map,"

perhaps during his visit to Bennington in April 1937, in which he spoke on his work at Phelps Dodge: "A map of the world used in the study was explained by Mr. Fuller as an 'air map'" ("Buckminster Fuller Speaks at College," *Bennington (VT) Evening Banner*, April 30, 1937). RBF lectured at Bennington again in April 1942 ("Technical Expert at College Tomorrow," *Bennington (VT) Evening Banner*, April 9, 1942), and it seems likely that he gave the Dymaxion Map to Park or his father on this later visit.

180 **"that they never knew were wrong"**: RBF to Wilber B. Larkin, March 26, 1942, B54-F3.

180 **five of his "igloos"**: Robert Colgate to Wilber B. Larkin, April 1, 1942, B53-F5.

180 **a custom tool kit**: Marks, 119.

180 **cooling effect**: Baldwin, *BuckyWorks*, 37.

180 **Martha Graham**: Martha Graham to RBF, January 30, 1964, B125-F8.

180 **hundreds were made**: Marks, 35.

180 **"immaculate discipline"**: RBF, "I Figure," in *BFR*, 125.

181 **"respective twenty-five-year martyrdom"**: RBF to Lincoln Pierce, telegram, July 10, 1942, B53-F10.

SIX: THE DWELLING MACHINE

182 **"Physical points"**: RBF with Applewhite, *Synergetics*, 258.

182 **speech on postwar industry**: RBF to Avery Pierce, October 1, 1943, B57-F2. A manuscript draft of the speech, "Flow Diagram into Tomorrow," appears in B54-F5.

183 **became the D-45**: Technical details drawn from Pawley, *Buckminster Fuller*, 79–84, and Foster and Glancey, *Dymaxion Car*, 14–15 and 72–73.

183 **"three-cornered deal"**: RBF to John Breck, January 26, 1943, B55-F1.

183 **Frank Lloyd Wright**: RBF, diary entry, February 25, 1943, B49-F4.

183 **"amateur mathematician"**: Edgar J. Applewhite, "An Account by R. Buckminster Fuller of His Relations with Scientists During the Development of His Energetic Geometry," Series 8, B12-F23, 4.

183 **Cynthia Lacey**: Biographical information drawn from "Particulars and Experience Record," 1947, B67-F16.

184 **cook breakfast**: Lacey to RBF, December 4, 1944, B60-F2.

184 **"functional" strategy**: RBF to Alex Taub, "Only One Substitution to Be Made," December 31, 1942, B54-F4.

184 **"Where a man is sinking"**: Conversion Conference transcript, January 14, 1943, B55-F1.

184 **"If you are in a shipwreck"**: RBF, *Operating Manual*, 9.

184 "an extraordinary engineer": Alex Taub to Henry Kaiser, January 27, 1943, B55-F1.

185 read secret intercepts: RBF to Alex Taub, March 8, 1943, B55-F1, and RBF, *CP*, 104.

185 "pasted together": RBF to William Phillip Churchill, October 19, 1943, B57-F2.

185 "pure invention": RBF, *Inventions*, 90.

185 "particularly fine": RBF to Kenneth R. Deardorf, March 19, 1943, B55-F1.

185 ignoring his grid: RBF to Martin Gardner, August 26, 1975, quoted in "Dymaxion Airocean World Map," in *SD*, 4314.

185 "irregular solid": "*Life* Presents R. Buckminster Fuller's Dymaxion World," *Life*, March 1, 1943, 42.

185 greatest success to date: Although the issue was said later to have been the first to sell 3 million copies (Hatch, 167), the magazine's circulation had really peaked at 4.2 million prior to the institution of wartime paper rationing ("An Announcement to *Life* Readers," *Life*, February 15, 1943, 34).

185 Winston Churchill: RBF, *EIK*, session 6.

185 "Bucky, in life": Dialogue reconstructed from Marks, "Bucky Fuller's Dymaxion World," *Science Illustrated*, November 1948; RBF, "Inexorability," in *SD*, 8100; and RBF, *EIK*, session 6.

186 copper hemispheres: RBF to A. J. Benson, December 14, 1943, B57-F2.

186 "the research and development": RBF to Alex Taub, memo, March 9, 1943, B55-F1.

186 "I'm going to make your hair": Memo by RBF, March 24, 1943, B55-F1.

186 spoke of "sunset": RBF and Cynthia Lacey, "Motion Economics," B537-F1, 52.

186 "from their enemy": Ibid., B537-F2, 82.

186 *up* and *down*: Ibid., B537-F1, 46.

187 "most financially successful": Ibid., B537-F2, 63.

187 "inferior citizenry": Ibid., B537-F2, 105.

187 "charges of paternalism": RBF and Cynthia Lacey, "Contact Economy," B537-F8, 5.

187 Taub had covertly assumed control: Alex Taub designed engines for Chevrolet, Vauxhall Motors, and the Churchill tank, but there is no evidence that he consulted for Kaiser during the period in question. In Hatch, 171, Taub is described only as "the man who had designed the current Chevrolet engine," and no mention is made of the fact that he was RBF's supervisor.

187 "I would rather": RBF to Avery H. Pierce, October 1, 1943, B57-F2.

187 Office of Strategic Services: Barry M. Katz, "The Arts of War: 'Visual Presentation' and National Intelligence," *Design Issues*, Summer 1995, 3–21.

188 **Charles Fenn:** Charles Fenn, *At the Dragon's Gate: With the OSS in the Far East* (Annapolis, MD: Naval Institute Press, 2004), 3–6, and correspondence in B55-F2.

188 **"slowly erected extension":** RBF to Alex Taub, memo, February 1, 1943, B54-F4.

188 **Getúlio Vargas:** RBF, *CP*, 288–89.

188 **Louis I. Kahn:** For Kahn and RBF's earlier encounters, see Sarah Williams Goldhagen, *Louis Kahn's Situated Modernism* (New Haven, CT: Yale University Press, 2001), 231n3.

188 **"Dymaxion Sleep":** "Science: Dymaxion Sleep," *Time*, October 11, 1943, 27.

189 **map by Irving Fisher:** "Airways to Peace" master checklist, Museum of Modern Art.

189 **"outside-in" globe:** Monroe Wheeler, "A Note on the Exhibition," in "Airways to Peace: An Exhibition of Geography for the Future," *Bulletin of the Museum of Modern Art*, August 1943, 22, caption on 9.

189 **would be "disgusted":** RBF, "Fluid Geography," in *BFR*, 148.

189 **Brazil report:** RBF, "A Compendium of Certain Engineering Principles Pertinent to Brazil's Control of Impending Acceleration in Its Industrialization," in RBF, *CP*, 293–308.

189 **move directly to "trackless":** Ibid., 302.

189 **"the leading skyport of the world":** Ibid., 294.

189 **"unofficial channels":** Memo, RBF meeting with Arthur Paul, July 4, 1944, B59-F4.

189 **"twenty times as big":** RBF, "Cost of Various Industries for China," August 30, 1943, B57-F1.

189 **running joke:** RBF, *EIK*, session 9.

189 **"scientific" house:** RBF to W. J. McGoldrick, December 10, 1943, B57-F2.

190 **eight tetrahedra:** RBF, "Dymaxion Comprehensive System: Introducing Energetic Geometry," in *Discourse*, 163.

190 **"comprehensive system":** RBF with Applewhite, *Synergetics*, 84.

190 **"energetic geometry":** RBF, "Dymaxion Comprehensive System," 160–68.

190 **By adding up the balls:** "Made experiment composing spheres—discovered 92 elements," RBF, diary entry, August 8, 1943, B59-F4.

191 **"my life's policy":** Gilbert A. Harrison, *The Enthusiast: A Life of Thornton Wilder* (New Haven, CT: Ticknor & Fields, 1983), 320.

191 **"as a joke, like numerology":** Applewhite, "Account by R. Buckminster Fuller of His Relations with Scientists," Series 8, B12-F23, 3.

191 **John Wolfenden:** Ibid., 2–3.

191 **Vannevar Bush:** RBF claimed later to have met Bush (Hatch, 172), but no such meeting was recorded in his diary.

191 **"relating the discovery":** RBF, office diary entry, October 13, 1943, B59-F4.

191 "powerfully effective": RBF and Dil, *Humans in Universe*, 125.

191 "unanimous surprise": RBF, "Earth, Inc.," in *BFR*, 232n.

191 "She aids me": RBF to Watson Davis, June 5, 1947, B67-F14.

191 "Dymaxion Comprehensive System": RBF, "Dymaxion Comprehensive System," 160–68.

192 attempt at "lucidity": RBF to Kenneth Beirn, March 13, 1944, B58-F2.

192 "It may be that we": RBF, office diary entry, October 26, 1943, B59-F4.

192 "The 'higher-ups' have decided": Alex Taub to RBF, memo, December 15, 1943, B57-F2.

192 German magazine: According to Dave Cort, the original article ("Air Transport," *Life*, November 30, 1942, 75–85) was reprinted in *Die Wehrmacht* (Cort to RBF, December 7, 1943, B57-F2).

192 "an aluminum bell": RBF to J. E. Haines, December 14, 1943, B57-F2.

192 "Fluid Geography": RBF, "Fluid Geography," *American Neptune* 4, no. 2 (April 1944): 118–36. Reprinted in *BFR*, 133–52.

192 "I prayed so hard last night": Lacey to RBF, undated (1944), B58-F2.

193 "the uncomfortable position": RBF, office diary entry, June 21, 1944, B59-F4.

193 "all our experience": "What Is a House?," *Arts & Architecture*, July 1944. Reprinted in Charles Eames and Ray Eames, *An Eames Anthology: Articles, Film Scripts, Interviews, Letters, Notes, Speeches*, ed. Daniel Ostroff (New Haven, CT: Yale University Press, 2015), 19.

193 "All three agreed": RBF office diary, July 12, 1944, B59-F4.

193 "I think this is the most important news": Ibid., July 13, 1944, B59-F4.

193 "round house with central chassis": Ibid., August 10, 1944, B59-F4.

193 "promote individual liberty": William F. Buckley Jr., "My Own Secret Right-Wing Conspiracy," in *Miles Gone By: A Literary Autobiography* (Washington, DC: Regnery, 2004), 503.

194 "Well, what are we waiting for?": "Fuller's House," *Fortune*, April 1946, 176. Although the article does not have a byline, it is attributed to George Nelson in Stanley Abercrombie, *George Nelson: The Design of Modern Design* (Cambridge, MA: MIT Press, 2000), 76.

194 They returned "triumphant": RBF, office diary entry, August 19, 1944, B59-F4.

194 "Bucky, you may save": Ibid., September 12, 1944, B59-F4.

195 "More than busy": Lacey to RBF, October 23, 1944, B59-F3.

195 "First Day Alone": Ibid.

195 he sounded "wistful": Lacey to RBF, November 5, 1944, B60-F1.

195 "Actually, our having two apartments": Ibid., December 1, 1944, B60-F2.

195 "Darling," she wrote, "I hope": AHF to RBF, October 26, 1944, B59-F3.

195 Wasserman embarrassed: "Notes by H. Wolf," April 15, 1945, B61-F5.

196 "a card-carrying journeyman-machinist": RBF, "Continuous Man," in *BFR*, 350.

196 underwriter's "fear of Bucky's suicide": Herman Wolf, year-end report, December 31, 1944, B60-F2, 9.

196 "too cramped": William Wasserman to Herman Wolf, February 24, 1945, B61-F3.

196 overhead carousel: Marks, 178.

196 "huge skies": Lacey to RBF, June 3, 1946, B64-F3.

196 train together to Washington: Michaelis, *Best of Friends*, 66.

197 "There will be plenty of time": RBF to Noguchi, January 19, 1945, B62-F4.

197 "but not if I win": Frank Lloyd Wright, "Ideas for the Future," *Saturday Review of Literature*, September 17, 1938, 14.

197 triangulated framework: Marks, 129.

197 a wind tunnel: Ibid., 130–31.

197 "little ships": "Washington Background," *Philadelphia Inquirer*, March 25, 1946.

198 Dymaxion Car #3: For the car's afterlife, see Foster and Glancey, *Dymaxion Car*, 115, and Herman Wolf, interviewed by Terry Hoover and Bill Pretzer, October 25, 2001, Benson Ford Research Center.

198 Japanese city of Hiroshima: The essayist Lewis H. Lapham claimed that RBF noted in his calendar that August 6, 1945, was "the day that humanity started taking its final exam" ("Spoils of War," *Harper's*, March 2002). No such entry has been found in the Chronofile.

198 no "practical application": RBF, marginalia in *Physical Science* by Sir James Jeans, quoted in Wong, 194n160.

198 "No matter how hard": Hatch, 179.

198 "As Bucky so aptly puts it": Lacey to Owen Smith, September 14, 1945, B61-F6.

199 "liaison between management": "Fuller's House," *Fortune*, April 1946, 174.

199 first projection to receive a patent: RBF, *Education Automation*, 19.

200 "The stacks of mudguards": RBF, "Designing a New Industry," in *BFR*, 198.

200 "mortality point": Ibid., 165.

201 "You can sit here": Ibid., 220.

201 "Negro press": Policy meeting memo, February 18, 1946, B63-F1.

202 "stationary airplane": RBF, "Designing a New Industry," 223.

202 Wichita House: The following description is based largely on Marks, 128–41.

202 "ovolving" conveyor shelves: Neder, *Fuller Houses*, 79–91. One year earlier, *Life* had publicized a "storage wall" system, without conveyor shelves, designed by RBF's friend George Nelson and the architect Henry Wright ("Storage Wall," *Life*, January 22, 1945, 64–71).

202 total price: Mary Roche, "50,000 Homes in '47 Aim of Company," *New York Times*, March 22, 1946, 18.

202 **"likely to produce"**: "Fuller's House," *Fortune*, April 1946, 167.

202 **3,500 orders**: Gorman, *Designing for Mobility*, 70.

203 **predicted in 1927**: "Dream House Goes into Production," *Nesosho (MO) Daily News*, March 22, 1946.

203 **"When the fin"**: "Notes of Harry L. Linsky," November 29, 1946, B66-F1.

203 **"Dear Bucky has worked"**: RBF to Edward D. Stone, November 15, 1946, B67-F14.

203 **"the Ides of March"**: RBF to J. Arch Butts, January 11, 1948, B67-F22.

203 **her "beau"**: Alexandra Snyder May, interviewed by author, June 7, 2019.

203 **return to art school**: Allegra Fuller Snyder, interviewed by author, March 26, 2019.

203 **"I just feel"**: AHF to RBF, "Friday" (March 29, 1946), B63-F3.

204 **"just a kid"**: Ibid.

204 **"You have a tendency"**: Allegra Fuller to RBF, "Thursday" (March 28, 1946), B63-F3.

204 **"*I* know that"**: AHF to RBF, "Sunday" (March 31, 1946), B63-F3.

204 **"giving a surprise party"**: Ibid.

204 **"I think you should know"**: Hatch, 181. Herman Wolf provided a somewhat different account of the incident: "So, around that time, Bucky's wife came out for the first time. And, unfortunately, I was so mad at Bucky that I made some insulting remarks, which she took very personally." Herman Wolf, interviewed by Terry Hoover and Bill Pretzer, October 25, 2001, Benson Ford Research Center.

204 **"I just hate that man"**: Tony Huston, interviewed by author, April 11, 2019.

205 **"It was wonderful"**: AHF to RBF, Sunday night, March 31, 1946, B63-F3.

205 **"express desire"**: Herman Wolf to staff, memo, April 8, 1946, B532-F4.

205 **"false agreement"**: RBF, note, undated (1948), B532-F4.

205 **"love and admiration"**: Herman Wolf to RBF, April 7, 1946, B532-F4.

205 **"He never once"**: Lacey to RBF, June 17, 1946, B64-F3.

205 **"opposed as chief policy maker"**: RBF to Safford Colby, June 26, 1946, B65-F2.

205 **"I am experienced"**: Ibid.

205 **"What I was really developing"**: Hatch, 180.

206 **"just not honest"**: Ibid.

206 **"I was forced"**: Ibid.

206 **"Upon completion"**: RBF to P. S. Barrows, April 29, 1946, B64-F1.

207 **"It may be that one"**: Ibid.

207 **"I welcome the opportunity"**: RBF to George Nelson, May 9, 1946, B64-F2.

207 **"a stooge"**: RBF to Edward D. Stone, November 15, 1946, B67-F14.

207 **"People just aren't"**: William Harris to RBF and Herman Wolf, May 17, 1946, B64-F3.

208 **"arranged to do battle"**: RBF to Harris, May 24, 1946, B64-F2.

208 "There is no feud": Ibid.

208 offered Wolf his "blessing": Ibid.

208 "put Herman where he belongs": RBF to Lacey, May 29, 1946, B64-F3.

208 "[We] were making small headway": Lacey to RBF, June 5, 1946, B64-F3.

208 rowing catamaran: "Dymaxion Projects," December 30, 1946, B66-F1.

209 lots throughout Wichita: "Ignoring the Past, Embracing Today, Will This Round Prefab Capture Tomorrow?," *Architectural Forum*, March 1945, 32.

209 "just a little": Lacey to RBF, "Monday night" (June 24, 1946), B64-F3.

209 aviation plants out west: Wong, 140.

209 "which they may deem necessary": RBF to Leland O'Neal, June 26, 1946, B65-F2.

209 "Remember that almost everyone": Lacey to RBF, "Monday night" (June 24, 1946), B64-F3.

209 "impractical" and "unreasonable": Ibid., December 13, 1946, B66-F1.

209 "The April debacle": Ibid., November 5, 1946, B66-F1.

210 "Is it still round?": AHF to RBF, "Thursday" (June 27, 1946), B64-F2.

210 "You had better notify them": RBF, "Notes of Understanding of R. Buckminster Fuller of Events of and Subsequent to January 25, 1947," February 15, 1947, B66-F4, 7.

210 California client: RBF and Lacey met in Santa Barbara with the executive Burton Tremaine, who was interested in building a house in Arizona. Tremaine introduced them to the architect John Tracey, who met with bankers to discuss production in California or Mexico (correspondence in B65-F3).

210 "one nebulous 'big deal' after another": Lacey to Lieutenant Robert C. Strobell, February 19, 1947, B66-F3.

211 "an alternate plan": RBF to Oliver F. Billingsley, February 25, 1947, B66-F3.

211 "I'm a very patient guy": Herman Wolf, interviewed by Terry Hoover and Bill Pretzer, October 25, 2001, Benson Ford Research Center.

211 "We may not be able": Lacey, introduction to "Earth, Inc.," B527-F3, 3.

211 "philosophy resolved": RBF, "Earth, Inc.," in *BFR*, 241.

211 "Unity is inherently two": RBF, "Appendix I: Energetic Geometry," in *Discourse*, 170.

211 "someday integrate them": RBF, notes on College of the Ozarks, 1947, B67-F10.

212 Dartmouth Hall: RBF, "Education for Comprehensivity," in RBF, Eric A. Walker, and James R. Killian Jr., *Approaching the Benign Environment* (New York: Collier Books, 1970), 18–19.

213 **"This little man"**: Christian William Miller to Monroe Wheeler, April 22, 1947, Glenway Wescott Papers, Yale University, Beinecke Rare Book and Manuscript Library, Series I, folder 1148. In *When Brooklyn Was Queer* (New York: St. Martin's Press, 2019), Hugh Ryan suggests that Bill Miller and RBF had "a brief, intense relationship" (226). A review of Miller's letters suggests that their interactions were confined to this lecture, although he later mentioned "reading some things Mr. Fuller has sent to me" (Miller to Wheeler, July 15, 1947, Glenway Wescott Papers). Miller made no further reference to RBF in his diary (Christian William Miller Papers, Yale University, Beinecke Rare Book and Manuscript Library, Series II—Appointment Books—1944–1969) or his other writings from this period (Christian William [Bill] Miller Photographs and Papers, ONE National Gay & Lesbian Archives, USC Libraries, boxes 44–46).

213 **"It was an amphibious plane"**: Cochrane, *Sense of Significance*, 126.

213 **"probably the only time"**: Allegra Fuller Snyder, author interview, March 26, 2019.

213 **"I must rather count on you"**: Lacey to RBF, May 26, 1947, B68-F5.

213 **"The advertising man"**: Mike Goldgar press release, 1947, B67-F16.

214 **"Your voice on the phone"**: Lacey to RBF, May 27, 1947, B68-F5.

214 **"His intentions are honorable"**: Ibid.

214 **"Please be a little more detailed"**: Lacey to RBF, June 4, 1947, B67-F14.

214 **George Balanchine**: Hatch, 183. Details confirmed by Jaime Snyder, email to author, April 27, 2021.

215 *The Secret Life of Walter Mitty*: RBF granted permission for a model of the house to be featured as a novelty hat—designed by Irene Sharaff—in the musical number "Anatole of Paris" (Neder, *Fuller Houses*, photo section following 191).

215 **"forever grounded this airplane"**: Marks, 141.

215 **"the most important prefabricated"**: Pawley, *Buckminster Fuller*, 13.

215 **never even managed to patent**: According to Richard Hamilton, a patent for the "Fuller House" (No. 655,601) was filed on March 20, 1946 (Hamilton, "R. Buckminster Fuller: Comprehensive Designer" [unpublished manuscript], "Patents Held by Fuller," B529-F11). No filing date or reason for abandoning the claim is given in RBF, *Inventions*, 95.

215 **"When I return"**: Lacey to RBF, August 19, 1947, B68-F6.

215 **"So often while here"**: Ibid., August 25, 1947, B68-F7.

216 **in his wallet**: Wallet items, photographs, Series 7, B19-F10.

PART THREE: GREAT CIRCLES

217 **"One single picture"**: RBF with Applewhite, *Synergetics*, 87.

SEVEN: GEODESICS

219 "When we say 'we think'": Ibid., 238.

220 "How blind people can be": Charles Eames to RBF, October 24, 1947, B68-F8.

220 "like lungs": Charles Eames to Cynthia Floyd, December 5, 1947, B68-F10.

220 "Ray of course": Charles Eames to RBF, October 24, 1947, B68-F8.

220 considered as a candidate to succeed: Thomas Dyja, *The Third Coast: When Chicago Built the American Dream* (New York: Penguin, 2013), 144.

220 László Moholy-Nagy: László Moholy-Nagy to RBF, March 3, 1939, and RBF to Moholy-Nagy, March 6, 1939, B39-V63.

220 exhibiting the bathroom: RBF to Serge Chermayeff, February 10, 1944, B58-F1.

220 "I had a long talk": Serge Chermayeff to RBF, July 14, 1947, B67-F20.

220 "a public appearance": RBF to Serge Chermayeff, November 17, 1947, B68-F9.

221 "from the hundreds": Cynthia Floyd to RBF, January 5, 1948, Series 18, B122-F10. Floyd's letter confirmed RBF's earlier encounter with Einstein: "When [Straus] mentioned the house Einstein remembered you."

221 "a woeful bit": RBF to Cynthia Floyd, January 20, 1948, B68-F2.

221 "In all humility": RBF to Albert Einstein, January 17, 1948, in *Discourse*, 171.

221 After a standing ovation: Harold and Mary Cohen, interviewed in Al Gowan, *Shared Vision: The Second American Bauhaus* (Cambridge, MA: Merrimack Media, 2012), 26.

221 "an unsolved social problem": RBF to "Smoky" (A. H. Mansbach), February 9, 1948, B67-F18.

221 "seating tool": Abercrombie, *George Nelson*, 110–11.

221 "I don't know to this day": Ibid., 111. For an alternative account, which attributes the ball clock to the designer Irving Harper, see Sadao, 117.

222 "what Archimedes sought": RBF, "Eureka—Eureka," April 23, 1948, in *Discourse*, 174.

222 "a railroad station": RBF, "The Comprehensive System," January 6, 1944, B58-F1, 13.

223 "identical axial rotations": RBF, "Dymaxion Kinetic Vector Structure," December 15, 1947, reproduced as the frontispiece to *Synergetics*.

224 "Atomic Buckalow": Wong, fig. 2.05a, and Wigley, *Buckminster Fuller Inc.*, 152.

225 "triangular intertension": Ibid.

225 "spherical mast": RBF, *Inventions*, 127.

226 he tripped, shattering it: Cochrane, *Sense of Significance*, 128–29.

226 "[Straus] didn't believe": Cynthia Floyd to RBF, May 24, 1948, B68-F18.

227 "a greenhouse enclosure": Serge Chermayeff to RBF, April 28, 1948, B68-F17.

227 "private sky": RBF to Chermayeff, May 26, 1948, B68-F17.

227 Black Mountain College: Sources on RBF's first visit to Black Mountain include Hatch, 187–92; Martin Duberman, *Black Mountain: An Exploration in Community* (Garden City, NY: Anchor Press, 1973), 292–304; and Ruth Asawa, "Black Mountain College," in *Anthology*, 201–4.

227 Theodore Dreier: Dreier, an engineer, should not be confused with the novelist Theodore Dreiser, whom RBF had encountered separately at Romany Marie's.

227 "the Oedipus complex": Duberman, *Black Mountain*, 31.

227 "The summer artists": Ibid., 291–92.

228 July 12, 1948: "Will arrive Monday July 12th" (RBF, notes, 1948, B68-F16).

228 "Call me Bucky": Asawa, "Black Mountain College," 202.

228 "cardboard polyhedra": Kenneth Snelson, "Snelson on the Tensegrity Invention," *International Journal of Space Structures* 11, no. 1/2 (1996): 43.

228 "unknown, humble": Sande-Friedman, "Kenneth Snelson and the Science of Sculpture in 1960s America" (doctoral thesis, Bard Graduate Center, 2012), 54.

228 "He looks stuffy": Elaine de Kooning, "de Kooning Memories," in *Black Mountain College: Sprouted Seeds—An Anthology of Personal Accounts*, ed. Mervin Lane (Knoxville: University of Tennessee Press, 1990), 246.

228 tossed it gently: RBF to Snelson, March 1, 1980, Series 8, B184-F4.

228 "the numbers nine and three": de Kooning, in *Black Mountain College*, 247.

229 "I knew you would": Ibid.

229 Max Dehn: Snelson, "Tensegrity Invention," 44.

229 "a great idealist": Duberman, *Black Mountain*, 295.

229 "I love you very much": Natasha Goldowski to RBF, October 24, 1948, B69-F9.

229 "for the least logical way": de Kooning, in *Black Mountain College*, 248.

229 "the eyes of a visionary": Ibid., 247.

229 "Zoroaster speaking Islamic": Prefatory material to Asawa, "Black Mountain College," in *Anthology*, 201.

229 affair with Philip Johnson: Mark Lamster, *The Man in the Glass House: Philip Johnson, Architect of the Modern Century* (New York: Little, Brown, 2018), 109.

229 caravan school: Dana Miller, "Thought Patterns," in *Starting with the Universe*, ed. Hays and Miller, 27.

229 "liveliness and optimism": Kenneth Silverman, *Begin Again: A Biography of John Cage* (Evanston, IL: Northwestern University Press, 2012), 74.

229 "From the beginning": John Cage, *Anarchy: New York City—January 1988* (Middletown, CT: Wesleyan University Press, 2001), v.

229 "Oh, isn't that marvelous": Duberman, *Black Mountain*, 296.

230 "Bucky Fuller and his magic show": Ibid.

230 "to rehouse Black Mountain": George Nelson to RBF, July 16, 1948, B69-F7.

231 "a giant's plate": Asawa, "Black Mountain College," in *Anthology*, 203.

231 chromosomes of a fruit fly: Sande-Friedman, "Snelson and Science of Sculpture," 57.

231 "critical point": Duberman, *Black Mountain*, 298.

231 "It was predicted to fall down": Ibid.

231 "supine dome": Prefatory material to Asawa, "Black Mountain College," 201.

231 "Let's put it together anyhow": de Kooning, in *Black Mountain College*, 248.

231 "I only learn what to do": John Cage, "An Autobiographical Statement," 1990, https://johncage.org/autobiographical_statement.html (accessed January 2021).

231 taping wooden struts: RBF to Snelson, March 1, 1980, Series 8, B184-F4.

231 "gently collapsed": Marks, 182.

231 Rudolf Laban: Asawa, "Black Mountain College," 202, and Lionel Wolberger, email to author, January 26, 2021. See also Caspar Schwabe, "Eureka and Serendipity: The Rudolf von Laban Icosahedron and Buckminster Fuller's Jitterbug," *Proceedings of Bridges 2010: Mathematics, Music, Art, Architecture, Culture*, 271–78.

231 James Fitzgibbon: Edward S. Popko, *Divided Spheres: Geodesics and the Orderly Subdivision of the Sphere* (Boca Raton, FL: CRC Press, 2012), 26.

232 "I wish you": RBF to Snelson, March 1, 1980, Series 8, B184-F4.

232 "to skip over a discrepancy": Sande-Friedman, "Snelson and Science of Sculpture," 56.

232 "When I listen to Bucky talk": Snelson, "Snelson on the Tensegrity Invention," *International Journal of Space Structures* 11, no. 1/2 (1996): 44.

232 "I'm afraid of making a damn fool": Duberman, *Black Mountain*, 302.

232 "We skipped around": Ibid., 302–3.

233 "let me learn to be myself": Ibid., 303.

233 Robert Rauschenberg: Calvin Tomkins, *Off the Wall: A Portrait of Robert Rauschenberg* (New York: Picador, 2005), 23 and 31.

233 "tinker toy": Snelson to RBF, September 9, 1948, B69-F8.

233 "Einstein has been searching": RBF to Snelson, September 18, 1948, B69-F8.

233 "a wonderful young protégé": Cynthia Floyd to RBF, September 22, 1948, B69-F3.

233 "might turn out to be": Sande-Friedman, "Snelson and Science of Sculpture," 59.

234 **"He seems to be conducting":** Peter Blake, *No Place Like Utopia: Modern Architecture and the Company We Kept* (New York: W. W. Norton, 1993), 93.

234 **"What I have just discovered":** Ibid., 94.

234 **"an incomprehensible sequence":** Ibid.

234 **"your bebop representative":** RBF to John and Jano Walley, telegram, March 13, 1949, Institute of Design Collection, University of Illinois at Chicago, B7-F216.

234 **"The next time":** Peter Blake, "The World of Buckminster Fuller," *Architectural Forum*, January 1972, 49.

234 **"because he was mixing":** Dyja, *Third Coast*, 149.

234 **"a Merlin's cave":** Ibid., 148.

234 **Mies van der Rohe:** Ibid.

234 **Konrad Wachsmann:** Wong, 267–72.

235 **"packaged house":** Fiona MacCarthy, *Gropius: The Man Who Built the Bauhaus* (Cambridge, MA: The Belknap Press of Harvard University Press, 2019), 403.

235 **Harold Young:** Michael Reid, email to author, May 10, 2021. According to Reid, Young later worked for the Chicago Housing Authority and the Department of Housing and Urban Development in Maryland. In 1963 Young asked RBF for information on domes while conducting planning work in Uganda. See correspondence in B125-F7.

235 **"kenning space":** Institute of Design project description, spring 1949, B70-F3.

235 **fog gun:** Marks, 99–100.

235 **"the worst thing":** RBF, *EIK*, session 11.

235 **venetian blinds covered in cement:** Ibid.

235 **aluminum pans:** RBF, "Architecture Out of the Laboratory," *Student Publication*, University of Michigan, College of Architecture and Design, Spring 1955, 16.

235 **Necklace Dome:** Wigley, *Buckminster Fuller Inc.*, 157–60.

236 **eight thousand square feet:** RBF, "Preview of Building," in *BFR*, 298–99.

236 **"The winner will be":** RBF, "Design for Survival—Plus," in *BFR*, 278.

236 **Reginald E. Gillmor:** Wong, 236.

236 **"He seemed to be a thinker":** Popko, *Divided Spheres*, 27.

236 **"a civilian stranger":** RBF to General Follett Bradley, March 25, 1949, B70-F3.

236 **"as good as a tent":** Ibid.

237 **"shot to the moon":** RBF, "Moon Structures," in *SD*, 10974.

237 **"to dodge widely":** RBF, "Preview of Building," 294.

237 **"His car or his train":** Ibid., 295.

237 "a super-camping structure": Ibid., 300.

237 "We need only revolve our charts": Ibid., 287.

237 reunited with Cage: Allegra Fuller Snyder, "John Cage Tribute," https://
     www.streamingmuseum.org/post/john-cage-tribute-allegra-fuller-snyder
     (accessed April 2021).

237 "I felt as though": Ruth Asawa to RBF, November 1, 1948, B69-F12.

237 Fuller designed her a silver ring: Marilyn Chase, *Everything She Touched:
     The Life of Ruth Asawa* (San Francisco: Chronicle, 2020), 70.

238 "How did you solve?": Snelson to RBF, December 5, 1948, B69-F12.

238 "a concentration camp": Natasha Goldowski to RBF, October 24, 1948,
     B69-F9.

238 instructors from the Institute of Design: The teachers from Chicago
     included John and Jano Walley; Emerson and Dina Woelffer; and Nataraj
     Vashi and Praveena Mehta, who taught Hindu philosophy and dance.

238 Masato Nakagawa: Dan Brogan, "Decorated WWII vet Masato
     Nakagawa, 64," *Chicago Tribune*, May 27, 1983. Nakagawa later worked
     with the graphic designer Herbert Bayer on the *World Geo-Graphic Atlas*,
     which featured the Dymaxion Map.

238 "I shall be there": Snelson to RBF, April 28, 1949, B70-F11.

238 "Student Dymaxion Designer": License for Joseph Manulik, May 27,
     1949, B70-F2.

238 "somewhat escapist": John and Jano Walley to Daisy Igel and "Polita,"
     August 2, 1949, quoted in Wong, 240.

238 "The Comprehensive Harvester": RBF, "Comprehensive Designer," in
     Discourse, 243.

238 "Man has now completed the plumbing": Ibid., 244.

239 "by indirection": Ibid., 245.

239 Necklace Dome: The account of the dome's construction is based on
     Wigley, *Buckminster Fuller Inc.*, 140–41.

240 fiberglass panels: Pawley, *Buckminster Fuller*, 120.

241 "When I showed him the sculpture": Valentín Gómez-Jáuregui, *Tensegrity
     Structures and Their Application to Architecture* (Santander, Sp.:
     Cantabria University Press, 2020), 160–61.

241 "As I photographed him": Snelson, "Snelson on the Tensegrity Invention,"
     International Journal of Space Structures 11, no. 1/2 (1996): 46.

242 "a special case demonstration": RBF to Snelson, March 1, 1980, Series 8,
     B184-F4.

242 novel development: Similar structures had been exhibited by the Latvian-
     Soviet artist Karl Johansson in 1920. Although his work was known to
     László Moholy-Nagy and Josef Albers, there is no evidence that either
     Fuller or Snelson were exposed to it during this period. See R. Motro,
     "Structural Morphology of Tensegrity Systems," *International Journal*

of Space Structures 11, no. 1/2 (1996): 233–34, and Lionel Wolberger at https://tensegritywiki.com/wiki/Ioganson,_Karl (accessed January 2021).

242 **"No one else in the world":** Snelson, "Tensegrity Invention," 46.

242 **stole it from his room:** Ibid.

242 **two hundred pages long:** Duberman, *Black Mountain*, 333.

242 **Anne had tea:** "Bill Treichler Remembers Black Mountain College," March 2004, http://www.blackmountaincollegeproject.org/MEMOIRS /TREICHLERwilliamMEMOIR.htm (accessed May 2021).

242 **model airplane wheels:** Sande-Friedman, "Snelson and Science of Sculpture," 140. RBF's version appears in RBF with Applewhite, *Synergetics*, 207.

242 **"Christ's Dymaxion disciples":** Sande-Friedman, "Snelson and Science of Sculpture," 83.

242 **stopped speaking to Fuller:** Duberman, *Black Mountain*, 333.

242 **"I can see him still":** Hatch, 195.

242 **"in the selective mutations":** RBF, "Total Thinking," in *BFR*, 310.

243 **"not informative but provocative curves":** Ibid., 312.

243 **"Every night he sleeps":** Charles Olson, *Kyklops II* manuscript, B71-F9, 1.

243 **"This is not a trip":** Ibid., 2.

243 **"that filthiest of all":** Charles Olson, *Charles Olson & Robert Creeley: The Complete Correspondence*, vol. 10, ed. Richard Blevins (Santa Rosa, CA: Black Sparrow Press, 1996), 68.

243 **"In what sense":** Ibid.

244 **"As nearly as I can tell":** Myron Goldsmith to RBF, August 16, 1949, B70-F19.

244 **Walter Gropius and László Moholy-Nagy:** Martino Peña Fernández-Serrano and José Calvo López, "Projecting Stars, Triangles, and Concrete," *Architectura* 47 (2017): 35.

244 **"If the Zeiss engineer":** RBF to Snelson, March 1, 1980, Series 8, B184-F4.

245 **"Rule I: Never show half-finished work":** RBF, "Universal Requirements of a Dwelling Advantage," in *BFR*, 269.

245 **"any tool-up expense whatsoever":** RBF, "Noah's Ark #2," in *Discourse*, 213.

245 **purchased "spontaneously":** Ibid.

245 **"I am absolutely confident":** Condensed from Hatch, 196.

246 **"Had she kept it":** Ibid.

246 **the "overhead":** RBF, "Noah's Ark #2," 213.

246 **"by borrowing on life insurance":** Ibid.

246 **put up a stunning $32,000:** McAtee, "Geodesic Dreams," in *Montreal's Geodesic Dreams,* ed. McAtee, 33.

246 **"an interlaced network":** Wigley, *Buckminster Fuller Inc.*, 146.

246 **"involute" shell:** Don and Robert Richter and Louis J. Caviani to RBF, May 15, 1950, B71-F11.

246 **Snelson left:** Sande-Friedman, "Snelson and Science of Sculpture," 65.

246 "I knew that if I went to work": Harold Cohen, author interview, March 6, 2021.

246 established by Lindsay: Details of Lindsay's background and the Montreal branch are based on McAtee, "Geodesic Dreams," 11–13 and 19–21.

247 "I have determined": Jeffrey Lindsay to Lee A. Hukar, September 13, 1949, B71-F13.

247 "You will doubtless understand": Hukar to RBF, September 21, 1949, B71-F13.

247 "Have you lost faith?": RBF to Hukar, July 11, 1949, B71-F7.

247 "thoroughly chastened": Hukar to RBF, July 12, 1949, B71-F7.

247 his reputation as a "promoter": Notes on phone conversation among RBF, Harold Young, and Hukar, September 25, 1949, B71-F13.

247 "It is unwise to present": RBF to Hukar, November 22, 1949, B71-F12.

248 "Bucky knows each one of us": Lindsay to RBF, December 19, 1949, B71-F12.

248 underwrite the Necklace Dome: McAtee, "Geodesic Dreams," 18.

248 Cyril Stanley Smith: According to a report on the meeting, Smith "fully understood Bucky's basic assumptions and nowhere refuted fundamental hypotheses" (Series 18, B129-F7). RBF later claimed that "Smith and others" introduced him to the physical chemist Harold Urey and the physicist Leo Szilard (Applewhite, "Account by R. Buckminster Fuller of His Relations with Scientists," Series 8, B12-F23, 5).

248 "peaceful types": Lindsay to "Mama" (Hukar), September 9, 1949, B71-F13.

248 "firecracker umbrella": Lindsay to RBF, March 7, 1955, B87-F1.

248 he and Fuller installed: Cochrane, Sense of Significance, 137.

248 invisible gaps, or "windows": Shoji Sadao, "A Brief History of Geodesic Domes," in Anthology, 25.

249 "the psychological factors": RBF, "Noah's Ark #2," in Discourse, 215.

249 "Beyond certain magnitudes": Ibid., 207.

249 "the brightest man alive": Hiram Hadyn, The American Scholar, Spring 1966, 190. Hadyn dated this discussion to "some twenty years ago (a little more, a little less)."

249 "chuckle indulgently": "Bucky Inc.," Time, November 7, 1949, 57.

249 "Continuous Tension": Itinerary, January 19, 1950.

249 recently published book Dianetics: John Moehlman to RBF, May 25, 1940, Series 18, B148-F11.

249 "an intensive professional course": W. Bradford Shank to RBF, June 7, 1950, B72-F7.

250 lecture the next day: L. Ron Hubbard, The Technical Bulletins of Dianetics and Scientology, vol. 1, 1950–1953, 14 (provided to the author by Chuck Beatty). RBF may also have met the science-fiction editor John

W. Campbell, who was involved at the foundation on a daily basis (Alec Nevala-Lee, *Astounding: John W. Campbell, Isaac Asimov, Robert A. Heinlein, L. Ron Hubbard, and the Golden Age of Science Fiction* [New York: Dey Street, 2018], 270–71).

250 **"I cannot continue"**: Don Richter to RBF, August 21, 1950, quoted in Wong, 294.

250 **Eleanor and Victor Cannon**: Wong, 189.

251 **"If sometimes"**: McAtee, "Geodesic Dreams," in *Montreal's Geodesic Dreams*, ed. McAtee, 27.

251 **Weatherbreak Dome**: For the dome's afterlife in the Hollywood Hills, see Dan MacMasters, "A Bubble on a Hilltop," *Los Angeles Times Home Magazine*, July 1, 1962.

251 **"the first convincing proof"**: Lindsay to RBF, March 7, 1955, B87-F1.

251 **"a dream come true"**: McAtee, "Geodesic Dreams," 29.

251 **"When the structure was completed"**: RBF, "Architecture Out of the Laboratory," *Student Publication*, University of Michigan, College of Architecture and Design, Spring 1955, 17.

251 **"a message"**: Sadao, "Brief History of Geodesic Domes," in *Anthology*, 25.

252 **triacon grid**: Ibid.

252 **Kidd Brewer**: Wong, 292–94.

252 **Zane Yost**: Marks, 187.

252 **three-quarter sphere**: Ibid.

252 **Patrick Baird**: McAtee, "Geodesic Dreams," 23–25.

252 **"totally skeptical"**: Abercrombie, *George Nelson*, 110.

252 **Bubble House**: Wong, 278–83.

252 **"Every problem"**: Charles Eames to RBF, August 13, 1950, quoted in Wong, 283.

253 **"What would Bucky Fuller say?"**: Eames and Eames, *Eames Anthology*, 339.

253 **another Dymaxion Car**: Foster and Glancey, *Dymaxion Car*, 210.

253 **Richard Hamilton**: Hamilton worked on the book between 1951 and 1957, but died before it could be completed. A partial unpublished draft of "R. Buckminster Fuller: Comprehensive Designer" can be found in B529-F11.

253 **"We sparked right away"**: Allegra Fuller Snyder, author interview, March 26, 2019.

253 **"He went to her"**: Hatch, 203.

253 **casually prejudiced remarks**: See, for example, her reference to "the little Jew boss" at the shop where she worked briefly (AHF to RBF, November 15, 1948, B69-F9).

253 **"Mr. Fuller"**: *AMS*, 93.

254 **"Anne gave Bob"**: Hatch, 203.

254 around 1970: "An Engineer Sees a Bright Future," *Independent Record* (Helena, MT), July 8, 1952.

254 "imagining that one": "Fuller Lectures on Motion Trend," *Kingston (NY) Daily Freeman*, July 25, 1951, 9.

254 Spaceship Earth: RBF, *EIK*, session 9. For an early use of the metaphor, see RBF to Harold E. Watson, April 19, 1955, Series 10, B13-F21: "Many a youngster says today: 'I wonder what it would feel like to be rocketing through Universe in a Space Ship?' The answer is: 'What *does* it feel like?'"

254 "to assert claims": Donald Robertson to RBF, July 12, 1951, quoted in Wong, 230.

254 "substantially uniform": RBF, *Inventions*, 130.

255 "patented error": T. C. Howard in Wong, 234n311.

255 "a paranoia that seemed": Wong, 233n309.

255 "If I had not taken out patents": RBF, *CP*, 147.

255 "The expense in this one": "Geodesic Dome," *Architectural Forum*, August 1951, 147.

255 Charles Eames: "We are often apt to think there are no Leonardos today, and as usual we are wrong, because there are. . . . Among such great original thinkers, we must certainly all be grateful for Buckminster Fuller" (Charles Eames, quoted in Richard Hamilton, "R. Buckminster Fuller: Comprehensive Designer," B529-F11, iii-iv).

255 "to develop a portable": J. T. Lawler to RBF, September 27, 1951, quoted in Wong, 303.

256 cotton mill: RBF, *CP*, 323–27.

256 put together a test section: Sherman Pardue Jr., *Student Publications of the School of Design*, North Carolina State College, Fall 1952, 7–8.

256 Alexander Graham Bell: Wong, 160–62.

256 isotropic vector matrix: The octet truss at North Carolina State was described as "the isotropic vector matrix or the tetrahedron truss" (Sherman Pardue Jr., *Student Publications of the School of Design*, North Carolina State College, Fall 1952, 8).

256 full octet truss: RBF with Applewhite, *Synergetics* 2, 232.

257 construct a plywood dome: Popko, *Divided Spheres*, 27–28, and Katrina Howard Fairley, email to author, January 5, 2021.

257 "Anne and I feel": McAtee, "Geodesic Dreams," in *Montreal's Geodesic Dreams*, ed. McAtee, 33.

257 "You achieved a far greater effort": Ibid.

257 For his first class: Sadao, 141–43.

257 Shoji Sadao: Biographical sources include Neil Genzlinger, "Shoji Sadao, Quiet Hand Behind Two Visionaries, Dies at 92," *New York Times*, November 13, 2019, and Shoji Sadao, interview with Thomas Devaney,

Amy Sadao, Elizabeth Asawa, Marty Griggs, and Tsuneko Sadao, November 22, 2015 (provided to the author by Devaney).

258 **an ignominious fate on Halloween:** Wigley, *Buckminster Fuller Inc.*, 245–48.

258 **revision of the Dymaxion Map:** Sadao, 149.

258 **a wire overlay:** *AMS*, 160.

258 **"self-effacing":** Shoji Sadao, interview with Devaney et al.

258 **Niels Diffrient:** Sadao, 145.

259 **"Mr. Industry himself":** Marks, 61.

259 **"a very fancy automobile":** RBF, *EIK*, session 8.

259 **worked them out on an envelope:** Robert W. Marks, "The Breakthrough of Buckminster Fuller," *New York Times Magazine*, August 23, 1959, 42.

259 **Bubble House:** Fred M. Taylor, "The 'Bubble House' and Other Summer Projects," *Student Publications of the School of Design*, North Carolina State College, Fall 1952, 20–22.

259 **"He may have been a machinist":** Wong, 271.

259 **"[The students] as much as dared":** Wolfe, *From Bauhaus to Our House*, 57.

259 **"Personally, I think":** George Howe to RBF, December 15, 1952, quoted in Wong, 252.

259 **"Roam Home to a Dome":** RBF, *EIK*, session 11.

260 **"I think this was very much":** K. Michael Hays, "Fuller's Geological Engagement with Architecture," in *Starting with the Universe*, ed. Hays and Miller, 19n18.

260 **"aesthetic nonsense":** RBF, "Octet Truss in Yale Art Gallery," in *SD*, 12293.

260 **"Lou and Bucky":** Daniel López-Pérez, "Postscript: R. Buckminster Fuller and Louis I. Kahn," in RBF, *World Man*, ed. López-Pérez (New York: Princeton Architectural Press, 2014), 136.

260 **"until the whole dome":** Charles Correa, "Bucky," in *Anthology*, 254.

260 **dome greenhouse:** Wigley, *Buckminster Fuller Inc.*, 182–94.

260 **Growth House:** RBF, *CP*, 327–29.

260 **a wooden truss dome:** Wigley, *Buckminster Fuller Inc.*, 186–90.

260 **University of Oregon:** Ibid., 184–86.

260 **Lincoln Laboratory:** Wong, 324–26.

261 **alternate grid:** Popko, *Divided Spheres*, 38.

261 **Fuller and his students piled into a taxi:** Fred M. Taylor, "The 'Bubble House,'" 23–24.

261 **granted the commission:** Sources on the Ford Rotunda include Marks, 196–98; Sadao, 145–47; Wong, 255–57; and Gorman, *Designing for Mobility*, 124–25.

261 **floating it up:** Leo Donovan, "World of Wheels," *Detroit Free Press*, June 2, 1953.

262 **supporting her husband's work:** In an interview earlier that year, AHF claimed that RBF had once conducted an argument "entirely in digits" ("Personality," *Time*, January 19, 1953, 39).

262 **Queen Elizabeth II:** Allegra Fuller Snyder, author interview, March 26, 2019.

263 **haul away the pieces:** RBF, *EIK*, session 8.

263 **"Ford's son and grandson":** RBF, *Grunch of Giants*, 56.

263 **atomic bombs:** "Tape, Plastic and Aluminum: Ford Builds a 'Geodesic Dome,'" *Life*, June 8, 1953, 67.

263 *Architectural Forum*: "Bucky Fuller Finds a Client," *Architectural Forum*, May 1953, 108–12.

263 **equivalent of a permanent roof:** Wong, 258.

263 **"should be prepared to compete":** Lindsay to RBF, March 7, 1955, B87-F1.

264 **"were now friends":** Ibid.

264 **"a goddamned shame":** Sadao, 147.

EIGHT: CONTINUOUS TENSION

265 **"The most economical":** RBF with Applewhite, *Synergetics*, 656.

265 **Fuller drove west:** Sadao, 147–49.

265 **"vast amounts of dollars":** RBF, *CP*, 225.

266 **Walter Paepcke:** Paepcke had introduced RBF to the designer Herbert Bayer, who met with him in Aspen from 1951 to 1953 (*From the Library of R. Buckminster Fuller*, 32–33). RBF advised Bayer on his *World Geo-Graphic Atlas: A Composite of Man's Environment* (Chicago: Container Corporation of America, printed privately, 1953), which included a Dymaxion Map to display the distribution of "energy slaves" (Eva Díaz, "Dome Culture in the Twenty-first Century," *Grey Room* 24 [Winter 2011]: 86).

266 **Woods Hole:** Details of the project are drawn from Robert A. Mohr and Joseph M. Swerdlin, "Making the Woods Hole Dome: The Story of R. Buckminster Fuller's Oldest Geodesic Structure," *Proceedings of the IASS Symposium 2018*.

267 **bomb shelter:** Itinerary, May 10–May 20, 1953, and May 25, 1953.

267 **"like Douglas Fairbanks":** RBF, *EIK*, session 11.

267 **"It cast shadows":** Jack Kniskern, quoted in Mohr and Swerdlin, "Making the Woods Hole Dome."

267 **Bear Island:** By this time, Jim Hardie had retired as caretaker, and he died the following year (Marvel, "Hardie Family," in *Bear Island Centennial Book*, 83).

267 **"discontinuous compression":** "Building Problems May Be Solved by Princeton Sphere," *Central New Jersey Home News* (New Brunswick), November 15, 1953, 24.

267 Lee Hogden: Marks, 167.

268 struck by a snowplow: RBF, *World Design Science Decade 1965–1975, Phase 1, Document 2*, 39.

268 atomic bomb: "Building Problems May Be Solved by Princeton Sphere," 24.

268 "I wasn't there": RBF and Dil, *Humans in Universe*, 49.

269 rise into the air: Marks, 168.

269 hung from the ceiling: Gorman, *Designing for Mobility*, 183–85.

269 *Architectural Forum*: Wong, 305. According to Manuel Bromberg, Lane's interest was aroused by a model of a dome in James Fitzgibbon's office (Popko, *Divided Spheres*, 33).

269 Colonel Lane's plans: Details on the Marine dome program are drawn from Wong, 303–24, and Henry C. Lane, "A Study of Shelter Logistics for Marine Corps Aviation," Marine Corps (Air Station, Cherry Point, NC), c. August 1955.

269 "It was no troubles at all": "Copter Nabs House; Away It Goes," *Chicago Tribune*, January 29, 1954, 2.

270 *New York Times*: "Marines Try Out Flyable Shelter," *New York Times*, January 29, 1954, 1.

270 "air-deliverable housing": Itinerary, January 28, 1954.

271 "a position of authority": RBF to Walter Paepcke, March 19, 1954, Series 4, B2-F3.

271 "You and your university": Henry Lane to RBF, May 28, 1954, quoted in Wong, 314.

271 ordered to fly out: *AMS*, 135–37.

271 "I already know": Marks, "Breakthrough of Buckminster Fuller," *New York Times Magazine*, August 23, 1959, 44.

271 The Sky Eye: RBF, *Artifacts of R. Buckminster Fuller*, vol. 3, 206–7, and Baldwin, *BuckyWorks*, 228–29. For its intended use, see Peter Robertson, "An Australian Icon—Planning and Construction of the Parkes Telescope," proceedings of *Science with Parkes @ 50 Years Young*, October 2012.

271 "the most dramatic": Olga Gueft to Peter Muller, May 3, 1954, quoted in Wong, 341.

272 Shoji Sadao was dispatched: Sadao, 155.

272 "bathing cap": Olga Gueft, "2 Cardboard Geodesics from the USA," *Interiors*, November 1954.

272 Gran Premio: After the award was announced, the Commandant of the US Marines planned to present it to RBF on behalf of the Italians. To ensure that he retained the primary association with the project, RBF contacted Clare Boothe Luce, who arranged for the prize to be presented by Italian representatives (Wong, 345–46).

272 pool in Aspen: Marks, 191.

272 Minni-Earth: Wigley, *Buckminster Fuller Inc.*, 241–45.

272 **Flying Seed Pod:** "An Experiment in Quick Construction," *St. Louis Post-Dispatch*, January 15, 1956.

273 **radome group:** Sources on the radome project include Wong, 324–35, and Sadao, 150–52.

273 **"opening [his] mail":** T. C. Howard, quoted in Wong, 327.

273 **sign a guest book:** RBF to James Fitzgibbon, September 29, 1954, B128-F7.

273 **"beat him down":** Bernie Kirschenbaum, quoted in Wong, 329.

273 **consulted on the Sky Eye:** RBF to "Mr. X," Series 4, B2-F4. A model of the telescope was built by Jeffrey Lindsay & Associates ("Plastics in Building," *Architectural Forum*, January 1955, 120).

274 **"Your reaction was such":** Lindsay to RBF, March 7, 1955, B87-F1. In the same letter, Lindsay wrote, "I also had time to consider and analyze your relationships with Ysidore Martinez, Serge Chermayeff, Don Richter, Duncan Stuart, Ken Snelson, George Nelson, Herman Wolf, George Welch, Charles Eames, John Dixon, Jim Fitzgibbon, all of whom I knew more than casually. I subsequently decided that it would be advisable for me not to become directly associated with you in any particular project."

274 **"The Bauhaus international school":** RBF, "Influences," 32–33. The essay originated as a January 7, 1955, letter from RBF to John McHale (Series 4, B2-F1).

274 **"Let Bucky Fuller":** Philip Johnson, *Writings* (New York: Oxford University Press, 1979), 103.

274 **"the negative stimulus":** RBF, "Influences," 23.

274 **"one of the most influential":** RBF, *Inventions*, 302.

274 **"I've always felt":** Adapted from RBF, "Later Development of My Work," in *BFR*, 95. For Salk's version of the encounter, see Charlotte DeCroes Jacobs, *Jonas Salk: A Life* (New York: Oxford University Press, 2015), 311–12.

275 **"pattern integrity":** RBF, "A Comprehensive Anticipatory Design Science," in *No More Secondhand God*, 106. The original poem was finished in July 1956 (Series 4, B2-F5).

275 **"grateful silence":** Ibid., 116.

275 **concrete block:** Sadao, 152.

275 **contract simultaneously:** RBF, *EIK*, session 11.

275 **"Oh, Buckling":** Cochrane, *Sense of Significance*, 142.

275 **Morley had died:** Marks, 35.

276 **"the first major basic improvement":** Henry C. Lane, "Study of Shelter Logistics for Marine Corps Aviation," 1.

276 **Fuller and Noguchi:** Sadao, 158. Sadao incorrectly dates the airlift to 1956.

276 **The marines ordered several hundred:** James Fitzgibbon to P. E. Peters, March 28, 1959, Series 22, B5-F14.

276 **underground bunkers:** Wong, 323.

276 **"advantage over our competitors":** RBF to Major George J. King, September 24, 1955, Series 4, B2-F4.

276 **"The cool barrel":** Ibid., July 10, 1956, Series 4, B2-F5.

277 **the "Eskimos":** RBF, "Continuous Man," in *BFR*, 361.

277 **"Luckily, in 1927":** Ibid.

278 **circumvent his claims:** Wong, 332.

278 **"superbly drawn":** Hatch, 198.

278 **policing his rights:** RBF draft letter to T. C. Ryerson, undated (1958), Series 8, B15-F15.

278 **"truncatable" grid:** Sadao, "Brief History of Geodesic Domes," in *Anthology*, 26.

278 **"We generated enough money":** Wong, 334.

278 **$2 million:** Ibid.

278 **flurry of spending:** Hatch, 211.

279 **"By coincidence":** RBF to Walter O'Malley, June 14, 1955, reproduced in Brent Shyer, "The O'Malley-Fuller Connection," https://www.walter omalley.com/en/features/buckminster-fuller/Page-1 (accessed January 2021).

279 **Denver and Minneapolis:** Wong, 400.

279 **"skinned with translucent fiberglass":** RBF to Walter O'Malley, June 14, 1955, reproduced in Shyer, "O'Malley-Fuller Connection."

279 **"built along the lines":** "The Dodgers' Dome," *Sports Illustrated*, October 31, 1955.

279 **"one of the wonders":** "Model of Dodgers' Proposed Stadium Shown at Princeton," *Central New Jersey Home News* (New Brunswick, NJ), November 23, 1955, 3.

279 **Robert Moses:** Shyer, "O'Malley-Fuller Connection."

280 **"energetic-synergetic geometry":** "Item—0: An Introduction to the Energetic-Synergetic Geometry of R. Buckminster Fuller," reproduced in Daniel López-Pérez, *R. Buckminster Fuller: Pattern-Thinking* (Zurich: Lars Müller, 2020), 72.

280 **tensional integrity, or "tensegrity":** The word *tensegrity* was RBF's preferred term by 1958, when he wrote an essay of the same name (RBF, "Tensegrity," in *Discourse*, 226).

280 **analyze defense systems:** Duncan Stuart to Grady Cox, undated, Series 18, B148-F10.

280 **aircraft industry:** John Dixon to "Dean X," June 30, 1956, reprinted in López-Pérez, *Pattern-Thinking*, 338.

280 **retaining all the rights:** Ibid., 339, and Alex Soojung-Kim Pang, "Dome Days: Buckminster Fuller in the Cold War," in *Cultural Babbage: Technology, Time and Invention,* ed. Francis Spufford and Jenny Uglow (London: Faber and Faber, 1996), 171.

280 "Space Ship": RBF to Harold E. Watson, April 19, 1955, Series 10, B13-F21.

280 United Nations: Wigley, *Buckminster Fuller Inc.*, 241, and RBF, *CP*, 174–75.

280 "catenary" dome: RBF, *Inventions*, 162.

281 Nelson Rockefeller: Wong, 349.

281 "excited as hell": Jack Masey and Conway Lloyd Morgan, *Cold War Confrontations: US Exhibitions and Their Role in the Cultural Cold War* (Baden, Switzerland: Lars Müller, 2008), 65.

281 exhibition in Sweden: Wong, 318–19.

282 Canadian pavilion: McAtee, "Geodesic Dreams," in *Montreal's Geodesic Dreams*, ed. McAtee, 41–42.

282 "a very poor show": Pang, "Dome Days," 186.

282 "talking chickens": Jack Masey, quoted in Emma Brockes, "Wait Till You See Our Talking Chickens," *Guardian* online, last modified September 17, 2008, https://www.theguardian.com/artanddesign/2008/sep/18/design.jackmasey.

282 "came inside": Wong, 352.

282 "The natives often thought": RBF, "Continuous Man," in *BFR*, 348–49.

282 "primitive" techniques: Pang, "Dome Days," 192.

282 Sir Halford Mackinder: RBF to Major George J. King, July 10, 1956, Series 4, B2-F5.

283 "a logistical Iwo Jima": Ibid.

283 slept on the floor: Roger VanDam, interviewed in Gowan, *Shared Vision*, 86.

283 dome on campus: "Huge Dome Studied for SIU," *Southern Illinoisan* (Carbondale), July 15, 1956, 2.

283 Union Tank Car Company: Details of project derived from Wong, 375–86.

283 Lake Merritt: "Giant Bird Cage for Lake Merritt," *Oakland (CA) Tribune*, August 9, 1956, 27E.

284 passing Richter's desk: RBF, *EIK*, session 11.

284 Hawaiian Village Hotel: "Kaiser Auditorium to Be 5-Day Job," *Honolulu Advertiser*, January 5, 1957, 1.

285 too late for an inspection: "Kaiser's Dome Rises Short of 36 Hours," *Honolulu Star-Bulletin*, January 14, 1957, 13. For the apocryphal version, see Hatch, 206.

285 minimum annual royalty: Wong, 393.

285 more than a dozen domes: Along with the Honolulu and Moscow domes discussed in the text, the Kaiser domes included auditoriums in Virginia Beach and Borger, Texas (1957); the Vacu-Blast industrial plant in Abilene, Kansas, the Citizens State Bank in Oklahoma City, the Casa

Mañana Theatre in Fort Worth, and the physical education building at Palomar College in San Marcos, California (1958); the Graham Memorial Auditorium in Pryor, Oklahoma (1959); the Miami Seaquarium, the Palais Des Sports in Paris, the National Orange Show Event Center in San Bernardino, California, and a storage facility at T. F. Green Airport in Warwick, Rhode Island (1960); a Lutheran church in Florida (1961); the Stepan Center at the University of Notre Dame and the Valley National Bank in Tempe, Arizona (1962); and the Aquatarium in St. Petersburg, Florida (1964). List based on online searches, slides in RBF's archive, and photo sections in Marks and in RBF, *Ideas and Integrities.*

285 **"We want to be geodesic":** Donald W. Robertson, *The Mind's Eye of Buckminster Fuller* (New York: St. Martin's Press, 1974), 54.

285 **"I don't want to mess":** T. C. Howard, quoted in Wong, 378.

286 **"dynamic developments":** Alvin Miller to RBF, December 6, 1955, quoted in Wong, 368.

286 **entire service division:** James Fitzgibbon to RBF, May 8, 1956, Series 22, B5-F12.

286 **Fletcher Jennings:** Lillian McLaughlin, "A Dome-Shaped House, That's What It Is," *Des Moines (IA) Tribune,* May 17, 1957.

286 **curtain of warm air:** Robertson, *Mind's Eye of Buckminster Fuller,* 56–57.

286 **"pinecone" version:** John McHale, *R. Buckminster Fuller* (New York: George Braziller, 1962), photo caption 73.

286 **Korea and the Philippines:** Marks, 218.

286 **Matrix Structures:** Lee Jones, "The Playhouse Pop Built," *This Week,* in *Semi-Weekly Spokesman Review* (Spokane, WA), March 3, 1957, 35.

286 **"Am I subversive?":** Wong, 355.

287 **"the simplest and the best way":** Ibid.

287 **USS *Leyte*:** "Down Comes Dome," *Green Bay (WI) Press-Gazette,* July 13, 1957. Footage appears in *The Love Song of R. Buckminster Fuller* by Sam Green.

287 **"The first line of defense":** AHF to Peter Floyd, September 26, 1957, quoted in Wong, 355–56.

287 **gala showing:** "'Around the World in 80 Days' Lives Up to Opening Fanfare," *Honolulu Star-Bulletin,* November 2, 1957, 2.

288 **"master architect":** Wong, 394.

288 **"the biggest ham actor":** Jane King Hession and Debra Pickrel, *Frank Lloyd Wright in New York: The Plaza Years, 1954–1959* (Salt Lake City: Gibbs Smith, 2007), 74.

288 **"gentler line of curvature":** Ibid.

288 **visited Wright in Scottsdale:** Indira Berndtson, email to author, August 4, 2020.

288 **twenty feet across:** RBF, "Later Development of My Work," in *BFR,* 83–84.

288 "first of his subsequently multiannual circuits": RBF, *CP*, 389. Toward the end of his life, RBF claimed that he had traveled more than 3.5 million miles and gone around the world forty-seven times (ibid., 131).

288 Spitz Laboratories: "50 to Visit Planetarium," *News Journal* (Wilmington, DE), July 12, 1957, 6, and Marks, 192.

288 Berger Brothers: Marks, 202.

288 "moon hut": Advertisement for American Legion and Swift & Company Space Show, *Boston Globe*, June 10, 1958.

288 domed shopping district: Wong, 399–400.

289 panels kept shattering: Wigley, *Buckminster Fuller Inc.*, 206.

289 "Now, I've been talking": Edgar W. Ray, *The Grand Huckster: Houston's Judge Roy Hofheinz, Genius of the Astrodome* (Memphis: Memphis State University Press, 1980), 304.

289 Fitzgibbon flew out: James Fitzgibbon to RBF, August 29, 1958, B99-F11.

289 Richard Lewontin: Popko, *Divided Spheres*, 44.

290 operational issues: Wong, 376. After the dome became obsolete, it was sold to Kansas City Southern and demolished in 2007 (*A Necessary Ruin*).

290 Graver Tank: Wong, 381–86.

290 American Society for Metals: W. L. Russell, "'Space Dome' 250 Feet in Diameter," *Pittsburgh Press*, March 6, 1958, and Marks, 230.

290 "buildings are trending": Strum, *Ideal of Total Environmental Control*, 216.

291 hopes of pitching Henry Kaiser: Jefferson Davis Brooks III to William Parkhurst, April 8, 1958, Series 22, B5-F15.

291 Chao Phraya River: RBF, *CP*, 21–22.

291 he acquired a toy ball: *Indlu Geodesic Research Project*, 9 and 22.

291 Nehru was her husband: RBF to AHF, April 16, 1958, B99-F10.

291 "He doesn't like people": RBF, *EIK*, session 7. RBF's exact words were "and don't go see [*sic*] the Taj."

291 "nearly five hours": Indira Gandhi to Dorothy Norman, April 17, 1958, quoted in Jairam Ramesh, *Indira Gandhi: A Life in Nature* (London: Simon & Schuster India, 2019, EPUB ed.), 74.

292 after twenty minutes: RBF, *EIK*, session 7.

292 suffering from exhaustion: "Nehru Sets Stage for Stepping Down?," *Cincinnati Enquirer*, April 13, 1958, sec. 2, 30.

292 hour and a half: Hatch, 215.

292 whenever he was in the country: RBF worked around this time on a proposal for a factory "for housing the poor people of India," with domes produced on an assembly line and flown by helicopter (Bob Stubenrauch, letter to *New York Times*, October 18, 1993). For Monsanto's interest in domes for India, see Wong, 397 and 442.

292 "They were all Zulu boys": RBF, *EIK*, session 11.

292 "If the technician": *Indlu Geodesic Research Project*, 85.

292 "demonstrably effective": Itinerary, June 4, 1958.

293 Reyner Banham: In *Theory and Design in the First Machine Age,* 2nd ed. (New York: Praeger, 1967), 326, Banham wrote that the Dymaxion House would have rendered the Villa Savoye by Le Corbusier and Pierre Jeanneret "technically obsolete before design had even begun."

293 "We find that industrialization": RBF, "Later Development of My Work," in *BFR,* 86–87. RBF's speech, which was originally delivered at the June 5, 1958, Annual Discourse to the Royal Institute of British Architects, was slightly reworked for publication. The original transcript was printed in the *Journal of the Royal Institute of British Architects* (October 1958).

293 "to build in between the earthquakes": Ibid., 88.

293 "You cannot stay out there": Ibid., 90.

293 "The minute you were not concerned": Ibid., 94.

293 "I am not afraid": Ibid., 95.

293 John Huston: Michaelis, *Best of Friends,* 70.

293 "I hurry to Bear Island": Allegra Fuller Snyder, "Bear Island Guest Book," in *Bear Island Centennial Book, 1904–2004,* 54.

293 Graver Tank was in trouble: James Fitzgibbon to RBF, March 9, 1959, B102-F8.

294 "refined blacksmithing": Wong, 383.

294 manager of the dome program: The Graver Tank division later contracted to build a geodesic aviary for Busch Gardens in Tampa ("Geodesic Dome Built for Birds at Gardens Here," *Tampa [FL] Tribune,* March 20, 1960, 5-G).

294 William Zeckendorf: Wong, 386 and 398–99.

295 Lucilla and her husband, Tom Marvel: Lucilla Fuller Marvel, interviewed by author, April 18, 2019.

295 Frobisher Bay: Wong, 425–26.

295 Missouri Botanical Garden: Ibid., 404–10.

295 took credit for it: RBF to Ernest G. Friesem, November 11, 1963, quoted in ibid., 414.

295 wagered privately that it would collapse: Ibid., 329.

295 "a nervous breakdown": T. C. Howard, quoted in ibid., 329.

295 "hollowed out balloon": RBF, "Tensegrity," in *Discourse,* 238.

295 "Cloud Nine": RBF was using the term "Cloud Nine" to refer to these structures by the midsixties (Ben Gelman, "Will We Live in Sky, on Water or Under Sea?," *Southern Illinoisan* [Carbondale], October 31, 1966, 3).

295 "launching platforms for missiles": Calvin Tomkins, "Umbrella Man," *Newsweek,* July 13, 1959, 87.

296 "to those who lack": Alfred North Whitehead, *Science and the Modern World* (Cambridge: Cambridge University Press, 1929), 245.

296 "the self-commissioned architect": RBF, "The Comprehensive Man," in *BFR,* 338.

296 "omnidirectional universal maelstrom": RBF, "Introduction to Omni-Directional Halo," in *No More Secondhand God*, 128.

296 "zone of lucidly": RBF, "Omni-Directional Halo," in ibid., 137.

296 systems theory: RBF was familiar with the work of the systems theorist Ludwig von Bertalanffy, whom he met at the Conference on World Affairs in Colorado in 1961 (RBF, *Education Automation*, 69).

296 rebranded it as "synergetics": This was RBF's preferred term by the time of his lecture "The Music of the New Life" (*Music Educators Journal*, June 1966, 68, reprinted in *UOO*, 77), which he delivered on December 10, 1964.

296 "My mother was really": "R. B. Fuller Lecture at San Quentin Prison," January 31, 1959, Series 8, B15-F11, 4.

297 "because I was almost": RBF, "*Playboy* Interview," *Playboy*, February 1972, 199.

297 "They were just going to sit": Ibid.

297 "turning point in our affairs": James Fitzgibbon to RBF, March 12, 1959, Series 22, B5-F6.

297 "some of us in Raleigh": Ibid.

297 American National Exhibition: Sources include Masey and Morgan, *Cold War Confrontations*, 162–78, and Pang, "Dome Days," in *Cultural Babbage*, ed. Spufford and Uglow, 188–90.

298 "the big blockbuster": Masey, quoted in Brockes, "Wait Till You See Our Talking Chickens," *Guardian* online, last modified September 17, 2008, https://www.theguardian.com/artanddesign/2008/sep/18/design.jackmasey.

298 "information machine": Masey and Morgan, *Cold War Confrontations*, 162.

298 "I am thinking": "Nikita Likes US Exhibit's Dome," *Baltimore Sun*, May 5, 1959.

299 "This dome is very interesting": Marks, "Breakthrough of Buckminster Fuller," *New York Times Magazine*, August 23, 1959, 14.

299 "counter-torquing tetrahedra": RBF, *Inventions*, 196.

299 visited Frank Lloyd Wright: Pedro E. Guerrero, *A Photographer's Journey*, (New York: Princeton Architectural Press, 2007), 87.

299 "I am an architect": RBF, "Frank Lloyd Wright," in *SD*, 20989.

299 "having a stream": Howard J. Brown, interviewed by author, April 29, 2019.

299 "the last of that era": RBF, "Frank Lloyd Wright," in *SD*, 20989.

299 "the instrument man": Ibid.

299 "with more absolute integrity": Marks, "Breakthrough of Buckminster Fuller," 14.

299 Johnson Wax Research Tower: David Leslie Johnson, *Frank Lloyd Wright Versus America: The 1930s* (Cambridge, MA: MIT Press, 1994), 168.

299 Price Tower: Patrick J. Meehan, *Frank Lloyd Wright Remembered* (Washington, DC: Preservation Press, 1991), 40.

299 **"I am one of Bucky's patron saints"**: "Berkshires Don't Know Wright from Wrong, Frank Lloyd (Who's the Best) Tells Us," *Berkshire Eagle* (Pittsfield, MA), March 14, 1957.

300 **"Of all the befriendings"**: Wright, *Shining Brow*, 133.

300 **"Frank Wright, as the ages go by"**: Ibid., 135.

300 **"There's no inherent"**: Calvin Tomkins, "Umbrella Man," *Newsweek*, July 13, 1959, 86.

301 **"a research laboratory"**: "R. Buckminster Fuller," *Form Givers at Mid-Century*, 60.

301 **a fire broke out**: RBF, *EIK*, session 12, and Marks, "Breakthrough of Buckminster Fuller," 44. Correspondence mentioning the fire appears in B103-F10.

301 **flames spread from the kitchen**: The original kitchens in the apartments at 6 Burns Street consisted of large metal wall units invented by the developer Guyon L. C. Earle, who had sold two to RBF for use in the Wichita House (Christopher Gray, "A Forest Hills Find for Fuller's Dymaxion House," *New York Times*, April 17, 1994, sec. 10, 7).

301 **"a ramshackle store"**: Marks, "Breakthrough of Buckminster Fuller," 44.

301 **"the most economical"**: "New Light on How Polio Starts," *Observer*, June 21, 1959.

301 **Klug and Finch**: Gregory J. Morgan, "Early Theories of Virus Structure," in *Conformational Proteomics of Macromolecular Architecture: Approaching the Structure of Large Molecular Assemblies and Their Mechanisms of Actions*, ed. R. Holland Cheng and Lena Hammar (Singapore: World Scientific, 2004), 25.

301 **"Some things never get out of date"**: Condensed from transcript of the Kitchen Debate, https://watergate.info/1959/06/24/kitchen-debate-nixon-khrushchev.html (accessed January 2021).

301 **twelve cultural representatives**: Roscoe Drummond, "US Plans Show in Moscow," *Decatur (IL) Daily Review*, May 29, 1959.

301 **"a lot of bows"**: Don Richter, quoted in Wong, 415n506.

302 **"You see, dear boy"**: Blake, *No Place Like Utopia*, 246.

302 **"a pair of enormous, thick lenses"**: Ibid.

302 **purchase the dome**: Masey and Morgan, *Cold War Confrontations*, 157.

302 **"We are going to have"**: RBF to K. Fulmer, November 23, 1959, quoted in Wong, 359.

302 **"in a big way"**: RBF, "Continuous Man," in *BFR*, 349.

302 **"a kid who had just gotten"**: Armand Re, interviewed in Gowan, *Shared Vision*, 106.

303 **"They said that I was recognized"**: "Fuller Revisited," *New Yorker*, October 10, 1959, 36.

303 **"Many a US worker citizen"**: RBF, "Continuous Man," 356.

303 "Khrushchev is glad": Ibid., 363–64.

303 "He's not a teacher": Harold Cohen, author interview, March 6, 2021.

304 appointment to the faculty: "SIU Appoints Fuller," *Southern Illinoisan* (Carbondale), September 20, 1959, 3.

304 salary of $12,000: Hatch, 218.

304 Museum of Modern Art: RBF was also featured in a touring show sponsored by *Time* magazine and the American Federation of Arts, "Form Givers at Mid-Century," which appeared at the Metropolitan Museum of Art (*Form Givers at Mid-Century* [New York: Time, 1959], 60–63).

304 modernist emphasis on function: Maria Gough, "Backyard Landing: Three Structures by Buckminster Fuller," in *New Views on R. Buckminster Fuller,* ed. Chu and Trujillo, 125–27.

304 offending Bernie Kirschenbaum: Wong, 238n122. A dome that Kirschenbaum built for Frank Safford in Wading River, Long Island, which became the headquarters of the Simulmatics Corporation, would also be erroneously attributed to RBF (Jill Lepore, *If Then: How the Simulmatics Corporation Invented the Future* [New York: Liveright, 2020], 13). A note in the Stanford finding aid refers to it as the "Kirchenbaum [sic] / Dr. Safford Wading River Residence" (Series 12, B22-F11).

304 Edison Price, a lighting consultant: Sadao, 163–66.

304 including Noguchi: Ibid., 166, and Herrera, *Listening to Stone,* 294.

304 Robert W. Marks: Marks had first covered RBF in "Bucky Fuller's Dymaxion World," *Science Illustrated,* November 1948.

304 "The domes do so much": Marks, "Breakthrough of Buckminster Fuller," *New York Times Magazine,* August 23, 1959, 15.

305 "weave" entire buildings: Gough, "Backyard Landing," 128.

305 "In effect, the city": Ada Louise Huxtable, "Future Previewed? Innovations of Buckminster Fuller Could Transform Architecture," *New York Times,* September 27, 1959, sec. 2, 21.

305 "the greatest advance": Ibid.

305 "the newest look": John Canaday, "Art: New Directions in Architecture," *New York Times,* September 22, 1959, 78.

305 "It could remain exactly where it is": Ibid.

305 "I had been successfully swallowed up": Snelson, "Tensegrity Invention," *International Journal of Space Structures* 11, no. 1/2 (1996): 46.

305 "cutting him to pieces": Sande-Friedman, "Snelson and Science of Sculpture," 67.

305 "I was suddenly gripped": Snelson, "Tensegrity Invention," 46.

306 "Pure jewelry": Ibid., 47. The dialogue that follows is based on Snelson's account.

306 second exhibit: Gough, "Backyard Landing," 137.

306 *Homage to New York*: Ibid., 131.

306 **"with the feeling"**: Gómez-Jáuregui, *Tensegrity Structures*, 163.

306 **"All of my structures"**: Gough, "Backyard Landing," 145. RBF included the frame of the 1928 Dymaxion House and a model of the Wichita House deck in a list of "experimental" tensegrity structures (Marks, 164–65). The debate over whether the dome embodied the principle of tensegrity has advocates on both sides. For the negative view, see Wong, 168–69; for a positive case, see Donald E. Ingber, "The Origin of Cellular Life," *BioEssays* 22 (2000): 1160–61.

307 **"They became his ideas"**: Wong, 173.

307 **"Ken, old man"**: Gómez-Jáuregui, *Tensegrity Structures*, 162.

NINE: INVISIBLE ARCHITECTURE

308 **"Only humans play"**: RBF with Applewhite, *Synergetics*, 661.

308 **"believed to be the largest"**: "Princeton Has Largest Global Map on Exhibition," *Central New Jersey Home News* (New Brunswick), April 7, 1960, 12.

309 **"the gnomes"**: Mary Cohen, interviewed in Gowan, *Shared Vision*, 14.

309 **custom sample cases:** Cary O'Dell and Thad Heckman, *Bucky's Dome: The Resurrection of R. Buckminster Fuller and Anne Hewlett Fuller's Dome Home in Carbondale, Illinois* (United Kingdom: Arcadia, 2020), 18–19.

309 **not a single one:** RBF's house eventually inspired the head of the local Student Christian Foundation to buy one as well ("Dome Home Number Two Is Going Up," *Southern Illinoisan* [Carbondale], May 10, 1960, 2), and several more were purchased by the SIU design department ("Domes for Designers," *Southern Illinoisan* [Carbondale], December 6, 1960, 12, photo caption).

309 **"We can do it in a day"**: Al Gowan, "R. Buckminster Fuller," in Gowan, *Shared Vision*, 25.

309 **construction workers:** O'Dell and Heckman, *Bucky's Dome*, 27–29.

310 **"coming out of a conch shell"**: "Antiques Fit Just Fine in Fullers' Dome Home," *Southern Illinoisan* (Carbondale), December 4, 1960. Additional details on the dome's dimensions and history are drawn from O'Dell and Heckman, *Bucky's Dome*, 24–57.

310 **"private motel"**: Ben Gelman, "Dome Home, Erected in Day, Is Innovation in Field of Dwelling Design," *Southern Illinoisan* (Carbondale), April 21, 1960.

310 **"would be just sort of"**: Hatch, 200.

310 **book in one hand:** Jay Baldwin in "Derelict Dome," *99% Invisible*, October 25, 2012, https://99percentinvisible.org/episode/episode-64-derelict-dome (accessed May 2021).

311 "Bucky was brilliant": Harold Cohen, interviewed in Gowan, *Shared Vision*, 15.

311 "an ego to match": Ibid.

311 "generic importance": Marks, 3.

311 "collaborate with local design": "Houston Due $15 Million Dome-Covered Stadium," *Victoria (TX) Advocate*, August 21, 1960. Additional information can be found in the Robert J. Minchew Houston Astrodome Architectural and Engineering Collection, Briscoe Center for American History, University of Texas at Austin, 94–274, box 4.

312 Sotaro Suzuki: Red McQueen, "Japan Plans Dome Stadium," Hoomalimali, *Honolulu Advertiser*, October 27, 1960, A14.

312 "ascending suspension": Sadao, 171, and RBF, *Inventions*, 201–13. See also "Fuller Ready to Put Dome over Baseball Park in Tokyo," *Southern Illinoisan* (Carbondale), April 9, 1961, 5.

312 begin work immediately: Robertson, *Mind's Eye of Buckminster Fuller*, 70–74.

313 laminar dome: RBF, *Inventions*, 227–40.

313 Monsanto Chemical Company shipped two hundred: Itinerary, September 7–September 11, 1961, and McHale, R. *Buckminster Fuller*, 34–35.

313 Describing himself as a "comprehensivist": RBF, *Education Automation*, 68.

313 "the children of the mobile people": Ibid., 33.

313 Robert Snyder: *Sketchbook: Three Americans* (1960) featured RBF, Igor Stravinsky, and Willem de Kooning. Snyder's later films included *Buckminster Fuller on Spaceship Earth* (1971) and *The World of Buckminster Fuller* (1974).

313 Francis Thompson: "Experimental Movies Lecture Set at SIU," *Southern Illinoisan* (Carbondale), November 9, 1960.

313 "There is no reason": RBF, *Education Automation*, 51.

313 "You are faced": Ibid., 48.

314 "the *infra*- and the *ultrasensorial* frequencies": Ibid., 62.

314 "control the invisible events": Ibid.

314 "A circus": Ibid., 85.

314 portable dome theater: The theater was developed by Synergetics Inc. for Ford and the corporate documentary producer Jam Handy (McHale, R. *Buckminster Fuller*, photo caption 87). In addition to its appearance at the Seattle World's Fair in 1962 (RBF, *EIK*, session 9), the Cine-Dome was used for short films at trade fairs ("In Focus," *Detroit Free Press*, January 7, 1960).

314 "delicate occulting membranes": RBF, *Education Automation*, 87.

314 "generalized computer setup": Ibid., 88.

314 "any dreamable vision": Ibid.

314 **"a comprehensive understanding"**: "Concepts for SIU Campus Discussed," *Alton (IL) Evening Telegraph*, June 3, 1961.

315 **"you or me"**: RBF, "The Architect as World Planner," in *Ideas and Integrities*, 254.

315 **Monica Pidgeon**: Wigley, *Buckminster Fuller Inc.*, 237.

315 **"stuff the results"**: Monica Pidgeon to RBF, December 17, 1962, quoted in Wong, 446.

315 **"a facility for displaying"**: RBF, "Proposal to the International Union of Architects," in *Discourse*, 248.

315 **"Al had run the math"**: Hideo Koike, interviewed in Gowan, *Shared Vision*, 96.

316 **not "efficient"**: RBF, "Aspension," in *SD*, 771.

316 **"deresonated" tensegrity dome**: For a similar basketry dome built at Carbondale, see Jerry Kinsman, interviewed in Gowan, *Shared Vision*, 112–13.

316 **Sunclipse Dome**: Allegra Fuller Snyder, "Bear Island Guest Book," in *Bear Island Centennial Book, 1904–2004*, 55.

316 **the misleading term *sunset***: According to RBF, Bill Wainwright proposed the term *sunclipse*, which suggested the corresponding word *sunsight* for *sunrise* (RBF, "Sunclipse," in *SD*, 17235).

316 **Yonkers Raceway**: Jim McCulley, "Zeckendorf Gives 14M for YR Control Stock," New York *Daily News*, October 11, 1961, 12.

316 **Princess Margaret**: The chandelier was coordinated by Cedric Price and built by James and Gillian Meller (RBF, *Artifacts of R. Buckminster Fuller*, vol. 4, 43). Price later worked on an aviary for the London Zoo influenced by RBF and conceived by Lord Snowdon (Noah Chasin, "Lord Snowdon Co-Designed the World's Strangest Aviary," *Garage*, March 5, 2018, https://garage.vice.com/en_us/article/gy8n9y/snowdon-aviary-cedric-price-london-zoo [accessed February 2021]).

316 **"poetry in its broadest sense"**: "Chair of Poetry at Harvard," *Baltimore Sun*, May 18, 1925, 2.

316 **"an analytical session"**: Duberman, *Black Mountain*, 303. Penn or Duberman related this session to a nonexistent "honorary degree from Harvard," which appears to refer to the Norton lectureship.

316 **"Medical science"**: RBF, "The Prospect for Humanity," in *World Game*, 175. The piece, which originally appeared as "Notes on the Future" in *Saturday Review* (August 29, September 19, and October 3, 1964), was adapted from the Norton lectures.

317 **Donald Caspar**: Morgan, "Early Theories of Virus Structure," in *Conformational Proteomics of Macromolecular Architecture*, ed. Cheng and Hammar, 27.

317 **toy called Geodestix:** Ian Menzies, "The Chinese Bandits with the Coolie Hats," *Boston Sunday Globe*, April 7, 1963, A-7.

317 **"The basic assumption":** D. L. Caspar and A. Klug, "Physical Principles in the Construction of Regular Viruses," *Cold Spring Harbor Symposia on Quantitative Biology* 17 (1962): 1–24.

317 **"Somebody told us":** "Fuller Gets Microbiology Idea Credit," *Southern Illinoisan* (Carbondale), February 17, 1964, 14.

317 **"comprehensive thinker":** RBF, "Prevailing Conditions in the Arts," in *UOO*, 104.

317 **"Any good virologist":** Ibid.

317 **"monohex" dome:** Wigley, *Buckminster Fuller Inc.*, 207–8, and RBF, *Inventions*, 214–26. A patent for the design was filed on December 19, 1961.

318 **Yomiuri Country Club:** Alfred Wright, "A Domed Club That Will Test the God of Golf," *Sports Illustrated*, November 7, 1966.

318 **"relatively uneconomical":** Ray, *Grand Huckster*, 304.

318 **"invade" his field:** RBF, *CP*, 392.

318 **"modified their designs":** RBF to Walter O'Malley, May 23, 1967, reproduced in Brent Shyer, "The O'Malley-Fuller Connection," https://www.walteromalley.com/en/features/buckminster-fuller/Page-1 (accessed January 2021).

318 **"dome building aspirants":** Ibid., September 7, 1965, reproduced in Shyer, "O'Malley-Fuller Connection."

318 **"an engineering blunder":** Al Reinert, "Greetings From the Eighth Wonder of the World," *Texas Monthly*, April 1975, 90.

318 **outlook on the whole:** Wong, 394–96. A plan around this time by Cedric Price for a geodesic auditorium in England was never realized ("Briton Picks Up Tips from Buildings Here," *Pittsburgh Post-Gazette*, October 12, 1962, and correspondence in B125-F1).

318 **"the number of licensees":** RBF to James Fitzgibbon, October 20, 1961, quoted in Wong, 396.

318 **International Union of Architects:** RBF, "Proposal to the International Union of Architects," in *Discourse*, 247.

319 **the entire roof ignited:** Robert Pearson and Don Beck, "Ford Rotunda Destroyed by $16-Million Fire," *Detroit Free Press*, November 10, 1962, 1.

319 **weren't written in English:** Mary Cohen, interviewed in Gowan, *Shared Vision*, 14.

319 **"problems of intellectual integrity":** RBF, "The Long Distance Trending in Pre-Assembly," in *Ideas and Integrities*, 271.

319 **"The Two Cultures":** C. P. Snow originally delivered the talk on May 7, 1959. He revisited the subject in *The Two Cultures: And a Second Look* (Cambridge: Cambridge University Press, 1965), 72–73, in which he

recommended the study of molecular biology, starting with "the analysis of crystal structure."

320 "I happen to have secret information": Jonathan Williams, *A Palpable Elysium* (New York: Dim Gray Bar, 1997), 134.

320 "their lack of economic experience": RBF to Luis G. Brugal, May 29, 1963, B121-F7.

320 "by the students' own uncoordinated efforts": RBF, preface to *World Design Science Decade 1965–1975, Phase 1, Document 1.*

320 they became "enthusiastic": Ibid.

320 "a little house": RBF, "The Designers and the Politicians," in *BFR*, 375.

320 "moon house": "'Moon House' Designed for US Astronauts," *St. Louis Post-Dispatch*, June 24, 1963. The study also developed computer models and reports on geodesic spheres and the jitterbug transformation, which Joseph Clinton saw as the first new technique for expanding rigid structures "since the time of Archimedes" (Joseph Clinton, interviewed by author, April 4, 2019).

321 "I am your disciple": RBF, "McLuhan, Marshall," in *SD*, 10229.

321 Fuller drank copiously: Wigley, *Buckminster Fuller Inc.*, 284.

321 "the glorious Bucky": *From the Library of Buckminster Fuller*, 80.

321 industry of the future: Marshall McLuhan to RBF, August 11, 1963, and RBF to McLuhan, August 25, 1963, B122-F6.

321 "technology as creator": McLuhan to RBF, September 17, 1964, in Marshall McLuhan, *Letters of Marshall McLuhan*, ed. Matie Molinaro, Corinne McLuhan, and William Toye (Toronto: Oxford University Press, 1987), 309.

321 "a necessity": Ibid.

321 "McLuhan has never": RBF, "McLuhan, Marshall," in *SD*, 10229.

322 "the message only": RBF, "Revolution in Wombland," in *Earth, Inc.*, 103.

322 "Bucky is my master": RBF, "McLuhan, Marshall," in *SD*, 10230.

322 the world's "have-nots": *World Design Science Decade 1965–1975, Phase 1, Document 1*, 27–28.

322 "all over": RBF, "New Forms vs. Reforms," in ibid., 53.

322 Fuller appealed unsuccessfully: RBF, *EIK*, session 12, and RBF to McGeorge Bundy, January 4, 1963, B121-F6.

322 "the positive potentials": RBF to August Heckscher, May 20, 1963, quoted in Wong, 448.

322 "was not a free citizen": RBF, *EIK*, session 12.

322 arranged for him alone: Ben Gelman, "He's Got Plans for the World," *Southern Illinoisan* (Carbondale), August 15, 1963, 1, and Auchincloss, "Dymaxion American," *Time*, January 10, 1964, 50.

322 the great "pirates": RBF, "World Design Initiative," in *Discourse*, 266.

323 "We could all live": Axel Madsen, *John Huston* (Garden City, NY: Doubleday, 1978), 207.

323 *The Bible*: Itinerary, July 21–23, 1964.

323 "She loved the idea too": Auchincloss, "Dymaxion American," 51.

323 "Your work and thought": Silverman, *Begin Again*, 213.

323 a series of public diaries: Donal Henahan, "He Is the American Optimist," *New York Times Book Review*, September 23, 1973, 34.

323 "without doubt the greatest living man": Silverman, *Begin Again*, 212.

323 "the most fascinating": "Stories Behind the Stories," *Time* press release, January 1964, B126-F2.

323 Fuller's eyes: *AMS*, 159.

323 "If my life provided": RBF, letter to the editor, *Time*, January 17, 1964.

324 "Today the world": Auchincloss, "Dymaxion American," 51.

324 "an untrue and arrogant picture": AHF to Naomi Smith, undated (1964), B126-F2.

324 "When dinner was finished": Harold Cohen, interviewed in Gowan, *Shared Vision*, 30.

324 annual income of $200,000: Rumwell, *Time* profile notes, final take, December 30, 1963, B126-F4. AHF thought that the profile's reference to his income was "embarrassing," since she often had trouble paying their mortgage on time (AHF to RBF, January 13, 1964, B126-F2).

324 wearing two watches: RBF, *World Man*, 42, and RBF, *Operating Manual*, 130.

324 "to protect clusters": "Students Experience Life in Ghana," *Chicago Tribune*, April 2, 1964, sec. 2D, 1.

324 physicist Vikram Sarabhai: Details of this unrealized deal appear in B126-F10. In 1958 RBF visited Sarabhai's brother Gautam (RBF to AHF, April 16, 1958, B99-F10), who built a geodesic showroom in Bombay (Marks, 223), followed four years later by the Calico Dome in Ahmedabad ("Plan to Resurrect Calico Dome Under Way," *Times of India*, November 22, 2017).

324 "I like your words": RBF, *EIK*, session 7.

324 "I read every word": Ibid.

325 "I would not think": Ibid.

325 "No one made clearer to me": RBF to Indira Gandhi, June 1, 1964, B130-F3.

325 Robert Moses: Robert Moses to RBF, August 5, 1970, B205-F4.

325 "a little automated spaceship": RBF, "The Prospect for Humanity," *Saturday Review*, August 19, 1964, quoted in "Spaceship Earth," in *SD*, 16472.

325 "I give up": Norman Cousins, "The Message Comes Across with Clarity and Love," *Architectural Forum*, January 1972, 60.

325 "absolutely intrigued": Neva Goodwin (Kaiser), interviewed by author, March 22, 2019.

325 invited to the family estate: Ibid.

325 "would turn into a twelve-year-old boy": Michael Denneny, interviewed by author, March 22, 2019.

326 "Everything I disapprove of": "White House Party Reveals Pre-Victory Lift," *Tennessean Sun* (Nashville), October 18, 1964.

326 Donald M. Wilson: Donald M. Wilson to RBF, August 14, 1964, B133-F5.

326 "Would Buckminster Fuller": Peter Chermayeff, email to author, January 7, 2022.

326 emerged from his "proximity": RBF to Shoji Sadao, March 16, 1965, Series 7, B19-F1.

326 "dictatorial authority": Ibid.

326 "You understand that you are": Ibid.

326 design for the pavilion: Sources include RBF, "World Game: How It Came About," in *World Game*, 18–19; Sadao 177–78; and Wong, 481–82.

327 a huge floor map: "Nations Will Light Up," *Southern Illinoisan* (Carbondale), February 2, 1964, 21.

327 "Let's Make the World Work": Ben Gelman, "Fuller Invents Game to 'Make the World Work' Properly," *Southern Illinoisan* (Carbondale), February 9, 1965, 14.

327 "its lowest point": RBF to Ben Hellman, November 29, 1966, quoted in Wong, 480.

327 Peter Chermayeff: Peter Chermayeff, email to author, January 7, 2022. See also Peter Chermayeff to RBF, December 29, 1964, B133-F6, and "Memorandum of Cambridge Meeting, December 29, 1964, on U.S. Pavilion for Montreal 1967 World's Fair," B133-F7.

327 "Peter, I am so tired": Peter Chermayeff, email to author, January 7, 2022.

327 "Let's do a big dome": Ibid. See also Masey and Morgan, *Cold War Confrontations*, 318, and Sadao, 178.

328 structurally impossible: Réjean Legault, "The Death and Life of Buckminster Fuller's U.S. Pavilion at Expo 67," in *Montreal's Geodesic Dreams*, ed. McAtee, 100.

328 "design creativity": Wong, 491.

328 "new establishment": Susan Sontag, *Against Interpretation*, 298. Originally published as "From New York: Susan Sontag," in *Mademoiselle*, April 1965.

328 components for a tensegrity truss: Robertson, *Mind's Eye of Buckminster Fuller*, 85–90, and RBF, *Inventions*, 241–47.

328 Fuller went "apeshit": Bill Perk, interviewed by author, March 22, 2019.

328 "He was abusive": Harold Cohen, author interview, March 6, 2021.

329 "He relied on other people": Ibid.

329 "He's not doing any patents": Robert Williams, interviewed by author, March 20, 2019.

329 **"I wasn't a person":** Ibid.

329 **"This is where I lived":** Studs Terkel with Sydney Lewis, *Touch and Go: A Memoir* (New York: New Press, 2007), 248.

329 **"Bucky Fuller, Bertrand Russell":** Ibid., 251.

329 **"filled with hatred":** June Jordan, *Civil Wars* (Boston: Beacon Press, 1981), x. Jordan appears under her married name, June Meyer, in RBF's itinerary for November 16 and 18, 1964.

330 **"a form of federal reparations":** Ibid., 24.

330 **"integration in reverse":** "Cone Sweet Home," *Southern Illinoisan* (Carbondale), April 18, 1965, 5.

330 **"Skyrise for Harlem":** Jordan, *Civil Wars*, 25.

330 **"Instant Slum Clearance":** June Meyer (Jordan), *Esquire*, April 1965, 108–11.

330 **"for participation by Harlem residents":** Jordan, *Civil Wars*, 26.

330 **"the most socialized":** RBF, "How to Maintain Man as a Success in Universe," in *UOO*, 243.

330 even **"fly-fishing":** RBF, "Prevailing Conditions in the Arts," in *UOO*, 110.

330 **"swiftly degeniused":** RBF, "Emergent Humanity," in *R. Buckminster Fuller on Education*, 86.

330 **"They call their project":** RBF, "How to Maintain Man as a Success in Universe," 254.

331 **"using their heads":** RBF, "Geosocial Revolution," in *UOO*, 204.

331 **"when everything else":** RBF, "Design Strategy," in *UOO*, 335.

331 **Arnold J. Toynbee:** "Someone suggested to me that *etherealization* may be a better word [than *ephemeralization*]. However, it is disqualified for my meaning because it is founded on the no longer physically accepted concept of *ether*" (RBF, *CP*, 133). RBF was probably referring to Toynbee, who developed the similar concept of "etherialization" in human society as early as 1934 (Toynbee, *A Study of History*, vol. 3 [London: Oxford University Press, 1934], 174–92).

331 **handed the platform over to his own daughter:** Tomkins, "In the Outlaw Area," *New Yorker*, January 8, 1966, 96.

331 **"Either war is obsolete":** RBF, "Utopia or Oblivion," in *UOO*, 282.

331 **"the more or less freewheeling":** Ibid., 291.

331 **"Because of the students' intuition":** Ibid.

331 **"intoxicated" him:** Calvin Tomkins, interviewed by author, March 7, 2019.

332 **"In the Outlaw Area":** Tomkins, "In the Outlaw Area," 35–97. According to Tomkins, RBF spent an "unprecedented" seventeen hours on the phone from New Zealand with the fact-checking department (Tomkins, interviewed in *Buckminster Fuller: Thinking Out Loud*).

332 **"an aesthetic"**: Tomkins, "In the Outlaw Area," 74.

332 **"the most perceptive"**: Ibid., 76.

332 **"I never work"**: Ibid., 85.

332 **"guinea pig"**: Ibid., 64.

332 **"At the beginning"**: Ibid., 90–91.

332 **"naked and helpless"**: RBF, *EIK*, session 3.

332 **beamed there from elsewhere**: RBF, "Brain and Mind," in *Intuition*, 170. He also raised the possibility of telepathic communication across space and time (ibid., 167–70).

333 **"disgustingly theorized"**: RBF to George N. Buffington, August 31, 1928, in *4DT*, 147.

333 **question Charles Darwin's assertion**: RBF, *EIK*, session 3, and RBF, *Tetrascroll: Goldilocks and the Three Bears—A Cosmic Fairy Tale* (New York: ULAE / St. Martin's Press, 1982), 66–67.

333 **"I gained twenty pounds"**: Tomkins, "In the Outlaw Area," 47.

333 **chalky low-calorie supplement Metrecal**: Jaime Snyder, "Experiments in Food and Festivity," afterword to *Synergetic Stew*, iii.

333 **"The cows are eating"**: Applewhite, *Cosmic Fishing*, 86.

333 **"The cut of beef"**: Hatch, 240.

333 **"Hot water would do"**: Ibid., 164.

334 **"After two sweltering hot nights"**: Applewhite, *Cosmic Fishing*, 90.

334 **diet caused "dismay"**: Ibid., 86.

334 **Free Speech Movement at the University of California at Berkeley**: RBF, "Learning Tomorrows: Education for a Changing World," in *R. Buckminster Fuller on Education*, 190.

334 **Stan VanDerBeek**: Dana Miller, "Thought Patterns," in *Starting with the Universe*, ed. Hays and Miller, 37.

334 **"the most astounding manner"**: Robert Smithson, "Entropy and the New Monuments," in *The Collected Writings*, ed. Jack Flam (Berkeley: University of California Press, 1996), 21.

334 **geodesic greenhouse**: Mark Matthews, *Droppers: The Story of America's First Hippie Commune* (Norman: University of Oklahoma Press, 2010), 64.

334 **"garbage of America"**: Felicity D. Scott, "Fluid Geographies: Politics and the Revolution by Design," in *New Views on R. Buckminster Fuller*, ed. Chu and Trujillo, 170.

334 **tempting Cadillac**: Peter Rabbit, *Drop City* (New York: Olympia Press, 1971), 45–46.

334 **the inventor of Zomes**: Steve Baer in Lloyd Kahn, *Shelter* (Bolinas, CA: Shelter, 1973), 134.

335 **"some Buck Rogers Indian village"**: William Hedgepeth, *The Alternative: Communal Life in New America* (New York: Macmillan, 1970), 153.

335 **"landed UFOs"**: Matthews, *Droppers*, 89.

335 "the Dymaxion Award": Ibid., 165–66.

335 "as your own local industry": John McHale to Peter L. Douthit (Peter Rabbit), November 9, 1966, quoted in Scott, "Fluid Geographies."

335 "was apparently home": Memorandum, October 16, 1965, FBI file on RBF.

335 Victor Lessiovski: In an undated letter from 1964, John McHale wrote to Bill Perk, "I have now forwarded at your suggestion copies of the two [World Design Science Decade] documents to Mr. Victor Lessiovsky [*sic*] at the U.N., and mentioned that you had asked that we might do so" (B151-F9). Lessiovski wrote to McHale on June 4, 1964, to confirm receipt of the reports and state that he was meeting with RBF (B130-F1). Their June 21, 1964, meeting in Carbondale is recorded in RBF's itinerary. RBF's contact is called "a KGB employee" in a May 27, 1965, memo in his FBI file, and Lessiovski was named publicly as a KGB officer in John Barron, *KGB: The Secret Work of Soviet Secret Agents* (New York: Reader's Digest Press, 1974), 19–20.

335 "influence operations": John Barron, *Operation Solo: The FBI's Man in the Kremlin* (Washington, DC: Regnery History, 1996), 265.

335 "inducing influential foreigners": Ibid.

335 "playing politics": Memorandum, October 16, 1965, FBI file on RBF.

335 alternative to Marxism: RBF claimed that if he had attended the UIA convention in Havana, his plans "might possibly have become adopted by Cuba in lieu of communist ideology" (RBF to W. R. Ewald Jr., November 6, 1963, quoted in Wong, 449), and he felt that his public appearances had ended "the political question of a communist Japan" (Rumwell, *Time* profile notes, final take, December 30, 1963, B126-F4).

336 "geosocial revolution": Memorandum, October 16, 1965, FBI file on RBF.

336 "by rocket ship": RBF, "Keynote Address at Vision 65," in *UOO*, 124.

336 "How to Make the World Work": Ibid., 125.

336 concentric spheres: Thomas Turman, interviewed by author, August 6, 2020.

336 "the biggest game": John Cage, *A Year from Monday: New Lectures and Writings* (Middletown, CT: Wesleyan University Press, 1969), 165.

336 "chilly machine": RBF, *Pound, Synergy, and the Great Design* (Moscow, ID: University of Idaho, 1977), 18.

336 lowered over him: Turman, author interview.

336 "To me, it is the most beautiful": RBF, *EIK*, session 10.

336 "critical path": "Next Step: 'Shaping' of Fair Site Islands," *Gazette* (Montreal), May 14, 1964.

336 The method directly guided the construction: Peter Floyd to Jack Masey, February 26, 1965, B133-F12.

336 "high performance per unit": RBF, "Aesthetics," in *SD*, 202.

336 "The aesthetics of such an undertaking": Ibid.

337 **"designed for visual impact":** Shoji Sadao to Samuel Spencer, April 20, 1966, B144-F4.

337 **hybrid design:** Sadao, 178–79, and Legault, "Death and Life of Buckminster Fuller's US Pavilion at Expo 67," in *Montreal's Geodesic Dreams*, ed. McAtee, 97–101.

337 **no construction drawings:** Hatch, 242. Many drawings for the Montreal Expo Dome appear in RBF, *Artifacts of R. Buckminster Fuller*, vol. 4, 103–38.

337 **"star tensegrity" truss:** RBF, *Inventions*, 248.

337 **"completely interchangeable":** RBF, *EIK*, session 11.

337 **"He was always careful":** Shoji Sadao, interview with Thomas Devaney, Amy Sadao, Elizabeth Asawa, Marty Griggs, and Tsuneko Sadao, November 22, 2015. Provided to the author by Devaney.

337 **Félix Candela:** Félix Candela, *Seven Structural Engineers: The Félix Candela Lectures*, ed. Guy Nordenson (New York: Museum of Modern Art, 2008), 10–11.

337 **"one of the prophetical voices":** Roszak, *From Satori to Silicon Valley*, 18.

337 **inadvertent pollination:** RBF, "Geosocial Revolution," in *UOO*, 165–66.

337 **"The drive in humanity":** RBF, "Sex," in *SD*, 16003. Baldwin stated that RBF believed that the "increase of homosexuality, and (especially) bisexuality, were natural evolutionary developments" (*BuckyWorks*, 82).

338 **"an array of multisensory enhancements":** Katz, *Make It New*, 128.

338 **"We're inside Bucky's brain!":** Ibid.

338 **"the final and most spectacular":** Roszak, *From Satori to Silicon Valley*, 19.

338 **Salvador Dalí:** The dome for the Dalí Theatre and Museum in Figueres, Spain, was designed by Emilio Pérez Piñero (José Calvo López and Juan Pedro Sanz Alarcón, "Folding Architecture for an Astonishing Decade," *EGA: Revista de expresión gráfica arquitectónica* [January 2011]: 124).

338 **"brief affair":** Linda Hamalian, *The Cramoisy Queen: A Life of Caresse Crosby* (Carbondale: Southern Illinois University Press, 2009), 187.

338 **"an international peace center":** Ibid., 188–89.

338 **"to cross the line":** CDNS, "Widow Tries Cyprus as Site for a World Peace Center," *Cincinnati Enquirer*, June 28, 1966, 14.

338 **persuading Cyprus:** RBF, *CP*, 394.

339 **"I feel our undertaking":** RBF to Caresse Crosby, January 26, 1967, B152-F6.

339 **"no physical device":** Sadao, 187.

339 **The final plan:** Pawley, *Buckminster Fuller*, 162, and RBF, *CP*, 337–40.

339 **"almost indecipherable":** Sadao, 188.

339 **annual fee of $25,000:** Hidetoshi Shibata to RBF, June 5, 1967, B158-F7.

340 **"big enough to entomb":** Lewis Mumford, "Autocratic Technocracy"

image caption, *The Pentagon of Power* (New York: Harcourt Brace Jovanovich, 1970).

340 **Minnesota Experimental City:** Steve Rose, "Minnesota Experimental City: The 1960s Town Based on a Comic Strip," *Guardian* online, November 5, 2018, https://www.theguardian.com/cities/2018/nov/05 /minnesota-experimental-city-the-1960s-town-based-on-a-comic-strip.

340 **politically frustrating:** RBF to Otto A. Silha, February 14, 1978, B360-F8.

340 *Who's Who in America:* Applewhite, *Cosmic Fishing*, 2.

340 **no longer informed:** For example, Sadao was obliged to tell RBF that his understanding of the truss used in the dome was incorrect: "Sepias for your signature will be arriving soon, which will give you a complete description of the structure" (Shoji Sadao to RBF, March 2, 1966, B143-F8).

340 **Tetrahelix Inc.:** "3 Firms Chartered by State," *Asheville (NC) Citizen-Times*, February 25, 1967, 17.

340 **"perhaps a prototype":** David Jacobs, "An Expo Named Buckminster Fuller," *New York Times Magazine*, April 23, 1967, 237.

340 **Fuller told reporters:** "Bubble Stays Put," Across Canada, *Ottawa Journal*, April 24, 1967, 14.

341 **"after almost no personal involvement":** Peter Chermayeff, email to author, January 7, 2022.

341 **"treated badly":** Ibid.

341 **"It was a great privilege":** Ibid.

341 **"skybreak bubble":** Fuller & Sadao report, "The United States Pavilion for the Montreal World's Fair," 1967, B136-F4.

342 **sunshade system:** Wong, 484, and Baldwin, *BuckyWorks*, 169.

342 **"should it ever have to be shown":** Jasper Johns to RBF, January 23, 1967, quoted in Dana Miller, "Thought Patterns," in *Starting with the Universe*, ed. Hays and Miller, 35.

342 **"of offering any of it to me":** RBF to Calvin Tomkins, October 6, 1982, quoted in ibid., 40.

342 **"an official acknowledgement":** Andy Warhol and Pat Hackett, *POPism: The Warhol Sixties* (Orlando, FL: Harvest Book / Harcourt, 1980), 277.

342 **Umberto Eco:** In *The Picture History of Inventions: From Plough to Polaris*, trans. Anthony Lawrence (New York: Macmillan, 1963), 258, Eco and G. B. Zorzoli had identified Joseph Paxton's Crystal Palace "as the main source of inspiration for such modern constructions as Fuller's famous cupolas." RBF challenged this association in a hostile review, in which he dismissed Eco as an "aesthete" ("Dissonant Chords on a Grand Piano," *Book Week*, February 9, 1964).

342 **"The dome was aesthetically":** Umberto Eco, *Travels in Hyperreality: Essays,* trans. William Weaver (San Diego: Harvest/Harcourt, 1986), 302.

342 "The US building at Expo": Blake Gopnik, *Warhol* (New York: Ecco, 2020), 553–54.

342 His office had lost money: Shoji Sadao to RBF, March 2, 1968, B169-F1.

343 "validation to the extraordinary backing": Hatch, 243.

343 "I have inadvertently": Ibid., 244.

343 "This gets continually worse": RBF to John Fistere, June 6, 1967, B158-F8.

344 "very slow but satisfactory": Itinerary, May 14, 1967.

344 carried in his briefcase: "Contents of BF's taupe suede briefcase," Series 4, B8-F7.

344 "Anne is a woman": Gene Fowler to RBF, June 13, 1967, B158-F8.

PART FOUR: WORLD GAME

345 "What is the minimum number?": RBF with Applewhite, *Synergetics*, 238–39.

TEN: WHOLE EARTH

347 "It is like the childhood game of Twenty Questions": Ibid., 268. RBF borrowed the use of Twenty Questions to approach design problems from Harold Cohen (Cohen, author interview, March 6, 2021, and Roger Van Dam and Steven Poster, interviewed in Gowan, *Shared Vision*, 86 and 122).

348 hated violent movies: Harold and Mary Cohen, interviewed in Gowan, *Shared Vision*, 26.

348 "the twentieth century's": Bruce Weber, "Priscilla Morgan, Cultural Matchmaker, Dies at 94," *New York Times*, April 3, 2014, 25.

348 local historical society: "Menotti Fete Opens on a Quieter Note," *Los Angeles Times*, July 1, 1967, pt. 3, 9.

348 documentary on Fuller: Robert Snyder, in *AMS*, 167.

348 complimentary reference to "Buckie": *Canto XCVII*: "The artigianato bumbles into technology, / 'Buckie' has gone in for structure (quite rightly) / but consumption is still done by animals."

348 "a parallel manifestation": J. Gocar to RBF, January 24, 1967, quoted in Wong, 450.

348 Victor Papanek and Armi Ratia: Yrjö Sotamaa, email to author, March 16, 2019.

349 Bloomsbury Square: Daniel López-Pérez, *Fuller en México: La Iniciativa Arquitectónica* (Mexico City: Arquine, 2015), 162–63.

349 thickened arteries: Dr. Douglas Rossdale to RBF, July 19, 1967, B159-F7.

349 raw sea urchins: Jaime Snyder, "Experiments in Food and Festivity," afterword to *Synergetic Stew*, iv. RBF's great-niece Deirdre Newman

recalled that John Cage and Merce Cunningham hunted for mushrooms on a separate visit to Bear Island (*New York Mycological Society Newsletter*, Summer 2013).

349 **"rowing needles":** RBF, *Inventions*, 257–58. Moore is credited for the prototype in Lord, *Pilot for Spaceship Earth*, 150.

349 **moving the Expo Dome:** "World Center Plan Awaits OK," *Southern Illinoisan* (Carbondale), July 16, 1967, sec, 5, 1.

349 **$4 million of state bonds:** "Kerner Approves Bill for Resources Center," *Southern Illinoisan* (Carbondale), September 11, 1967.

349 **"It would be pretty much a detective game":** RBF to John Huston, September 27, 1967, Calvin Tomkins Papers, Museum of Modern Art, folder I.4.

349 **"Maoris sailing between the islands":** Calvin Tomkins to John Huston, October 9, 1967, Calvin Tomkins Papers, Museum of Modern Art, folder I.4.

350 **American Institute of Planners:** E. J. Applewhite in RBF, *SD*, xxvi.

350 **"I've often heard people say":** RBF, *Operating Manual for Spaceship Earth*, 46–47.

350 **"the number of forward days":** Ibid., 86.

350 **"You may very appropriately":** Ibid., 132.

351 **"at the outset":** RBF to Dale Klaus, February 9, 1970, quoted in *SD*, xxxiii.

351 **USCO, an experimental collective:** Stewart Brand interview, *The Tim Ferriss Show*, February 3, 2018, https://tim.blog/2018/02/03/the-tim-ferriss-show-transcripts-stewart-brand (accessed January 2021).

351 **"an Indian freak":** Tom Wolfe, *The Electric Kool-Aid Acid Test* (New York: Bantam Books, 1969), 10.

352 **"I was bored":** Stewart Brand interview, *The Tim Ferriss Show*, February 3, 2018.

352 **"Why haven't we seen":** Ibid.

352 **only one to reply:** Stewart Brand interview with Massive Change Radio, https://www.yumpu.com/en/document/view/5742799/bruce-mau-design-stewart-brand-interview-25-massive-change (accessed January 2021).

352 **"powerful tools technology":** Stewart Brand to RBF, November 21, 1967, quoted in Wong, 458.

352 **"Oh, yes, I wrote to that guy":** Stewart Brand interview with Massive Change Radio.

353 **Fuller advised him:** Stewart Brand to RBF, January 3, 1968, B152-F5, and April 10, 1968, B172-F6.

353 **"You're right":** Peter Collier, "Drop-Out's How-To," *New York Times Book Review*, March 7, 1971, 8.

353 **"The program is operational":** Stewart Brand to RBF, January 3, 1968, B152-F5. RBF had previously received a report from Don Moore that

described Engelbart's research on using computers to navigate "conceptual space" (Moore to RBF, January 22, 1965, B134-F5).

353 **South Carolina Tricentennial Commission:** Margaret Shannon, "Like Nothing Else in the World," *Atlanta Journal and Constitution Magazine,* October 3, 1971; drawings in RBF, *Artifacts of R. Buckminster Fuller,* vol. 4, 181–91.

353 **his patron died:** RBF, *CP,* 333.

353 **"He wants to do it better":** David Jacobs, "An Expo Named Buckminster Fuller," *New York Times Magazine,* April 23, 1967, 237.

353 **"several grave misunderstandings":** Shoji Sadao, "Buckminster Fuller's Floating City," *Futurist,* February 1969.

354 **Triton City:** Sadao, 190–91, and RBF, *CP,* 332–35.

354 **estimated cost per resident:** "U.S. Agency Comments on 'Floating Cities,'" *Southern Illinoisan* (Carbondale), December 27, 1968, 2.

354 **Nixon's election:** RBF, *CP,* 334.

354 **Gerald Gladstone:** Mark Osbaldeston, *Unbuilt Toronto: A History of the City That Might Have Been* (Toronto: Dundurn Press, 2008), 39.

355 **"second order metropolitan area":** Ibid., 42. RBF and Sadao revisited some of their concepts for Toronto in an unrealized proposal for a megastructure in Pearland, Texas (RBF, *Artifacts of R. Buckminster Fuller,* vol. 4, 233–34).

355 **Cinerama dome:** Don Kirkland, "Freeway Signs Built at Torrance," *Independent Press-Telegram* (Long Beach, CA), January 24, 1965.

355 **"to reform the environment":** RBF, "What I Am Trying to Do," *Saturday Review,* March 2, 1968, 13.

355 **American Association of Neurological Surgeons:** RBF, "Brain and Mind," 77–171. After meeting the psychologist B. F. Skinner, RBF noted, "I observe that he does not differentiate between brain and mind" (RBF, "Skinner, B. F.," in *SD,* 16235). He owed the introduction to Harold Cohen, who went on to conduct research on operant conditioning at the Institute for Behavioral Research in Maryland, where RBF served in "an unpaid, intermittent consult. function" (Itinerary, March 19, 1965). RBF discussed these experiments in "The Music of a New Life," in *UOO,* 31–32, and in his foreword to Harold Cohen and James Filipczak, *A New Learning Environment: A Case for Learning* (San Francisco: Jossey-Bass, 1971), xi–xvi.

355 **Dick Cavett joked:** "Yuk Among the Yaks," *Time,* March 22, 1968.

355 **Sun Dome:** Charles E. Rhine, "Amazing Sun Dome You Can Build," *Popular Science,* May 1966, 108–12.

356 **"the work of small time ingrates":** Steve Baer, *Dome Cookbook,* 5th printing (Corrales, NM: Lama Foundation, 1970), 8.

356 **"long dying and funeral":** Brand, *Last Whole Earth Catalog,* 439.

356 **Barbara Ward:** Ward credited the book's title to RBF: "I borrow the comparison from Professor Buckminster Fuller" (*Spaceship Earth* [New York: Columbia University Press, 1966], 15).

356 **"Amid the fever I was in":** Brand, *Last Whole Earth Catalog*, 439.

356 **"Techniques and tools":** Ibid.

356 **"I dunno, *Whole Earth Catalog*":** Ibid.

356 **"with manufacturers":** Stewart Brand to RBF, April 10, 1968, quoted in Wong, 454.

357 **"We *are* as gods":** Stewart Brand, *Whole Earth Catalog*, Fall 1968, 3.

357 **"access to tools":** John Markoff, *Whole Earth: The Many Lives of Stewart Brand* (New York: Penguin, 2022), 138.

357 **"People who beef about Fuller":** Stewart Brand, *Whole Earth Catalog*, Fall 1968, 3.

357 **"one of the most original":** Ibid.

357 **"baling wire hippies":** J. D. Smith, quoted in Andrew G. Kirk, *Counterculture Green: The* Whole Earth Catalog *and American Environmentalism* (Lawrence: University Press of Kansas, 2007), 76.

357 **On December 9, 1968, Brand assisted:** Fred Turner, *From Counterculture to Cyberculture: Stewart Brand, the Whole Earth Network, and the Rise of Digital Utopianism* (Chicago: University of Chicago Press, 2010), 110.

358 **center for futures studies:** Jenny Andersson, *The Future of the World: Futurology, Futurists, and the Struggle for the Post–Cold War Imagination* (New York: Oxford University Press, 2018), 202.

358 **"the Spaceship Earth Game":** John McHale, *The Future of the Future* (New York: Ballantine Books, 1969), 279.

358 **World Game:** "CU Confab on World Affairs Set Monday," *Colorado Springs (CO) Gazette-Telegraph*, April 9, 1967. The term had been used in RBF's circle for at least two years (Don Moore to RBF, January 22, 1965, B134-F5).

358 **"realizing a lifestyle":** RBF, *CP*, 191.

358 **"The young will lead the old":** RBF, "Self-Debiasing," in *SD*, 15750.

359 **"They are wondering if you":** Pat White to RBF, February 17, 1969, Series 18, B133-F9.

359 **Paul McCartney:** "The Mad Day Out: Location Seven," *The Beatles Bible*, July 23, 2018, https://www.beatlesbible.com/1968/07/28/the-mad-day-out-location-seven (accessed May 2021).

359 **would be "delighted":** RBF to John Lennon, telegram, February 26, 1969, Series 18, B133-F9.

359 **"Dr. Fuller found [it]":** Constance Abernathy to John Lennon and Yoko Ono, April 4, 1969, Series 18, B133-F9. RBF's office was later in touch with Yannis "Magic Alex" Mardas, the head of technology for the Beatles, but no project resulted (Series 18, B133-F9).

359 "I have learned": RBF testimony before the Senate Subcommittee on Intergovernmental Relations, March 4, 1969, reprinted in RBF, *World Game*, 25.

359 Herbert Matter: Itinerary, various dates, February 6–April 4, 1969.

360 found the experience "transformative": Schlossberg, author interview, March 27, 2019.

360 stealing his ideas: RBF to John Lloyd, March 21, 1969, B184-F1, and Amei Wallach, introduction to RBF, *Tetrascroll*, xii.

360 "That seems really interesting": Schlossberg, author interview.

360 thought that he was "insane": Ibid.

360 "I don't know who you are": Ibid.

360 "That I could survive it": Ibid.

360 had forgotten Schlossberg completely: Ibid.

361 he was "exaggerating": Alexandra Snyder May, author interview, June 7, 2019.

361 "an incredible list of freaks": Peter Rabbit, *Drop City*, 131.

361 "He brought it all together": Ibid., 144.

361 "my continued interest": Edwin Schlossberg to RBF, undated (1968), B175-F6.

361 New York Studio School: Details of the World Game workshop are derived primarily from *World Game Report*, New York Studio School of Painting and Sculpture, 1969.

362 "You're in charge": Schlossberg, author interview.

362 "It was really fun": Ibid.

362 for the "disaster": Mercedes Matter to RBF, July 6, 1969, B188-F3.

363 audio message: Itinerary, June 27, 1969.

363 "as the degree of gravitational hazard": RBF, "Vertical Is to Live, Horizontal Is to Die," in *Discourse*, 281.

363 Kurt Vonnegut and Arthur C. Clarke: *10:56:20 PM EDT 7/20/69*, 102.

363 enter environmental design: Stephen E. Selkowitz, interviewed by author, March 13, 2019.

363 "They looked like": John Cage, "Diary: How to Improve the World (You Will Only Make Matters Worse)," in *Liberations: New Essays on the Humanities in Revolution*, ed. Ihab Hassan (Middletown, CT: Wesleyan University Press, 1971), 20.

363 "bare maximum": "I. Pre-Scenario Facts," *World Game Report*, New York Studio School of Painting and Sculpture, 1969.

363 "not his minimum potential": Ibid.

363 "our first move": "II. Scenario," *World Game Report*, New York Studio School of Painting and Sculpture, 1969.

363 "I feel": Edwin Schlossberg, "World Game Diary," Series 18, B39-F2, 19.

364 "My whole strategy": RBF, "Rearrange the Scenery," in *SD*, 14920.

364 efforts by other futurists: Andersson, *Future of the World*, 179–81.

364 "When you can't go any further": RBF, "Rearrange the Scenery," in *SD*, 14921.

364 water tower in Kuwait: Zahra Ali Baba and Jaber Al-Qallaf, "Dynamic Heritage," *World Heritage*, December 2016, 58.

364 kibbutz in Israel: RBF, *Artifacts of R. Buckminster Fuller*, vol. 4, 219–22. Sadao noted later that their involvement in the Kibbutz Kfar Menachem project would be "very limited" (Sadao to RBF, October 23, 1970, B209-F6).

364 religious center at SIU Edwardsville: "Architect Fuller Shows Sketches of Religious Center," *Alton (IL) Evening Telegraph*, March 22, 1969, 6. Photos and drawings in RBF, *Artifacts of R. Buckminster Fuller*, vol. 4, 192–215.

364 Francis Warner: In 1966 Warner had tried unsuccessfully to arrange a meeting between RBF and Beckett, who had been receptive but unfamiliar with the other man's work: "If you could lighten a little my ignorance concerning him I should be grateful" (Beckett to Warner, June 9, 1966, reprinted in David Tucker, *A Dream and Its Legacies: The Samuel Beckett Theatre Project, Oxford c. 1967–76* [Gerrards Cross, UK: Colin Smythe, 2013], 47).

364 Fuller's representative: Gillian Meller, email to author, July 30, 2021.

364 Norman Foster: Norman Foster, "Richard Buckminster Fuller," in *Anthology*, 1.

365 "It must have been like hearing": Foster, author interview, October 30, 2020.

365 "This is what it would look like": Tucker, *Dream and Its Legacies*, 13.

365 "an underground submarine": "An Historic and Unique Conference," B188-F1, 4.

365 "while slightly quailing": Samuel Beckett to Francis Warner, June 28, 1969, reprinted in Tucker, *A Dream and Its Legacies*, 49.

365 "He's a skeptical playwright": Bernard Weinraub, "Fuller Is Designing Theater at Oxford," *New York Times*, March 13, 1970, 28.

365 "to the ultimate: invisibility": "An Historic and Unique Conference," B188-F1, 4.

365 "The principle of mobility": Ibid.

365 "To Indira, in whose integrity": Indira Gandhi to Dorothy Norman, May 26, 1969, quoted in Ramesh, *Indira Gandhi*, 161.

366 "deeply touched": Ibid.

366 "the usual drive": Hatch, 217.

366 "a remarkable person": Indira Gandhi, opening speech before Nehru Lecture by RBF, reprinted in *Planetary Planning* booklet, B450-F3, 1.

366 "I have done so simply": RBF, "Planetary Planning," *Jawaharlal Nehru Memorial Lectures, 1967–1972*, 74.

366 **airports in New Delhi:** Sadao, 203.

366 **Edgar J. Applewhite:** Biographical information in Applewhite, *Cosmic Fishing*, 1–21.

366 **computer memory:** Trevor Blake, "Buckminster Fuller's Ultra-Micro Computer," Synchronofile.com, last modified July 12, 2014, https://synchronofile.com/2014/07/12/buckminster-fullers-ultra-micro-computer.

366 **"become the whole curriculum":** RBF interview in *The World of Buckminster Fuller.*

367 **John Entenza:** Wong, 272–73.

367 **"He is the only truly great man":** Huston, author interview, April 11, 2019.

367 **architectural group Archigram:** Simon Sadler, *Archigram: Architecture Without Architecture* (Cambridge, MA: MIT Press, 2005), 6, 37–38, 100–103.

367 **"kidnapped" Fuller:** Constance M. Lewallen and Steve Seid, *Ant Farm, 1968–1978* (Berkeley: University of California, 2004), 44.

367 **Lloyd Kahn:** Biographical information derived from Lloyd Kahn, interviewed by author, March 21, 2019, and Kahn, *Domebook 2*, 16.

368 **"I've found this to be true":** J. Baldwin, ed., *The Essential Whole Earth Catalog: Access to Tools and Ideas* (Garden City, NY: Doubleday, 1986), 21.

368 **he was "very straight":** Lloyd Kahn, author interview.

368 **"There's someone you should meet":** Ibid.

368 **"Persons in their late twenties":** Brand, *Last Whole Earth Catalog*, 112.

368 **Abandoning the "second-rate":** Jay Baldwin, interviewed in Gowan, *Shared Vision*, 73.

368 **Pacific High School:** Kahn, *Domebook 2*, 32–34, and Alastair Gordon, *Spaced Out: Radical Environments of the Psychedelic Sixties* (New York: Rizzoli, 2008), 204–11.

369 **Charles Hall:** Edward M. Wysocki Jr., *Out of This World Ideas—And the Inventions They Inspired* (Scotts Valley, CA: CreateSpace, 2018), 102.

369 **"The youth of today":** Elizabeth Barlow, "The New York Magazine Environmental Teach-In," *New York*, March 30, 1970, 30.

370 **He planned to hand out leaflets:** Tucker, *Dream and Its Legacies*, 25.

370 **presence on campus was controversial:** Martin Pawley describes a fraught encounter with RBF on this trip in *Buckminster Fuller*, 20–31.

370 **"Welcome to the Buckminster Führer Show":** Ibid., 26.

370 **"Just as ephemeralization":** RBF, "Education for Comprehensivity," in RBF, Eric A. Walker, and James R. Killian Jr., *Approaching the Benign Environment*, 106–7.

371 **"everything with nothing":** Ibid., 108.

371 **"gratifying their own personal discontent":** Ibid., 107.

371 **"Because almost everyone":** Ibid., 107–8.

371 bamboo domes—or "bambassadors": *Industrial Design*, February 1961, 65–68.

371 solution to Vietnam: Keats, *You Belong to the Universe*, 85.

371 Abbie Hoffman: David Farber, *Chicago '68* (University of Chicago Press, 1988), 47.

371 "agitating simply because": RBF, "Education for Comprehensivity," 109–110.

371 "Whites are bleached out": RBF, "Whites," in *SD*, 20656.

371 "swiftly eradicated": RBF, "Raison d'Etre," in *SD*, 14786.

371 "I have never met you before": George Boesel, interviewed in Gowan, *Shared Vision*, 167.

371 "I am confident": RBF to Francis Warner, April 20, 1970, quoted in Tucker, *Dream and Its Legacies*, 58n36.

372 "Marines eat rice too": Scott, "Fluid Geographies," in *New Views on R. Buckminster Fuller*, ed. Chu and Trujillo, 164.

372 until late at night: Gene Youngblood, interviewed by author, January 30, 2020.

372 "A concrete scientific alternative": Gene Youngblood, "Buckminster Fuller's World Game," in *Whole Earth Catalog*, March 1970, 30.

372 dome in Edwardsville: Ibid., 32, and Wigley, *Buckminster Fuller Inc.*, 267.

372 federal legislation: *Congressional Record* extract in RBF, *World Game*, 39.

372 close the campus: "SIU Shut Down After Protests," *Southern Illinoisan* (Carbondale), May 13, 1970, 1.

372 "brilliantly exploited": RBF, *CP*, 236.

373 "the determination of the students": Ibid.

373 "sitting ducks": Ibid.

373 "He was supposedly coming in": Syeus Mottel, *Charas: The Improbable Dome Builders* (New York: Drake, 1973), 150. For another account of the lecture, see Allan Katzman, "When Will John V. Lindsay Meet Buckminster Fuller?," *East Village Other*, April 12, 1968, 4.

373 "who basically spent time": Mottel, *Charas*, 22.

373 "completely crazy situation": Michael Ben-Eli, interviewed by author, April 1, 2019.

374 "They could not see": Mottel, *Charas*, 109.

374 "It's more about involvement": Ibid., 73.

374 "a New York City Lower East Side gang": RBF, *Inventions*, 145.

374 Agel had met: Itinerary, April 24, 1964.

374 "turn of mind": Jeffrey T. Schnapp and Adam Michaels, *The Electric Information Age Book: McLuhan/Agel/Fiore and the Experimental Paperback* (New York: Princeton Architectural Press, 2012), 131.

374 "I always say to myself": RBF, with Agel and Fiore, *I Seem to Be a Verb*, 6.

374 "reigning sex symbols": Laurie Johnston, "Elderly Are Told of the Power in a May-December Coalition," *New York Times*, May 1, 1973, 22. Another

older rebel adopted by the counterculture was Henry Miller, whom Robert Snyder videotaped for two days with RBF in California (Itinerary, June 15–16, 1970). This unlikely pairing may have inspired Norman Mailer to lament the comparative lack of mainstream interest in Miller: "On any given night in any lecture hall of America, Buckminster Fuller—another genius—would draw an audience five to, would it be, ten times the size, and Fuller cannot write a sentence which does not curdle the sap of English at its root" (Norman Mailer, *Genius and Lust: A Journey Through the Major Writings of Henry Miller*, comp. Mailer [New York: Grove Press, 1976], 9).

374 **"his contribution to mankind's environment":** Francis Warner to Calvin Tomkins, June 19, 1970, Calvin Tomkins Papers, Museum of Modern Art, folder I.7. RBF was also nominated for the Nobel Peace Prize that year by the psychiatrist Francis J. Braceland (B204-F5).

375 **"the strongest":** Eugene Garfield, "A Tribute to R. Buckminster Fuller—Inventor, Philosopher," *Current Contents*, November 2, 1981, 16.

375 **Alzheimer's disease:** Al Gowan, "Delyte W. Morris," in Gowan, *Shared Vision*, 20.

375 **"a broken man":** Harold Cohen, interviewed in ibid., 16.

375 **"strife-free center":** RBF to Farah Pahlavi, undated (fall 1970), B211-F4.

375 **found it "feasible":** RBF, *CP*, xxxi.

375 **Ezra Pound:** RBF and Dil, *Humans in Universe*, 20–25.

376 **"the gold from the pine cones":** Ibid., 24.

376 **"tremendously hurt man":** RBF, *Pound, Synergy, and the Great Design*, 2.

376 **"That conversation":** Guy Davenport to Hugh Kenner, December 30, 1970, reprinted in Guy Davenport and Hugh Kenner, *Questioning Minds: The Letters of Guy Davenport and Hugh Kenner*, vol. 2, ed. Edward M. Burns (Berkeley, CA: Counterpoint, 2018), 1321.

376 **World Game centers:** Hal Aigner, "Sustaining Planet Earth: Researching World Resources," *Mother Earth News*, November 1970.

376 **"egocentric soul-searching":** RBF to Edwin Schlossberg, October 13, 1970, B209-F7.

376 **participants were "exploited":** RBF to John Cage, September 14, 1970, B209-F3.

376 ***Whole Earth Catalog:*** RBF to Edwin Schlossberg, October 13, 1970, B209-F7.

376 **"I thought if I exaggerated":** Youngblood, author interview, January 30, 2020.

376 **"greatly weakening":** RBF to Gene Youngblood, October 13, 1970, B209-F7.

376 **"Humanity is acquiring":** RBF, "The Earthians' Critical Moment," *New York Times*, December 11, 1970, 47.

376 "I'm the only man": RBF, "*Playboy* Interview," *Playboy*, February 1972, 198.

377 "Something hit me very hard once": Ibid., 199–200.

377 "demountable" auditorium: Norman Foster to the Master of St. Peter's College, October 26, 1970, B209-F6.

378 5 percent of the royalties: Dale Klaus to Lloyd Kahn, June 8, 1970, quoted in Scott, "Fluid Geographies," in *New Views on R. Buckminster Fuller,* ed. Chu and Trujillo, 172.

378 "Your engineer": Lloyd Kahn, author interview, March 21, 2019.

378 Fuller was excessively "pissed": Ibid.

378 "everyone concerned": Joseph D. Clinton, email to author, January 31, 2021.

378 "What's good about 90° walls": Kahn, *Shelter,* 117.

379 "*less* with less": Ibid., 117.

379 "Our work, though perhaps smart": Ibid., 112.

379 "demonstrative urban conservation park": Wong, 433.

379 "Bucky, I think you've been in Africa": RBF, *EIK,* session 11.

379 "There is nothing in Calcutta": RBF, *Education Automation,* 50.

379 "raceless and classless": Wong, 437.

379 Old Man River's City: Sources include ibid., 433–40, and RBF, *CP,* 315–23.

380 "might be part of its social enemy's conspiracy": RBF, *CP,* 322.

380 "I said that what I would design": Ibid., 322.

380 "our design solution": Ibid., 321.

380 searched by the CIA: *Your Private Sky,* ed. Krausse and Lichtenstein, 37.

380 "using my phone": RBF, "Carbondale Office," in *SD,* 1742.

380 basketball stadium: O'Dell and Heckman, *Bucky's Dome,* 59.

380 maintaining his archives: Ibid., 60.

380 event of a state audit: Bill Perk, author interview, March 15, 2019.

380 "these two fighters": Tony Pugh, interviewed in Gowan, *Shared Vision,* 77. Pugh believed that this referred to the "Rumble in the Jungle" between Ali and George Foreman on October 30, 1974, but the chronology points to the Ali-Frazier fight on March 8, 1971.

380 "Every time humanity": RBF and Dil, *Humans in Universe,* 21.

380 "friend of the universe": Ibid., 20.

381 "You could not meet Maharishi": RBF press conference with Maharishi Mahesh Yogi, https://www.organism.earth/library/document/amherst-press-conference (accessed January 2021). RBF contributed an introduction to Harold H. Bloomfield, Michael Peter Cain, Dennis T. Jaffe, and Robert B. Kory, *TM: Discovering Inner Energy and Overcoming Stress* (London: Unwin Paperbacks, 1976), xiii, in which he related Transcendental Meditation to his own techniques. Two domes were later built at Maharishi International University in Fairfield, Iowa, "as a place to practice levitation" (Sam Farling to RBF, February 10, 1980, B411-F4).

381 sent Senator Edmund Muskie: RBF, "Telegram to Senator Edmund Muskie," in *Earth, Inc.*, 155–71.

381 "All those years": Hatch, 79.

381 Hexa-Pent Dome: RBF, "My New Hexa-Pent Dome," *Popular Science*, May 1972, 128–31, and RBF, *Inventions*, 264–68.

381 "the time or skills": Norman Foster, "Buckminster Fuller—File for Future." Manuscript provided to the author by Norman Foster.

381 "bizarre antics": Thomas Turner to RBF, July 21, 1971, quoted in Hsiao-Yun Chu, "The Archive of R. Buckminster Fuller," in *New Views on R. Buckminster Fuller*, ed. Chu and Trujillo, 20.

381 "complete strangers to me": Written statement by RBF, quoted in ibid.

382 "I live on planet Earth": "Fuller Plans Office on SIU-Edwardsville Campus," *Southern Illinoisan* (Carbondale), September 24, 1971, 1.

382 Herman Wolf: According to Wolf, he and RBF reconciled after a chance encounter at Yale in 1957 (Herman Wolf, interviewed by Terry Hoover and Bill Pretzer, October 25, 2001, Benson Ford Research Center).

382 "I could conk out": Boyce Rensberger, "New Institute to Promote Buckminster Fuller's Pioneer Ideas: Reform of Environment Long Goal of Maverick," *New York Times*, June 29, 1972, 41.

382 "I would have assumed": Hatch, 223.

382 "a joint manifesto": Ibid.

382 "are in fact the feedback": RBF, "The Lord's Prayer," in *Intuition*, 184.

382 "a financial intervention": Medard Gabel, interviewed by author, January 22, 2020.

382 tired and irritable: Hatch, 272.

382 put his grandson through college: Chu, "Archive of R. Buckminster Fuller," 21.

383 moving to Philadelphia: Gabel, author interview.

383 annual funds of $50,000: Martin Meyerson to RBF, October 4, 1978, B374-F9.

383 "a coup roughly of the same magnitude": William C. Lyon, "Buckminster Fuller: Our New Ben Franklin," *Philadelphia Inquirer*, November 1, 1972, 1.

383 keep her husband at home: Creed C. Black, "Reflections: Congress Retards Nixon Domestic Policy," *Philadelphia Inquirer*, November 26, 1972, 5-H.

383 "the interaccommodative": Shirley Sharkey, email to author, March 18, 2019. For the Scheherazade numbers, see RBF with Applewhite, *Synergetics*, 771–83.

384 Cam Smith: Biographical information derived from "Public Invited to Hear Dr. Buckminster Fuller Feb. 1, Green Auditorium," *Vincennes (IN) Sun-Commercial*, January 29, 1973, 3.

384 "a tremendous tool": Cam Smith to RBF, September 15, 1968, B176-F5.

384 "Things could and should be": Ibid.

384 "in line with each other": Charlene Sprang to RBF, October 16, 1971, B231-F3.

384 "Dr. Fuller is familiar": Shirley Swansen to Charlene Sprang, October 28, 1971, B231-F3.

384 "dedicated to reducing crime": Copyright page of RBF, *Buckminster Fuller to Children of Earth*, comp. and photographed by Cam Smith (Garden City, NY: Doubleday, 1972).

384 a vehicle for Hubbard's theories: Lawrence Wright, *Going Clear: Scientology, Hollywood, and the Prison of Belief* (New York: Vintage Books, 2013), 241–43.

384 "the leader of something": Alvin Toffler, *The Futurists* (New York: Random House, 1972), 6.

384 Libre and Red Rockers: Amy Azzarito, "Libre, Colorado, and the Hand-Built Home," in *West of Center: Art and the Counterculture Experiment in America, 1965–1977*, ed. Elissa Auther and Adam Lerner (Minneapolis: University of Minnesota Press, 2012), 95–108.

384 youth center in the Bronx: John L. Hess, "No Youth Center in Spuyten Duyvil, So Youngsters Construct Their Own," *New York Times*, November 19, 1972, 45.

385 "It's beautiful, isn't it?": Mottel, *Charas*, 61.

385 "a triumph": Ibid., 63.

385 "Hey, is this man important?": Ibid.

ELEVEN: SYNERGY

386 "Fictional history": RBF with Applewhite, *Synergetics*, 661.

386 "I seem to be a verb": Stephen Schwartz, "Prologue / Tower of Babble," *Godspell*, https://www.allmusicals.com/lyrics/godspell/prologuetowerof babble.htm (accessed January 2021).

387 "a visionary trance": Sean Kelly, Joe Orlando, and John Costanza, "Utopia Four Comix," *National Lampoon*, June 1971, 32. "Buckminster Fuller's Repair Manual for the Entire Universe" by Henry Beard, Harry Fischman, and Jeffrey Prescott ran in *National Lampoon*, January 1972, 43–46.

387 episode at the Expo Dome: The same site later served as a location for Robert Altman's 1979 film *Quintet* (Legault, "Death and Life of Buckminster Fuller's US Pavilion at Expo 67," in *Montreal's Geodesic Dreams*, ed. McAtee, 105).

387 *Earth II*: The pilot ran as a movie on CBS and in theaters overseas. RBF's role included "reading and checking the scientific accuracy and plausibility of stories and scripts," proposing his own ideas, and granting permission for the use of the Dymaxion Map (F. C. Houghton to Frank Rosenfelt, MGM memo, September 15, 1970, B211-F1).

387 *Brave New World*: Milton Sperling to RBF, March 15, 1967, B155-F6.

387 **Amundsen-Scott South Pole Station**: Frank A. Blazich Jr., "Rendezvous with Penguins: Seabee Construction of the South Pole Dome," *Seabee Magazine*, https://seabeemagazine.navylive.dodlive.mil/2014/08/27/rendezvous-with -penguins-construction-of-the-south-pole-dome (accessed May 2021).

387 **Richter published a computer analysis**: Don L. Richter, "Geodesic Domes: The Rationale and the Reality," *Architectural Forum*, January 1972, 84–86.

388 **his staff had been reduced**: Hiring history provided by Tim Wessels, interviewed by author, March 21, 2019.

388 **despite back pain**: Kenner, *Bucky*, 47. A decade earlier, RBF had learned that his left leg was an inch shorter than his right, resulting in leg pain that required him to wear special shoes (Rumwell, *Time* profile notes, Take 6, December 26, 1963, B126-F4).

388 **Fuller's standard quote**: Shirley (Swansen) Sharkey, interviewed by author, March 18, 2019.

388 **"I have absolutely no budget"**: Applewhite, *Cosmic Fishing*, 31.

389 **a floatable breakwater**: RBF, *Inventions*, 269–73. An improved version was patented in 1979.

389 **Fuller advised Windworks**: AP, "Answer to Energy Problems May Be Blowing in Wind," *Oshkosh (WI) Northwestern*, October 3, 1974, 9, and RBF, "Wind Power Sequence," in *SD*, 20735–47.

389 **convince the Rockefellers**: RBF, "Wind Power Sequence," in *SD*, 20735–47.

389 **"when the sun"**: RBF, "Energy Through Wind Power," *New York Times*, January 17, 1974, 39.

389 **"A spaceship is built by man"**: Rasa Gustaitis, *Wholly Round* (New York: Holt, Rinehart and Winston, 1973), 158–59.

389 **"one has an obligation"**: Kenner to Guy Davenport, July 2, 1963, reprinted in Davenport and Kenner, *Questioning Minds*, vol. 1, 358.

390 **"[Fuller] reiterates"**: Kenner, *Bucky*, 181.

390 **"some sharp kids from the streets"**: Ibid.

390 **"I wouldn't have"**: Ibid., 182.

390 **lost research and speaking time**: Hatch, 248–49.

390 **"If anyone were to ask you"**: RBF, "Airport," in *SD*, 290.

390 **"I hope no irrevocable commitments"**: Indira Gandhi to Karan Singh, March 1, 1973, quoted in Ramesh, *Indira Gandhi*, 163.

391 **"visibly had little to do with Bucky"**: Sadao, 203.

391 **U Thant**: Itinerary, March 5, 1969, and October 30, 1970, and RBF, *CP*, 183–84.

391 **compiled from his earlier work:** Michael Ben-Eli, quoted in Jaime Snyder, *"And It Came to Pass—Not to Stay*: Background," appendix to RBF, *And It Came to Pass—Not to Stay* (Zurich: Lars Müller, 2008), 182–83.

391 **Marks felt "terribly hurt":** Robert W. Marks to RBF, July 14, 1973, B261-F4.

391 **second book by Kenner:** Hugh Kenner's *Geodesic Math and How to Use It* (Berkeley: University of California Press) was published in 1976. RBF was originally slated to write an introduction, but the result was rejected. "Bucky [Fuller] has thrown a mini-tantrum. The introduction that he promised turned out to be 57 pages long, full of cadenzas and insinuations that the book (which he hasn't read) might better have been produced by for instance himself, if his patent attorney (he now says) hadn't counseled against publishing methods of design" (Kenner to Guy Davenport, June 10, 1975, reprinted in Davenport and Kenner, *Questioning Minds*, vol. 2, 1580).

391 **"He had proposed":** Applewhite, *Cosmic Fishing*, 5.

391 **"thinking out loud":** Ibid., 1.

391 **"Address! That isn't the right question":** RBF, "Planet Earth," in *SD*, 13478.

392 **"an unexpected shipment":** Applewhite, *Cosmic Fishing*, 91.

392 **"I am feeling myself":** Ibid., 95.

392 **"confronting the author":** Edgar J. Applewhite, "A Note on Collaboration," in RBF with Applewhite, *Synergetics*, vii.

392 **"For Fuller, nature":** Applewhite, *Cosmic Fishing*, 78.

392 **Kahn's associate Anne Tyng:** Ibid., 108.

392 **"the bridge builder":** RBF to Arthur L. Loeb, January 6, 1967, quoted in ibid., 111.

392 **"digressions into metaphysics":** Loeb to J. Pack, September 3, 1971, quoted in Wong, 467.

393 **"My own lay conclusion":** Siobhan Roberts, *King of Infinite Space: Donald Coxeter, the Man Who Saved Geometry* (Toronto: Anansi, 2006), 179.

393 **Gerald Dickler:** Applewhite, *Cosmic Fishing*, 123–24.

393 **went to Hugh Kenner for comments:** Kenner to Guy Davenport, May 7, 1972, reprinted in Davenport and Kenner, *Questioning Minds*, vol. 2, 1404.

393 **impact on Fuller was "galvanic":** Applewhite, *Cosmic Fishing*, 127.

393 **"His imagination is triggered":** Ibid., 129.

393 **"Now we know that it is right":** RBF, "Fuller, R. B.: On Galley Proofs," in *SD*, 6342.

394 **"a book about models":** Applewhite, *Cosmic Fishing*, 62.

394 **H. S. M. Coxeter, "*the* geometer":** Dedication page in RBF with Applewhite, *Synergetics*.

394 **"after much deliberation"**: Applewhite, "Account by R. Buckminster Fuller of His Relations with Scientists," Series 8, B12-F23, 9.

394 **"foolhardy freedom to explore"**: Ibid.

394 **"a remarkable discovery"**: Roberts, *King of Infinite Space*, 179.

394 **whom he praised for his intuition**: Charles Panati, "Almighty Tetrahedron," *Newsweek*, April 21, 1975.

394 **"a brilliant architect and engineer"**: Roberts, *King of Infinite Space*, 178.

395 **"a lot of nonsense"**: Ibid., 181.

395 **"dare to be naïve"**: RBF with Applewhite, *Synergetics*, xix. The statement recalls "Dare to Be True," the motto of Milton Academy (Sadao, 26).

395 **"Synergy means behavior"**: RBF with Applewhite, *Synergetics*, 3.

395 **"You may say that we had no right"**: Ibid., 4.

395 **"When we draw a triangle"**: Ibid., 443.

395 **"Only the triangle produces structure"**: Ibid., 318.

396 **"the minimum thinkable set"**: Ibid., 239.

396 **"the number of telephone lines"**: Ibid., 60.

396 **a triangular number**: RBF, "Whole Number or Rational Basic Relationship of Dymaxion Comprehensive System to Light Quanta Particle Growths," May 7, 1948, reproduced in ibid., 816–17.

397 **"Tetrahedron Discovers Itself and Universe"**: RBF with Applewhite, *Synergetics*, 209–15.

397 **"a business of stark homage"**: Applewhite, *Cosmic Fishing*, 143.

397 **mass of other material**: Other topics discussed in *Synergetics* include the concept of "triangling" or "tetrahedroning" instead of squaring and cubing for finding the second and third powers of numbers; the A and B quanta modules, which RBF derived by dividing the tetrahedron and octahedron into the minimum units of his geometry; and the Scheherazade numbers, which he obtained by multiplying primes into enormous integers that would allow computers to perform trigonometric calculations in whole number increments.

397 **prose recalled Gertrude Stein**: Arthur L. Loeb, preface to RBF with Applewhite, *Synergetics*, xvi.

397 **"James Joyce to a mild extent"**: RBF, "Conceptuality of Fundamental Structures," in *Structure in Art and in Science*, ed. Kepes, 81.

397 **"Make it new"**: Michael North, "The Making of 'Make It New,'" *Guernica*, August 15, 2013, https://www.guernicamag.com/the-making -of-making-it-new (accessed January 2021).

397 **"There are to this day certain passages"**: Applewhite, *Cosmic Fishing*, 48.

397 **"I find certain passages sublime"**: Ibid., 49.

397 **"Maybe the reader"**: Ibid., 73.

397 **"If we had only two days"**: Ibid., 93.

398 **"I always assume"**: Ibid.

398 "intimations of mortality": Ibid.

398 "Boy, this must be a nice job": Sharkey, author interview, March 18, 2019.

398 Fuller became distressed: Tim Wessels, author interview, March 21, 2019.

398 an even larger pattern: RBF with Applewhite, *Synergetics 2*, 122–23.

398 "Lady Anne and Her Little Old Man": Jaime Snyder, email to author, April 27, 2021.

398 world president of the high-IQ society Mensa: RBF later offered to resign in the wake of criticism from certain American members of Mensa, but he elected to retain the honorary position, in part because of a dispute with its international chairman, Victor Serebriakoff (B377-F5). He also began to serve around this time as a tutor in design science for International College, which paired students with mentors for private study (Bruce Winters, "The World for a Classroom," *Baltimore Sun*, April 7, 1974).

399 "give the reviewers": Applewhite, *Cosmic Fishing*, 141.

399 "You're very . . . *universe*": Annie Dillard, "Buckminster Fuller," in *Brushes with Greatness: An Anthology of Chance Encounters with Celebrities*, ed. Russell Banks, Michael Ondaatje, and David Young (Toronto: Big Bang, 1989), 29.

399 "After he sat on me": Annie Dillard, email to author, February 25, 2021.

399 "scores of incomprehensible letters": Dillard, "Buckminster Fuller," 29.

399 "carry real weight in the universe": Annie Dillard, *Living by Fiction* (New York: Harper Perennial, 1988), 174.

400 "A dome really might not leak": Stewart Brand, *Whole Earth Epilog: Access to Tools* (San Francisco: Point, 1974), 467.

400 "As a major propagandist": Stewart Brand, *How Buildings Learn: What Happens After They're Built* (New York: Viking, 1994), 59–60.

400 "How do you convince people": Stewart Brand, afterword to Gabel, *Energy, Earth, and Everyone*, 155.

400 Brown invoked Spaceship Earth: Joseph Lelyveld, "Jerry Brown's Space Program," In America, *New York Times Magazine*, July 17, 1977, 55.

400 used Brand as a matchmaker: Ibid.

400 "*Synergetics* excludes": Applewhite, *Cosmic Fishing*, 66.

401 "All humanity has always been born naked": RBF, *EIK*, session 1.

401 "tremendously tender": Ibid., session 9.

401 "People would say": Ibid.

401 "When you can't find the logical way": Ibid., session 5.

401 "When the kids can really get information": Ibid., session 12.

401 "I don't see any pure males": Ibid., session 2.

401 "When the bound galleys": Denneny, author interview, March 22, 2019.

402 "the Leonardo da Vinci": Ibid.

402 "You sent me a world": Annie Dillard to RBF, May 12, 1975, B375-F4.

402 thirty thousand copies: Denneny, author interview.

402 "many flaws": RBF, *EIK*, session 11.

402 Tatyana Grosman: The following account is based largely on Wallach, intro. to RBF, *Tetrascroll*, ix–xxvii.

402 "The stone invites me": Polly Lada-Mocarski, "Book of the Century: Fuller's *Tetrascroll*," *Craft Horizons*, October 1977, 18.

402 "the empirical, the scientific way": Applewhite in RBF, *SD*, xxvii.

403 "The Red Sea is still open": Wallach, intro. to *Tetrascroll*, xxi.

403 Fuller proudly called it *Tetrascroll*: The original book closed with a photo of RBF's granddaughter, Alexandra, captioned as the "only known picture of Goldy" (E. Maurice Bloch, *Words and Images: Universal Limited Art Editions, Frederick S. Wight Art Gallery, March 14–May 7, 1978* [Los Angeles: The Gallery, 1978], 6). An epilogue by Schlossberg, "Epilever," played on RBF's notion that the principle of the lever had been discovered by a man standing on a log (*Tetrascroll*, 128–29).

403 archaeological site in Thailand: RBF, *Artifacts of R. Buckminster Fuller*, vol. 4, 353–59.

403 "all kinds of undemocratic": Indira Gandhi to RBF, July 10, 1975, Series 7, B19-F5.

403 "treachery of information": RBF, *EIK*, session 12.

403 printing "propaganda": Ibid.

403 World Society of Ekistics: Itinerary, November 11, 1975.

404 "human unsettlement": RBF, "Accommodating Human Unsettlement," September 20, 1976, presented to the House Committee on Banking, Currency and Housing (*The Rebirth of the American City*, US Government Printing Office, 1976).

404 Richard Burton: Tucker, *Dream and Its Legacies*, 8.

404 Burton paid another £17,500: Richard Burton, *The Richard Burton Diaries*, ed. Chris Williams (New Haven, CT: Yale University Press, 2012), 613.

404 "very diminutive": RBF to Francis Warner, August 13, 1976, B325-F4.

404 "What I had surmised": Sadao, 213. For Noguchi's use during this period of a "post-tensioning" sculptural technique inspired by RBF, see Herrera, *Listening to Stone*, 328.

404 renewed for two more years: Martin Meyerson to RBF, October 4, 1978, B374-F9.

404 International Conference on the Unity of the Sciences: James Robison, "Charges Cloud Moon's Conference," *Chicago Tribune*, September 6, 1975, sec. 2, 11.

405 "vectors and valences": Tom Patton, *The Synanon Philosophy* booklet, B211-F2, 28.

405 "gluing their fingers": Ibid., 20.

405 "Chuck has installed": David Cole Gordon to RBF, September 18, 1968, B176-F6.

405 "allowed for individual expression": Rod Janzen, *The Rise and Fall of Synanon: A California Utopia* (Baltimore: Johns Hopkins University Press, 2001), 59.

405 "The process of World Game": Mark Victor Hansen to Steve Meyers, June 17, 1970, B280-F6.

405 "an extraordinary philosophic treatise": RBF to Steve Meyers, October 26, 1973, B280-F6.

405 "I am an admirer of Chuck Dederich": Ibid., July 29, 1971, B280-F6.

405 repressive practices: See correspondence between RBF and Meyers in B280-F6.

406 they agreed to work together: Roger Stoller, interviewed by author, May 13, 2019.

406 "Grandpa," Jaime said: Jaime Snyder, interviewed by author, April 23, 2019.

406 often referring to trainees as "assholes": Luke Rhinehart, *The Book of est* (Hypnotic I Media, 2010, EPUB ed.), 39.

407 "Jaime, I don't really": Jaime Snyder, author interview, April 23, 2019.

407 "a great conversation": Barbara Stevenson, email to author, February 9, 2022.

407 conventional notions of gender: RBF's views on women's rights were largely mediated by his sense of domestic labor, which he framed as a matter of design ("Dymaxion House," in *Discourse*, 92). He supported the Equal Rights Amendment (RBF, *Cosmography*, 263), believed that abortion should be legalized (RBF, *Synergetics 2*, 366, and RBF, "Abortion," in *SD*, 16), and proposed that the futurist Barbara Marx Hubbard run for president (Hubbard, "Path of Social Evolution," in *Anthology*, 140), but in "Why Women Will Run the World" (*McCall's*, March 1968), he envisioned a new set of gender roles, rather than their removal: "Women will convert man's scientific findings to industrial production."

407 "My brother": Alexandra Snyder May, author interview, May 17, 2019.

407 "Grandpa, how do you deal?": Jaime Snyder, author interview, April 23, 2019.

408 The North Face: Account based on Bruce Hamilton, "The North Face and R. Buckminster Fuller Special Connection," https://www.outinunder.com/content/north-face-and-r-buckminster-fuller-special-connection-part-1 (accessed January 2021).

408 "I am writing to you now": Don Butts to RBF, April 10, 1970, reproduced at "Bruce Hamilton's The North Face and R. Buckminster Fuller Archive," https://www.outinunder.com/documents/bruce-hamiltons-north-face-and-r-buckminster-fuller-archive (accessed January 2021).

408 "the single occupant": RBF to Don Butts, June 4, 1970, reproduced at ibid.

408 **"It is no aesthetic accident"**: Bruce Hamilton, "The North Face and R. Buckminster Fuller Special Connection—Part 2," https://www.outinunder .com/content/north-face-and-r-buckminster-fuller-special-connection -part-2 (accessed May 2021).

409 **Philadelphia folk club:** At the Cherry Tree Folk Club in Philadelphia on January 18, 1976, RBF sang and danced onstage with a band led by Pete Stampfel of the psychedelic folk group the Holy Modal Rounders. The live album *Dymaxion Ditties* was produced in a limited edition of fifty copies (Series 17, Mixed Materials 71, Reel 292). An item on the concert ran in *The Village Voice* ("The Geodesic Drone," March 1, 1976).

409 **Lim Chong Keat:** Lim later funded the publication of *Synergetics Folio*, a limited edition of ten posters that were printed in Singapore and completed in 1979. Reduced versions were reproduced in *Synergetics 2*.

409 **plans for Komtar:** Soon-Tzu Speechley, "Komtar: Malaysia's Monument to Failed Modernism," *Failed Architecture*, June 6, 2016, https:// failedarchitecture.com/komtar-malaysias-monument-to-failed-modernism (accessed May 2021).

409 **"his best friend"**: Shirley Sharkey to Lim Chong Keat, telegram, July 2, 1983, B489-F6.

409 **"in awesome review"**: "Campuan Society Meeting, January 10–16, 1977," B347-F1, 1.

409 **dome out of bamboo:** Daniel Carleton Gajdusek, *Stumbling Along the Tortuous Road to Unanticipated Nobility: Melanesian, Indonesian, and Malaysian Expedition* (Bethesda, MD: Study of Child Growth and Development and Disease Patterns in Primitive Cultures, National Institute of Neurological and Communicative Disorders and Stroke, 1996), 313.

409 **"Carl, I have a good friend"**: Carl Solway, letter to author, April 1, 2019.

409 **Ohio Building Authority:** Ibid.

410 **"I've seen time and again"**: RBF, *EIK*, session 12.

410 **rechristened the Biosphere:** "Son of Expo," *Courier-Journal & Times* (Louisville, KY), April 14, 1968.

410 **caught fire:** "US-Built Geodesic Dome in Montreal Gutted in Fire," *Palm Beach Post*, May 21, 1976.

410 **"It almost seemed"**: RBF, "Accommodating Human Unsettlement," September 20, 1976, presented to the House Committee on Banking, Currency and Housing (*The Rebirth of the American City*, US Government Printing Office, 1976, 1019).

410 **"What would have happened?"**: Guy Hulot, letter in Kahn and Easton, *Shelter II*, 203.

410 **Erhard, Ruth Asawa, and the neuroscientist John Lilly:** Jaime Snyder, author interview, April 23, 2019.

410 **"I don't see this":** John Poppy, "Bringing the Universe Home," *est Graduate Review*, December 1976, 7.

411 **Now House:** RBF, "Accommodating Human Unsettlement," 1026.

411 **North Face donated two tents:** Bruce Hamilton, "The North Face and R. Buckminster Fuller Special Connection—Part 2," April 19, 2018, https://www.outinunder.com/content/north-face-and-r-buckminster-fuller-special-connection-part-2 (accessed May 2021). RBF later discussed consulting for The North Face on a tensegrity tent, which was never produced.

411 **Pushcart Prize:** Ben Reuven, "Leviathans of the Little Magazines," *Los Angeles Times Book Review*, July 25, 1976, 13.

411 **"The young people":** RBF, "Politics & Property," in *SD*, 13688.

411 **Peter Senge:** Poppy, "Bringing the Universe Home," 6.

411 **"In other words":** John Mankiewicz, "Shoot-Out at the I'm OK/You're OK Corral," in *Very Seventies: A Cultural History of the 1970s, from the Pages of* Crawdaddy, ed. Peter Knobler and Greg Mitchell (New York: Fireside, 1995), 299.

411 **"Computers can be run":** Ibid., 300.

412 **suspicious of the relationship:** Tim Wessels, author interview, March 21, 2019.

412 **"Bucky hated Werner Erhard":** Brown, author interview, April 16, 2019.

412 **"a good human being":** RBF, "Fuller, R. B.: On Werner Erhard & est," in *SD*, 6327.

412 **"New York was one of many":** Mankiewicz, "Shoot-Out at the I'm OK/You're OK Corral," 301.

412 **"in a surprising and brilliant way":** Sadao, 213.

412 **Guy Nordenson:** Ibid.

412 **"He did one of his freakouts":** Roger Stoller, author interview, May 13, 2019.

412 **"I would have gone on loving you":** RBF to Noguchi, June 4, 1975, B297-F6.

413 **"Despite our best efforts":** Sadao, 213.

413 **"planetary revolution":** RBF, "And It Came to Pass (Not to Stay)," in *And It Came to Pass—Not to Stay* (New York: Macmillan, 1976), 113.

413 **"edifice complex":** Robert Whymant, "A Conjugal Dictatorship Rules in the Philippines," *Gazette* (Montreal), May 10, 1976, 9. Reprint from the *London Sunday Times*.

413 **"Within one of the great dictatorships":** RBF, *CP*, 138.

413 **"not pursued":** Ibid.

413 **"an only barely audible":** Diplomatic cable from US embassy in Manila, October 28, 1976, archived at https://wikileaks.org/plusd/cables/1976MANILA16743_b.html (accessed January 2021).

413 **"geodesic communities":** Samuel Lanahan, "Dome Communities for Southeast Asia" report, B343-F4.

414 **"definite fear"**: Tom Vinetz, interviewed by author, August 5, 2020.

414 **"This is not to say"**: Ibid.

414 **"memorably delightful"**: RBF to Imelda Marcos, April 3, 1977, B343-F4.

414 **"dome farm"**: Lanahan, "Dome Communities for Southeast Asia" report, B343-F4.

414 **fee of $320,000**: Draft agreement with the government of the Philippines, 1977, B343-F4.

414 **her advisors were concerned**: "Comments on the Proposed Agreement," B343-F4.

414 **"only when asked"**: RBF, "Promote: I Don't Promote," in *SD*, 14329.

414 **"fuselage cartridge"**: RBF, "Curricula and the Design Initiative," in *UOO*, 295–97.

415 **reminded Neva Kaiser of a coffin**: Goodwin (Kaiser), author interview, March 22, 2019.

415 **"a spoke in the wheels"**: Ibid.

415 **precursor organizations**: Ibid. One predecessor was an early incarnation of the Buckminster Fuller Institute, which was run by Don Moore for twelve years (Henry E. Strub to RBF, December 17, 1973, B267-F7).

415 **"Making sense"**: RBF, "Technology: Enchantment vs. Disenchantment," in *SD*, 17852.

415 **"the world's politicos"**: RBF to Neva Kaiser, June 10, 1974, quoted in "Cosmic vs. Terrestrial Accounting," in *SD*, 3143.

415 **"I thought that perhaps"**: Goodwin (Kaiser), author interview.

415 **"One never knows"**: Glenn A. Olds, "R. Buckminster Fuller: Cosmic Surfer," in *Anthology*, 64.

415 **"Everybody sat down"**: Jaime Snyder, email to author, April 27, 2021.

415 **"The necessary restrictions"**: Neva Kaiser to Glenn A. Olds, April 24, 1977, provided to the author by Neva Goodwin (Kaiser).

416 **In the end, Norman Cousins**: Olds, "Cosmic Surfer," 64.

416 **intended to live to be 120**: John Cage and Morton Feldman, *Radio Happenings I–V,* trans. Gisela Gronemeyer (Cologne, Ger.: MusikTexte, 1993), 149.

416 **He talked about dying"**: Alexandra Snyder May, author interview, June 7, 2019.

TWELVE: EQUILIBRIUM

417 **"We can put"**: RBF with Applewhite, *Synergetics*, 441.

417 **a "context" for ending hunger**: Werner Erhard, *The Hunger Project* (San Francisco: Hunger Project, 1977), 1.

417 **"after considerable research"**: Ibid., 39.

417 **"anything dreamable"**: RBF, "The Music of the New Life," in *UOO*, 52.

417 "What can the little individual do?": Erhard, *Hunger Project*, 26.

418 "a morally religious idea": Schlossberg, author interview, March 27, 2019.

418 "If you could see it": Ibid.

418 "You and I want to contribute": Erhard, *Hunger Project*, 3.

418 it raised more than $800,000: Suzanne Gordon, "Let Them Eat est," *Mother Jones*, December 1978, 42.

418 "a thinly veiled recruitment arm": Ibid.

418 Douglas Engelbart: Markoff, *What the Dormouse Said*, 210–11.

418 Carl Sagan and . . . Arnold Schwarzenegger: Classified ad in a 1977 issue of *Saturday Review*, https://books.google.com/books?id=SPseAQAAMAAJ (accessed December 2020). RBF, Schwarzenegger, and Werner Erhard were listed as speakers at an event organized by Jerry Rubin (John Leonard, "Private Lives," *New York Times*, November 8, 1978, sec. C, 12).

418 lectures on campuses: Susan Heller Anderson, "Stars of Today's Lecture Circuit," *New York Times*, April 8, 1982, sec. C, 1. The firm was Brian Winthrop International.

419 North Perry, Ohio: RBF, *Artifacts of R. Buckminster Fuller*, vol. 4, 326–35.

419 "revolutionary courage": RBF to Jane Fonda, draft letter (unsent), 1978, B360-F9.

419 Prince Shahram Pahlavi-Nia: James Parks Morton, "Bucky Fuller—Nine Epiphanies," in *Anthology*, 77–78, and correspondence in B360-F3.

419 Arthur C. Clarke: RBF also corresponded and made appearances with the science-fiction author Isaac Asimov (for example, "Five Noted Thinkers Explore the Future," *National Geographic*, July 1976, 68–74). For Robert A. Heinlein's interest in the first version of the Dymaxion House, which predated his career as a writer, see Heinlein to RBF, December 8, 1934, B33-V52. In the unpublished 1963 essay "If You Don't See It, Just Ask," Heinlein described his attempt to order a Wichita House (Heinlein Archives, UC Santa Cruz, Box 72, Opus 148). In July 1979 both men also testified separately before the House of Representatives on applications of space technology for the elderly and handicapped. See Itinerary and William H. Patterson, *Robert A. Heinlein: In Dialogue with His Century*, vol. 2, *1948–1988: The Man Who Learned Better* (New York: Tor Books, 2016), 412–14.

419 "one of the important exceptions": RBF to A. C. Spectorsky, October 11, 1968, B175-F8.

419 "one of the world's": RBF, "Fuller, R. B.: Described as Engineer-Saint," in *SD*, 6326.

419 "I think he's the world's first engineer-saint": Ibid.

419 "traffic vertically ascending": Arthur C. Clarke, "The Fountains of Paradise," in *Anthology*, 134.

419 "many alternate plantings": Morton, "Fuller—Nine Epiphanies," 79.

419 "wildly impractical": Ibid., 81.

420 **"I'm glad you liked":** Cam Smith to RBF, June 20, 1978, B370-F1.

420 **"people in my environment":** Mary to RBF, May 22, 1978, B370-F7. At Mary's request, her last name has been omitted from the text.

420 **accused of stealing ideas from Hubbard:** Since 1976 Hubbard had been convinced that est had copied his theories and techniques, which led to an ongoing campaign by Scientology to discredit Erhard. See Robert W. Welkos, "Founder of est Targeted in Campaign by Scientologists," *Los Angeles Times,* December 29, 1991, 1, and Peter Haldeman, "The Return of Werner Erhard, Father of Self-Help," *New York Times,* November 28, 2015, ST1.

420 **"I have learned, however":** Mary to RBF, June 26, 1978, B370-F7.

420 **"intellectual curiosity":** Mary, email to author, April 19, 2021.

420 **swallowing difficulties:** Allegra Fuller Snyder, "Notes on Anne and Bucky Fuller's Deaths," Buckminster Fuller Institute, https://www.bfi.org/about-fuller/biography/notes-anne-and-bucky-fullers-deaths (accessed January 2021).

421 **"various seaweed things":** Roger Stoller, author interview, May 13, 2019.

421 **diagnosed with pericarditis:** Shirley Sharkey to Jeremy Bradford, August 28, 1978, B371-F5.

421 **inscribed a copy of** *Synergetics*: Mary, email to author, April 19, 2021.

421 **"an engineer and architect":** L. Ron Hubbard, "Positioning, Philosophic Theory," January 30/1979, reprinted at http://suppressiveperson.org/1979/01/30/hcopl-positioning-philosophic-theory (accessed January 2021).

421 **"Its handling feels better":** 1978 Honda Civic advertisement, The Henry Ford Museum, https://www.thehenryford.org/collections-and-research/digital-collections/artifact/367351 (accessed December 2020). RBF also appeared in commercials for *Funk & Wagnalls New Encyclopedia* (Philip H. Dougherty, "B. & W. Offering a Rolls-Royce," *New York Times,* October 7, 1975, 55).

421 **Fuller looked "adorable":** Valerie Harper, "Bucky: Citizen of the Universe," in *Anthology,* 151.

421 **dinner at Trader Vic's:** Jody Jacobs, "Mae Enters—Slowly, Deliberately," *Los Angeles Times,* March 30, 1977.

421 **kissing him on the cheek:** "Rocky Mountain 'Hi,'" *Times-News* (Twin Falls, ID), September 24, 1978, 2, photo caption.

421 **visited the office:** Ann Mintz, interviewed by author, March 8, 2019.

421 **skepticism toward solar power:** Associated Press, "Sun Day Yields Vows of New Era," *Minneapolis Star,* May 4, 1978, 10.

421 **"You brought me through":** RBF to John Denver, October 5, 1978, B374-F11.

422 **Control of his office:** Edward Schumacher, "Area Colleges End Funds for Buckminster Fuller," *Philadelphia Inquirer,* November 16, 1978, 1.

422 **"a whole new life program"**: RBF to Martin Meyerson, October 5, 1978, B374-F9.

422 **"I feel a deep responsibility"**: Ibid.

422 **"of marginal benefit"**: Edward Schumacher, "The Likely Exit of Fuller: Do These Things Just Happen?," *Philadelphia Inquirer*, November 19, 1978, 54-M.

422 **"not the blue star"**: Ibid.

422 **"two very attractive offers"**: Schumacher, "Area Colleges End Funds for Buckminster Fuller," 1.

422 **financed with a large loan**: "Buckminster Fuller Gets Loan to Continue Work Here," *Philadelphia Inquirer*, February 1, 1979, 2-B. RBF eventually moved his office into the Institute for Scientific Information in the University City Science Center (Eugene Garfield, "A Tribute to R. Buckminster Fuller—Inventor, Philosopher," *Current Contents*, November 2, 1981, 5).

422 **"He has not made a penny"**: Thomas Hine, "Fuller Keeps, for Now, Science Center Office," *Philadelphia Inquirer*, December 7, 1978, 5D.

422 **"He expected a house"**: Joe Fowler, quoted in Neal Gabler, *Walt Disney: The Triumph of the American Imagination* (New York: Vintage Books, 2007), 609.

423 **the Houston Astrodome**: Steve Mannheim, *Walt Disney and the Quest for Community* (Aldershot, UK: Ashgate, 2002), 45.

423 **"giant golden geodesic dome"**: Epcot press kit, c. 1978, B380-F9.

423 **Disney park**: Legault, "Death and Life of Buckminster Fuller's US Pavilion at Expo 67," in *Montreal's Geodesic Dreams*, ed. McAtee, 104.

423 **"familiar" with Fuller**: Interview with John Hench, *Disney Parks Blog*, https://disneyparks.disney.go.com/blog/2012/09/the-scientist-who-inspired-the-name-of-epcots-spaceship-earth (accessed January 2021).

423 **a belated invitation**: Martin A. Sklar to RBF, November 8, 1978, B380-F8.

424 **Margaret Mead**: In an undated letter from 1971, Mead questioned one of RBF's key assumptions: "Do you think it would be possible to say that with the introduction of the square, man became conscious of creating an artificial world, and that this was necessary if we were to have the development of technology?" (B226-F6).

424 **"woman anthropologist"**: RBF, office diary entry, June 9, 1944, B59-F4.

424 **event in her memory**: Panayis Psomopoulos to RBF, November 18, 1978, B377-F4.

424 **"I know she can hear me"**: Jane Howard, *Margaret Mead: A Life* (New York: Fawcett Columbine, 1984), 428.

424 **1982 World's Fair in Knoxville, Tennessee**: Norman Foster, "Buckminster Fuller—File for Future." Manuscript provided to the author by Norman Foster. A rendering is reproduced at "International Energy Expo '82, Knoxville, USA," Norman Foster Foundation Archive, https://archive

.normanfosterfoundation.org/en/consulta/registro.do?id=3463 (accessed May 2021).

424 "obviously too radical": Foster, author interview, October 30, 2020.

424 "hip pocket": "Take 5 with Jeff Jarvis," *San Francisco Examiner,* February 23, 1979.

424 "some kind of charity": Clyde Haberman and Albin Krebs, "No Alms, Please, for R. Buckminster Fuller," Notes on People, *New York Times,* February 21, 1979, sec. C, 12.

424 "I have quite a few million": Ibid.

424 "Great Grandfather": Sue Fernicola, "5,000 Take Responsibility for Making the World Work," *News* (Paterson, NJ), February 28, 1979, 12.

424 "a huge influx of support": Jaime Snyder, author interview, April 23, 2019.

424 earned him $63,000: RBF to David Travers, February 11, 1980, B411-F6.

425 "He's eighty-three": Andy Warhol, *The Andy Warhol Diaries,* ed. Pat Hackett (New York: Warner Books, 1991), 207.

425 "Would Mr. Fuller": Alexandra Snyder May, author interview, June 7, 2019.

425 "like little mirror images": Ibid.

425 "Who was that little man, darling?": Ibid.

425 "All of humanity": RBF, "Learning Tomorrows," in *R. Buckminster Fuller on Education,* 187–88.

425 Fly's Eye Dome: RBF, *CP,* 310–15.

426 "I've been thinking about you": *AMS,* 177.

426 buy his own apartment: Jaime Snyder, author interview, February 12, 2019.

426 "Reorganizing all its strategies": RBF, *Grunch of Giants,* xvii.

426 In May Fuller toured China: RBF corresponded with the novelist Pearl S. Buck about their shared interest in China (Thomas Zung in *Anthology,* 149 and 376–78). Anwar Dil stated incorrectly that *Operating Manual for Spaceship Earth* was "number one on the list of the ten most widely read American books in the People's Republic of China today" (RBF and Dil, *Humans in Universe,* 15). In fact, it was mentioned in *Publishers Weekly* (July 31, 1981) as one "of the ten books the Chinese showed the most interest in" at a publishing trade fair (RBF, *Grunch of Giants,* 77).

426 China had extended: RBF and Dil, *Humans in Universe,* 14–15, and RBF, *EIK,* session 12.

426 "Chinese psycho-guerrilla warfare": RBF, "*Playboy* Interview," *Playboy,* February 1972, 200.

426 "anxious" to use: RBF, *CP,* 408

426 "grand strategy": RBF and Dil, *Humans in Universe,* 15. Gao Yuan is identified in a list of Chinese officials in RBF's itinerary (page following entry for May 10, 1979).

426 "You are China": Ibid.

426 "I get so preoccupied": Applewhite, *Cosmic Fishing*, 90.

426 astronaut diaper system: Wigley, *Buckminster Fuller Inc.*, 123.

426 discreetly urinate: Ibid.

427 "busman's leg": Casey Mack, email to author, January 15, 2020. One student at SIU was told that RBF used "a bladder bag" (George Boesel, interviewed in Gowan, *Shared Vision*, 167).

427 "standing on head after meals": Kenner to Guy Davenport, September 6, 1979, reprinted in Davenport and Kenner, *Questioning Minds*, vol. 2, 1701.

427 "would be willing to put a price": RBF to Martin Meyerson, October 5, 1978, B374-F9.

427 "I do have the know-how": RBF to Charles Sachs, January 14, 1980, quoted in Chu, "Archive of R. Buckminster Fuller," in *New Views on R. Buckminster Fuller*, ed. Chu and Trujillo, 21.

427 owed to Werner Erhard: Jonathan Stoller, interviewed by author, May 14, 2019.

427 "That's fine": Ibid.

427 bank sculpture in Dayton, Ohio: Wessels, author interview, March 21, 2019, and Rob Grip, interviewed by author, December 26, 2019.

428 "a large manned stratospheric station": E. C. Okress and R. K. Soberman, "Spherical Tensegrity Atmospheric Research Station (STARS)," June 6, 1978, B370-F3.

428 unforgivable sin of "promoting" him: Wessels, author interview.

428 plotting features by hand: Christopher Kitrick, interviewed by author, March 10, 2019.

428 a tensegrity truss: RBF, *Inventions*, 286–93.

428 miscalculated the relative volumes: RBF with Applewhite, *Synergetics 2*, 245.

428 "The music of John Cage": Ibid., 434. The book also featured one of RBF's more memorable poems: "Universe to each must be / All that is, including me. / Environment in turn must be / All that is, excepting me" (ibid., 3).

428 Thomas Zung: Biographical details from Thomas T. K. Zung, "Fuller Today: The Legacy of Buckminster Fuller," in *Anthology*, 361–62.

428 space frame vault: According to Zung, RBF called it "the caterpillar dome" (*Anthology*, 384). Synergetics did the calculations for the vault, but after a windstorm, "structural inadequacies" caused it to sag ("People," *Daily Reporter* [Dover, OH], May 1, 1971).

429 "We've been doing some things": Associated Press, "R. Buckminster Fuller Heads New Corporation," *Newark (OH) Advocate*, February 28, 1980, 24.

429 "I shall never let you down": Thomas Zung to RBF, February 26, 1980, B411-F3.

429 **"Even though Thomas"**: Shoji Sadao, interview with Thomas Devaney, Amy Sadao, Elizabeth Asawa, Marty Griggs, and Tsuneko Sadao, November 22, 2015. Provided to the author by Devaney.

429 **"original demonstration"**: RBF to Snelson, December 22, 1949, Series 8, B184-F4.

429 **"You even resorted"**: Snelson to RBF, December 31, 1979, Series 8, B184-F4.

429 **"I would ask you please"**: Ibid.

429 **"hundreds of deceitfully"**: RBF to Snelson, March 1, 1980, Series 8, B184-F4.

429 **"He has concluded that I am a charlatan"**: Ibid.

430 **"You used to tell me"**: Ibid.

430 **"Until then"**: RBF to Snelson, draft letter (unsent), February 11, 1980, B411-F9.

430 **"all you will have to do"**: Ibid., March 1, 1980, Series 8, B184-F4.

430 Jaime proposed: Jaime Snyder, author interview, April 23, 2019.

431 to present the report: RBF, *CP*, 292.

431 Winooski, Vermont: AP, "A Town in Vermont Is Bursting at a Bubble Idea," *New York Times*, March 30, 1980, 26.

431 packaging toilet: RBF had looked into a packaging toilet for the Wichita House, which William Wasserman remembered as his "self-contained, shit-packaging john" (Hatch, 179), and worked on a design for a "waterless" toilet in George Nelson's office in the 1950s (Abercrombie, *George Nelson*, 116). He later collaborated on a more finished version with Don Moore (Baldwin, *BuckyWorks*, 30). RBF described its design in RBF, *CP*, 314–15. A prototype for a "dry toilet" is listed in his office inventory (Dymaxion Index, "Section 20. Non-Archive Materials," Series 4, B8-F9).

431 **"completely irrational"**: Roger Stoller, author interview, May 13, 2019.

431 **"discourage the volume of requests"**: Bob Baker, "'Planetary Planner' in West Hollywood Stays in Orbit with Buckminster Fuller," *Los Angeles Times*, February 26, 1981.

431 force the ArtCenter: Roger Stoller, author interview.

431 **"I love you, Bucky"**: "Well, Kiss My Dome—It's a Heady 85th Birthday Bash for Plucky Bucky Fuller," *People*, July 21, 1980.

432 mesostic poem: Reprinted in *Your Private Sky*, ed. Krausse and Lichtenstein, 324.

432 *Themes and Variations*: John Cage, *Composition in Retrospect* (Cambridge, MA: Exact Change, 1993), 85–86, 154.

432 randomness in art: John Cage to Jerry Novesky, February 9, 1972, in *The Selected Letters of John Cage*, ed. Laura Kuhn (Middletown, CT: Wesleyan University Press, 2016), 417.

432 **"We'll be remembered":** Peder Anker, *From Bauhaus to Ecohouse: A History of Ecological Design* (Baton Rouge: Louisiana State University Press, 2010), 68.

432 **more recent acquaintances:** The journalist Richard J. Brenneman arranged for Fuller to explain the principles of synergetics in conversations with three schoolchildren, which were published as *Fuller's Earth*. The children were Rachel Myrow, Benjamin Mack, and Jonathan Nesmith, the son of musician-songwriter Michael Nesmith of the Monkees (Brenneman, *Fuller's Earth*, xx).

432 **talks "on tensegrity and cosmic energy":** Thomas T. K. Zung, "Fuller Today: The Legacy of Buckminster Fuller," in *Anthology*, 370.

432 **Los Angeles Bicentennial:** Alice Rawsthorn, "The Prescient Vision of a Gentle Revolutionist," *New York Times*, May 12, 2013.

433 **Kiyoshi Kuromiya:** Biographical details drawn from "Life of Kiyoshi Kuromiya," *NBC News*, March 7, 2015, https://www.nbcnews.com /storyline/527elma-50th-anniversary/527elma-marcher-aids-activist-life -steven-kuromiya-n318876 (accessed May 2021).

433 **"positive personality":** "Positively Good Guys Honored," *Miami Herald*, April 4, 1973, 18-A.

433 **typist at the office:** Sharkey, author interview, March 18, 2019.

433 **"adjuvant," or writing assistant:** "I have served as adjuvant, a term Fuller borrowed from medicine (specifically immunology) in 1980 to designate my role in the writing of *Critical Path*" (Kiyoshi Kuromiya, in RBF, *Cosmography*, vii). *Merriam-Webster* defines *adjuvant* as "one that helps or facilitates"—for instance, by "enhancing the immune response to an antigen."

433 **"fucking impossible":** Denneny, author interview, March 22, 2019.

433 **Fuller was "totally" involved:** Ibid.

433 **"best and most important book":** RBF, *SD*, xxviii.

433 **"would feel happy":** RBF, *CP*, xiii.

433 **critical path method:** The term was used in RBF's circle by 1974, when a World Game workshop "surveyed all known sources of energy and devised ten-year 'Critical Path' strategies for getting there from here" (Kenner, "Bucky Fuller and the Final Exam," *New York Times Magazine*, July 6, 1975, 10). The results were published by Medard Gabel in *Energy, Earth, and Everyone*.

433 **"Two million things":** RBF, *CP*, xviii.

434 **"a very reluctant player":** Ibid., 116.

434 **"get out their guns":** Ibid., 217.

434 **"The 'greenhouse' effect":** Ibid., 112.

434 **Fuller framed his proposals as an "icebreaker":** Ibid., 249.

434 **rendering existing systems obsolete:** "World Game is Anti-Obnoxico and commits itself to making Obnoxico and allied activities obsolete rather

than attacking it directly" (ibid., 226). RBF claimed to have made a similar statement at his first meeting with Nehru: "I must actually find a tool that solves the problem—makes what is going on obsolete" (RBF, *EIK*, session 7). Over time, this would be refined into a more elegant but apocryphal quote: "You never change things by fighting the existing reality. To change something, build a new model that makes the existing model obsolete" (for example, Rick Ingrasci in Sieden, *Fuller View*, 236).

434 **"friendliness" to "moon rocks":** RBF, *CP*, 221.

434 **"that the greatest luxury":** Ibid., 263.

434 **"The larger the number":** Ibid., 125. A nearly identical passage appeared in 1967 in the essay collected as "Buckminster Fuller Chronofile" (*BFR*, 17).

434 **"What you want for yourself":** RBF, *CP*, 269.

435 **"no change":** RBF, *Nine Chains*, xvii.

435 **"That's not the case":** Denneny, author interview.

435 **"Oh, I don't mind":** Emily Gaudette and Zach Schonfeld, "Ellen Burstyn on 'House of Tomorrow,' Bad Horror Movies, and Why #MeToo Is 'the Biggest Thing That's Happened in My Life,'" *Newsweek* online, last modified April 27, 2018, https://www.newsweek.com/ellen-burstyn-house-tomorrow-patriarchy-buckminster-fuller-903463. A similar version appears in Burstyn, *Lessons in Becoming Myself* (New York: Riverside Books, 2006), 281–82.

435 **"mystically inspired":** RBF to Elizabeth Channing Fuller, December 10, 1976, quoted in Hsiao-Yun Chu, "R. Buckminster Fuller's Model of Nature" (doctoral thesis, University of Brighton, 2014), 100.

435 **Jerry Brown, Jonas Salk, and Woody Allen:** RBF had contributed an illustrated introduction to Woody Allen and Stuart Hample, *Non-Being and Somethingness* (New York: Random House, 1978), a collection of the comic strip *Inside Woody Allen*. "I thought it was wonderful, fascinating, imaginative, and presented me as a far deeper and more interesting character than I really am" (Woody Allen to RBF, April 11, 1979, B389-F8).

435 **"It's criminal that he doesn't":** John Duka, "Buckminster Fuller: Celebrating the 86th," *New York Times*, July 9, 1981, sec. 3, 1.

436 **"a famous surfer":** Roger Stoller, author interview, May 13, 2019.

436 **"I did not like the way":** Alexandra Snyder May, author interview, May 17, 2019.

436 **"She could really teach you":** Ibid.

436 **"sweet, supportive granddaughter":** Ibid.

436 **"I would like to take advantage":** Amy Edmondson, *A Fuller Explanation: The Synergetic Geometry of R. Buckminster Fuller* (Boston: Birkhäuser, 1987), 5.

437 **"I hadn't yet seen Bucky Fuller":** Amy Edmondson, quoted in Baldwin, *BuckyWorks*, 141.

437 **had met with Fuller:** Itinerary, September 23, 1980.

437 **"Money and You":** Margery Eagan, "On the Righteous Road to Riches," *Burlington (VT) Free Press*, August 13, 1978, sec. C, 1.

437 **Robert Kiyosaki:** Robert Kiyosaki, *Second Chance: For Your Money, Your Life and Our World* (Plata Publishing, 2015, EPUB ed.), 78.

437 **Mark Victor Hansen:** Brochure for Dome East Corporation, 1973, B268-F4.

438 **"You invited me to talk":** Morton, "Fuller—Nine Epiphanies," in *Anthology*, 83.

438 **hexagonal hanging bookshelf:** RBF, *Inventions*, 294–98.

438 **Gigundo Dome:** Photo caption, *Anthology*, 391.

438 **"double deresonated dome":** Foster, "Richard Buckminster Fuller," in *Anthology*, 1.

438 **"self-isolation":** Applewhite, *Cosmic Fishing*, 153.

438 **"You never know":** *The Buckminster Fuller Institute Newsletter*, October 1983, quoted in Pawley, *Buckminster Fuller*, 171.

438 **"I think I'd like":** Allegra Fuller Snyder, author interview, March 26, 2019.

439 **"a mild but definite encephalopathy":** Dr. John C. Wilson, "To Whom This May Concern," August 2, 1982, B489-F9.

439 **dreamed about an award:** Jaime Snyder, author interview, April 23, 2019.

439 **songs for a television special:** Mark Stouffer, teleplay for *What One Man Can Do*, B496-F7.

439 **"It does not seem to me":** RBF to John Denver, January 3, 1982, B450-F8.

439 **"operating only as a beneficiary":** Ibid.

439 **"longest time in one place":** Jaime Snyder in *Synergetic Stew*, 111.

439 **DeLand, Florida:** Al Truesdell, "Fuller Made Volusia a Part of His Dream," *Orlando (FL) Sentinel*, July 10, 1983.

439 **Friends of Buckminster Fuller:** Jaime Snyder, author interview, April 23, 2019.

440 **"There's one thing":** Brown, author interview, April 16, 2019.

440 **"When we went back":** Ibid.

440 **"in an intelligent way":** Jaime Snyder, email to author, April 27, 2021.

440 **United States House of Representatives:** Itinerary, December 16, 1982. At the talk, RBF's staff members dropped fifty thousand bingo markers on the map to illustrate the devastation of a nuclear war ("Dymaxion Map Displayed," *Buckminster Fuller Institute Newsletter*, February 1983, Series 7, B10-F13).

440 *Spruce Goose*: John Davies, "Hoopla Marks Start of Dome for Flying Boat," *News-Pilot* (San Pedro, CA), April 30, 1981.

441 **"It's not ours":** Thomas Zung, interviewed by author, March 12, 2019.

441 **Spaceship Earth:** Construction details based on John P. Grossman and Glenn R. Bell, "Spaceship Earth: Epcot Center's Gateway to Tomorrow,"

Modern Steel Construction, 4th Quarter, 1982. A second geodesic dome at Epcot, the Wonders of Life dome, was engineered by Don Richter and Temcor (RBF, "Epcot," *Buckminster Fuller Institute Newsletter*, February 1983, Series 7, B10-F13). Richter and Thomas Zung also contributed to the tetrahedral design of the external panels for Spaceship Earth (Zung, author interview).

441 **"We regret to inform you":** Disney form letter, c. 1983, Series 7, B10-F13.

441 **"my Cambridge, Massachusetts, engineers":** RBF, "Epcot," *Buckminster Fuller Institute Newsletter*, February 1983, Series 7, B10-F13.

441 **"It must have been their business people":** Ibid.

441 **Home of the Future:** "'Home of the Future' Fair's Ice Cream Soda?," *World's Fair Guide*, in *Kingsport (TN) Times-News*, April 29, 1982, 51.

441 **"Well, old man?":** Baldwin, *BuckyWorks*, 224.

441 **"in case I was expecting":** Goodwin (Kaiser), author interview, March 22, 2019.

442 **"He manipulated the model":** John Litweiler, *Ornette Coleman: A Harmolodic Life* (New York: Da Capo, 1994), 186.

442 **"number one hero":** Syd Fabio, "Ornette Coleman—*Prime Design/Time Design*," *Rock Salted*, December 19, 2017, http://rocksalted.com/2017/12 /ornette-coleman-prime-design-time-design (accessed January 2021). Coleman made the remark in the documentary *Ornette: Made in America*.

442 **"I have always had the experience":** RBF and Dil, *Humans in Universe*, 135.

442 **her husband's interpreter:** Kenner to Guy Davenport, December 27, 1972, reprinted in Davenport and Kenner, *Questioning Minds*, vol. 2, 1433.

443 **"You know, that Bucky":** Jaime Snyder, author interview, April 23, 2019.

443 **"the very few":** Archibald MacLeish to President Jimmy Carter, February 24, 1978, Series 7, B10-F10.

443 **"If there is a living contemporary":** Ibid.

443 **"a true Renaissance man":** "Remarks at the Presentation Ceremony for the Presidential Medal of Freedom," *The American Presidency Project*, https://www.presidency.ucsb.edu/documents/remarks-the-presentation -ceremony-for-the-presidential-medal-freedom-0 (accessed January 2021).

443 **"A farce":** Bob Baker, "'Planetary Planner' in West Hollywood Stays in Orbit with Buckminster Fuller," *Los Angeles Times*, February 26, 1981, sec. 9, 1.

443 **"a dumb actor":** Robert Anton Wilson, "Interview: Buckminster Fuller," *High Times*, May 1981, 37.

443 ***Grunch of Giants:*** RBF dedicated the book to three people: Margaret Fuller; Marilyn Ferguson, who named RBF as one of the thinkers frequently mentioned as an influence on the New Age in *The Aquarian Conspiracy: Personal and Social Transformation in the 1980s* (Los Angeles: J. P. Tarcher, 1980), 420; and the futurist Barbara Marx Hubbard.

443 "Gross Universe Cash Heist": RBF, *Grunch of Giants*, 1.

443 "the most negatively momentous": Ibid., 50.

443 "the Reagan government's reduction of taxes": Ibid., 72.

444 "the Republican Party of Lincoln": Ibid., 73.

444 "It would take another fifty years": Ibid., 86.

444 known as a realist: *AMS*, 197.

444 he would resign: RBF, "Epilogue," in *UOO*, 343.

444 more than a thousand attendees: Jaime Snyder, author interview, April 23, 2019.

444 "reduces the credibility": RBF, "Guinea Pig B," in *Inventions*, xii.

444 "An individual": Jaime Snyder, email to author, February 16, 2021.

444 "a curriculum vitae": Michael Denneny, "Bucky's Apologia," in *Anthology*, 280.

445 "experimental data": Ibid., 282.

445 *Inventions: The Patented Works of R. Buckminster Fuller*: RBF had used a similar title for a boxed portfolio, *Inventions: Twelve Around One*, that he produced with Carl Solway in an edition of sixty in 1981 (Buckminster Fuller Institute catalog, undated, provided to the author by Solway).

445 "publicly declaring yourself": RBF, "Guinea Pig B," vii.

445 "miracles that occur": Ibid., xxviii.

445 "Fortunately, this gambling": Ibid., xxx.

445 line of praise "to God": RBF, *Cosmography*, 169.

445 "Grandma, you have to spit that out": Jaime Snyder, author interview, April 23, 2019.

445 "was basically starving to death": Allegra Fuller Snyder, "Notes on Anne and Bucky Fuller's Deaths," https://www.bfi.org/about-fuller/biography/notes-anne-and-bucky-fullers-deaths (accessed January 2021).

445 stalled escalator: Sieden, *Buckminster Fuller's Universe*, 415.

445 "ran out of wind": RBF to Henry Gomperts, June 18, 1983, B488-F5.

446 "Bucky did *not* ask me": Shirley Sharkey to Calvin Tomkins, June 13, 1983, B487-F7.

446 "instructions from the master": Foster, author interview, October 30, 2020.

446 "very much bigger": "Transcript of Meeting/Conversation Between Norman Foster, Buckminster Fuller, and James Meller," June 1983, 1. Provided to the author by Norman Foster.

446 "I think it is your world": Ibid., 2.

446 "pull the plug": Norman Foster, "Richard Buckminster Fuller," https://www.fosterandpartners.com/media/546507/essay12.pdf (accessed January 2021).

446 "the longest, most exhausting day": Sharkey, author interview, March 18, 2019.

446 "the sweetest": Ibid.

446 **punctured her esophagus:** Allegra Fuller Snyder, "Notes on Anne and Bucky Fuller's Deaths."

446 **Fuller wondered aloud:** Jaime Snyder, author interview, April 23, 2019.

447 **"Do you see that?":** Allegra Fuller Snyder, author interview, March 26, 2019.

447 *Mister God, This Is Anna:* Jaime Snyder, author interview, April 23, 2019.

447 **"I think she's worried":** Allegra Fuller Snyder, author interview.

447 **"She's squeezing my hand":** Ibid.

447 **"After he had died":** Allegra Fuller Snyder, "Notes on Anne and Bucky Fuller's Deaths."

448 **"so beautifully":** John Cage to Norman O. and Beth Brown, July 14, 1983, in *Selected Letters of John Cage*, 533.

448 **"It said to me things":** Alexandra Snyder May, author interview, May 17, 2019.

448 **letter to Annie Dillard:** Edgar J. Applewhite to Gerald Dickler, June 22, 1983, B488-F10.

448 **"Dear reader":** RBF, *Cosmography*, 264.

448 **"a thoroughgoing original":** Albin Krebs, "R. Buckminster Fuller Dead: Futurist Built Geodesic Dome," *New York Times*, July 2, 1983, 1.

448 **"a somewhat curious cult figure":** Paul Goldberger, "An Apostle of Technology: An Appreciation," *New York Times*, July 3, 1983, 17.

448 **Mount Auburn Cemetery:** "Fullers Buried in Cambridge," *Boston Globe*, July 9, 1983, 27.

448 **"sentimental and small":** Silverman, *Begin Again*, 345.

449 **Fuller's most spectacular memorial:** A separate memorial service was hosted at the American Society for Metals dome in Ohio by Governor Richard F. Celeste (Thomas Zung to Allegra Fuller Snyder, October 13, 1983, B496-F4, and Thomas T. K. Zung, "A Fuller Family," in *Anthology*, 413–14).

449 **"I go right on seeing them":** Morton, "Fuller—Nine Epiphanies," in *Anthology*, 82.

449 **"Almighty and everlasting God":** Ibid., 83.

449 **"Eternal Father, strong to save":** Ibid.

449 *CALL ME TRIMTAB:* "R. Buckminster Fuller," Find a Grave, https://www.findagrave.com/memorial/1403/r.-buckminster-fuller (accessed January 2021).

449 **"The behavior of 'Universe'":** RBF with Applewhite, *Synergetics*, 86.

EPILOGUE: TETRAHEDRON DISCOVERS ITSELF AND UNIVERSE

451 **"It could be that by traveling":** Ibid., 288.

451 **"Because life *is* three-dimensional design":** Adapted from Donald E. Ingber, "Spanning the Divide," SciArt, last modified October 2018, https://www.sciartmagazine.com/spanning-the-divide.html.

452 who had been "wowed": Donald E. Ingber, interviewed by author, March 12, 2019.

452 "bounced up off the dish": Ingber, "Spanning the Divide."

452 "The tensegrity must have changed!": Donald E. Ingber, "Tales of Discovery from a Life Without Borders" (lecture, Wyss Institute for Biologically Inspired Engineering, Harvard University, June 13, 2009), available at https://wyss.harvard.edu/news/ingber-dartmouth-medical -school-class-day-lecture.

452 "Those who are versed": RBF, "Conceptuality of Fundamental Structures," in *Structure in Art and in Science*, ed. Kepes, 88. The images of geodesic structures in tissue were by Dr. Arthur von Hochstetter (ibid., 76–77).

453 article about his work: D. E. Ingber and J. D. Jamieson, "Tumor Formation and Malignant Invasion: Role of Basal Lamina," in *Tumor Invasion and Metastasis*, ed. L. A. Liotta and I. R. Hart (The Hague: Martinus Nijhoff, 1982), 335–57.

453 "Cancer may then be viewed": Donald E. Ingber to RBF, April 5, 1983. Provided to the author by Ingber.

453 "I would just like": Ibid.

453 "The whole encounter": Ingber, "Spanning the Divide."

454 "elated" by his ideas: RBF to Donald E. Ingber, April 11, 1983. Provided to the author by Ingber.

454 "setting up a network": Undated staff member note to RBF (1983), B489-F4.

454 "You are the first one": RBF to Donald E. Ingber, May 18, 1983. Provided to the author by Ingber.

454 "I believe he was telling me": Ingber, "Spanning the Divide."

454 explored these constructions: Donald E. Ingber and James D. Jamieson, "Cells as Tensegrity Structures: Architectural Regulation of Histodifferentiation by Physical Forces Transduced over Basement Membrane," in *Gene Expression During Normal and Malignant Differentiation*, ed. Leif C. Anderson, Carl G. Gahmberg, and P. Ekblom (Orlando, FL: Academic Press, 1985), 13–32.

455 "My presentation": Ingber, "Spanning the Divide."

456 "If you stop using the word": Ingber, author interview.

456 scientific claims of tensegrity: A related but distinct line of argument conceives of anatomical features on a larger scale as tensegrity structures. See Randel L. Swanson II, "Biotensegrity: A Unifying Theory of Biological Architecture with Applications to Osteopathic Practice, Education, and Research—A Review and Analysis," *Journal of the American Osteopathic Association* 113, no. 1 (January 2013): 34–52.

456 paper from the seventies: P. C. Rathke, M. Osborn, and K. Weber, "Immunological and Ultrastructural Characterization of Microfilament

Bundles: Polygonal Nets and Stress Fibers in an Established Cell Line," *European Journal of Cell Biology* 19 (January 1979): 40–48. Ingber also drew on E. Lazarides, "Actin, alpha-Actinin, and Tropomyosin Interactions in the Structural Organization of Actin Filaments in Nonmuscle Cells," *Journal of Cell Biology* 68, no. 2 (February 1976): 202–19.

457 "It was like touching the heartstrings": "Sculpting Life," *Boston Globe*, July 26, 2004.

457 R. Buckminster Fuller Professorship of Design Science: Harvard Graduate School of Design announcement, https://www.gsd.harvard.edu/2021/04 /harvard-gsd-announces-establishment-of-the-r-buckminster-fuller -professorship-of-design-science (accessed May 2021).

458 "middle zone": David Edwards, *ArtScience: Creativity in the Post-Google Generation* (Cambridge, MA: Harvard University Press, 2008), 132.

458 "In the end": Donald E. Ingber, email to author, December 26, 2020.

458 "What the hell am I doing?": Ingber, author interview.

458 "Imagine human beings": RBF, *EIK*, session 3.

458 "a kind of pilgrimage": Margaret Kroto, email to author, September 6, 2020.

459 "on the organized growth": Harold W. Kroto, "C_{60}: Buckminsterfullerene, the Celestial Sphere That Fell to Earth," *Angewandte Chemie* 31, no. 2 (February 1992): 117.

459 Rice University: Details of the team's work at Rice are based on Hugh Aldersey-Williams, *The Most Beautiful Molecule: The Discovery of the Buckyball* (New York: John Wiley & Sons, 1995), 52–90, and Jim Baggott, *Perfect Symmetry: The Accidental Discovery of Buckminsterfullerene* (New York: Oxford University Press, 1994), 34–76.

459 "cockamamie theory": Richard E. Smalley, oral history, interviewed by Robbie Davis Floyd, January 22, 2000, 17, http://www.davis-floyd.com /smalley6.

459 "I thought Harry": Ibid.

459 "What's the worst?": Richard E. Smalley interview in "Race to Catch a Buckyball," *Nova* television episode, December 19, 1995.

460 "the mother wodge": Aldersey-Williams, *Most Beautiful Molecule*, 67.

460 "Who was that guy?": Smalley, interviewed by Robbie Davis Floyd, 9.

461 "Here, after all, we had a hexagonal sheet": Richard E. Smalley interview in "Race to Catch a Buckyball," *Nova* television episode, December 19, 1995.

461 "Juicy Fruit gum balls": Richard E. Smalley, "Great Balls of Carbon," *The Sciences*, March 1991, 24.

462 "ecstatic and overtaken by its beauty": Harold W. Kroto, "C_{60}: Buckminsterfullerene, the Celestial Sphere," 117.

462 "I could explain this to you": Aldersey-Williams, *Most Beautiful Molecule*, 76.

462 **soccer ball's familiar pattern:** In 1978 RBF explored an unproduced design for a "Dymaxion Soccer Ball" (Series 19, Mixed Materials 86), and he would sometimes be credited incorrectly with the standard arrangement of a soccer ball's panels (Lloyd Alter, "Did Bucky Fuller Really Design a Soccer Ball?," October 11, 2018, archived at https://web.archive.org/web/20201112041208/https://www.treehugger.com/did-bucky-fuller-really-design-soccer-ball-4855430).

462 **The proposals included "ballene":** Aldersey-Williams, *Most Beautiful Molecule*, 76.

462 **"Your name sucks":** Margaret Kroto, email to author, January 21, 2021.

463 **"an unforgettable name":** Robert Curl, email to author, September 7, 2020.

463 **"Only a spheroidal structure":** H. W. Kroto et al., "C_{60}: Buckminsterfullerene," *Nature* 318 (November 1985): 162–63, https://doi.org/10.1038/318162a0.

463 **might yield a "super-lubricant":** Ibid.

463 **like the "inadvertencies":** RBF, "Geosocial Revolution," in *UOO*, 166.

464 **references to "buckyballs":** Among the earliest print references was an Associated Press article, "Rice Researchers Find Carbon 60 'Buckyballs,'" *Tyler (TX) Courier-Times*, December 23, 1985, 9.

464 **"We refined":** Smalley, interviewed by Robbie Davis Floyd, 9.

464 **"commit suicide":** Harold W. Kroto, "C_{60}: Buckminsterfullerene, the Celestial Sphere," 119.

464 **"The geodesic ideas":** E. J. Applewhite, "The Naming of Buckminster-fullerene," in *Anthology*, 332.

464 **Fuller knew well:** RBF, "Twelve Pentagons," in *SD*, 19255.

465 **Lone Ranger and Tonto:** Baggott, *Perfect Symmetry*, 56.

465 **Bucky and Mrs. Bucky:** Ibid., 269.

465 **"round" chemistry":** Ibid., 214.

466 **Because a "buckytube":** *Merriam-Webster* defines "buckytube" as "a nanotube composed of pure carbon with a molecular arrangement similar to that of a fullerene."

466 **Bruce Sterling called "buckyjunk":** Bruce Sterling, "Dawn of the Carbon Age," *Wired*, December 2005.

466 **"The name has given it":** Harold W. Kroto, blurb for *Anthology*.

466 **unsuitable as lubricants:** Aldersey-Williams, *Most Beautiful Molecule*, 252.

466 **a spectrum of C_{60} ions:** E. K. Campbell et al., "Laboratory Confirmation of C_{60}^+ as the Carrier of Two Diffuse Interstellar Bands," *Nature* 523 (2015): 322–23, https://doi.org/10.1038/nature14566.

466 **Hubble Space Telescope:** Evan Gough, "Hubble Finds Buckyballs in Space," Universe Today, last modified June 26, 2019, https://www.universetoday.com/142644/hubble-finds-buckyballs-in-space.

467   **"If fullerenes have been present"**: Richard E. Smalley, "Great Balls of Carbon," *The Sciences*, March 1991, 28.

467   **"Buckminster Fuller would have loved it"**: Ibid.

467   **"One guy"**: Sonia Rao, "Tag, You're It!", *Washington Post*, January 6, 2019, https://www.washingtonpost.com/arts-entertainment/2019/01/07/tag-youre-it-heres-full-transcript-jeff-bridgess-wacky-golden-globes-speech (accessed January 2021).

467   **"When he dies"**: Kenner, *Bucky*, 262.

468   **conflict erupted over the World Game**: Among its successors was the Global Energy Network Institute, which was founded in 1986 to promote RBF's idea of a global energy grid ("About Us," http://www.geni.org/globalenergy/about_us/index.shtml [accessed December 2020]).

468   **"warring factions"**: Alexandra Snyder May, author interview, June 7, 2019.

468   **sold Fuller's archives**: Chu, "Archive of R. Buckminster Fuller," in *New Views on R. Buckminster Fuller*, ed. Chu and Trujillo, 22.

468   **tetrahelix memorial**: Sadao, 225. In the course of this project, it was discovered that RBF had described the repeating structure of the tetrahelix incorrectly.

468   **occupied by raccoons**: Baldwin, *BuckyWorks*, 56. Another alleged detail of the Wichita House's afterlife was even more ignominious: "It appears to have become, for a short time, a house of ill-repute, and was even painted pink to entice the airmen as they flew into Wichita" (Gorman, *Designing for Mobility*, 76).

468   **explored onstage and in fiction**: Works inspired by RBF's life included the one-person play *R. Buckminster Fuller: The History (and Mystery) of the Universe* by D. W. Jacobs (2000); *The Love Song of R. Buckminster Fuller*, a multimedia project by the filmmaker Sam Green and the alt-rock band Yo La Tengo (2012); and the novel *The House of Tomorrow* by Peter Bognanni (2010), which was adapted into a feature film in 2017. The movie starred Ellen Burstyn, who also appeared in archive footage with RBF.

468   **unveiled a postage stamp**: Thomas T. K. Zung, "Fuller Today: The Legacy of Buckminster Fuller," in *Anthology*, 370–71.

468   **"Piles of them"**: Donald E. Ingber, "Utopia or Oblivion Revisited." Manuscript provided to the author by Ingber.

468   **Carlos Castaneda**: In the nineties, Castaneda promoted "Tensegrity," which was described as a system of "ancient, energy-gathering movements" (advertisement in "Bulletin Board Annex," *LA Weekly*, March 2, 1995).

468   **Oliver Stone**: In an introduction to L. Fletcher Prouty's *JFK: The CIA, Vietnam, and the Plot to Assassinate John F. Kennedy* (New York: Birch

Lane Press / Carol, 1992), xiv, Stone commended the author for "bringing back the ghost of Buckminster Fuller and his great book, *The Critical Path* [*sic*]."

468 "colossuses in China": Nicolai Ouroussoff, "Fixing Earth One Dome at a Time," *New York Times*, July 4, 2008, E23.

468 "a privilege to spend": Foster, author interview, October 30, 2020.

468 "diagrid" framework: Ian Volner, "Dissecting Diagrid," *Architect*, October 5, 2011, https://www.architectmagazine.com/technology/dissecting-diagrid_o (accessed January 2021).

469 "The idea that somebody": Cynthia Haven, "Stanford Libraries Acquire the Archives of Leading Environmentalist William McDonough," *Stanford News Service*, November 15, 2012, https://news.stanford.edu/pr/2012/pr-libraries-mcdonough-archives-111512.html (accessed January 2021).

469 "in a very free": Renzo Piano, *Logbook* (New York: Monacelli Press, 1997), 22.

469 "Fuller did everything wrong": Thom Mayne, *ArtForum*, November 2008, https://www.artforum.com/print/200809/thom-mayne-41980 (accessed January 2021).

469 "He will put the data": RBF, *Education Automation*, 85.

469 Medusa: Frank Gehry, *Gehry Talks: Architecture + Process*, ed. Mildred Friedman (New York: Rizzoli, 1999), 214–15.

469 James Dyson: "My History Hero: Buckminster Fuller (1895–1983)," July 6, 2012, https://www.historyextra.com/period/20th-century/my-history-hero-buckminster-fuller-1895–1983 (accessed May 2021).

469 Yves Béhar: Yves Béhar interview, "Buckminster Fuller: Utopia Rising," https://www.nowness.com/story/buckminster-fuller-utopia-rising (accessed February 2021).

469 Roman Mars: "Bucky Fuller is also responsible for the title of this radio show. He said, '99% of who you are is invisible and untouchable'" (Roman Mars in "Derelict Dome," *99% Invisible*, October 25, 2012, https://99percentinvisible.org/episode/episode-64-derelict-dome [accessed May 2021]). Various versions of this quotation exist, including "Reality is more than 99 percent invisible" (RBF, *Earth, Inc.*, 62).

469 "He was organizing his present": Olafur Eliasson, interviewed by author, August 29, 2019.

469 Bill McKibben: "Marshall McLuhan . . . coined the term 'global village.' Fuller's 'Spaceship Earth' is the other great description of our planet from the moment of new insight that coincided with the first views back from outer space" (Bill McKibben, ed., *American Earth: Environmental Writing Since Thoreau* [New York: Penguin, 2008], 464).

469 Al Gore: Gore served on the Congressional Clearinghouse on the Future, which invited RBF to speak on April 19, 1978 (Itinerary, and Al Gore, *The*

Future: Six Drivers of Global Change [New York: Random House, 2013],
xvi). He later drew on RBF's metaphors of Earth, Inc. (*The Future*, 4) and
Spaceship Earth (Al Gore, "Al Gore Weighs In on a Long-Delayed Earth
Observatory Launch," *Scientific American* online, last modified February
6, 2015, https://www.scientificamerican.com/article/al-gore-weighs-in-
on-sunday-s-long-delayed-earth-observatory-launch).

469 "A tremendous mind": "Drilling Deeper," *PB Oil & Gas*, October 1, 2013,
https://pboilandgasmagazine.com/drilling-deeper-october-2013 (accessed
January 2021).

470 "What are you going to do about it?": Russell Gold, *The Boom: How
Fracking Ignited the American Energy Revolution and Changed the
World* (New York: Simon & Schuster, 2014), 95.

470 "the father of fracking": Tom Fowler, "'Father of Fracking' Dies at 94,"
Wall Street Journal, July 26, 2013.

470 Richard Dawkins: "I was once privileged to hear [RBF], in his nineties [*sic*],
lecturing for a mesmerizing three hours without respite" (Richard Dawkins,
Climbing Mount Improbable [New York: W. W. Norton, 1996], 234).

470 its basic "epiphany": John P. Allen and Kathelin Gray interview with
Hans Ulrich Obrist, https://www.synergeticpress.com/wp-content/uploads
/2018/08/Mousse2017allenhoffmanulrich.pdf (accessed January 2021).

470 "The idea that Bucky had": Ibid.

470 "They are visionaries": Marc Cooper, "The Profits of Doom: The
Biospherians Lure Scientists to a High-Price Feast Under Glass," *Phoenix
New Times*, June 19, 1991.

470 appeared in NASA proposals: "The US government is specifying in its
bids for space structures the octet truss (my copyrighted trademark name)
for all the main structuring for space stations" (RBF, *Inventions*, 167).
A more accurate characterization would be that tetrahedral trusses were
seen as "an excellent configuration for preliminary studies" (Martin M.
Mikulas Jr., Harold G. Bush, and Michael F. Card, "Structural Stiffness,
Strength and Dynamic Characteristics of Large Tetrahedral Space Truss
Structures," NASA technical memorandum, March 1977, 1), but the octet
truss was rarely mentioned by name. RBF had a more direct influence on
the Super Ball Bot, a NASA prototype based on a tensegrity icosahedron
("Super Ball Bot," September 10, 2014, https://www.nasa.gov/content
/super-ball-bot).

470 Hawaii Space Exploration Analog and Simulation: Marina Koren, "When
a Mars Simulation Goes Wrong," *Atlantic*, June 22, 2018.

470 Mars Science City: Poppy Koronka, "Architects Have Designed a
Martian City for the Desert Outside Dubai," June 10, 2020, https://www
.cnn.com/style/article/mars-science-city-design-spc-scn/index.html (accessed
May 2021).

470 **poorly suited for Mars:** Casey Handmer, "Domes Are Over-Rated," https://caseyhandmer.wordpress.com/2019/11/28/domes-are-very-over -rated (accessed May 2021).

471 **technology company Foxconn:** Josh Dzieza, "The 8th Wonder of the World," The Verge, October 19, 2020, https://www.theverge.com /21507966/foxconn-empty-factories-wisconsin-jobs-loophole-trump (accessed May 2021).

471 **"geodesic domes":** "Welcome to Fyre Festival," April 15, 2017, https:// medium.com/@fyrefestival/welcome-to-fyre-festival-c8a565cd401 (accessed January 2021).

471 **Hurricane Katrina:** Alice Rawsthorn, "Can Fuller Be Rehabilitated as a 21st Century Design Hero?," *New York Times*, June 20, 2008.

471 **overflow morgues:** Photo in Khaleda Rahman, "New York City Builds Makeshift Morgue to Handle Expected Rise in Coronavirus Deaths," *Newsweek*, March 26, 2020, https://www.newsweek.com/nyc -erects-temporary-morgue-handle-rise-coronavirus-deaths-1494423 (accessed May 2021).

471 **"We're talking about domes":** Soleil Ho, "SF Restaurant's $200-Per-Person Dome is America's Problems in a Plastic Nutshell," *San Francisco Chronicle*, August 12, 2020, https://www.sfchronicle.com/restaurants /article/The-200-per-person-fine-dining-dome-is-15475555.php (accessed January 2021).

471 **"We want to be known":** Katrina Brooker, "WeFail," *Fast Company* online, last modified November 15, 2019, https://www.fastcompany .com/90426446/wefail-how-the-doomed-masa-son-adam-neumann -relationship-set-wework-on-the-road-to-disaster.

471 **"future cities initiative":** Niall Patrick Walsh, "The We Company Launches Future Cities Initiative with Studio Dror," *ArchDaily*, March 22, 2019, https://www.archdaily.com/913746/the-we-company-launches-future -cities-initiative-with-studio-dror (accessed January 2021).

471 **"The universe does not allow waste":** Reeves Wiedeman, *Billion Dollar Loser: The Epic Rise and Spectacular Fall of Adam Neumann and WeWork* (New York: Little, Brown, 2020), 136.

471 **"maybe in the way":** Lanier, author interview, March 12, 2019.

471 **geodesic domes on Mars:** John Brownlee, "Elon Musk Wants to Build a Martian Colony out of Geodesic Domes," *Fast Company* online, last modified October 24, 2016, https://www.fastcompany.com/3064918 /elon-musk-wants-to-build-a-martian-colony-out-of-geodesic-domes.

472 **"where employees can think":** The Spheres, https://www.seattlespheres .com (accessed January 2021).

472 **"Ideally," Brand said:** Stewart Brand, "The Clock and Library Projects," https://longnow.org/about (accessed January 2021).

472 **"Industrialization must be recognized"**: RBF, "Industrialization," in *SD*, 8047.

473 **Mayor Rudolph Giuliani:** Allegra Hobbs, "Gentrification's Empty Victory," *New York Times*, June 1, 2018, MB1.

473 **geodesic cabin:** Luke Winkie, "What's It Like to Own the Most-Visited Airbnb in the World," Vox, last modified September 20, 2019, https://www.vox.com/the-goods/2019/9/20/20857633/airbnb-most-visited-popular-mushroom-dome-cabin.

473 **replacing doorknobs:** RBF, *EIK*, session 9.

474 **Andrew Yang:** While Yang's platform was the closest to his ideas, the 2020 Democratic presidential candidate who expressed admiration for RBF most openly was Marianne Williamson, who called him "one of those world historic geniuses who reminded us of the extraordinary things that are possible" (Williamson, blurb for Sieden, *Fuller View*).

474 **"We should not be haunted":** Adi Robertson, "Alexandria Ocasio-Cortez Says 'We Should Be Excited About Automation,'" The Verge, last modified March 10, 2019, https://www.theverge.com/2019/3/10/18258134/alexandria-ocasio-cortez-automation-sxsw-2019.

475 **compared to "blood clots":** RBF, *CP*, 217.

475 **"I feel that his lack":** RBF, "United States: One of the Most Difficult Sovereignties to Break Up," in *SD*, 19534.

475 **Critical Path Project:** Douglas Martin, "Kiyoshi Kuromiya, 57, Fighter for the Rights of AIDS Patients," *New York Times*, May 28, 2000, 34.

475 **"Marx's ideas changed history":** William Kuhns, *The Post-Industrial Prophets: Interpretations of Technology* (New York: Weybright and Talley, 1971), 244.

476 **"I do want you to understand":** RBF, "Man as a Function of Universe," in *SD*, 9988.

BIBLIOGRAPHY

Abercrombie, Stanley. *George Nelson: The Design of Modern Design.* Cambridge, MA: MIT Press, 2000.

Aldersey-Williams, Hugh. *The Most Beautiful Molecule: The Discovery of the Buckyball.* New York: John Wiley & Sons, 1995.

Allen, Woody, and Stuart Hample. *Non-Being and Somethingness: Selections from the Comic Strip* Inside Woody Allen. Introduction by R. Buckminster Fuller. New York: Random House, 1978.

Andersson, Jenny. *The Future of the World: Futurology, Futurists, and the Struggle for the Post–Cold War Imagination.* New York: Oxford University Press, 2018.

Andersson, Leif C., Carl G. Gahmberg, and P. Ekblom, eds. *Gene Expression During Normal and Malignant Differentiation.* Orlando, FL: Academic Press, 1985.

Anker, Peder. *From Bauhaus to Ecohouse: A History of Ecological Design.* Baton Rouge: Louisiana State University, 2010.

Applewhite, E. J. *Cosmic Fishing: An Account of Writing* Synergetics *with Buckminster Fuller.* New York: Macmillan, 1977.

Auther, Elissa, and Adam Lerner, eds. *West of Center: Art and the Counterculture Experiment in America, 1965–1977.* Minneapolis: University of Minnesota Press, 2012.

Baer, Steve. *Dome Cookbook.* 5th printing. Corrales, NM: Lama Foundation, 1970.

Baggott, Jim. *Perfect Symmetry: The Accidental Discovery of Buckminsterfullerene.* New York: Oxford University Press, 1994.

Baldwin, J. *BuckyWorks: Buckminster Fuller's Ideas for Today.* New York: John Wiley & Sons, 1996.

Baldwin, J., ed. *The Essential Whole Earth Catalog: Access to Tools and Ideas.* New York: Doubleday, 1986.

Banham, Reyner. *Theory and Design in the First Machine Age.* 2nd ed. New York: Praeger, 1967.

Banks, Russell, Michael Ondaatje, and David Young, eds. *Brushes with Greatness: An Anthology of Chance Encounters with Celebrities.* Toronto: Big Bang, 1989.

Barron, John. *KGB: The Secret Work of Soviet Secret Agents.* New York: Reader's Digest Press, 1974.

———. *Operation Solo: The FBI's Man in the Kremlin.* Washington, DC: Regnery History, 1996.

Bear Island Centennial Book, 1904–2004. Published privately, April 2011.

Birmingham, Stephen. *The Late John Marquand: A Biography.* Philadelphia: J. B. Lippincott, 1972.

Blake, Peter. *No Place Like Utopia: Modern Architecture and the Company We Kept.* New York: W. W. Norton, 1996.

Blake, Trevor. *Buckminster Fuller Bibliography.* Synchronofile.com, 2016.

Bloch, E. Maurice. *Words and Images: Universal Limited Art Editions, Frederick S. Wight Art Gallery, March 14–May 7, 1978.* Los Angeles: The Gallery, 1978.

Bloomfield, Harold H., Michael Peter Cain, Dennis T. Jaffe, and Robert B. Kory. *TM: Discovering Inner Energy and Overcoming Stress.* Introduction by R. Buckminster Fuller. London: Unwin Paperbacks, 1976.

Boyce, Robert. *Keck and Keck.* New York: Princeton Architectural Press, 1993.

Brand, Stewart. *How Buildings Learn: What Happens After They're Built.* New York: Viking, 1994.

Brand, Stewart, ed. *The Last Whole Earth Catalog: Access to Tools.* Menlo Park, CA: Portola Institute, 1971.

———. *Whole Earth Epilog: Access to Tools.* San Francisco: Point, 1974.

Braun, Frederick Augustus. *Margaret Fuller and Goethe.* New York: Henry Holt, 1910.

Brenneman, Richard J. *Fuller's Earth: A Day with Buckminster Fuller and the Kids.* New York: New Press, 2009.

Buckley, William F., Jr. *Miles Gone By: A Literary Autobiography.* Washington, DC: Regnery, 2004.

Buckminster Fuller: Thinking Out Loud. Directed by Karen Goodman and Kirk Simon. Simon & Goodman Picture Company, 1996.

Burstyn, Ellen. *Lessons in Becoming Myself.* New York: Riverhead Books, 2006.

Burton, Richard. *The Richard Burton Diaries.* Edited by Chris Williams. New Haven, CT: Yale University Press, 2012.

Cage, John. *Anarchy: New York City—January 1988*. Middletown, CT: Wesleyan University Press, 2001.

———. *Composition in Retrospect*. Cambridge, MA: Exact Change, 1993.

———. *The Selected Letters of John Cage*. Edited by Laura Kuhn. Middletown, CT: Wesleyan University Press, 2016.

———. *A Year from Monday: New Lectures and Writings*. Middletown, CT: Wesleyan University Press, 1969.

Cage, John, and Morton Feldman. *Radio Happenings, I–V*. Translated by Gisela Gronemeyer. Cologne, Ger.: MusikTexte, 1993.

Candela, Félix. *Seven Structural Engineers: The Félix Candela Lectures*. Edited by Guy Nordenson. New York: Museum of Modern Art, 2008.

Chase, Marilyn. *Everything She Touched: The Life of Ruth Asawa*. San Francisco: Chronicle, 2020.

Cheney, Ednah Dow. *Reminiscences of Ednah Dow Cheney*. Boston: Lee & Shepard, 1902.

Cheng, R. Holland, and Lena Hammar, eds. *Conformational Proteomics of Macromolecular Architecture: Approaching the Structure of Large Molecular Assemblies and Their Mechanisms of Actions*. Singapore: World Scientific, 2004.

Chu, Hsiao-Yun. "R. Buckminster Fuller's Model of Nature." Doctoral thesis, University of Brighton, 2014.

Chu, Hsiao-Yun, and Roberto G. Trujillo, eds. *New Views on R. Buckminster Fuller*. Stanford, CA: Stanford University Press, 2009.

Cochrane, Louise. *The Sense of Significance*. Great Britain: Sidaway and Barry, 2015.

Cohen, Harold, and James Filipczak. *A New Learning Environment: A Case for Learning*. Foreword by R. Buckminster Fuller. San Francisco: Jossey-Bass, 1971.

Cort, David. *The Sin of Henry R. Luce: An Anatomy of Journalism*. Secaucus, NJ: Lyle Stuart, 1974.

Davenport, Guy, and Hugh Kenner. *Questioning Minds: The Letters of Guy Davenport and Hugh Kenner*. Edited by Edward M. Burns. Berkeley, CA: Counterpoint, 2018.

Dawkins, Richard. *Climbing Mount Improbable*. New York: W. W. Norton, 1996.

De Forest, Lee. *Father of Radio: The Autobiography of Lee de Forest*. Chicago: Wilcox & Follett, 1950.

Dickenson, Donna. *Margaret Fuller: Writing a Woman's Life*. London: Palgrave Macmillan, 1993.

Dillard, Annie. *Living by Fiction*. New York: Harper Perennial, 1988.

Dormehl, Luke. *The Apple Revolution: The Real Story of How Steve Jobs and the Crazy Ones Took Over the World.* London: Virgin, 2012.

Drake, Harold L. "Alfred Korzybski and Buckminster Fuller: A Study in Environmental Theories." Doctoral thesis, Southern Illinois University, 1972.

Drucker, Peter. *Adventures of a Bystander.* New York: John Wiley & Sons, 1997.

Duberman, Martin. *Black Mountain: An Exploration in Community.* Garden City, NY: Anchor Press, 1973.

Dyja, Thomas. *The Third Coast: When Chicago Built the American Dream.* New York: Penguin, 2013.

Eames, Charles, and Ray Eames. *An Eames Anthology: Articles, Film Scripts, Interviews, Letters, Notes, Speeches.* Edited by Daniel Ostroff. New Haven, CT: Yale University Press, 2015.

Eastham, Scott. *American Dreamer: Bucky Fuller and the Sacred Geometry of Nature.* Cambridge, UK: Lutterworth Press, 2007.

Eco, Umberto. *Travels in Hyperreality: Essays.* Translated by William Weaver. San Diego: Harvest/Harcourt, 1986.

Eco, Umberto, and G. B. Zorzoli. *The Picture History of Inventions: From Plough to Polaris.* Translated by Anthony Lawrence. New York: Macmillan, 1963.

Edmondson, Amy. *A Fuller Explanation: The Synergetic Geometry of R. Buckminster Fuller.* Boston: Birkhäuser, 1987.

Edwards, David. *ArtScience: Creativity in the Post-Google Generation.* Cambridge, MA: Harvard University Press, 2008.

Emerson, Ralph Waldo, W. H. Channing, and James Freeman Clarke, eds. *Memoirs of Margaret Fuller Ossoli.* Boston: Phillips, Sampson, 1852.

Erhard, Werner. *The Hunger Project.* San Francisco: Hunger Project, 1977.

Farber, David. *Chicago '68.* University of Chicago Press, 1988.

Fenn, Charles. *At the Dragon's Gate: With the OSS in the Far East.* Annapolis, MD: Naval Institute Press, 2004.

Ferguson, Marilyn. *The Aquarian Conspiracy: Personal and Social Transformation in the 1980s.* Los Angeles: J. P. Tarcher, 1980.

Floyd, Margaret Henderson. *Architecture After Richardson: Regionalism Before Modernism—Longfellow, Alden, and Harlow in Boston and Pittsburgh.* University of Chicago Press, 1994.

Form Givers at Mid-Century. Exhibition catalog. New York: Time, 1959.

Foster, Norman, and Jonathan Glancey. *Dymaxion Car.* Edited by David Jenkins and Hsiao-Yun Chu. London: Ivorypress, 2010.

Fraase, Michael. *Macintosh Hypermedia.* Vol. 1. *Reference Guide.* Glenview, IL: Scott, Foresman, 1990.

Friedman, B. H. *Gertrude Vanderbilt Whitney: A Biography*. Garden City, NY: Doubleday, 1978.

From the Library of R. Buckminster Fuller. New York: Glenn Horowitz Bookseller, 2004.

Fuller, Arthur Buckminster. *Historical Notices of Thomas Fuller and His Descendants, with a Genealogy of the Fuller Family, 1638–1902*. Cambridge, MA: 1902.

Fuller, Margaret. *The Letters of Margaret Fuller: 1845–1847*. Vol. 4. Edited by Robert N. Hudspeth. Ithaca, NY: Cornell University Press, 1987.

———. *The Portable Margaret Fuller*. Edited by Mary Kelley. New York: Penguin Books, 1994.

Fuller, R. Buckminster. *And It Came to Pass—Not to Stay*. New York: Macmillan, 1976.

———. *And It Came to Pass—Not to Stay*. Edited by Jaime Snyder. Zurich: Lars Müller, 2008.

———. *Anthology for the Millennium*. Edited by Thomas T. K. Zung. Carbondale: Southern Illinois University Press, 2014. First published 2001 by St. Martin's Press (New York).

———. *The Artifacts of R. Buckminster Fuller: A Comprehensive Collection of His Designs and Drawings*. Edited by James Ward. New York: Garland, 1985.

———. *An Autobiographical Monologue/Scenario*. Documented and edited by Robert Snyder. New York: St. Martin's Press, 1980.

———. *The Buckminster Fuller Reader*. Edited by James Meller. Harmondsworth, UK: Penguin, 1972.

———. *Buckminster Fuller to Children of Earth*. Compiled and photographed by Cam Smith. Garden City, NY: Doubleday, 1972.

———. *Cosmography: A Posthumous Scenario for the Future of Humanity*. Kiyoshi Kuromiya, adjuvant. New York: Macmillan, 1992.

———. *Critical Path*. Kiyoshi Kuromiya, adjuvant. New York: St. Martin's Press, 1981.

———. *4D Time Lock*. Albuquerque, NM: Lama Foundation, 1972.

———. *Earth, Inc.* Garden City, NY: Anchor Press, 1973.

———. *Education Automation: Freeing the Scholar to Return to His Studies—A Discourse Before the Southern Illinois University, Edwardsville Campus Planning Committee, April 22, 1961*. Carbondale: Southern Illinois University Press, 1962.

———. *Everything I Know*. Amazon Kindle edition, 2019.

———. *Grunch of Giants*. Santa Barbara, CA: Design Science, 2008. First published 1983 by St. Martin's Press (New York).

————. *Ideas and Integrities: A Spontaneous Autobiographical Disclosure*. Edited by Robert W. Marks. New York: Collier Books, 1969. First published 1963 by Prentice-Hall (Englewood, NJ).

————. *Intuition*. Garden City, NY: Doubleday, 1972.

————. *Inventions: The Patented Works of R. Buckminster Fuller*. New York: St. Martin's Press, 1983.

————. *Nine Chains to the Moon*. Philadelphia: J. B. Lippincott, 1938.

————. *No More Secondhand God, and Other Writings*. Carbondale: Southern Illinois University Press, 1963.

————. *Operating Manual for Spaceship Earth*. New York: E. P. Dutton, 1978. First published 1969 by Southern Illinois University Press (Carbondale).

————. *Pound, Synergy, and the Great Design*. Moscow: University of Idaho, 1977.

————. *R. Buckminster Fuller on Education*. Edited by Peter H. Wagschal and Robert D. Kahn. Amherst: University of Massachusetts Press, 1979.

————. *R. Buckminster Fuller Sketchbook*. Philadelphia: University City Science Center, 1981.

————. *Synergetics: Explorations in the Geometry of Thinking*. In collaboration with E. J. Applewhite. New York: Macmillan, 1975.

————. *Synergetics 2: Explorations in the Geometry of Thinking*. In collaboration with E. J. Applewhite. New York: Macmillan, 1979.

————. *Synergetics Dictionary: The Mind of Buckminster Fuller*. 4 vols. Edited by E. J. Applewhite. New York: Garland, 1986.

————. *Tetrascroll: Goldilocks and the Three Bears—A Cosmic Fairy Tale*. New York: ULAE / St. Martin's Press, 1982.

————. *Untitled Epic Poem on the History of Industrialization*. New York: Simon and Schuster, 1962.

————. *Utopia or Oblivion: The Prospects for Humanity*. New York: Bantam Books, 1969.

————. *The World Game: Integrative Resource Utilization Planning Tool*. Carbondale: Southern Illinois University Press, 1971.

————. *World Man*. Edited by Daniel López-Pérez. New York: Princeton Architectural Press, 2014.

————. *Your Private Sky: Discourse*. Edited by Joachim Krausse and Claude Lichtenstein. Zurich: Lars Müller, 2001.

Fuller, R. Buckminster, with Jerome Agel and Quentin Fiore. *I Seem to Be a Verb*. New York: Bantam Books, 1970.

Fuller, R. Buckminster, and Anwar Dil. *Humans in Universe*. New York: Mouton, 1983.

Fuller, R. Buckminster, and Robert Marks. *The Dymaxion World of Buckminster Fuller*. Garden City, NY: Anchor Books, 1973. First published 1960 by Reinhold (New York), with Marks credited as the sole author.

Fuller, R. Buckminster, Eric A. Walker, and James R. Killian Jr. *Approaching the Benign Environment*. New York: Collier Books, 1970.

Fuller, Richard F. *Chaplain Fuller: Being a Life Sketch of a New England Clergyman and Army Chaplain*. Boston: Walker, Wise, 1863.

Gabel, Medard. *Energy, Earth, and Everyone: A Global Energy Strategy for Spaceship Earth*. With the World Game Laboratory. Foreword by R. Buckminster Fuller. San Francisco: Straight Arrow Books, 1975.

———. *Ho-Ping: Food for Everyone*. With the World Game Laboratory. Garden City, NY: Anchor Press, 1979.

Gabler, Neal. *Walt Disney: The Triumph of the American Imagination*. New York: Vintage Books, 2007.

Gajdusek, Daniel Carleton. *Stumbling Along the Tortuous Road to Unanticipated Nobility: Melanesian, Indonesian, and Malaysian Expedition*. Bethesda, MD: Study of Child Growth and Development and Disease Patterns in Primitive Cultures, National Institute of Neurological and Communicative Disorders and Stroke, 1996.

Gehry, Frank. *Gehry Talks: Architecture + Process*. Edited by Mildred Friedman. New York: Rizzoli, 1999.

Geithner, Timothy. *Stress Test: Reflections on Financial Crises*. New York: Crown, 2014.

Gerber, Alex, Jr. *Wholeness: On Education, Buckminster Fuller, and Tao*. Kirkland, WA: Gerber Educational Resources, 2001.

Gerst, Cole. *Buckminster Fuller: Poet of Geometry*. Portland, OR: Overcup Press, 2013.

Gillette, King C. *The Human Drift*. Boston: New Era, 1894.

Gold, Russell. *The Boom: How Fracking Ignited the American Energy Revolution and Changed the World*. New York: Simon & Schuster, 2014.

Goldhagen, Sarah Williams. *Louis Kahn's Situated Modernism*. New Haven, CT: Yale University Press, 2001.

Gómez-Jáuregui, Valentín. *Tensegrity Structures and their Application to Architecture*. Santander, Spain: Cantabria University Press, 2020.

Goode, James M. *The Outdoor Sculpture of Washington, DC*. Washington, DC: Smithsonian Institution, 1974.

Gopnik, Blake. *Warhol*. New York: Ecco, 2020.

Gordon, Alastair. *Spaced Out: Radical Environments of the Psychedelic Sixties*. New York: Rizzoli, 2008.

Gore, Al. *The Future: Six Drivers of Global Change.* New York: Random House, 2013.

Gorman, Michael John. *Buckminster Fuller: Designing for Mobility.* Milan, It.: Skira Editore, 2005.

Gowan, Al. *Shared Vision: The Second American Bauhaus.* Cambridge, MA: Merrimack Media, 2012.

Guerrero, Pedro E. *A Photographer's Journey.* New York: Princeton Architectural Press, 2007.

Gustaitis, Rasa. *Wholly Round.* New York: Holt, Rinehart and Winston, 1973.

Hamalian, Linda. *The Cramoisy Queen: A Life of Caresse Crosby.* Carbondale: Southern Illinois University Press, 2009.

Harrison, Gilbert A. *The Enthusiast: A Life of Thornton Wilder.* New Haven, CT: Ticknor & Fields, 1983.

Hassan, Ihab, ed. *Liberations: New Essays on the Humanities in Revolution.* Middletown, CT: Wesleyan University Press, 1971.

Hatch, Alden. *Buckminster Fuller: At Home in the Universe.* New York: Crown, 1974.

Hays, K. Michael, and Dana Miller, eds. *Buckminster Fuller: Starting with the Universe.* New York: Whitney Museum of Modern Art, 2008.

Hedgepeth, William. *The Alternative: Communal Life in New America.* New York: Macmillan, 1970.

Hein, Hilde. *The Exploratorium: The Museum as Laboratory.* Washington, DC: Smithsonian Institution Press, 1990.

Herrera, Hayden. *Arshile Gorky: His Life and Work.* New York: Farrar, Straus & Giroux, 2003.

———. *Listening to Stone: The Art and Life of Isamu Noguchi.* London: Thames & Hudson, 2015.

Hession, Jane King, and Debra Pickrel. *Frank Lloyd Wright in New York: The Plaza Years, 1954–1959.* Salt Lake City: Gibbs Smith, 2007.

Howard, Jane. *Margaret Mead: A Life.* New York: Fawcett Columbine, 1984.

Howland, Llewellyn, III. *No Ordinary Being: W. Starling Burgess, Inventor, Naval Architect, Poet, Aviation Pioneer, and Master of American Design.* Jaffrey, NH: David R. Godine, 2015.

Hubbard, L. Ron. *Adventurer/Explorer: Daring Deeds & Unknown Realms.* Commerce, CA: Bridge, 1996.

Indlu Geodesic Research Project. Durban, South Africa: School of Architecture, University of Natal, 1958.

Jacobs, Charlotte DeCroes. *Jonas Salk: A Life.* Oxford, 2015.

James, William. *The Varieties of Religious Experience.* New York: Modern Library, 1994. First published 1902 by Longmans, Green & Co. (London).

Janzen, Rod. *The Rise and Fall of Synanon: A California Utopia.* Baltimore: Johns Hopkins University Press, 2001.

Jawaharlal Nehru Memorial Lectures, 1967–1972. New Delhi: Jawaharlal Nehru Memorial Fund, 1973.

Johnson, David Leslie. *Frank Lloyd Wright Versus America: The 1930s.* Cambridge, MA: MIT Press, 1994.

Johnson, Philip. *Writings.* New York: Oxford University Press, 1979.

Johnson, Philip, and Robert A. M. Stern. *The Philip Johnson Tapes: Interviews by Robert A. M. Stern.* New York: Monacelli Press, 2008.

Jordan, June. *Civil Wars.* Boston: Beacon Press, 1981.

Jumsai, Sumet. *Naga: Cultural Origins in Siam and the West Pacific.* With contributions by R. Buckminster Fuller. New York: Oxford University Press, 1988.

Kahn, Lloyd. *Domebook One.* Los Gatos, CA: Pacific Domes, 1970.

———. *Domebook 2.* Bolinas, CA: Pacific Domes, 1971.

———. *Shelter.* Bolinas, CA: Shelter, 1973.

Kahn, Lloyd, and Bob Easton. *Shelter II.* Bolinas, CA: Shelter, 2010.

Kahney, Leander. *Inside Steve's Brain.* New York: Portfolio, 2008.

Katz, Barry. *Make It New: The History of Silicon Valley Design.* Cambridge, MA: MIT Press, 2015.

Keats, Jonathon. *You Belong to the Universe: Buckminster Fuller and the Future.* New York: Oxford University Press, 2016.

Kenner, Hugh. *Bucky: A Guided Tour of Buckminster Fuller.* New York: William Morrow, 1973.

———. *Geodesic Math and How to Use It.* Berkeley: University of California Press, 1976.

Kepes, György, ed. *Structure in Art and in Science.* New York: George Braziller, 1965.

Kettenmann, Andrea. *Frida Kahlo, 1907–1954: Pain and Passion.* Cologne, Ger.: Taschen, 2003.

Kirk, Andrew G. *Counterculture Green: The* Whole Earth Catalog *and American Environmentalism.* Lawrence: University Press of Kansas, 2007.

Kiyosaki, Robert T. *Second Chance: For Your Money, Your Life and Our World.* Plata Publishing, 2015. EPUB edition.

Knight, Damon. *Charles Fort: Prophet of the Unexplained.* Introduction by R. Buckminster Fuller. London: Victor Gollancz, 1971.

Knobler, Peter, and Greg Mitchell, eds. *Very Seventies: A Cultural History of the 1970s, from the Pages of* Crawdaddy. New York: Fireside, 1995.

Krausse, Joachim, and Claude Lichtenstein, eds. *Your Private Sky: R. Buckminster Fuller—The Art of Design Science.* Zurich: Lars Müller, 2017.

Kuhns, William. *The Post-Industrial Prophets: Interpretations of Technology.* New York: Weybright and Talley, 1971.

Kurian, George Thomas, and Graham T. T. Molitor, eds. *Encyclopedia of the Future.* New York: Macmillan, 1996.

Lamster, Mark. *The Man in the Glass House: Philip Johnson, Architect of the Modern Century.* New York: Little, Brown, 2018.

Lane, Mervin, ed. *Black Mountain College: Sprouted Seeds—An Anthology of Personal Accounts.* Knoxville: University of Tennessee Press, 1990.

Lanier, Jaron. *Dawn of the New Everything: Encounters with Reality and Virtual Reality.* New York: Henry Holt, 2017.

Lant, Jeffrey L., ed. *Our Harvard: Reflections on College Life by Twenty-two Distinguished Graduates.* New York: Taplinger, 1982.

Le Corbusier. *Towards a New Architecture.* Translated by Frederick Etchells. New York: Dover, 1986. First published 1927 by Architectural Press (London).

Lepore, Jill. *If Then: How the Simulmatics Corporation Invented the Future.* New York: Liveright, 2020.

Levy, Steven. *Hackers: Heroes of the Computer Revolution.* Sebastopol, CA: O'Reilly Media, 2010.

Lewallen, Constance M., and Steve Seid. *Ant Farm, 1968–1978.* Berkeley: University of California, 2004.

Liotta, L. A., and I. R. Hart, eds. *Tumor Invasion and Metastasis.* The Hague: Martinus Nijhoff, 1982.

Litweiler, John. *Ornette Coleman: A Harmolodic Life.* New York: Da Capo, 1994.

Lonsdale-Bryans, James. *The Curve of Fate.* London: Andrew Dakers, 1941.

López-Pérez, Daniel. *Fuller en México: La Iniciativa Arquitectónica.* Mexico City: Arquine, 2015.

———. *R. Buckminster Fuller: Pattern-Thinking.* Zurich: Lars Müller, 2020.

Lorance, Loretta. *Becoming Bucky Fuller.* Cambridge, MA: MIT Press, 2009.

———. "Building Values: Buckminster Fuller's Dymaxion House in Context." Doctoral thesis, City University of New York, 2004.

Lord, Athena V. *Pilot for Spaceship Earth.* New York: Macmillan, 1978.

The Love Song of R. Buckminster Fuller. Directed by Sam Green. 2012.

Luce, Clare Boothe. *The Women.* New York: Dramatists Play Service, 1966.

MacCarthy, Fiona. *Gropius: The Man Who Built the Bauhaus.* Cambridge, MA: The Belknap Press of Harvard University Press, 2019.

Madsen, Axel. *John Huston*. Garden City, NY: Doubleday, 1978.

Mailer, Norman. *Genius and Lust: A Journey Through the Major Writings of Henry Miller*. Compiled by Norman Mailer. New York: Grove Press, 1976.

Mannheim, Steve. *Walt Disney and the Quest for Community*. Aldershot, UK: Ashgate, 2002.

Markoff, John. *What the Dormouse Said*. New York: Penguin, 2006.

——. *Whole Earth: The Many Lives of Stewart Brand*. New York: Penguin, 2022.

Marshall, Megan. *Margaret Fuller: A New American Life*. New York: Houghton Mifflin Harcourt, 2013.

Masey, Jack, and Conway Lloyd Morgan. *Cold War Confrontations: US Exhibitions and Their Role in the Cultural Cold War*. Baden, Switzerland: Lars Müller, 2008.

Matthews, Mark. *Droppers: The Story of America's First Hippie Commune*. Norman: University of Oklahoma Press, 2010.

McAtee, Cammie, ed. *Montreal's Geodesic Dreams: Jeffrey Lindsay and the Fuller Research Foundation Canadian Division*. Halifax, NS: Dalhousie Architectural Press, 2017.

McHale, John. *The Future of the Future*. New York: Ballantine Books, 1969.

——. *R. Buckminster Fuller*. New York: George Braziller, 1962.

McKibben, Bill, ed. *American Earth: Environmental Writing Since Thoreau*. New York: Penguin, 2008.

McLuhan, Marshall. *Letters of Marshall McLuhan*. Edited by Matie Molinaro, Corinne McLuhan, and William Toye. Toronto: Oxford University Press, 1987.

Meehan, Patrick J. *Frank Lloyd Wright Remembered*. Washington, DC: Preservation Press, 1991.

Metz, Jerred. *Halley's Comet, 1910: Fire in the Sky*. St. Louis: Singing Bone, 1985.

Mezzrow, Mezz, and Bernard Wolfe. *Really the Blues*. New York: New York Review Books, 2016.

Michaelis, David. *The Best of Friends: Profiles of Extraordinary Friendships*. New York: William Morrow, 1983.

Moore, Joe S. *Basic Bucky*. Salinas, CA: J. S. Moore, 1992.

Morris, Sylvia Jukes. *Rage for Fame: The Ascent of Clare Boothe Luce*. New York: Random House, 1997.

Morshed, Adnan. *Impossible Heights: Skyscrapers, Flight, and the Master Builder*. Minneapolis: University of Minnesota Press, 2015.

Mottel, Syeus. *Charas: The Improbable Dome Builders*. New York: Drake, 1973.

Mumford, Lewis. *The Pentagon of Power*. New York: Harcourt Brace Jovanovich, 1970.

A Necessary Ruin. Directed by Evan Mather. Hand Crafted Films, 2010.

Neder, Federico. *Fuller Houses: R. Buckminster Fuller's Dymaxion Dwellings and Other Domestic Adventures.* Baden, Switzerland: Lars Müller, 2008.

Nef, Evelyn Stefansson. *Finding My Way: The Autobiography of an Optimist.* Washington, DC: Francis Press, 2002.

Nevala-Lee, Alec. *Astounding: John W. Campbell, Isaac Asimov, Robert A. Heinlein, L. Ron Hubbard, and the Golden Age of Science Fiction.* New York: Dey Street, 2018.

New Worlds in Engineering. Detroit: Chrysler Corporation, 1940.

The 1936 Book of Small Houses. New York: Simon & Schuster, 1936.

Noguchi, Isamu. *Isamu Noguchi: A Sculptor's World.* Foreword by R. Buckminster Fuller. London: Thames and Hudson, 1967.

O'Dell, Cary, and Thad Heckman. *Bucky's Dome: The Resurrection of R. Buckminster Fuller and Anne Hewlett Fuller's Dome Home in Carbondale, Illinois.* United Kingdom: Arcadia, 2020.

Olson, Charles. *Charles Olson & Robert Creeley. The Complete Correspondence.* Vol. 10. Edited by Richard Blevins. Santa Rosa, CA: Black Sparrow Press, 1996.

O'Neill, Eugene. *Selected Letters of Eugene O'Neill.* Edited by Travis Bogard and Jackson R. Bryer. New York: Limelight Editions, 1994.

Osbaldeston, Mark. *Unbuilt Toronto: A History of the City That Might Have Been.* Toronto: Dundurn Press, 2008.

Ouspensky, P. D. *A New Model of the Universe.* London: Routledge & Kegan Paul, 1960.

Patterson, William H., Jr. *Robert A. Heinlein: In Dialogue with His Century.* Vol. 2. *1948–1988: The Man Who Learned Better.* New York: Tor Books, 2016.

Pawley, Martin. *Buckminster Fuller.* London: Grafton / HarperCollins, 1992.

Perl, Jed. *Calder: The Conquest of Time—The Early Years, 1898–1940.* New York: Alfred A. Knopf, 2017.

Piano, Renzo. *Logbook.* New York: Monacelli Press, 1997.

Popko, Edward S. *Divided Spheres: Geodesics and the Orderly Subdivision of the Sphere.* Boca Raton, FL: CRC Press, 2012.

Potter, Robert R. *Buckminster Fuller.* Englewood Cliffs, NJ: Silver Burdett Press, 1990.

Prouty, L. Fletcher. *JFK: The CIA, Vietnam, and the Plot to Assassinate John F. Kennedy.* Introduction by Oliver Stone. New York: Birch Lane Press / Carol, 1992.

Rabbit, Peter. *Drop City.* New York: Olympia Press, 1971.

Ramesh, Jairam. *Indira Gandhi: A Life in Nature.* London: Simon & Schuster India, 2019. EPUB edition.

Ray, Edgar W. *The Grand Huckster: Houston's Judge Roy Hofheinz, Genius of the Astrodome.* Memphis: Memphis State University Press, 1980.

Reid, Michael. *Convergence/Divergence: Exploring Black Mountain College + Chicago's New Bauhaus/Institute of Design.* Asheville, NC: Black Mountain College Museum, 2015.

Rhinehart, Luke. *The Book of est.* Hypnotic I Media, 2010. EPUB edition.

Rich, Doris L. *Amelia Earhart: A Biography.* Washington, DC: Smithsonian Institution, 1989.

Roberts, Siobhan. *King of Infinite Space: Donald Coxeter, the Man Who Saved Geometry.* Toronto: Anansi, 2006.

Robertson, Donald W. *The Mind's Eye of Richard Buckminster Fuller.* New York: St. Martin's Press, 1974.

Romig, D. K. *The United States Ship Great Northern.* Brooklyn: Eagle, 1919.

Rosen, Sidney. *Wizard of the Dome: R. Buckminster Fuller, Designer for the Future.* Boston: Little, Brown, 1969.

Rosenberg, Samuel. *The Confessions of a Trivialist.* Foreword by R. Buckminster Fuller. Baltimore: Penguin, 1972.

Roszak, Theodore. *From Satori to Silicon Valley.* San Francisco: Don't Call It Frisco, 1986.

Ryan, Hugh. *When Brooklyn Was Queer.* New York: St. Martin's Press, 2019.

Sadao, Shoji. *Buckminster Fuller and Isamu Noguchi: Best of Friends.* Long Island City, NY: The Isamu Noguchi Foundation and Garden Museum, 2011.

Sadler, Simon. *Archigram: Architecture Without Architecture.* Cambridge, MA: MIT Press, 2005.

Sande-Friedman, Amy. "Kenneth Snelson and the Science of Sculpture in 1960s America." Doctoral thesis, Bard Graduate Center, 2012.

Schnapp, Jeffrey T., and Adam Michaels. *The Electric Information Age Book: McLuhan/Agel/Fiore and the Experimental Paperback.* New York: Princeton Architectural Press, 2012.

Schulman, Robert. *Romany Marie: The Queen of Greenwich Village.* Louisville, KY: Butler Books, 2006.

Schwantes, Carlos A. *Vision & Enterprise: Exploring the History of Phelps Dodge Corporation.* Tucson: University of Arizona Press, 2000.

Sheed, Wilfrid. *Clare Boothe Luce.* New York: Berkley Books, 1984.

Sieden, Lloyd Steven. *Buckminster Fuller's Universe: An Appreciation.* Cambridge, MA: Perseus, 2000. First published 1989 by Plenum Press (New York).

Sieden, Lloyd Steven, ed. *A Fuller View: Buckminster Fuller's Vision of Hope and Abundance for All.* Studio City, CA: Divine Arts, 2011.

Sigstedt, Cyriel Odhner. *The Swedenborg Epic*. London: Swedenborg Society, 1981.

Silverman, Kenneth. *Begin Again: A Biography of John Cage*. Evanston, IL: Northwestern University Press, 2010.

Smithson, Robert. *The Collected Writings*. Edited by Jack Flam. Berkeley: University of California Press, 1996.

Snow, C. P. *The Two Cultures: And a Second Look*. Cambridge: Cambridge University Press, 1965.

Sontag, Susan. *Against Interpretation and Other Essays*. New York: Picador / Farrar, Straus & Giroux, 1966.

———. *Alice in Bed: A Play in Eight Scenes*. New York: Farrar, Straus & Giroux, 1993.

———. *Styles of Radical Will*. New York: Farrar, Straus & Giroux, 1969.

Spaceship Earth. Directed by Matt Wolf. Impact Partners, RadicalMedia, and Stacey Reiss Productions, 2020.

Spufford, Francis, and Jenny Uglow, eds. *Cultural Babbage: Technology, Time and Invention*. London: Faber and Faber, 1996.

Stone, Brad. *Amazon Unbound*. New York: Simon & Schuster, 2021. EPUB edition.

Stott, Timothy. *Buckminster Fuller's World Game and Its Legacy*. New York: Routledge, 2022.

Strum, Suzanne. *The Ideal of Total Environmental Control: Knud Lönberg-Holm, Buckminster Fuller, and the SSA*. London: Routledge, 2017.

Synergetic Stew: Explorations in Dymaxion Dining. Zurich: Lars Müller, 2020. Originally published in 1982.

10:56:20 PM EDT 7/20/69. New York: Columbia Broadcasting Network, 1970.

Terkel, Studs, with Sydney Lewis. *Touch and Go: A Memoir*. New York: New Press, 2007.

Thompson, D'Arcy Wentworth. *On Growth and Form*. 1st ed. Cambridge, 1917.

Toffler, Alvin, ed. *The Futurists*. New York: Random House, 1972.

Tomkins, Calvin. *Off the Wall: A Portrait of Robert Rauschenberg*. New York: Picador, 2005.

Toynbee, Arnold J. *A Study of History*. Vol. 3. London: Oxford University Press, 1934.

Tucker, David. *A Dream and Its Legacies: The Samuel Beckett Theatre Project, Oxford c. 1967–76*. Gerrards Cross, UK: Colin Smythe, 2013.

Turner, Fred. *From Counterculture to Cyberculture: Stewart Brand, the Whole Earth Network, and the Rise of Digital Utopianism*. University of Chicago Press, 2010.

Upham, Charles W. *Salem Witchcraft*. Boston: Wiggin and Lunt, 1867.

US Bureau of the Census. *The Statistical History of the United States, from Colonial Times to the Present*. New York: Basic Books, 1976.

Wallace, Amy. *The Prodigy*. New York: E. P. Dutton, 1986.

Ward, Barbara. *Spaceship Earth*. New York: Columbia University Press, 1966.

Warhol, Andy. *The Andy Warhol Diaries*. Edited by Pat Hackett. New York: Warner Books, 1991.

Warhol, Andy, and Pat Hackett. *POPism: The Warhol Sixties*. Orlando, FL: Harvest Book / Harcourt, 1980.

Weber, Nicholas Fox. *Patron Saints: Five Rebels Who Opened America to a New Art, 1928–1943*. New Haven, CT: Yale University Press, 1995.

Whitehead, Alfred North. *Science and the Modern World*. Cambridge: Cambridge University Press, 1929.

Wiebé, Edward. *The Paradise of Childhood*. Springfield, MA: Milton Bradley, 1869.

Wiedeman, Reeves. *Billion Dollar Loser: The Epic Rise and Spectacular Fall of Adam Neumann and WeWork*. New York: Little, Brown, 2020.

Wigley, Mark. *Buckminster Fuller Inc.: Architecture in the Age of Radio*. Zurich: Lars Müller, 2015.

Williams, Jonathan. *A Palpable Elysium*. New York: Dim Gray Bar, 1997.

Wolf, Amy. *On Becoming an Artist: Isamu Noguchi and His Contemporaries, 1922–1960*. Long Island City, NY: Isamu Noguchi Foundation and Garden Museum, 2010.

Wolfe, Tom. *The Electric Kool-Aid Acid Test*. New York: Bantam Books, 1969.

———. *From Bauhaus to Our House*. New York: Farrar, Straus & Giroux, 1981.

Wong, Yunn Chii. "The Geodesic Works of Richard Buckminster Fuller, 1948–1968 (The Universe as a Home of Man)." Doctoral thesis, Massachusetts Institute of Technology, 1999.

World Design Science Decade 1965–1975, Phase 1, Document 1: Inventory of World Resources, Human Trends and Needs. Carbondale: Southern Illinois University Press, 1963.

The World of Buckminster Fuller. Directed by Robert Snyder and Baylis Glascock. Robert Snyder Films, 1974.

Wozniak, Steve, with Gina Smith. *iWoz: Computer Geek to Cult Icon—How I Invented the Personal Computer, Co-founded Apple, and Had Fun Doing It*. New York: W. W. Norton, 2006.

Wright, Frank Lloyd. *Collected Writings*. Vol. 2. Edited by Bruce Brooks Pfeiffer. New York: Rizzoli, 1992.

Wright, Lawrence. *Going Clear: Scientology, Hollywood, and the Prison of Belief.* New York: Vintage Books, 2013.

Wright, Olgivanna Lloyd. *The Shining Brow.* New York: Horizon, 1960.

Wysocki, Edward M., Jr. *Out of This World Ideas—And the Inventions They Inspired.* Scotts Valley, CA: CreateSpace, 2018.

INDEX

A and B quanta modules, 583n397
Abbot, George, 84, 89
Abernathy, Constance, 359
abortion, 586n407
Abstraction in Almost Discontinuous Tension (Noguchi), 510n120
Accra Dome, 336
Acheson, Dean, 322
ACT UP Philadelphia, 475
Adams Brothers Company, 54
Aéroports de Paris, 366
Agel, Jerome, 374
Ahern, William, 260–261, 327
"air map," 156, 164, 527–528n179
Airbarac, 197, 198
Airflow sedan, 149
"Airways to Peace" (Bayer), 189, 213
Albers, Anni, 229
Albers, Josef, 227, 228, 229, 238, 540n22
Alcott, Amos Bronson, 38
Alcott, Louisa May, 38
Aldrin, Buzz, 363, 432, 468
Alfa Romeo Castagna Aerodinamica, 142
Ali, Muhammad, 380
Allen, John P., 470
Allen, Woody, 435
Alloy, 368
alternate grid, 261

alternative energy, 389, 421, 470
Altman, Robert, 580n387
aluminum, 117, 118, 120
Aluminum Company of America (Alcoa), 205, 210
Aluminum Limited, 304
Amazon, 472
American Association of Neurological Surgeons, 355
American Geographical Society, 185
American Humanist Association, 365
American Institute of Architects, 80, 109, 110–111, 116, 122, 165, 353, 375
American Museum of Natural History, 424
American National Exhibition (Moscow), 297–299, 301–303, 302*fig*
American Neptune, 192
American Radiator and Standard Sanitary Corporation, 126, 132, 165
American Scholar, 249
American Society for Metals, 290–291, 601n449
Amherst College, 422
Amundsen-Scott South Pole Station, 387–388
And It Came to Pass—Not to Stay (Denneny), 413
Anderson, Marian, 201

Andrews, Caroline Wolcott
    (grandmother), 27, 32
Andrews, Caroline Wolcott (mother).
    See Fuller, Caroline Wolcott Andrews
    (mother)
Andrews, Martin, 35
Ant Farm, 367
anti-Semitism, 253
antiwar movement, 371
Apaka, Alfred, 285
Apollo space program, 363, 433
Apple Computer, 1–4, 6–8, 7fig
Apple Industrial Design Group, 5
Apple Marina Bay Sands, 485n8
Applewhite, Edgar J.
    airport project and, 390
    in Carbondale, 366–367, 382
    Design Science Institute and, 414
    Designing a New Industry and, 200
    on Fuller, 438
    on Fuller's alcohol use, 334
    on Fuller's fears, 415
    Presidential Medal of Freedom and,
        443
    recruitment of, 199
    resignation of, 205
    on solar energy and Fuller's diet,
        333
    Synergetics and, 367, 391–392,
        393–394, 397–398, 399, 400–401,
        433
Archigram, 367
Architects Club, 103
Architectural Design, 315
Architectural Forum, 173, 235, 255,
    263, 269
Architectural League of New York, 58,
    80, 117
Architectural Record, 160
Armour & Company, 51–52, 61,
    70–71, 74
Armstrong, Edwin Howard, 171
Armstrong, Neil, 363
Arnold, John E., 5
Arnstein, Karl, 508n112

Around the World in 80 Days, 287–288
Arrow Plane, 142
ArtCenter College of Design, 431
Arts & Architecture, 193
Artzybasheff, Boris, 323
Asawa, Ruth, 231–232, 237, 310, 410,
    432
Asbury, Keith, 441
Asimov, Isaac, 590n419
Aspen Design Conference, 469
aspension dome, 312–313, 316
Astor, John Jacob, IV, 76
Astor, Vincent, 76–77, 81, 114, 499n70
Atkinson, Brooks, 43
Atkinson, Edward W., 46–47, 48
Atkinson, Haserick & Co., 46
"Atomic Buckalow," 224
Atwood, Leland, 112, 152, 199, 227
Auchincloss, Douglas, 323–324
Auchincloss, Lily, 323
"auto-airplane," 113
Automatic Machine Company, 140
automation, 474
Autonomous House, 438, 446
Autonomous Living Package, 235, 260

Baer, Steve, 334–335, 355–356, 368
Baird, Bil, 161, 164, 183, 435
Baird, Patrick, 248, 252, 408
Baker, Newton D., 69
Balanchine, George, 214
Baldwin, Jay, 368–369, 441, 468,
    514n132
ball clock, 221
"Ballistics of Civilization" (Fuller), 169
Banham, Reyner, 293
Barcroft, Taylor, 1–3
Bardacke, Gregory, 193, 194, 205
"bare maximum" concept, 363
Barr, Alfred, 175–176
Basing, Charles, 80
Bass, Ed, 470
Bassett, John, 354
bathroom project, 160–161, 160fig,
    162–164, 220

Battiste, Roy, 373
Battlestar Galactica, 387
Bauersfeld, Walther, 244
Bauhaus school, 81, 115, 129, 220, 229, 274
Bayer, Herbert, 189, 540n238, 564n266
Bear Island, Maine
  Alexandra and, 72
  Allegra and, 117
  ecological threats to, 389
  Fuller on, 27–28, 28*fig*, 29, 30, 32–33, 34, 62, 71, 76, 117, 123, 128, 293, 349, 381
  Lacey on, 214
  loss of share of, 154
  renting out of, 173
  Tomkins on, 331–332
Beard, Marjorie, 62
Beatty, Warren, 347–348
Beaux Arts Ball, 58, 124
bebop, 234
Becker, Sal, 373
Beckett, Samuel, 364, 365, 370
Beckett Theatre, 364–365, 377–378, 404
Bedford, Gladys, 72
Beech, Walter, 195–196
Beech Aircraft, 193–194, 197, 207, 210
Béhar, Yves, 12, 469
"Being with Bucky," 408, 410
Beirn, Ken, 183, 192
Bel Geddes, Norman, 154, 278–279
Bell, Alexander Graham, 256–257
Bell Telephone, 157
Bellinger, Patrick, 64, 65, 66, 137
Ben-Eli, Michael, 336, 373–374, 384, 391
Benshetrit, Dror, 471
Benson Manufacturing of Missouri, 186
Berger Brothers, 288
Bernofsky, Gene, 334, 335
Bertalanffy, Ludwig von, 554n296
Better Homes and Gardens, 286
Beyond the Horizon (O'Neill), 70
Bezos, Jeff, 472

Biddle, Anna "Nannie," 136, 138–139, 140, 141, 143–144, 148, 154
Big House, Bear Island, Maine, 30
biohacking, 11
biology, Fuller's interest in, 316–317
Biosphere, 410
Biosphere 2, 470
Black Mountain College, 227–233, 230*fig*, 237–243, 239*fig*
Blake, Peter, 234, 301–302, 328
Blouke, Pierre, 110
Blue Hills Reservation state park, 25
Board of Economic Warfare (BEW), 180–181, 182, 183, 192
Bodky, Erwin, 232
Boggs, Samuel Whittemore, 185
Bonenkamp, Nettie, 94, 95
Bonnie and Clyde, 347–348
Borland, John Nelson, II, 93
Boudreau Machine and Tool Company, 139
Boy-Ed, Karl, 60–61
Bragdon, Claude, 111
Braggiotti, Mario, 168
brain/mind distinction, 355
Brancusi, Constantin, 120, 121
Brand, Stewart
  Clock of the Long Now and, 472
  on dome, 400
  in Fuller's classes, 338
  Fuller's legacy and, 366
  on Great Depression, 120
  Joint Computer Conference and, 357–358
  Kahn and, 368
  meeting with in California, 338
  as outlier, 359
  Whole Earth Catalog and, 4, 356–357
  whole earth photo and, 352–353
Brave New World (Huxley), 387
Brazil, 188, 189
Breath (Beckett), 370
Breck, John, 183
Breines, Simon, 129
Brenneman, Richard J., 596n432

Brewer, Kidd, 252
Brezhnev, Leonid, 375
Bridges, Jeff, 467
British Commonwealth Scientific Office, 191
Brokaw, Clare Boothe (later Luce)
    Hale and, 167–168
    later interactions with, 183, 193, 438
    Milan Triennale and, 271, 547n272
    Presidential Medal of Freedom and, 443
    relationship with, 149–151, 169
    Vanity Fair and, 134–135
Bromberg, Manuel, 236, 257, 547n269
Brook Farm, 38
Brooklyn Dodgers, 278–280
Brooks, Jefferson Davis, III, 282
Brooks, Louise, 98
Browall, Harold, 154
Brown, Howard, 388, 412, 440, 468
Brown, Jerry, 400, 435
Brownjohn, Robert, 234
Bryan, Joseph "J.," III, 504n95
Bryn Mawr College, 383, 449
Bubble House, 252, 259
Buck, Pearl S., 593n426
Buckley, William F., Jr., 382
Buckminster Fuller: An Autobiographical Monologue/Scenario (Fuller), 433
Buckminster Fuller: At Home in the Universe (Hatch), 398
Buckminster Fuller Institute, 468, 589n415
Buckminster Fuller Reader, The (Fuller), 370
Buckminster Fuller to Children of Earth (Fuller), 384
Buckminster Fuller's Universe (Sieden), 16
buckminsterfullerene, 9, 462–467, 465*fig*
Bucky (Kenner), 389–390
buckyball. *See* buckminsterfullerene
Buffington, George N., 114
Bundy, McGeorge, 316, 322

Burgess, Frederick V. "Bunny," 85, 86
Burgess, Starling, 34, 132, 139–144, 148, 152
Burklyn Business School, 437
Burning Man, 471
Burns, Robert, 32
Burr, Laurence Hewlett, 74, 75
Burr, Robert, 75, 79
Burroughs, William S., 320
Burstyn, Ellen, 435–436, 449
Burton, Isabel, 92
Burton, Richard, 322, 404
Burton, Richard Francis, 92
Bush, Vannevar, 191
Butler Manufacturing of Missouri, 173–174
Butts, Don, 408
Butts, J. Arch, Jr., 198, 209, 210, 408
Byron, Lord, 94

Cage, John
    on Bear Island, 570n349
    at Black Mountain, 229, 231, 232, 233
    Fuller's birthday and, 432, 435
    Fuller's death and, 448
    Fuller's longevity and, 416
    impact of Fuller on, 323
    music of, 428
    "new establishment" and, 328
    in New York, 237
    Schlossberg and, 360
    Solway and, 409
    Studio School and, 363
    Warhol and, 342
    World Game and, 336, 376
Cahill, Bernard J. S., 177, 200
Calder, Alexander, 119, 149–150, 241, 242, 410
Cambridge Seven Associates, 326
Campbell, John W., 542–543n250
Campbell, Joseph, 237
Campuan Conference on Global Issues, 409
Canaday, John, 305

Canadian Connecticut Cotton Mills Company, 46–47
Candela, Félix, 316, 337
Canfield, Jack, 437
Cannon, Eleanor and Victor, 250
Cantos, The (Pound), 233, 348, 375–376
capitalism, 165
Capone, Al, 88
Capote, Truman, 425
Carbondale house, 309–310, 310fig
Carlyle Hotel, 133
Carnegie, Andrew, 412
Carpenter, Rue Winterbotham, 117
Carrick, Bruce, 399
Carter, Jimmy, 418
Caspar, Donald, 317, 461
Castaneda, Carlos, 468
"casting out nines," 54
catenary dome, 280–283, 281fig, 287
Cates, Louis S., 157, 163
Cavett, Dick, 355
Celebrity Center Mission, 384
Celeste, Richard F., 601n449
Celotex Corporation, 102, 506n106
Central Intelligence Agency (CIA), 380
Chalkley, Lyman, 184
Challenger, 468
Chamberlain, Martin, 93, 95
CHARAS, 373–374, 384–385, 473
Chassis Fountain, 167
Chavez, Cesar, 405
Chermayeff, Peter, 326–327, 341
Chermayeff, Serge, 220–221, 226–227, 233–234, 237
Cherry Tree Folk Club, 587n409
Chicago, architecture of, 86
Chicago Architects Exhibition League, 117
Chicago Sunday Evening Club, 92, 95
Chicago World's Fair, 116, 123, 126, 134, 145, 152, 258
Cholmondeley, Lady Margot, 422
chord factors, 11, 249, 328, 355, 369, 378

Chrysler, 172–173
Chrysler, Walter, 144, 149
Chrysler Building, 122, 125
Churchill, Winston, 185
Cinerama, 323, 355
City Gone Wild, The, 98
civil rights movement, 329
Civil War, 37
Clarke, Arthur C., 363, 382, 419, 466
Clarke, James Freeman, 41
climate change, 434
Climatroffice, 381
Climatron, 295, 387
Clinton, Joseph, 320, 369, 378, 391, 561n320
Clock of the Long Now, 472
clothes, experiments with, 117
"Cloud Nine" structures, 294fig, 295
Clow, Francise, 155, 156, 158, 159, 166, 168
Club of Rome, 383
Coffyn, Frank T., 144
Cohen, Harold
  on Fuller's relationships, 15, 246, 311, 328–329, 430
  Skinner and, 571n355
  at Southern Illinois University, 283, 303, 309
  Time article and, 324
  Twenty Questions approach and, 569n347
Cohen, Mary, 309
Colby, Safford, 205
Coleman, Ornette, 442
Colgate, Robert, 173, 174
College of the Ozarks, 211–212
Colstad, Ken, 5
Columbia Broadcasting System, 169
Common Brick Manufacturers' Association, 91
Community Memory, 5
Compass Island, Maine, 27
comprehensive anticipatory design science, 275, 277
comprehensive designer, 3, 357

"Comprehensive Designer, The"
   (Fuller), 238
computers, 316, 350–351, 353,
   357–358, 366, 401
Conference on World Affairs, 334, 335
Congressional Clearinghouse on the
   Future, 606n469
"conning tower," 133
Constant Nymph, The (Kennedy), 128
"Contact Economy" (Fuller), 187
Container Corporation of America, 266
Convoy (magazine), 68
Cooper, Merian C., 187
Cooper, Viola Irene, 161–162, 164
Cooper Hewitt Museum, 412–413
copper, 157, 159–160, 202, 258
Corbett, Harvey, 116, 117, 121, 122,
   126, 179–180
Cornell University, 257
coronavirus pandemic, 473
Correa, Charles, 260, 366
Cort, David, 176, 282, 286–287,
   531n192
Cosmic Fishing (Applewhite), 333, 393
Cosmography (Fuller), 448
Cosmopolitan Homes, 107. *See also*
   Fuller Houses
Cosmos Club, 184, 189, 191, 192
cotton mill project, 256
Council on Science and Technology, 426
Cousins, Norman, 325, 355, 391, 404,
   416
Cousteau, Jacques, 419
Cox, G. W., 269
Coxeter, H. S. M., 394–395
Crane, Margarett, 36
Crawley, Calvin, 498–499n66
Creeley, Robert, 243
Crespo, Humberto, 373
Critchlow, Keith, 336
critical path, 336, 433
Critical Path (Fuller), 433–435, 444
Critical Path Project, 475
Crocodile, 176
Cronkite, Walter, 340, 436

Crosby, Caresse, 338–339, 370
Crosby, Harry, 338
Crosland, Anthony, 411
Crystal House, 152
Crystal Palace, 568n342
cuboctahedron, 177–178, 178*fig*, 222.
   See also vector equilibrium
Cummings, E. E., 119, 172, 433
Cunningham, Merce, 229–230, 231,
   237, 360, 370n349, 435
Curl, Robert, 459, 462–463, 465
Curtiss-Wright aviation company, 144
Curve of Fate, The (Lonsdale-Bryans),
   169

D-45 (car), 183, 187
Daley, George, 457
Dalí, Salvador, 338
Dalton School, 158
Dancing House, 469
"Dare to Be True" (Milton Academy
   motto), 31
Dartmouth College, 212
Dartmouth Conference, 325, 335, 358
Darwin, Charles, 333
Davenport, Guy, 376
Davenport, Russell, 168, 171–172
Davis, Emma Lu, 140
Davison, Robert L., 126–127, 131
Dawes, Rufus C., 134
Dawkins, Richard, 470
de Forest, Lee, 64, 115
de Kooning, Elaine, 228–229, 230–231,
   232
de Kooning, Willem, 228–229, 232, 338
*Death and Life of Great American Cit-
   ies, The* (Jacobs), 330
"debunking," 107
decentralization, 107, 110, 236
Dederich, Charles E., Sr., 405–406
Dehn, Max, 229
Delano, William, 122
Delos Symposium, 321, 340, 403
demonstrative urban conservation park,
   379

Denneny, Michael, 401–402, 413, 433, 435, 443, 444
Denver, John, 407, 421, 422, 423, 427, 431, 439, 448
Department of Commerce, 209, 283
Deren, Maya, 193, 237
Design Science Institute, 382, 414–416
Designing a New Industry, 200
desovereignization, 475
"diagrid" framework, 468
Dial, 39
Dianetics (Hubbard), 249–250
Dickler, Gerald, 393
Dieges, Jon, 360
Diffrient, Niels, 258–259
Digressionist Society, 80
Dil, Anwar, 442, 593n426
Dillard, Annie, 399, 402, 448
Dirigo Island, Maine, 28
discontinuous compression, continuous tension, 249, 267, 268*fig*, 269, 280. *See also* tensegrity
discontinuous compression domes, 267–269
Disney, Walt, 422–423
Disney World, 422–423, 441
Distant Early Warning Line, 273, 277, 277*fig*
Dixon, John, 269, 282, 306
Dollfus, Charles, 145–146, 147–148
Dome Cookbook (Baer), 355–356
Dome over Manhattan. *See* Manhattan, dome for
Domebook One (Kahn), 369, 378
Donavin, Kirkwood H., 75
Douglas, John W., 108
Doxiadis, Constantinos, 321, 331, 403
Doyle, Stafford, 68
Draper, Muriel, 123
Dreier, Theodore, 227, 229, 238
Dreiser, Theodore, 119
Drexler, Arthur, 251, 304, 305, 306, 429
Drop City, 356, 361, 461
Droppers, 334

Drucker, Peter, 171, 188
"dry toilet," 595n431
Duberman, Martin, 227
Ducas, Robert, 192, 194
Dunaway, Faye, 348
Dunham, Katherine, 379
duralumin, 104
dwelling machine, 192–194, 195–206, 214–215
Dymax, 4
Dymaxion, 4, 10, 115–116, 484n6
Dymaxion Bathroom. *See* bathroom project
Dymaxion Car. *See also* 4D Transport
   #1, 141–148, 142*fig*
   #2, 142*fig*, 148–150, 153, 431
   #3, 152–154, 198
   Arnstein and, 508n112
   building of, 140*fig*
   C. Eames and, 220
   Felsenstein and, 4
   new design for, 253
   New York World's Fair and, 170
   as part of shelter, 10
   pattern of failures and, 207
   plans to sell, 159
   replica of, 469
   supercharged, 176
Dymaxion Chronofile, 13, 167, 427
Dymaxion Company, 174
Dymaxion Comprehensive System, 191–192
Dymaxion Deployment Unit (DDU), 174–176, 175*fig*, 180, 240
Dymaxion Ditties, 587n409
Dymaxion Dwellings Inc., 194
Dymaxion Dwelling Machine. *See* Wichita House
Dymaxion House. *See also* Wichita House
   Delano on, 122
   Heinlein and, 590n419
   hexagonal structure and, 177
   Hotel Winthrop and, 131
   lack of patent for, 444
   MacLeish on, 133

Dymaxion House. *See also* Wichita House (*cont.*)
  model of, 120*fig*, 121, 125–126
  MOMA exhibition and, 130
  potential building of, 134
  tensegrity and, 307
Dymaxion Map
  Brand and, 356
  exhibition of, 188–189
  Gandhi and, 365–366
  House of Representatives and, 440
  icosahedron and, 248
  illustration of, 179*fig*, 225*fig*
  infographics and, 10
  mathematical transformation for, 428
  Nehru and, 291
  patent for, 199–200, 225–226
  publication of, 181, 182, 185–186
  revision of, 258
  World Game and, 361, 362*fig*
Dymaxion Sleep, 188
Dymaxion World of Buckminster Fuller, The (Marks), 311, 317, 327, 391, 461
Dyson, James, 469

Eagle 15, USS, 75
Eagle's Eye, The, 61
Eames, Charles, 193, 205, 211, 219–220, 252–253, 255, 266, 271, 301
Eames, Ray, 193, 219–220, 252–253, 266, 301
Earhart, Amelia, 153, 154
Earle, Guyon L. C., 555n301
Early X Piece (Snelson), 240*fig*, 242
Earth, Inc. (Fuller), 211, 381
Earth II, 387
Earth Metabolic Design, 388
Ebbets Field, 278, 285
Eberle, Edward W., 66
Echevarria, James, 374
Eco, Umberto, 343
ecology, Fuller's definition of, 133
Edison, Charles, 176

Edmondson, Amy, 436–437, 445, 453, 454, 457
education, Fuller on, 313–315
Education Automation, 313, 319
Edwards-Davies, Harold, 149
Ehrlich, Matthew, 514n132
Eight Chains to the Moon (Fuller), 156–157
Einstein, Albert, 35, 123, 158–159, 161–162, 198, 221, 233, 268
Eisenhower, Dwight, 281
ekistics, 321
Electric Kool-Aid Acid Test, The (Wolfe), 352
elements, discovery of, 191, 211, 212*fig*
Eliasson, Olafur, 469
Elizabeth II, 262–263, 411
emergence through emergency, 166
Emerson, Ralph Waldo, 38–39, 41–42
energetic geometry, 190, 195, 233, 241, 248, 317, 319, 328
"energetic-synergetic geometry," 280
Energy, Earth, and Everyone (Gabel), 596n433
Engelbart, Douglas, 353, 357–358, 418
Entenza, John, 193, 367
environment, shaping of, 33
Epcot, 9, 423, 441
ephemeralization, 11–14, 131, 157, 165, 169, 255, 270, 365, 379, 418, 471–474
Equal Rights Amendment, 586n407
Erdman, Jean, 237
Erhard, Ludwig, 406
Erhard, Werner, 406–407, 410–412, 415–416, 417–418, 420–422, 424–425, 427, 444, 448
Erhard Seminars Training, 406
Erickson, Mark, 408
Esalen Institute, 5, 351, 352
Escher, M. C., 392–393
est, 406–407, 411, 417, 420, 424, 437
Everything I Know, 401, 433
evolution, 332–333

Experimental Prototype Community of Tomorrow, 423
Exploratorium children's museum, 4
Explorers Club, 170

Fales, DeCoursey, 89
Farrell, Barry, 376–377
Federal Housing Administration, 199
Felsenstein, Lee, 4–5
Fenn, Charles, 188
Ferguson, Marilyn, 599n443
Ferriss, Hugh, 119
Festival of Two Worlds, 348
Field, William Lusk Webster, 34
Fifth Discipline, The (Senge), 411
Figueroa, Anthony, 373
Finch, John, 301
Finland, USS, 67
Fiore, Quentin, 374
Fisher, Howard, 116, 132
Fisher, Irving, 189
Fisher, Paul, 207
Fistere, John, 176
Fitzgerald, F. Scott, 85
Fitzgibbon, James
  catenary dome and, 283
  financial difficulties and, 295
  Graver Tank and, 294
  in Houston, 289
  Lane and, 547n269
  Montreal proposal and, 288
  North Carolina State College and, 231, 236, 251–252
  Old Man River's City, 379
  plydomes and, 286
  plywood dome and, 257, 269
  Synergetics Inc. and, 278, 280, 297
flat plane, absence of, 108
Floyd, Cynthia (née Lacey). *See* Lacey, Cynthia (later Floyd)
Floyd, David, 214, 215, 220, 236
Floyd, Peter, 267, 273, 287, 327, 336–337, 354, 441
"Fluid Geography" (Fuller), 192
Flying Fish (biplane), 34

flying fish logo, 139
Flying Seed Pod, 272, 281, 320
Fly's Eye Dome, 425–426, 431, 432–433, 432*fig*
fog gun, 235
Fogg Museum, 121
Fonda, Jane, 419
Food and Agriculture Organization, 215
Forbes-Sempill, William Francis, 145–147, 148
Ford, Henry, 108, 111, 152, 200, 259
Ford, Henry, II, 263
Ford, Walter B., 258
Ford Foundation, 315
Ford Industrial Exposition, 97, 98, 104, 263
Ford Motor Company, 72, 127
Ford Rotunda, 258–259, 261–264, 262*fig*, 269, 319
Foreign Economic Administration, 192, 194
Fort, Charles, 510n119
Fortune, 167, 168–169, 170–171, 174, 202
Foster, Norman, 364–365, 381, 404, 424, 438, 446, 468–469, 485n8
Foundation for Contemporary Performing Arts, 360
Fountains of Paradise, The (Clarke), 419, 466
Four Saints in Three Acts, 149
4D Company, 108
4D Dymaxion Corporation, 139
4D Time Lock (Fuller), 109–114, 113*fig*, 246
4D Transport, 127, 133, 135–141. *See also* Dymaxion Car
fourth dimension, 42, 108
Fowler, Gene, 344
Foxconn, 471
fracking, 470
Frangipani, 439
Franklin Institute Science Museum, 435
Franklin Research Institute, 428
Frazier, Joe, 380

Free Speech Movement, 334
French, Gordon, 5
Friends of Buckminster Fuller, 439
Frobisher Bay, 295
Froebel, Friedrich, 22–23, 23fig
Fugitive Slave Clause, 36
Fuller, Alexandra Willets (daughter), 68,
  71–74, 75–77, 75fig, 78–79, 381
Fuller, Anna Allegra (daughter)
  advice from, 170
  on Bear Island, 117, 123
  birth of, 94, 95
  Black Mountain College and, 227
  car accident and, 153
  career of, 407
  childhood of, 103
  in Connecticut, 141
  correspondence from, 204
  early work life of, 253–254
  education of, 158, 214
  family finances and, 220
  4D Transport and, 139–140
  on Fuller, 213
  marriage of, 253–254
  mother and, 203
  mother's health and, 445, 447
  photographs of, 109fig
  The Proud Shirtfront and, 496n53
  Stearns and, 155
  in Washington, 191, 194
Fuller, Anne Hewlett (wife)
  Alexandra and, 71–74
  Allegra and, 94, 253–254
  attempt at reconciliation with, 132
  on Bear Island, 123
  car accident and, 153, 343–344,
    347
  CHARAS project and, 384
  in Chicago, 93–94
  in Connecticut, 141
  correspondence from, 195
  as creative partner, 105
  death of, 447–449
  early relationship with, 52–58
  estrangement from, 149

  family finances and, 102, 108–109,
    117–118, 245–246, 562n324
  on Fuller, 442–443
  Fuller Houses and, 203–205, 210
  Fuller's alcohol use and, 181
  on Fuller's alleged silence, 101
  Fuller's housing ideas and, 111–112
  Geodesics Inc. and, 272–273
  health of, 398, 420, 445, 446–448
  Lacey and, 194–195, 214
  in London, 292–293
  married life of, 76–77, 85, 88, 90–93,
    98, 139
  in Michigan, 262–263, 265
  in New York, 203
  in North Carolina, 236, 238, 240, 242
  "partial estrangement" from, 128–129
  in Philadelphia, 383
  photographs of, 53fig, 109fig, 310fig
  Pound and, 375–376
  pregnancy of, 67, 68
  reunion with, 213
  Schwartz and, 124
  Snelson and, 430
  Stearns and, 155–156
  on success of dome, 287
  in Virginia, 64
  in Washington, 191, 194
  Weatherbreak Dome and, 250–251
  wedding to, 61–62
Fuller, Arthur Buckminster, 36–37
Fuller, Caroline Lesley "Leslie" (sister),
  22, 25, 33, 43, 44, 52, 92, 93, 274
Fuller, Caroline Wolcott Andrews
  (mother)
  death of, 149
  Fuller's childhood and, 24–25, 30–31,
    32
  Fuller's engagement and, 56–58, 61
  Fuller's finances and, 33–34, 46, 47,
    48
  Fuller's housing ideas and, 111
  marriage of, 35
  Schwartz and, 124
Fuller, Emma Lucilla Reeves, 36

Fuller, Margaret, 37–41, 40*fig*, 95, 96, 310, 395, 435, 599n443
Fuller, Margarett Crane, 36
Fuller, Richard Buckminster, Jr. *See also individual colleagues and relatives; individual projects and works*
  alcohol and, 57, 79, 82, 90–91, 135, 180–181
  alleged silence of, 101–102
  ancestry of, 35–41
  at Annapolis, 66–67
  book deal for, 161–162
  in Chicago, 86–87, 88–91, 92–94
  childhood and early life of, 21–50
  control and, 14–15
  death of, 447–449
  diet of, 333–334
  early work life of, 51–52
  education of, 22–23, 31–32, 34–35, 42–46, 48–50
  farsightedness of, 21–22, 24, 25–26, 59
  flying of, 76
  health of, 349, 383–384, 421, 426–427, 438–439, 445–446
  in Illinois, 303–304
  interwar years and, 70–77
  legacy of, 8–13, 16–17, 467–476
  mental turning point for, 95–98
  misinformation and falsehoods from, 15
  "partial estrangement" and, 128–129
  patent law and, 114–115, 254–255
  photographs of, 7*fig*, 22*fig*, 28*fig*, 46*fig*, 53*fig*, 63*fig*, 109*fig*, 230*fig*, 266*fig*, 270*fig*, 298*fig*, 310*fig*, 362*fig*, 432*fig*
  in Quebec, 46–48, 46*fig*
  sleep experiments of, 112
  World War I and, 57–60, 62–70
  writing priorities of, 397–398
Fuller, Richard Buckminster, Sr. (father), 22*fig*, 24, 26–27, 30, 32, 35, 36
Fuller, Richard Frederick, 493n36

Fuller, Rosamond (sister), 28*fig*, 29, 33, 58, 124, 149, 154, 214, 309, 432
Fuller, Sarah Williams, 36
Fuller, Thomas, 35–36
Fuller, Timothy, 36
Fuller, Timothy, Jr., 36, 37, 44
Fuller, Waldo, 42
Fuller, Wolcott (brother)
  childhood of, 22, 25, 29, 30–31
  correspondence from, 52
  death of, 309
  Fuller's housing ideas and, 111, 117
  Fuller's wedding and, 61–62
  mother's death and, 149
  Stockade Midwest and, 86
  World War I and, 59–60
Fuller & Sadao, 326
Fuller Houses, 102, 104, 106–107, 199–206, 207–211, 214–215. *See also* Cosmopolitan Homes
Fuller Research Foundation, 211, 246–249, 250–252
fullerenes, 465–467
Fuller's Earth (Fuller and Brenneman), 596n432
Fuller's Law of Economics, 108
fuselage cartridge, 414–415
Future World, 423
Fyodorov, E. K., 325
Fyre Festival, 471

Gabel, Medard, 361, 376, 388, 400, 440, 443, 468, 596n433
Game (Synanon), 405
Gandhi, Indira, 291–292, 324–325, 365–366, 390–391, 403, 414, 437
Gao Yuan, 426
García, Carlos "Chino," 373
Garden of Eden, 237, 252
Garibaldi, Giuseppe, 39
Gaty, Jack, 193–194, 196
Gay Liberation Front, 433
Gehry, Frank, 469
Geithner, Timothy, 493n36
General Electric, 111

General Houses, 132
general semantics, 116
geodesic dome. *See also* individual
  domes
  attempt to patent, 254–255
  Bauersfeld and, 244
  discovery of, 10
  hacking and, 11
  number of, 14
  origins of, 222–226
  patent for, 271
geodesic lines, 176–177
Geodesic Math and How to Use It
  (Kenner), 581n391
geodesic cotton mill, 256
Geodesics Inc., 272–273, 275, 278
Geometrics Inc., 278, 318, 355
George Washington, USS, 68–69
Geoscope, 308, 315, 320, 328, 364,
  391, 472
Gershwin, George, 121
Giacometti, Alberto, 362
gig economy, 473
Gigundo Dome, 438
Gila River Indian Reservation, 257
Gillette, King C., 511n122
Gillette Corporation, 511n122
Gillis, Bob, 408, 411
Gillmor, Reginald E., 172, 236
Gilmour, Léonie, 119
Giuliani, Rudolph, 473
Gladstone, Gerald, 354
Gleaves, Albert, 67, 69–70, 72
Glenn L. Martin aircraft company, 171
Glimpses of the U.S.A., 301
global dwelling service, 425
global energy network, 434
Global Energy Network Institute,
  605n468
Godspell (Schwartz), 386
Gold Medal of the American Institute of
  Architects, 375
Goldberg, Bertrand, 227
Goldberger, Paul, 448
Goldgar, Mike, 213, 220, 221

Goldowski, Natasha, 229, 238, 242
Goldsmith, Myron, 244
Gonzalez, Angelo, Jr, 373
Gordon, David Cole, 405
Gore, Al, 469
Gorky, Arshile, 170
Graf Zeppelin, 145
Graham, Billy, 443
Graham, Martha, 119, 121, 180,
  412–413
Graham, Paul, 12
Graham, William, 209, 215
Graham Foundation, 367
Grand Central Palace exhibition hall,
  84, 136, 148–149, 403
Grand Central Station, 80
Graver Tank, 293–294
Great Depression, 120, 139
Great Gatsby, The (Fitzgerald), 85
"Great Lawsuit, The" (M. Fuller), 39
Great Northern, USS, 67
Greeley, Horace, 39
Greenwood, Marion, 187–188
Grip, Rob, 412, 428
Gropius, Walter, 161, 187, 235, 244,
  249, 332
Grosman, Tatyana, 402–403, 442
Growth House, 260
Grunch of Giants (Fuller), 443–444
Guaranty Trust Company, 132
Gueft, Olga, 271
"Guinea Pig B" (Fuller), 445
Gulf Refining Company, 144
Gurdjieff, George, 114, 123, 407
Gustaitis, Rasa, 389
Gutenberg Galaxy, The (McLuhan),
  321
Gwilliam, Tony, 411

Haar, Charles, 353–354
*Hackers: Heroes of the Computer Revo-
  lution* (Levy), 5
Hale, Dorothy, 149, 151, 167–168
Hall, Charles, 369
Halle, Kay, 121, 520n153

Halstead, Alexander S., 67
Halsted, Samuel, 146, 147
Hamilton, Bruce, 408, 411
Hamilton, Richard, 253
Hansen, Mark Victor, 376, 405, 437
Hardie, Jim, 33, 59, 76, 564n267
Harper, Valerie, 407, 421
Harrah, William F., 431, 520n153
Harris, Ruth Green, 163
Harris, William B., 207–208
Harrison, Richard Edes, 185
Harvard, 27, 34–35, 36, 42–46, 48–50,
    457
Harvard Graduate School of Business
    Administration, 501n76
Harvard Society for Contemporary Art,
    116–117, 121
Hasty Pudding Club, 36
Hatch, Alden
    Anne and, 111–112
    on Anne's family, 53
    bathroom project and, 163–164
    biography by, 398
    on family finances, 246
    first meeting with, 52
    on Fuller as mystic, 14
    Fuller's alleged silence and, 101
    on Fuller's character, 43
    on Fuller's diet, 333–334
    Fuller's engagement and, 61
    on Fuller's fame, 8–9
    on Fuller's marital problems, 85
    on housing project, 205–206
    misinformation and, 15–16
    on Taub, 529n187
    on Wolf, 204
Hauben, Ed, 376
Hauer, Erwin, 451–452
Haverford, 383
Hawaii Space Exploration Analog and
    Simulation, 470
Hawaiian Village Hotel, 284–285,
    284fig, 287–288
Hawes, Phil, 470
Hawthorne, Nathaniel, 38

Hazard, William A., 52, 57, 71
HCKP Club, 44
Hearst Tower, 468
Heath, Jim, 459–460, 461, 463
Heger, Frank, 337, 340
Heinlein, Robert A., 590n419
Heisenberg, Werner, 406
Held, Isidore W., 162
helicopter airlifts, 269–270, 270fig, 287
Helleu, Paul César, 80
Hench, John, 423
Henry, Frank, 152, 161
Herbert, George, 31
Herman Miller, 219
Herrera, Hayden, 121
Hertzfeld, Andy, 484n6
Hewlett, Anglesea "Anx," 52–53, 53fig,
    62, 72, 76, 432
Hewlett, Anna Willets, 53, 56–57, 71
Hewlett, Anne (later Fuller). See Fuller,
    Anne Hewlett (wife)
Hewlett, Arthur, 199
Hewlett, Carman, 86, 87, 90
Hewlett, Estelle Rodgers Wilbur, 84
Hewlett, James Monroe
    background of, 80
    connections of, 58, 70
    death of, 175
    death of Willets and, 71
    Fuller and Anne's marriage and, 57, 61
    homage to, 403
    influence of, 78–79
    patents and, 111
    Stockade Building System and, 81–84,
        85, 87
    work of, 53
Hewlett, Jim, 62, 118
Hewlett, Laurence (later Burr), 74, 75
Hewlett, Roger, 124
Hewlett, Willets, 62, 71
hexagonal structure, 114, 177
Hexa-Pent Dome, 381, 464, 465fig
Hill, Martha, 253
Hinkley, Clair, 105–106
Hiroshima bombing, 198

Hirschfeld, Al, 170
Hitchcock, Henry-Russell, 130
HIV/AIDS, 475
Ho, Soleil, 471
Ho Chi Minh, 188
Hoffman, Abbie, 361, 371
Hofheinz, Roy "Judge," 289, 311, 318
Hogan, John V. L., 191
Hogden, Lee, 267
Hollein, Hans, 412
Holy Modal Rounders, 587n409
Homage to New York (Tinguely), 306
Home of the Future, 441
home versus house, 105
Homebrew Computer Club, 5
Honda Civic, 421
Hood, Raymond, 93
Hopkins, Farley, 87, 89, 90, 92, 97,
    106
Horne, Robert, 317
Hotel Champlain, 57
Hotel Wentworth, 56
Hotel Winthrop, 131, 132–133
housing. *See also* Dymaxion House;
    Fuller Houses; Wichita House
  aircraft facilities and, 192–194,
    195–196
  DDU and, 173–176, 180
  Dymaxion Car and, 10
  Fuller's early ideas on, 29, 31
  Fuller's focus on, 9
  Institute of Design projects on,
    235–236
  mass production and, 105
  State Department and, 189
  Stockade Building System and, 81–84
  tension as support for, 106–107
Houston Astrodome, 289, 311–312,
    318, 423
"How to Make the World Work," 349,
    358
How to Work Wonders with Words,
    102
How to/Let's Make the World Work, 3,
    327, 336

Howard, Thomas C.
  Astrodome and, 312
  catenary dome and, 283
  Climatron and, 295
  dome design by, 290–291
  Ford Rotunda and, 259, 261
  Manhattan dome and, 300*fig*
  MOMA exhibition and, 306
  New York World's Fair and, 325
  octet truss and, 304
  plywood dome and, 257, 269
  Synergetics Inc. and, 280
Howe, Fuller & Trunkett, 26–27
Howe, George, 130, 131, 259
Howe & Goodwin, 35
Hubbard, Barbara Marx, 586n407,
    599n443
Hubbard, L. Ron, 170, 249–250, 384,
    407, 420, 421
Hubble Space Telescope, 466
Hughes, Howard, 440
Hughes Aircraft Company, 209
Hukar, Lee A., 246–247
Humans in Universe (Fuller), 442
Humphrey, Hubert, 340
Hunger Project, 417–418, 467
Hunter, Joe, 230
Hunter Douglas, 226, 230
Huntington dome, 275, 279
Hurlbut, Byron, 48
Huston, John, 293, 322–323, 332, 349,
    359, 387, 437
Huston, Tony, 367, 422
Huxley, Aldous, 387
Huxtable, Ada Louise, 305
hypertext (term), 484n6

I Seem to Be a Verb (Fuller), 374
"ice cream cone" clause, 138
icosahedron, 177–178, 189, 222*fig*,
    223–224, 226, 230, 244, 248, 258,
    301, 455
Ideas and Integrities (Fuller), 319, 323
Illinois Institute of Technology, 236
Imperial Hotel, 45–46

Inca, USS, 63, 63*fig*, 64, 65, 67, 76
India airport project, 390–391
Industrial Engineering Division, 183
infographics, 10
information technology, 313–314, 316
Ingber, Donald E., 451–458, 468
Inside Woody Allen, 597n435
Institute for Advanced Study in Princeton, 221
Institute for Behavioral Research, 571n355
Institute for Contemporary Arts, 365
Institute for Scientific Information, 592n422
Institute of Design, 220–221, 226–227, 233–236, 238
Institute of Ecotechnics, 470
Integrity Day, 439, 444, 446
Interiors, 252
International and Universal Exposition (Montreal Expo). *See* Montreal Expo
International College, 584n398
International Conference on the Unity of the Sciences, 404
International Design Conference, 265–266
International Film Foundation, 253–254
International Machinists Union, 193
International Merchant Marine, 75
International Style, 130
International Union of Architects, 318
International University of Art, 375
Interview, 425
Intuition (Fuller), 382
Inventions Board of the British Purchasing Commission, 174
Inventions: The Patented Works of R. Buckminster Fuller (Fuller), 445
Inventions: Twelve Around One, 600n445
isotropic vector matrix, 256, 366, 445, 456–457

J. P. Morgan Co., 74–75
Jackson, Jesse, 446

Jacobs, David, 340
Jacobs, Jane, 330, 379
James, William, 41
Japanese Pavilion, World's Columbian Exposition, 25
Jawaharlal Nehru Memorial Lecture, 365–366
Jeanneret, Charles-Édouard. *See* Le Corbusier
Jennings, Fletcher, 286
Jennings, Lois, 356
Jeshyn International Fair, 280–283
Jiménez, José Cha Cha, 329
jitterbug sculptures, 409–410
jitterbug transformation, 222*fig*, 223, 228, 232, 233, 409, 456
Jobs, Steve, 1–4, 8, 475
Johansen, John M., 326
Johansson, Karl, 540n22
Johns, Jasper, 343
Johnson, Al, 315
Johnson, Lyndon B., 326, 340, 354, 371
Johnson, Philip, 13, 116–117, 130, 131, 133, 229, 274, 326
Johnson Wax Research Tower, 299
Joint Computer Conference, 357–358
Jones, Dan, 490n29
Jones, David E. H., 463
Jordan, June, 329–330
Joyce, James, 156, 397
Jumsai, Sumet, 403

Kahlo, Frida, 168
Kahn, Albert, 258
Kahn, Ely Jacques, 509n117
Kahn, Lloyd, 10, 367–369, 378–379, 410, 472
Kahn, Louis I., 188, 260, 392, 398
Kahn, Robert D., 388
Kaiser, Henry J., 182–183, 187, 284–285, 288, 291, 440
Kaiser, Neva, 382, 411, 415–416, 441–442
Kaiser Aluminum, 283–284

Kaiser domes, 550–551n285
Kaliczak, Janek, 7
Kanter, Jonathan, 369, 378
Kay, Alan, 6
Keck, George Fred, 116, 152
Kelly-Springfield Motor Truck Company, 72–73, 74–75
Kenison, Alphonse, 214
Kennedy, Caroline, 437
Kennedy, Margaret, 128
Kenner, Hugh, 157, 389–390, 391, 393, 467, 492n35
Kent State shootings, 372
Kepes, György, 452
Kesey, Ken, 351
Khrushchev, Nikita, 298–299, 301, 303, 336
Kibbutz Kfar Menachem project, 574n364
Kiesler, Frederick, 129
Kiewitt, Gustel, 318
King, Alexander, 170
King, Andy, 28, 30–31, 86, 89, 91
King, George J., 276
King, MacGregor, 28, 30–31
King, Martin Luther, Jr., 433
King, Rockwell, 27, 29, 30
Kinzie, John, 492n35
Kirkwood conference, 436–437
Kirschenbaum, Bernie, 273, 304
Kirstein, Lincoln, 116
"Kitchen Debate," 301
Kitrick, Chris, 428
Kitty Foyle (Morley), 173
Kiyosaki, Robert, 437
Klaus, Dale, 329
Kleenex House, 271
Klopp, Kenneth "Hap," 408
Klug, Aaron, 301, 317, 461
Klutznick, Philip, 415
Knoxville World's Fair, 441
Koike, Hideo, 315
Korzybski, Alfred, 116, 242, 296
Kottke, Daniel, 2, 3
Krehbiel Plastic Products, 196

Kronprinzessin Cecilie, 48
Kroto, Harold W., 458–466
Kroto, Margaret, 458
Kubrick, Stanley, 419
Kuhns, William, 475
Kuromiya, Kiyoshi, 433–434, 445, 448, 475
Kwinter, Sanford, 488n16
Kyklops II (Olson), 243

La Guardia, Fiorello H., 74
Laban, Rudolf, 231
labor, students for, 10
Lacey, Cynthia (later Floyd)
    Anne and, 194–195, 204–205
    chemical elements chart and, 212fig
    contributions from, 191–192
    dwelling machine and, 198–199
    Eames and, 220
    Einstein letter and, 221
    end of relationship with, 215–216
    on finances, 212
    geodesic dome and, 226
    later interactions with, 183–184, 398, 432, 441
    "Motion Economics" and, 186
    Mount Kisco project and, 213
    Pentagon meeting and, 236
    photographs of, 199fig, 216fig
    relationship with, 195, 196, 208–211
    on Snelson, 233
Lacey, Ginnie, 208
Ladle, 523n163
Lady Anne, 117, 123
lag, concept of, 159
Lake View Hotel, 102
laminar dome, 313
Lane, Henry, 269–270, 271, 275–276
Lane, Jeff, 516n143
Lanier, Albert, 231, 232, 237
Lanier, Jaron, 6, 471
Lapham, Lewis H., 532n198
Larned, Edward, 43, 44, 83, 84, 92
Larned, William, 83, 92
Larson, Ted, 129, 209

laser supersonic cluster beam apparatus, 459–460
laundry, 110
Le Corbusier, 23, 104–105, 107, 133, 137, 138*fig*, 237
Le Petit Gourmet restaurant, 114
LeBrun, Marc, 4
Lehman, Herbert H., 189
Lehr, Mira, 363
Lehr, Richard A., 283, 294
Lennon, John, 359
Leonardo da Vinci, 80, 177, 255
LeSourd, Homer, 34
Lessiovski, Victor, 335
Let's Make the World Work. *See* How to/Let's Make the World Work
Levinson, Leon, 134
Levinson, Max, 130, 139, 148
Levy, Steven, 5
LeWitt, Sol, 334
Lewontin, Richard, 289
Lieber, Lillian Rosanoff, 185
Life, 171, 176, 181, 185, 263, 286–287
"Lightful Houses" (Fuller), 107–108
Lilly, John, 410
Lim Chong Keat, 409
Lincoln, Abraham, 103
Lincoln Laboratory, 260–261, 304
Lindbergh, Charles, 93
Lindsay, Jeffrey
  antenna project and, 261
  Astrodome and, 312
  Black Mountain College and, 238, 240
  break with, 257, 263–264, 273–274
  Canadian pavilion and, 282
  Institute of Design and, 235
  Montreal branch of business and, 246–248, 252
  tents and, 408
  Weatherbreak Dome and, 251, 255
Lipkin, Efrem, 5
Lippincott & Co., J. B., 161
Lippold, Richard, 229, 232
Little Dipper, 152, 155
Little Spruce Head Island, Maine, 27

Liu, Yuan, 460
Lloyd, Robert McAllister, 83
Locke, Edwin, 283, 285, 294
Locomobile Company of America, 141, 143
Loeb, Arthur L., 392–393, 397, 436
Lönberg-Holm, Knud, 127, 129, 131, 209, 290
Longfellow, Alexander Wadsworth, Jr., 25, 30
Lonsdale-Bryans, James, 169
Lorance, Loretta, 15
Lord, Athena, 16
Loring, Lindsley, 46, 48, 50
Los Angeles Bicentennial, 432–433, 432*fig*
Luce, Clare Boothe (née Brokaw). *See* Brokaw, Clare Boothe (later Luce)
Luce, Henry, 167, 168, 169, 172, 173, 185–186, 322
Ludin, Mohammed Kabir, 282
Lupu, Aurel, 147
Lusitania, 52

MacBride, Robert, 425
"Machine Tools" (Fuller), 173
Mack, Benjamin, 596n432
Mack Trucks company, 74
Mackinder, Halford, 282
MacLeish, Archibald, 133, 443
Macmillan, 393, 401–402
MacNaughton, Leslie M., 65, 76, 398–399
Macneil, Marie and Robert, 174
Madarasz, George, 7
Mailer, Norman, 577n374
Malisoff, William Marias, 166
Malthus, Thomas, 29, 315
Man, Alrick, H., Jr., 85, 203
man, Fuller's definition of, 164
Mango, Roberto, 271
Manhattan, dome for, 300, 300*fig*, 306
Manhattan Project, 191, 248
Manock, Jerry, 5–6

Marchand, Marie "Romany Marie," 70, 77, 118–119, 120, 123, 129
Marcos, Ferdinand, 413–414
Marcos, Imelda, 413–414
Mardas, Yannis "Magic Alex," 572n359
Marimekko textiles, 348
marine shelter program, 269–271, 275–276
Markkula, Mike, 2
Marks, Robert W., 134, 140, 304, 311, 391, 461
Marquand, John P., 43, 110
Mars, Roman, 469
Mars Science City, 470
Marshall, Herbert, 364
Marshall Field's, 115–116
Martinez, Ysidore, 247
Marvel, Lucilla Fuller, 295
Marvel, Tom, 295
Marx, Karl, 475
Masey, Jack, 281–282, 283, 298, 326–327
Mason, Dixie, 60–61
Massachusetts Constitutional Convention, 36
Massachusetts Volunteer Militia, 48
Matrix Structures, 286
Matter, Herbert, 359, 361
Matter, Mercedes, 359, 362
Mayne, Thom, 469
McCamy, James L., 180, 192, 193
McCartney, Paul, 359
McDonnell, Edward, 63, 72–73, 75
McDonough, William, 469
McHale, John, 274, 293, 301, 320, 322, 328, 335, 358
McHale, Magda, 358
McIntosh, Arthur T., 87
McKibben, Bill, 469
McKim, Mead & White, 80
McKim, Robert, 5, 338
McLuhan, Marshall, 321–322, 328, 352, 374, 387, 389, 404, 485n6
McNamara, Robert, 366

Mead, Margaret, 321, 382, 424
Mechanical Wing, 173, 235
Medium Is the Massage, The (McLuhan), 374
Meeker, Arthur, 52, 70–71
Meeks, Everett V., 122
Mehta, Praveena, 540n238
Meller, Gillian, 559n316
Meller, James, 364, 370, 559n316
Melnikov, Konstantin, 299
Menotti, Gian Carlo, 348
Mensa, 398
Mercury space capsule, 320
Merrimac Chemical Company, 46
Merry Pranksters, 351
Metropolitan Life Insurance Company, 125
Meyer, Hans, 381, 389
Meyerson, Martin, 383, 422, 427
Mezes, John, Sr., 520n153
Mezes, Theodore E., 520n153
Michigan lecture, 253–254
Mid-America Jubilee, 280
Mies van der Rohe, Ludwig, 234, 244, 305, 332
Milan Triennale, 271–272, 272fig, 283
Miller, Alvin, 286, 309
Miller, Christian William, 212–213
Miller, Donald C., 203, 205
Miller, Henry, 577n374
Miller, Marilyn, 44
Miller, Osborn Maitland, 185
Millett, Kate, 387
Milton Academy, 25, 31–32, 34
Milunić, Vlado, 469
Minnesota Experimental City, 340, 379
Minni-Earth, 272, 280
Mintz, Ann, 443
Miss Expanding Universe (Noguchi), 133
Missouri Botanical Garden, 295
Mister God, This Is Anna, 447
Mitchell, George P., 469–470
mobile inventory control, 200
Moehlman, John, 249

Moholy-Nagy, László, 163, 220, 244, 540n22
monohex dome, 317 425
Monsanto Chemical Company, 313
Montreal Biosphere, 410, 471. *See also* Montreal Expo Dome
Montreal Expo, 326, 336, 341, 433
Montreal Expo Dome
  Coxeter and, 394
  design for, 326–328, 336–337
  destruction of, 410
  Epcot and, 423
  Ingber and, 452
  Kroto and, 458
  photograph of, 341*fig*
  reception of, 15, 340–343, 347–348, 358
  Sadao and, 353
"moon house," 320
moon landing, 363
Moore, Don, 316, 349, 570–571n353, 589n415, 595n431
Morgan, Jasper, 161, 164
Morgan, Priscilla, 348, 375
Morley, Christopher
  amphibious plane and, 213
  bathroom project and, 161, 163
  "To a Child" and, 79, 151
  connections of, 174
  death of, 275
  geodesic dome and, 226
  introduction to, 151
  manifesto mailed to, 111
  in Midwest, 173
  The Proud Shirtfront and, 496n53
  relationship with, 156
Morley, Frank, 151
Morris, Delyte, 303–304, 372, 375
Morris, Si, 289
Morton, James Park, 419
Moscow Dome. *See* American National Exhibition (Moscow)
Moscow Palace of Labor, 106
Moses, Robert, 279, 325
Mother Jones, 418

Mother of All Demos, 358
Motherwell, Robert, 402
"Motion Economics" (Fuller and Lacey), 186–187
Mottel, Syeus, 373, 384
Mount Auburn Cemetery, 448
Mount Kisco project, 213
Muller, Franklyn R., 102
Mumford, Lewis, 107, 117, 339–340
Museum of Modern Art (MOMA)
  "Airways to Peace" exhibition at, 213
  Allegra and, 214
  Bubble House and, 259
  Chermayeff and, 220
  contemporary architecture show and, 130
  DDU and, 175–177
  Drexler and, 251
  Dymaxion Map and, 188–189
  "outside-in" globe from, 227
  thirtieth anniversary exhibition and, 304–306
"Music of a New Life, The" (Fuller), 571n355
Musk, Elon, 471, 474
Muskie, Edmund, 359, 381
Myers, Forrest, 334
Myrow, Rachel, 596n432

Naga story, 332, 349–350
Nakagawa, Masato, 238, 239, 239*fig*
nanotechnology, 466
Napier, Susan, 73–74
Narconon, 384
Nation, The, 125, 126, 282, 286–287
National Automobile Chamber of Commerce, 135–136
National Housing Agency, 209
National Institute of Arts and Letters, 355
National Lampoon, 386–387
Nature, 463
Nautilus Motor Inn and Dome Restaurant, 267

Necklace Dome, 235–237, 239–241,
    239*fig*, 248
Nehru, Jawaharlal, 291–292, 324,
    597n434
Nelson, George
    American National Exhibition (Mos-
        cow) and, 298
    Black Mountain College and, 230
    dome design and, 226
    Fortune article by, 202
    Fuller Research Foundation and, 211
    Fuller's engagement by, 252, 259
    Fuller's resignation and, 207
    Fuller's work with, 219–220, 221–222
    "storage wall" system and, 532n202
    "waterless" toilet and, 595n431
Nelson, Paul, 112
Nelson, Ted, 484n6
neo-futurism, 468
Nervi, Pier Luigi, 316
Nesmith, Jonathan, 596n432
Nesmith, Michael, 596n432
Neumann, Adam, 471
Neutra, Richard, 126, 127, 161
New Alchemy Institute, 441
New Dimensions, 408
"New Forms vs. Reforms" (Fuller), 322
New Learning Environment, A (Fuller),
    571n355
New Left, 357
New Worlds in Engineering (Fuller),
    172–173
New York Automobile Show, 148–149
New York Public Library, 188
New York School of Applied Design for
    Women, 53
New York Studio School, 359, 361–364
New York Times articles, 304–305, 376
New York World's Fair, 167, 170, 325
New Yorker profile, 17
New Yorker, The, 167, 332
Newman, Barnett, 402
Newman, Deirdre, 369–370n349
Newton, Isaac, 393
New-York Daily Tribune, 39

Nine Chains to the Moon (Fuller), 162,
    163, 164–167, 169, 197, 228, 263,
    321, 435
Nixon, Richard, 301, 354, 475
"No More Secondhand God" (Fuller),
    172
No More Secondhand God (Fuller),
    319, 321
Noguchi, Isamu
    background of, 119–121
    bathroom project and, 162–163
    Clow and, 154–155
    Cunningham and, 229
    dwelling machine and, 196–197
    Fuller's death and, 448
    Fuller's friendship with, 121–122
    Fuller's legacy and, 468
    Greenwood and, 187
    Hale and, 149–150, 168
    Huston and, 293
    in London, 141
    MOMA exhibition and, 304
    in New York, 131, 132, 133
    Pound and, 375–376, 380
    reconciliation with, 158–159, 167
    Romany Marie's and, 118
    Sadao and, 404, 428
    support from, 338
    "tension room" and, 509n117
    theater project and, 412–413
    at United States National Air Show,
        276
    Wright and, 135
Noguchi, Yone, 119
Nordenson, Guy, 412
Norfolk Country Club, 64
Norquist, E. E., 173, 174, 186
North American Aviation, 291
North Carolina State College, 236,
    251–252, 256, 260, 274
North Face, 408, 411
Nourmahal, 76
Now House, 411
Nowlan, Phil, 121
Noyes, David, 210

Obnoxico, 265
O'Brien, Sean, 459–460
Ocasio-Cortez, Alexandria, 474
octahedron, 23, 23*fig*, 222*fig*, 223–224
octet truss, 256–257, 261, 279, 304, 305, 470
Oelrich, Herman, 48, 76
Office of Economic Programs, 193
Office of Economic Warfare, 192
Office of Foreign Relief and Rehabilitation Operations, 189
Office of War Information, 183
Okress, Ernest C., 428
Old Man River's City, 379–380, 414
Olds, Glenn, 382, 415–416
Olson, Charles, 243
Olson, Ken, 286
O'Malley, Walter, 278–280, 311, 312
"Omni-Directional Halo" (Fuller), 296
On Growth and Form (Thompson), 177
One Flew Over the Cuckoo's Nest (Kesey), 351
One Ocean World Town Plan, 113, 113*fig*
O'Neill, Eugene, 70
Ono, Yoko, 359
operant conditioning, 571n355
Operating Manual for Spaceship Earth (Fuller), 350, 593n426
Order of the Knights of St. John of Jerusalem, 427
Ortega Park Teachers Laboratory, 356
O'Ryan, John F., 149
Osawa, Eiji, 463
Osborn, Peggy, 157
Osborn, William H., 157, 160, 164
Ossoli, Giovanni Angelo, 39
Ossoli, Nino, 39
Ouroussoff, Nicolai, 468
Ouspensky, P. D., 507n111
"outside-in" globe, 189, 227
Oval Intention, 408
Own Your Own Home Show, 91
Oxfam International, 418
Oxford University, 370, 371–372

Pacific High School, 368–369, 378
Packer Collegiate Institute, 53
Paepcke, Walter, 266, 271
Page, Ruth, 133
Pahlavi, Farah, 375
Pahlavi-Nia, Shahram, 419–420
Papanek, Victor, 348
Papen, Franz von, 60
paperboard dome, 270–272, 272*fig*, 281, 286
Paradise of Childhood, The, 23*fig*
Paris Peace Conference, 68–69
Park, David, 527–528n179
Park, Edwin, 179–180
Parker, Grace, 23
Parkhurst, William M., 152–153, 205, 209
Passing Show of 1913, The, 44–45
Patterson, James, 439
Pawley, Martin, 215
Paxton, Joseph, 568n342
Peabody, Elizabeth, 38
Pearce, Peter, 367, 470
Pearl Harbor, 176
Pearson, Philip C., 147, 149, 154
"peas work," 23, 23*fig*
Pease Woodworking Company, 295, 309
Penn, Arthur, 229, 230–231, 232–233, 316, 348, 449
People's Computer Company, 4
Perk, Bill, 328
Perky, Scott H., 83
Peterson, Gunnar, 266–267
Phelps Dodge, 157, 159, 160*fig*, 163, 528n179
Philip, Prince, 411
Phillips, Kenneth, 52–53, 54
Piano, Renzo, 469
Pidgeon, Monica, 315
Piel, Gerard, 182, 187
Pierce, John B., 126
Pierce, Lincoln, 25, 31, 42, 58, 59, 61–62, 68, 87, 398
Pierce, Vassar, 25, 32

Pierce Foundation, 126, 127, 131, 133, 160, 207
Pilgrim at Tinker Creek (Dillard), 399
Pillow Dome, 369
Piñero, Emilio Pérez, 567n338
Plattsburg military camp, 57–58
Playboy, 376–377
playdome, 286
Plumbe, John, 40*fig*
plydomes, 286
Plydomes Inc., 286
Poe, Edgar Allan, 173
polar air routes, 113, 176
polar warfare plan, 187, 192
poles, geography of, 119
Police Reserve Aviation Division, 74, 76
polio virus, 301, 317
Pope, Ted, 246
Popular Science, 355, 381
Portola Institute, 4, 352, 356
Post, Marion, 175*fig*
Pound, Ezra, 233, 348, 375, 380, 389, 397
Poznań International Trade Fair, 286–287
Prance, Ghillean, 470
precession, principle of, 172
Preparedness Movement, 57
Presidential Medal of Freedom, 443
Price, Cedric, 559n316, 560n318
Price, Edison, 304, 306
Price Tower, 299
Princeton University, 267–268, 268*fig*, 279, 308
Principia Mathematica (Newton), 393
Project One, 4–5
Proud Shirtfront, The (Morley), 496n53
Pullman railway coach, 127
Pushcart Prize, 411

Queen of the Sea, The, 62
Quintet, 580n387

Rabbit, Peter, 361
race/racism, 371, 390

Racquet and Tennis Club, 151
radio, 64, 69, 115
radome, 261, 273, 275, 277*fig*, 278, 279, 295, 304, 305–306, 387
Ragan, Philip, 172
Rahman, Habib, 390–391
Randolph Hall, 42, 45–46
Rasch, Bodo and Heinz, 106
Ratia, Armi, 348
Rauschenberg, Robert, 233
Raymond, Dick, 352, 356
Reagan, Ronald, 443–444
Real Great Society, 373
Reconstruction Finance Corporation, 192
Reeves, Alfred, 135–136
Reeves, Emma Lucilla, 36
Reich, Charles A., 387
Reichstag dome, 468
Reid, Ogden, 153
remote learning, 313
Rendleman, John, 382
Reuther, Walter, 192, 193
Rice, John Andrew, 227
Richards, Atherton, 187
Richards, Mary Caroline, 232
Richert, Clark, 334
Richter, Donald
  alternate grid and, 261
  American National Exhibition (Moscow) and, 301
  aviation and, 454
  Black Mountain College and, 238
  commissions and, 440–441
  Epcot and, 599n441
  Fuller Research Foundation and, 246, 250
  Hawaiian Village Hotel and, 284–285
  Institute of Design and, 235
  at Kaiser Aluminum, 283–284
  pragmatism of, 248
  prototypes and, 252
  South Pole dome and, 387
  Temcor and, 355
Rivera, Diego, 129–130, 143, 158

Rizzo, Frank, 433
Roadside Home, 503n88
Robbins, Tony, 437
Robertson, Donald W., 254–255, 285,
    286
Robeson, Paul, 119
Rock Hall, 54, 62, 253
Rockefeller, David, 325
Rockefeller, Nelson, 281
Roerich, Nicholas, 123
"Romany Marie." See Marchand, Marie
Roosevelt, Eleanor, 153, 167
Roosevelt, Franklin D., 134, 139, 185
Roosevelt, Theodore, 57
Root, Elihu, Jr., 140
Rosenberg, Jack. See Erhard, Werner
Roszak, Theodore, 6, 337, 338
Roth, Meyer, 145–146, 147–148
"rowing needles," 349
Royal Gold Medal for Architecture, 446
Royal Institute of British Architects,
    293, 355
Rubin, Jerry, 361
Rudge, Olga, 375–376
Rudolph, Paul, 326
Rumsfeld, Donald, 446
Rumwell, Miriam, 323, 324
Ruse of Medusa, The (Satie), 232–233,
    243
Russell, Bertrand, 111
Russell, George William, 156
Ryan, Hugh, 535n213

Saarinen, Eero, 258
Saarinen, Eliel, 103
Sachs, Charles, 427
Sadao, Shoji
    aspension dome and, 312
    background of, 257
    "Cloud Nine" structures and, 294fig
    commissions and, 364
    Cooper Hewitt Museum and, 412–413
    Festival of Two Worlds and, 348
    fire caused by, 301
    Ford Rotunda and, 263, 264

    in Fuller's classes, 257–258
    Fuller's legacy and, 468
    General Electric and, 325
    Hexa-Pent Dome and, 381, 464,
        465fig
    India airport project and, 390–391
    India commissions and, 366
    International Design Conference and,
        265–266
    jitterbug sculptures and, 409
    Jordan and, 330
    Milan Triennale and, 272
    Montreal Expo and, 336–337, 340,
        343, 353
    Morton and, 419
    Noguchi and, 404, 428
    partnership with, 326
    tetrahedral mast and, 304, 306
    Toronto revitalization project and, 355
    Tower of World Man and, 339
    Triton Foundation and, 354
Safford, Frank, 556n304
Sagan, Carl, 418
Saint-Gaudens, Augustus, 103
Salemme, Antonio, 70, 118, 119
Salemme, Betty, 70
Salk, Jonas, 274–275, 382, 404, 435
San Jose State, 337–338
San Quentin State Prison, 296–297
Sandburg, Carl, 301
Sanders, Walter, 173, 183
Sarabhai, Vikram, 324
Sargent, Ralph, 173
Satie, Erik, 232
Saylor, Henry, 117
Schaeffer, Ray, 115
Scheherazade numbers, 583n397
Schlossberg, Edwin, 15, 359–364,
    366–367, 373, 383, 402–403, 418,
    437
Schundler, Fritz E., 87
Schuster, Max, 169
Schwartz, Evelyn, 123–125, 127–129,
    161, 170, 398, 441
Schwartz, Stephen, 386

Schwarzenegger, Arnold, 418
Science Center, 422
Scientology, 384, 420, 421
Scott, Howard, 129
Scott, Ralph, 4
Sea Bird, 28–29
Sears, Roebuck, 174
Seattle World's Fair, 314
Secret Life of Walter Mitty, The, 215
Seldes, Gilbert, 169
Seldes, Marian, 169
Selkowitz, Stephen E., 363
Senge, Peter, 411
Serebriakoff, Victor, 584n398
Shah, Mohammed Zahir, 282
Shank, W. Bradford, 249
Shapley, Harlow, 249
Sharaff, Irene, 535n215
Sharkey, Shirley (née Swansen). *See*
    Swansen, Shirley (later Sharkey)
Shelter, 130–132, 133–134, 135,
    156
Shelter (Kahn), 378–379
Shire, Albert C., 183
Shirley, Jim, 408
Shōriki, Matsutarō, 312–313, 317, 339,
    353
Shrubb, Al, 42–43
Shubert Theatre, 44
Sidis, William James, 42
Sieden, Lloyd Steven, 16, 486n11
Silent Running, 387
Silhaus, Athelstan, 340
Simpson Gumpertz & Heger, 441
Sims, William, 76
Simulmatics Corporation, 556n304
Singh, Karan, 366, 390
Skinner, B. F., 571n355
Sklar, Martin A., 423
Sky Eye, 271, 273
Skybreak Carolina Corporation, 257
Skybreak House, 237
Slaughterhouse-Five (Vonnegut), 387
sleep experiments, 112, 188, 231
Smalley, Richard, 459–466

Smith, Cam, 384, 420
Smith, Charles Kingsford, 153
Smith, Cyril Stanley, 248
Smithson, Robert, 334
Snelson, Kenneth
    Allegra and, 253
    artwork of, 410
    Black Mountain College and, 228,
        232, 233
    credit for, 401
    Early X Piece by, 240*fig*, 241–242
    on Fuller, 307
    lack of credit and, 244, 246
    MOMA exhibition and, 305–306
    patent infringement and, 429–430
    relationship with, 237–238
Snow, C. P., 319–320
Snyder, Alexandra (granddaughter),
    274, 407, 412, 416, 425, 436, 448,
    449, 468
Snyder, Jaime (grandson)
    Alexandra and, 436
    Anne and, 443, 445, 446–447
    birth of, 274
    Buckminster Fuller Institute and, 468
    Design Science Institute and, 415–416
    Erhard and, 406–407, 411, 424
    events organized by, 410
    Fuller's death and, 447, 468
    Fuller's health and, 438–439
    I. Marcos and, 414
    Presidential Medal of Freedom and, 445
    relationship with, 407–408
    in Santa Monica, 430
    Stollers and, 406
    World Game and, 440
Snyder, Robert, 253–254, 274, 313,
    348, 407, 426, 433, 577n374
Society Hill Towers, 383
Solari, Tom, 384
Solway, Carl, 409–410, 429, 600n445
Somerset Club, 27, 93
Sonnenberg, Henry, 226, 230
Sontag, Susan, 37, 101, 328
Soundex, 506n106

South Carolina Tricentennial Commission, 353
South Dakota, USS, 67
Southern Illinois University
  book projects and, 319
  break with, 380, 381–382
  protests and, 372, 375
  Sadao and, 364
  Schlossberg and, 360
  teaching at, 283, 302, 303–304, 308–311
  World Resources Center and, 349
Spaceship Earth concept, 254, 356, 389, 400
Spaceship Earth (ride), 9, 423, 441, 599n441
Spaceship Earth (Ward), 356
Spaghetti (dog), 65
Spanish flu epidemic, 68
Sparks, William A., 62
Speakers Unlimited, 418
spectroscope, 458–459
Sperry Corporation, 172, 236
spheres, theory of, 108
Spheres Inc., 238
Spitz Laboratories, 288
Sprang, Charlene, 384
Spruce Goose, 440–441
Sputnik 1, 302–303
St. Chrysostom's Episcopal Church, 504n95
St. Nicholas, 31
Stampfel, Pete, 587n409
Stankiewicz, Richard, 312
Starrett-Lehigh Building, 128, 129–130, 131
State University of New York, 422
Stearns, Gilbert "Gee," 155
Stefansson, Vilhjalmur, 119, 125, 136, 143, 145–148, 151, 170, 333
Steichen, Edward, 301
Stein, Forbes & Co., 27
Stein, Gertrude, 397
Steinbeck, John, 338
Sterling, Bruce, 466

Stevenson, Adlai, 300
Stockade Blue Ridge Corporation, 84
Stockade Building System, 81–86, 82*fig*, 105–106
Stockade Midwest, 85–87, 89–90, 94–95
Stokowska, Evangeline, 149, 151–152, 158
Stokowski, Leopold, 136, 149
Stoller, Jonathan, 406, 412, 427
Stoller, Roger, 406, 412, 431, 436
Stone, Edward Durrell, 203
Stone, Oliver, 468
Storyk, John, 363
Stout, William B., 144, 160
Straus, Ernst, 221, 226
"Streamlining" (Fuller), 133
Structural Study Associates (SSA), 129, 131–132
Structure in Art and in Science (ed. Kepes), 452
Stuart, Duncan, 251–252, 255, 257, 280, 367
Submarisle, 299
substitutions, 184
Sullivan, Johanna, 26, 30, 33
Sun Dome, 355, 368
Sun Myung Moon, 404
Sunclipse Dome, 316
Super Ball Bot, 607n470
"supine dome," 230*fig*, 231
Suzuki, Sotaro, 312
Swansen, Shirley (later Sharkey), 382, 383–384, 388, 398, 422, 445–446
Swarthmore, 383, 422
Swedenborg, Emanuel, 41–42, 98
Swift, Narcissa, 163
Swiss Re Building (Gherkin), 468
Synanon, 405–406
synergetics, 296. *See also* energetic geometry
Synergetics (Fuller), 367, 391–397, 394*fig*, 398–399, 400–402, 408, 421, 433, 454
Synergetics 2 (Fuller), 428

Synergetics Folio, 587n409
Synergetics Inc., 278–280, 282–284,
    290, 294–295, 297, 306, 311–312,
    318, 353
synergy, 5, 11, 236, 395
systems theory, 296
Szilard, Leo, 542n248
Szpakowski, Mark, 5

Taliesin, 122
Taliesin Fellowship, 135
Taliesin West, 288, 299
Taub, Alex, 182, 184, 186, 187, 192
Taylor, Alfred J. T., 143, 145, 147, 148,
    151
Taylor, Elizabeth, 288, 322–323
technocracy movement, 129
television, 169, 313
Temcor, 355, 387, 426, 599n441
tensegrity. *See also* discontinuous com-
    pression, continuous tension
    asymmetrical, 377
    Carbondale project and, 327–328
    elongated, 424
    floating sphere and, 428
    Gandhi and, 291
    Ghana project and, 324
    Ingber and, 452–457
    Noguchi and, 510n120
    potential of, 295–296
    principle of, 12, 241, 280
    robotics and, 470
    scale and, 300, 305
    Snelson and, 306–307, 401, 429–430
    Synanon and, 405
    Tokyo projects and, 312, 316
tensegrity mast, 304, 305
tension, housing and, 106–107
"tension room," 509n117
tents, 408–409, 411
Terkel, Studs, 329
tetrahedron, 23, 23*fig*, 190, 223, 232,
    241, 296, 391, 395–397, 442
Tetrahedron City, 339–340, 342,
    353–354

Tetrahelix Inc., 340
tetrahelix memorial, 468
Tetrascroll, 403
Tetron, 353
Themes and Variations (Cage), 432
"Think Different" campaign, 7*fig*, 8
Thompson, D'Arcy Wentworth, 177
Thompson, Derek, 13
Thompson, Francis, 313
Thoreau, Henry David, 39
Thorsteinn, Einer, 469
"Three Structures by Buckminster
    Fuller" (exhibition), 305
Thurber, Marshall, 437
Tilton, Ellen, 161
Time, 132, 249, 323–324
"time binding," 116
Tinguely, Jean, 306
Titan, The, 253
Titanic, 76
"To a Child" (Morley), 79, 151
Todd, Mike, 287–288, 323
Toffler, Alvin, 384
Tomkins, Calvin, 16–17, 300, 331–333,
    349–350, 443, 445–446, 487n14
Toomer, Jean, 114, 119
Toronto revitalization project, 354–355
"Total Thinking" (Fuller), 242–243
Towards a New Architecture (Le Cor-
    busier), 104–105, 137, 138*fig*
Tower of World Man, 339
Toynbee, Arnold J., 331
Tracey, John, 534n210
transatlantic communications, 69
Transcendental Meditation, 380–381
transcendentalism movement, 39
Tremaine, Burton, 534n210
trend curves, 10
triacon grid, 252
trial balance, 54
triangles, 395
Tribune Tower, 103
trim tab, 322, 377, 467
Trips Festival, 351
Triton City, 354, 354*fig*

Triton Foundation, 354
Trudeau, Margaret, 411
Trudeau, Pierre, 375, 411
truncated icosahedron, 462
T-Square, 130
Tulane University, 270–271
tumor formation, 453
Turner, Francis T., 145–147
Turner, Tom, 372, 381
Turtle Dome, 426
Twain, Mark, 55
"Two Cultures, The" (Snow), 319
Two on the Town, 443
2001: A Space Odyssey, 419
Tyng, Anne, 260, 392, 398

U Thant, 335, 391
UIA Congress, 322, 330–331, 348
Ulysses (Joyce), 397
"Underlying Order in Randomness,"
    396*fig*
Unification Church, 404
Union Tank Car Company, 283, 285,
    289–290, 290*fig*, 295, 461
United Automobile Workers, 192
United Farm Workers of America, 405
United Nations, 215
United Nations Conference on Human
    Settlements, 411
United States Housing Corporation,
    91
United States Information Agency
    (USIA), 281, 283, 288, 326
United States National Air Show, 276
United States Student Press Association,
    361
"Universal Architecture" (Fuller), 130
universal basic income, 13, 363, 474
Universal Limited Art Editions, 402
"Universal Requirements for a Dwelling
    Advantage" (Fuller), 245
University City Science Center, 383,
    388
University of Houston, 289
University of Michigan, 260, 261, 274

University of Michigan School of Archi-
    tecture and Design, 254
University of Minnesota dome, 266,
    266*fig*, 269
University of Oregon, 296
University of Pennsylvania, 383, 422
University of Rochester, 427
university system, Fuller's use of, 10. *See
    also individual institutions*
"Untitled Epic Poem on the History of
    Industrialization" (Fuller), 172
*Untitled Epic Poem on the History of
    Industrialization* (Fuller), 320
Urey, Harold, 542n248
Ursa Minor, 29
US Army Signal Corps, 180
US Department of Housing and Urban
    Development, 354
US Festival, 7
"US Industrialization," 171
US National Security Agency, 278
US Naval Academy, 64, 66–67
US Navy Bureau of Aeronautics,
    255–256, 269, 270
US Office of Scientific Research and
    Development, 191
US Office of Strategic Services (OSS),
    187–188
US War Production Board, 193
Utopia or Oblivion (Fuller), 331

Van Alen, William, 122, 124–125
Van Judah, Nelson, 338
VanDerBeek, Stan, 334
Vanderbilt, William H., 188
Vanderlip, Frank A., 126
Vanity Fair, 134–135
Vargas, Getúlio, 188
Varieties of Religious Experience, The
    (James), 41
Vashi, Nataraj, 540n238
vector equilibrium, 178, 179*fig*, 190,
    222–225, 222*fig*, 223*fig*, 224*fig*
Veech, Bill, 462
Vesnin, Leonid and Viktor, 106

Vietnam war, 370–371
Vinetz, Tom, 414, 433
Virginia Hotel, 93, 102
Visual Presentation Unit, 187
Vogue, 176
Vonnegut, Kurt, 363, 387

Wachsmann, Konrad, 234–235
Wainwright, Bill, 261, 273, 278, 297,
    326
Walcott, Russell, 104, 108
Walker, Hugh McLean, 63
Walker, John, 116
Wallace, David, 441
Wallace, Henry A., 183, 194
Wallace, Naomi, 328, 360
Walley, John and Jano, 540n238
Wallis, Barnes, 508n112
Walt Disney Productions, 298. *See also*
    Disney World
Warburg, Edward, 116
Ward, Barbara, 321, 356
Warhol, Andy, 118, 343, 425
Warner, Francis, 364, 365, 371–372,
    375, 404
Warren, John, 426
Warren, Waldo, 115
Wasserman, William, 194, 195–196,
    198, 595n431
Waterhouse Company of Massachusetts,
    154
Weatherbreak Dome, 250*fig,* 251, 255,
    304
Webster, Clifford L., 65, 76
Wego, 58–60, 62–63
Weisenborn, Rudolph, 115
Welch, George, 265
Wellington, Persis, 149
Wells, H. G., 151
Wenner, Jann, 400
Wessels, Tim, 388, 428
Western Cold Storage Company, 27
Western Electric Company, 273,
    277–278
"What I Am Trying to Do" (Fuller), 355

Wheeler, Harvey, 374
Wheeler, Monroe, 189, 212–213
Whirl of the World, The, 45
Whirlpool Corporation, 316
Whistler, USS, 63
Whitall, Thomas, 42, 62
White, Pat, 359
white supremacism, 169
Whitehead, Alfred North, 296
Whitney, Gertrude Vanderbilt, 170
Whitney, Gloria, 170
Whitney Museum of American Art, 468
Whole Earth Catalog, 3–4, 356–357,
    368, 376, 400
"Why Women Will Run the World"
    (Fuller), 586n407
Wichita House, 194–198, 199*fig,*
    201*fig,* 202–203, 214–215, 225,
    237, 249, 278, 468, 472, 557n306,
    590n419. *See also* Dymaxion House
Wiener, Norbert, 494n42
Wigley, Mark, 246
Wilbur, Estelle Rodgers (later Hewlett),
    84
Wild West Festival, 368
Wilder, Thornton, 111, 150, 183, 187,
    191, 192, 226, 393
Willets, Anna (later Hewlett), 53, 56–57
Willets, Edward, 56
William J. Burns agency, 45
Williams, Alford F., Jr., 144, 146–147
Williams, John, 261, 273, 320
Williams, Paul, 231
Williams, Robert, 329
Williams, Sarah, 36
Williams & Co., 506n106
Williamson, Marianne, 609n474
Wilson, Donald M., 326
Wilson, Woodrow, 58, 68–69
winches, boat-mounted, 65–66
Windstar Foundation, 431
Windworks, 389
Winsey, Val R., 376
Winter Garden Theatre, 45
Woelffer, Emerson and Dina, 540n238

Wofford, Harris L., Jr., 449
Wolberger, Lionel, 504n95
Wolf, Emily, 194
Wolf, Herman, 193, 194, 196, 201, 203–206, 207–209, 210–211, 382
Wolfe, Tom, 161, 259, 351–352
Wolfenden, John, 191
Woman in the Nineteenth Century (M. Fuller), 38, 39, 395
Women, The (Brokaw), 167
women's rights, 586n407
Women's University Club, 122
Wonders of Life dome, 599n441
Woodruff, Helen Morley, 275
Woodruff, W. L., 269
Woods Hole, 266–267
Woolley, Clarence, 126
World, 391
World Congress of the International Union of Architects (UIA), 315
World Design Science Decade, 319, 320, 335, 348, 363
World Game. *See also* How to/Let's Make the World Work
computers and, 7
Congressional hearing and, 359
Expo Dome and, 358
Fuller lecturing at, 362*fig*
Fuller's death and, 468
furthering of, 366, 383
Gabel and, 400
as inheritance, 440
Kaiser and, 415
Lipkin and, 5
resource classification from, 434
workshops on, 361–364, 418
Youngblood and, 376, 405
World Geo-Graphic Atlas (Bayer), 564n266
World Man Center, 338, 370, 391
World Peace Prayer Tower, 339
World Resources Center, 349, 350–351, 353, 358, 366
World Resources Inventory, 328, 380
World Society of Ekistics, 331, 403–404

World War I, 48, 57, 58–61, 62–70, 81, 105
World War II, 176, 179–180, 184–185, 198
World's Columbian Exposition, 25
Wozniak, Steve, 5, 6–7
Wrather, Jack, 440
Wright, Frank Lloyd
book review by, 166–167, 197
Chicago and, 86
death of, 299–300
Erhard compared to, 407
Froebel and, 23
Hawaiian Village Hotel and, 288
"No More Secondhand God" and, 172
photograph of, 298*fig*
Taliesin Fellowship and, 135
visit with, 183
Women's University Club event and, 122
Wright, Henry, 126
Wright, Olgivanna, 123

Xerox PARC, 6

Yale, 259–260
Yang, Andrew, 474
Yeomans, Henry A., 45–46, 48, 49–50
Yogi, Maharishi Mahesh, 380–381
Yomiuri Country Club, 313, 318
Yost, Zane, 252
You Only Live Twice, 387
Young, Charles, 74
Young, Harold, 235, 237, 238, 246–247
Young Lords Organization, 329
Youngblood, Gene, 372, 376, 405
youth movement, 372–373

Zeckendorf, William, 294, 316
Zeiss Planetarium, 244
Zomes, 334–335
Zorzoli, G. B., 568n342
Zung, Thomas, 428–429, 599n441

ABOUT THE AUTHOR

Alec Nevala-Lee was born in Castro Valley, California, and graduated from Harvard University with a bachelor's degree in classics. He was a 2019 Hugo and Locus Awards finalist for *Astounding: John W. Campbell, Isaac Asimov, Robert A. Heinlein, L. Ron Hubbard, and the Golden Age of Science Fiction*, which was named one of the best books of the year by *The Economist*, and is the author of three novels, including *The Icon Thief*. His nonfiction has appeared in the *New York Times*, the *Los Angeles Times*, Salon, the Daily Beast, Longreads, the Rumpus, Public Books, and the *San Francisco Bay Guardian*, and his short stories have been published in *Analog Science Fiction and Fact, Lightspeed,* and *The Year's Best Science Fiction.* He lives with his wife and daughter in Oak Park, Illinois.